T0282396

Probability: The Classical Limit Theorems

The theory of probability has been extraordinarily successful at describing a variety of natural phenomena, from the behavior of gases to the transmission of information, and is a powerful tool with applications throughout mathematics. At its heart are a number of concepts familiar in one guise or another to many: Gauss' bell-shaped curve, the law of averages, and so on, concepts that crop up in so many settings that they are, in some sense, universal. This universality is predicted by probability theory to a remarkable degree. It is the aim of the book to explain the theory, prove classical limit theorems, and investigate their ramifications.

The author assumes a good working knowledge of basic analysis, real and complex. From this, he maps out a route from basic probability, via random walks, Brownian motion, the law of large numbers and the central limit theorem, to aspects of ergodic theorems, equilibrium and nonequilibrium statistical mechanics, communication over a noisy channel, and random matrices. Numerous examples and exercises enrich the text.

HENRY MCKEAN is a professor in the Courant Institute of Mathematical Sciences at New York University. He is a fellow of the American Mathematical Society and in 2007 he received the Leroy P. Steele Prize for his life's work.

Probability:
The Classical Limit Theorems

HENRY MCKEAN
New York University

CAMBRIDGE
UNIVERSITY PRESS

CAMBRIDGE
UNIVERSITY PRESS

University Printing House, Cambridge CB2 8BS, United Kingdom

Cambridge University Press is part of the University of Cambridge.

It furthers the University's mission by disseminating knowledge in the pursuit of education, learning and research at the highest international levels of excellence.

www.cambridge.org
Information on this title: www.cambridge.org/9781107053212

© Henry McKean 2014

First published 2014

A catalogue record for this publication is available from the British Library

Library of Congress Cataloguing in Publication data
McKean, Henry P. (Henry Pratt), 1930- author.
Probability : the classical limit theorems / Henry Mckean, New York University.
pages   cm
1. Limit theorems (Probability theory)   I. Title.
QA273.67.M35   2014
519.2–dc23
2014022165

ISBN 978-1-107-05321-2 Hardback
ISBN 978-1-107-62827-4 Paperback

Il y a des faussetés déguisées qui représentent si bien la verité, que ce serait mal juger de ne s'y laisser pas tromper.

La Rochefoucauld, Maximes no. 282

DEDICATION

To the memory of M. Kac, W. Feller, K. Itô, N. Levinson, and Gretchen Warren, who taught me so much about everything, and to my dear wife Rasa.

Contents

Preface   *page* xvii
Guide   xx

1   Preliminaries   1
1.1   Lebesgue measure: an outline   1
   1.1.1   Carathéodory's lemma   1
   1.1.2   Measurable sets   2
   1.1.3   The integral   4
1.2   Probabilities and expectations   5
   1.2.1   Reduction to Lebesgue measure   5
   1.2.2   Expectation   6
   1.2.3   Independence   7
1.3   Conditional probabilities and expectations   9
   1.3.1   Signed measures   9
   1.3.2   Radon–Nikodym   10
   1.3.3   Conditional probabilities and expectations   11
   1.3.4   Games, fair and otherwise   15
1.4   Rademacher functions and Wiener's trick   19
1.5   Stirling's approximation   22
   1.5.1   The poor man's Stirling   22
   1.5.2   Kelvin's method   24
1.6   Fast Fourier: series and integrals   25
   1.6.1   Fourier series   26
   1.6.2   Fourier integral   28
   1.6.3   Poisson summation   29
   1.6.4   Several dimensions   31
1.7   Distribution functions and densities   32
   1.7.1   Compactness   32

	1.7.2	Convolution	33
	1.7.3	Fourier transforms	34
	1.7.4	Laplace transform	35
	1.7.5	Some special densities	35
1.8		More probability: some useful tools illustrated	36
	1.8.1	Chebyshev's Inequality	36
	1.8.2	Mills's Ratio	37
	1.8.3	Doob's Inequality	37
	1.8.4	Kolmogorov's 01 Law	38
	1.8.5	Borel–Cantelli Lemmas	39

2 Bernoulli Trials — 42

2.1		The law of large numbers (LLN)	43
	2.1.1	The weak law	43
	2.1.2	The individual or strong law	43
	2.1.3	Application of the weak law: polynomial approximation	44
	2.1.4	Application of the strong law: empirical distributions	45
2.2		The Gaussian approximation (CLT)	48
2.3		The law of iterated logarithm	51
2.4		Large deviations	54
	2.4.1	Cramér's estimate	54
	2.4.2	Empirical distributions	55
	2.4.3	The descent	56
	2.4.4	Legendre–Fenchel duality	58
	2.4.5	General variables	58

3 The Standard Random Walk — 60

3.1		The Markov property	61
	3.1.1	The simple Markov property	61
	3.1.2	The strict Markov property	62
3.2		Passage times	64
	3.2.1	A better way: the reflection principle of D. André.	66
	3.2.2	Yet another take: stopping	68
	3.2.3	Passage to a distance	68
	3.2.4	Two-sided passage: the Gambler's Ruin	69
3.3		Loops	71
	3.3.1	Equidistribution	71
	3.3.2	The actual number of visits	73

	3.3.3	Long runs	74
3.4	The arcsine law		76
	3.4.1	The conditional arcsine law	81
	3.4.2	Drift and the conventional wisdom	82
3.5	Volume		83

4 The Standard Random Walk in Higher Dimensions 87

4.1	What RW(2) and RW(3) do as $n \uparrow \infty$	87
4.2	How RW(3) escapes to ∞	91
	4.2.1	Speed	91
	4.2.2	Direction	92
4.3	Gauss–Landen, Pólya and RW(2)	94
4.4	RW(2): loops and occupation numbers	97
	4.4.1	Duration of a large number of loops	99
	4.4.2	Long runs	101
4.5	RW(2): a hitting distribution	101
4.6	RW(2): volume	104
4.7	RW(3): hitting probabilities	105
	4.7.1	The meaning of G	106
	4.7.2	Comparison with $\mathbb{R}^3$	107
	4.7.3	Electrostatics	108
	4.7.4	Back to $\mathbb{Z}^3$	108
	4.7.5	Energy and capacity in $\mathbb{R}^3$	109
	4.7.6	Finale in $\mathbb{Z}^3$	110
	4.7.7	Grounding	111
	4.7.8	Harmonic functions	111
4.8	RW(3): volume	112
4.9	Non-negative harmonic functions	115
	4.9.1	Standard walks: first pass	115
	4.9.2	Standard walks: second pass	118
	4.9.3	Variants of RW(1)	120
	4.9.4	Space-time walks, mostly in dimension 1	126

5 LLN, CLT, Iterated Log, and Arcsine in General 132

5.1	LLN	132
	5.1.1	Another way	133
	5.1.2	Kolmogorov's 1933 proof	133
	5.1.3	Doob's proof	134
5.2	Kolmogorov–Smirnov statistics	135
5.3	CLT in general	139
	5.3.1	Conventional proof	140

	5.3.2	Second proof	141
	5.3.3	Errors	143
5.4	The local limit	144
5.5	Figures of merit	146
	5.5.1	Gibbs's lemma and entropy	146
	5.5.2	Fisher's information	148
	5.5.3	Log Sobolev	148
5.6	Gauss is prime	149
5.7	The general iterated log	152
5.8	Sparre-Andersen's combinatorial method	152
	5.8.1	Sparre-Andersen's combinatorial lemma	152
	5.8.2	Application to random walk	154
	5.8.3	Spitzer's identity	155
5.9	CLT in dimensions 2 or more	157
	5.9.1	Gaussian variables	157
	5.9.2	Gauss and independence	158
	5.9.3	CLT itself	159
	5.9.4	Maxwell's distribution	160
5.10	Measure in dimension $+\infty$	162
	5.10.1	A better way	163
	5.10.2	Curvature	164
	5.10.3	A more delicate description	165
5.11	Prime numbers	166
6	**Brownian Motion**	**170**
6.1	Preview	171
6.2	Direct construction of BM(1)	175
	6.2.1	P. Lévy's construction	175
	6.2.2	Wiener's construction	178
6.3	Markov property and passage times	179
	6.3.1	The simple Markov property	180
	6.3.2	The strict Markov property	180
	6.3.3	Passage times	181
	6.3.4	Two-sided passage or the Gambler's Ruin	183
6.4	The invariance principle	184
	6.4.1	Proof	185
	6.4.2	Reprise	187
6.5	Volume RW(1) (reprise)	188
6.6	Arcsine (reprise)	191
	6.6.1	Feynman–Kac (FK)	192

	6.6.2	Proof by Brownian paths	194
	6.6.3	Still another way by Brownian paths	196
6.7	Skorokhod embedding		199
	6.7.1	CLT	200
	6.7.2	The iterated log	201
6.8	Kolmogorov–Smirnov (reprise)		202
	6.8.1	Tied Brownian motion	203
	6.8.2	Tied Poisson walk	204
	6.8.3	Evaluations	205
6.9	Itô's lemma		206
	6.9.1	Brownian integrals and differentials	207
	6.9.2	An example	207
	6.9.3	Itô's lemma	208
	6.9.4	Robert Brown and Einstein	213
6.10	Brownian motion in dimensions ≥ 2		214
	6.10.1	Itô's lemma	214
	6.10.2	BM(2): some details	215
	6.10.3	BM(3) and how it goes to ∞	217
6.11	$S^\infty(\sqrt{\infty})$ revisited		221
	6.11.1	div, grad, and all that	221
	6.11.2	Hermite and polynomial chaos	223
	6.11.3	The Brownian format	224
	6.11.4	Back to Δ	226
	6.11.5	Drift and Jacobian	228
7	**Markov Chains**		**231**
7.1	Set-up and the Markov property		232
7.2	The invariant distribution		233
	7.2.1	Geometrical proof	233
	7.2.2	Analytical proof	235
	7.2.3	Probabilistic proof	235
7.3	LLN for chains		237
	7.3.1	LLN improved	238
	7.3.2	Mixing	240
	7.3.3	McMillan's theorem	240
7.4	CLT for chains		241
	7.4.1	Kubo's formula	243
	7.4.2	CLT improved	244
7.5	Real time		244
	7.5.1	The Markov property	245

	7.5.2	Loops and the invariant distribution	246
7.6		The standard Poisson process	248
7.7		Large deviations	250
	7.7.1	Setup and simplest examples of the main result	250
	7.7.2	Preliminaries about I	252
	7.7.3	Proof of the main result	255
	7.7.4	Legendre duality	257

8   The Ergodic Theorem   **260**

8.1		Hamiltonian mechanics	260
8.2		Gibbs, Birkhoff, and the statistical method	262
	8.2.1	Gibbs's canonical ensemble	262
	8.2.2	Time averages	263
	8.2.3	H. Weyl's example	264
8.3		A more general set-up	265
	8.3.1	Metric transitivity and mixing	265
	8.3.2	Poincaré recurrence	267
8.4		Riesz's lemma and Garsia's trick	268
8.5		Continued fractions	270
	8.5.1	The set-up	270
	8.5.2	Birkhoff applied	272
	8.5.3	Proof of metric transitivity	273
	8.5.4	Mixing	274
	8.5.5	Information rate (McMillan's theorem)	275
8.6		Geodesic flow	278
	8.6.1	Sphere	278
	8.6.2	Plane	279
	8.6.3	Poincaré's half-plane	279
	8.6.4	$\mathbb{H}^2/\Gamma$: the circle bundle	284
	8.6.5	How continued fractions enter	285
	8.6.6	CLT	288
	8.6.7	Back to $\mathbb{R}^2/\mathbb{Z}^2$	288
	8.6.8	Why $\mathbb{H}^2/\Gamma$ is better	289

9   Communication over a Noisy Channel   290

9.1   Information/Uncertainty/Entropy   291
    9.1.1   What Boltzmann said   294
    9.1.2   Information as a guide to gambling   296
    9.1.3   A dishonest coin   297
    9.1.4   Relative entropy   298
9.2   Noiseless coding   299
9.3   The source   300
    9.3.1   The rate   301
    9.3.2   McMillan's theorem (reprise)   302
9.4   The noisy channel: capacity   305
    9.4.1   Simplest example   306
9.5   The noisy channel: coding   308
9.6   Communication when $H > C$ 309
9.7   Communication when $H < C$ 310
9.8   The binary symmetric channel   313
    9.8.1   Shannon's idea for $H < C$ 313
    9.8.2   Garbage out   316

10   Equilibrium Statistical Mechanics   317

10.1   What Gibbs said   317
    10.1.1   Phase space and energy   317
    10.1.2   The microcanonical ensemble   318
    10.1.3   Heat bath and the canonical ensemble   319
    10.1.4   Large volume   320
    10.1.5   Thermodynamics: free energy   321
    10.1.6   Thermodynamics: pressure   323
10.2   Two simple examples   326
    10.2.1   Ideal gas   326
    10.2.2   Hard balls   326
10.3   Van der Waals' gas law: dimension 1   328
10.4   Van der Waals: dimension 3   331
    10.4.1 Z bounded below   331
    10.4.2   Scaling and the van der Waals limit   333
    10.4.3   Finishing the proof   333
10.5   The Ising model   334
    10.5.1   Overview   334
    10.5.2   Dimension 1   336
    10.5.3   Dimension 2   337
10.6   Existence of 3 338

10.7	Magnetization per spin: dimension 2	341
10.7.1	Shape of $\mathfrak{m}$, T fixed	341
10.7.2	Shape of $\mathfrak{m}$, K fixed	342
10.8	Change of phase: dimension 2	345
10.8.1	High temperature	345
10.8.2	Low temperature	348
10.9	Duality and the critical temperature	350

11 Statistical Mechanics Out of Equilibrium — 352

11.1	What Boltzmann said and what came after	354
11.2	The two-speed gas: chaos and the law of large numbers	363
11.2.1	The empirical distribution	364
11.2.2	Chaos and the law of large numbers	365
11.2.3	Why chaos propagates	367
11.3	The two-speed gas: fluctuations	369
11.4	More about Boltzmann's equation	371
11.5	The two-speed gas with streaming	374
11.5.1	Solving Boltzmann	375
11.5.2	Carleman's gas	376
11.5.3	The surprising equation	377
11.5.4	Velocity and displacement	380
11.6	Chapman–Enskog–Hilbert	381
11.6.1	First pass	382
11.6.2	Second pass	383
11.6.3	Making better sense of all that	384
11.6.4	Focusing	385
11.7	Kac's gas	385
11.7.1	Boltzmann's equation and Wild's sum	386
11.7.2	Entropy and the tendency to equilibrium	390
11.7.3	Fisher's information	391
11.7.4	CLT: Trotter's method	393
11.7.5	CLT: Grünbaum's method	395
11.7.6	A tagged molecule	397
11.7.7	CLT: $S^\infty(\sqrt{\infty})$ revisited	400

12   Random Matrices   402

12.1   The Gaussian orthogonal ensemble (GOE)   402

12.2   Why a semi-circle?   404

    12.2.1   Reduction to spec **x**   404

    12.2.2   Steepest descent   405

12.3   The semi-circle: a hands-on proof   408

    12.3.1   Wiener's recipe   409

    12.3.2   Traces   409

    12.3.3   Samples   410

    12.3.4   A better way   411

    12.3.5   Leading order   412

    12.3.6   Convergence of traces   413

    12.3.7   The semi-circle   414

12.4   Dyson's Coulomb gas   415

    12.4.1 2×2 hands-on   415

    12.4.2 $n \times n$ in general   416

    12.4.3   Coda   418

12.5   Brownian motion without crossing   419

    12.5.1   Crossing times   419

    12.5.2   Connection to spec **x**   420

12.6   The Gaussian unitary ensemble   422

    12.6.1   Reduction to spec **x**   422

    12.6.2   Scaling and the semi-circle   423

    12.6.3   Dyson's gas   423

12.7   How to compute   424

    12.7.1   Andréief's lemma   424

    12.7.2   Application to spec **x**: first pass   425

    12.7.3   Hermite polynomials   426

    12.7.4   Application to spec **x**: second pass   427

    12.7.5   Fredholm determinants   428

    12.7.6   Application to spec **x**: third pass   430

12.8   In the bulk   431

    12.8.1   A single gap   432

    12.8.2   Wigner's surmise   433

12.9   The ODE   434

12.10   The tail   439

    12.10.1   Wigner's surmise (corrected)   442

12.11   At the edge   443

12.12   Coda   444

    12.12.1   Some history old and new   444

12.12.2 What's happening   445

12.12.3 Riemann and the prime numbers   445

Bibliography   447

Index   458

Preface

The goal of this book is to present the elementary facts of classical probability, namely the law of large numbers (LLN) and the central limit theorem (CLT), first in the simplest setting (Bernoulli trials), then in more generality, and finally in some of their ramifications in, e.g. arithmetic, geometry, information and coding, and classical mechanics.

Let's talk about coin-tossing to illustrate the principal themes in the simplest way.

Let the coin be honest so that the probability of heads or tails is the same $(= 1/2)$. After a large number (n) of independent trials, you have $\# = \mathbf{e}_1 + \cdots + \mathbf{e}_n$ successes, meaning heads, let's say $\mathbf{e} = 1$ for heads, $\mathbf{e} = 0$ for tails. The law of large numbers states that $\#/n$ tends to $1/2$ as $n \uparrow \infty$ with probability $1 - \mathbb{P}\left[\lim_{n \uparrow \infty} \frac{\#}{n} = \frac{1}{2}\right] = 1$ as it is written. That's only common sense if you like. The central limit theorem is deeper. It says that if you center and scale $\#$ as in $(\# - n/2)$ over $\sqrt{n}/2$, then for large n you will see the celebrated bell-shaped curve of Gauss:

$$\lim_{n \uparrow \infty} \mathbb{P}\left[a \leq \frac{\# - n/2}{\sqrt{n}/2} < b\right] = \int_a^b \frac{e^{-x^2/2}}{\sqrt{2\pi}} \, dx \quad \text{for any } a < b.$$

I say it lies deeper, but it is only the proof that it is so. The phenomenon itself is easily illustrated in Nature. To do this, it is best to make a little change, making new es from the old by the rule $\mathbf{e} \rightarrow 2\left(\mathbf{e} - \frac{1}{2}\right) = \pm 1$. Then $\mathbf{x}(n) = \mathbf{e}_1 + \cdots + \mathbf{e}_n$ is the standard random walk, so-called, taking independent steps ± 1 with probabilities $\mathbb{P}(\mathbf{e} = \pm 1) = \frac{1}{2}$, and you have the more symmetrical law:

$$\lim_{n \uparrow \infty} \mathbb{P}\left[a \leq \frac{\mathbf{x}(n)}{\sqrt{n}} < b\right] = \int_a^b \frac{e^{-x^2/2}}{\sqrt{2\pi}} \, dx.$$

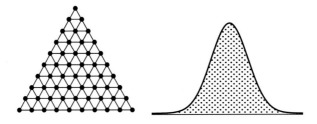

Figure 1

The walk is easily simulated. Take a board studded with nails as in Figure 1 (left) and incline it not too steeply (at 20° say), pour in bird shot from a little funnel at the top, and look to see what piles up at the bottom. You *should* see a scaled approximation to Gauss's celebrated bell-shaped curve $(2\pi)^{-1/2}e^{-x^2/2}$ as in Figure 1 (right) and that is just what happens: the individual shot, running down, hits a nail and is deflected, roughly half the time to the right and half the time to the left, imitating the random walk to produce a bell-shaped heap at the bottom in vindication of CLT.

Of course, that is not *exactly* what happens: the shot is not perfectly round, the nails are not perfectly placed, perfect statistical independence does not exist in Nature, and so on. But it is *approximately* so, which is why I have placed on the title page the maxim of La Rochefoucauld (who surely never thought about the standard random walk but has it just right): That there are certain deceptions, wrong in fact, but which come so close to the real truth that it would be a mistake of judgment not to let oneself be fooled by them. Or, to vary the *mot*: Probability is only a manner of speaking – not the real thing – but is wonderful how well it works.

Take for example, Gibbs's statistical mechanics of which you will get a glimpse in Chapter 10. It is based, of course, on ideas from Nature, but the language is probability, and it is successful beyond all dreams. Or again, take Shannon's ideas about the quantity of information and the means (coding) for its faithful communication over a noisy channel – an equally successful statistical picture of the thing, explained, in part, in Chapter 9.

Well, you get the idea of what I want to do and will judge at the end if I succeeded. Einstein said: "Everything should be made as simple as possible, but not more simple". I have tried to follow that.

Acknowledgement

My debt to Anne Boutet de Monvel is very great. She has typed the whole book from my poor writing with such care, both for how it looks and what it says. I do not know how I could have finished it without her friendly help.

Henry McKean

NYC and Essex, MA, December 2013

Guide

Prerequisities

Not much. I need a good *working* knowledge of the vector spaces $\mathbb{R}^d$ and $\mathbb{C}^d$, and of calculus in several variables; also the rudiments of probability and some knowledge of Lebesgue's measure and integral on the unit interval $[0,1]$. As to the last, I will sketch what I need and ask you to read up on it if you don't know it already, or just to believe what I tell you and use it (with care). Here are some books at about the right level: Munroe [1953], Breiman [1968], and/or the more advanced Billingsley [1979]. Any of these will tell what is wanted, through the little sketch in §1.1 is plenty if you fill it in. Besides, some complex function theory would be nice; it is used sparingly. Ahlfors [1979] is best for this. As to basic probability, Breiman [1968] is excellent and above all Feller [1966] which is full of verve, much information, and a variety of practical examples. In short, only a modest technical machinery is required so as to keep the probability to the fore.

Exercises

Please do these faithfully. It is the only known way to learn the tricks of the trade. Some are marked with a star ($\star$) as being unnecessary to the sequel and/or more difficult. Certain articles and sections are also starred for similar reasons.

References

References are indicated by a name followed by the year of publication in *square* brackets, as in Feller [1950: 33–37]: 1950 indicates the date of publication, 33–37 gives the paging. These are listed at the end. Things like Gauss (1800) or Jacobi (1820), with the year in *parentheses*, are not precise references, only historical indications: *who* and very roughly *when*.

Notations/Usage

Positive means > 0, non-negative ≥ 0, and similarly for negative and non-positive; x^+/x^- is the positive/negative part of the number x; $x \wedge y$ is the smaller of x and y, $x \vee y$ the larger. The symbol $\simeq$ means approximately equal, up to something really small or with a small percentage error, as the context will indicate; $\lesssim$ is used similarly. $\mathbb{Z}$ is the integers, $\mathbb{Z}^+$ the non-negative integers, $\mathbb{N}$ the whole numbers $(1, 2, 3, \ldots)$, $\mathbb{R}$ the line, $\mathbb{C}$ the complex plane. $\mathbf{X}$ is a (sample) space; A, B, C and the like are (mostly) events, i.e. subsets of $\mathbf{X}$. The symbol $\mathfrak{F}$ denotes a field (of events) meaning that it is closed under complements indicated by a prime $(')$, countable unions $(\cup)$, and countable intersections $(\cap)$ – not the common usage but more brief. German (fraktur) is reserved for these. Boldface $\mathbf{x}$ and the like is used, not wholly consistently, for random quantities. Italic x and the like is (mostly) for non-random things. $\mathbb{P}$ is probability. $\mathbb{E}$ is expectation as in $\mathbb{E}(\mathbf{x}) = \int \mathbf{x} \, d\mathbb{P}$; the shorthand $\mathbb{E}(\mathbf{x}, A)$ is for $\int_A \mathbf{x} \, d\mathbb{P}$. By $C[0, 1]$, $C^2[0, \infty)$, $C^\infty(\mathbb{R})$, are designated various classes of continuous functions. $C_\downarrow^\infty(\mathbb{R})$ is the class of smooth, rapidly vanishing functions. $L^1(\mathbb{R})$, $L^2(\mathbb{R})$, and so on are the usual Lebesgue spaces. $\#$ is for counting as in $\#(p \leq n : p$ a prime number$)$. $2+$ means a number a little bit more than 2, say; $2-$ means a number a little bit less. The natural logarithm to the base $e = 2.718+$ is written log; loglog means $\log(\log)$. The symbol $\mathbf{1}_K$ is an indicator function: 1 on K, 0 elsewhere. Traces are denoted by tr.

For those not so familiar with spherical polar coordinates in three dimensions, I remind you that $x \in \mathbb{R}^3$ may be written $x = |x|e$ in which $|x|$ is length and e is the (unit) direction $(\sin\phi\cos\theta, \sin\phi\sin\theta, \cos\phi)$. Here $x_3 = |x|\cos\phi$, $\frac{-\pi}{2} \leq \phi \leq \frac{\pi}{2}$ being co-latitude, and $0 \leq \theta \leq 2\pi$ is the longitude, measured counter-clockwise in the plane $x_3 = 0$ as in the picture.

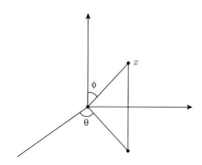

1

Preliminaries

This is a collection of more or less elementary topics you will need to be familiar with. Much of it you may know already but look it over anyhow. Munroe [1953] tells most of what you need – and nothing more, which I like.

1.1 Lebesgue measure: an outline

Let's begin with the unit interval $[0,1]$ and the class $\mathfrak{F}'$ of subsets $A, B, \ldots \subset [0,1]$ comprised of countable sums of non-overlapping intervals I: open, closed, half and half, I don't care. The measure of $A \in \mathfrak{F}'$ is declared to be the sum of their lengths: $\mathrm{m}(A) = \sum |I|$. Obviously, A can be written as such a sum of intervals in many ways, but I take it as geometrically obvious that $\mathrm{m}(A)$ is not changed thereby. It is plain that

(1) $\mathrm{m}(A) \le \mathrm{m}(B)$ if $A \subset B$ and also
(2) $\mathrm{m}(A \cup B) \le \mathrm{m}(A) + \mathrm{m}(B)$, or more precisely
(3) $\mathrm{m}(A \cap B) + \mathrm{m}(A \cup B) = \mathrm{m}(A) + \mathrm{m}(B)$.

1.1.1 Carathéodory's lemma

Lemma 1.1.1 *If the sets $A_n \in \mathfrak{F}'$ decrease to the empty set, then* $\mathrm{m}(A_n) \downarrow 0$.

Proof. If not, then $\mathrm{m}(A_n) \ge c > 0$ for any $n \ge 1$. Choose compact $B_n =$ a finite sum of closed intervals inside A_n with $\mathrm{m}(B_n) > \mathrm{m}(A_n) - c\, 2^{-n-1}$, and let $C_n = B_1 \cap \cdots \cap B_n$. These are compact, decreasing, and $\bigcap_1^\infty C_n \subset$

$\bigcap_1^\infty A_n = \varnothing$ so C_n is void from some n on, by Heine–Borel. But this cannot be. In fact,

$$m(C_n) = m(C_{n-1} \cap B_n)$$
$$= m(C_{n-1}) + m(B_n) - m(C_{n-1} \cup B_n) \qquad \text{by (3)}$$
$$> m(C_{n-1}) + m(A_n) - c\, 2^{-n-1} - m(A_{n-1})$$

since $C_{n-1} \cup B_n \subset A_{n-1}$. Now move $m(C_{n-1})$ to the left and add from $n = 2$ on to produce

$$m(C_n) - m(C_1) > m(A_n) - m(A_1) - \frac{c}{4}$$
$$> c - \left(m(B_1) + \frac{c}{4} \right) - \frac{c}{4}$$
$$= \frac{c}{2} - m(B_1)$$

for any $n \geq 2$. This is contradictory since $m(B_1) = m(C_1)$ and $m(C_n) = 0$.

1.1.2 Measurable sets

Any open set $G \subset [0,1]$ is the sum of a countable number of disjoint open intervals, so $m(G)$ is defined; the recipe $m(A) = \inf m(G)$ for open $G \supset A$ extends on from $A \in \mathfrak{F}'$ to any $A \subset [0,1]$; and you have $m(A \cup B) \leq m(A) + m(B)$ in general. Declare A to be "measurable" if $m(A)$ so defined coincides with $\sup m(K)$ for compact $K \subset A$ and write $A \in \mathfrak{F}''$ in that case. Obviously, $\mathfrak{F}' \subset \mathfrak{F}''$; in particular, any open G belongs to $\mathfrak{F}''$. I claim that any compact K does, too. It is easy to make open G_n decrease to K. Then $G_n \setminus K$ is open and decreases to the empty set, so $m(G_n \setminus K) \downarrow 0$ by Carathéodory's lemma, and $m(K) \leq m(G_n) \leq m(G_n \setminus K) + m(K)$ as $n \uparrow \infty$, i.e. $m(K) = \inf_{G \supset K} m(G)$, as required. Note, by the way, that G_n may be chosen as the sum of finitely many non-overlapping open intervals, by Heine–Borel, so that $1 - m(K) = \lim_{n \uparrow \infty} [1 - m(G_n)] = \lim_{n \uparrow \infty} m(G'_n)$. Here, the compact G'_n is contained in the open complement K', and $K' \setminus G'_n \in \mathfrak{F}'$ decreases to the empty set, with the result that $m(K') = m(K' \setminus G'_n) + m(G'_n)$ is very nearly $1 - m(G_n)$, i.e. $m(K') = 1 - m(K)$ or, equivalently, $m(G') = 1 - m(G)$.

The rest is easy enough.

(1) $\mathfrak{F}''$ is closed under complements, generalizing the fact that $G' = K \in \mathfrak{F}''$: indeed, if $A \in \mathfrak{F}''$, then

$$\inf_{G \supset A'} m(G) = \inf_{K' \supset A'} m(K') = 1 - \sup_{K \subset A} m(K)$$

and likewise

$$\sup_{K \subset A'} \mathrm{m}(K) = \sup_{G' \subset A'} \mathrm{m}(G') = 1 - \inf_{G \supset A} \mathrm{m}(G),$$

and as these match, so $A' \in \mathfrak{F}''$; in particular, $\mathrm{m}(A') = 1 - \mathrm{m}(A)$.

(2) If $A_n \in \mathfrak{F}''$ and $A_i \cap A_j = \varnothing$ for $i \neq j$, then $A = \bigcup_1^\infty A_n \in \mathfrak{F}''$. Here's the proof. Choose $c > 0$ and $G_n \supset A_n \supset K_n$ so that $\mathrm{m}(G_n) - c\,2^{-n} < \mathrm{m}(A_n) < \mathrm{m}(K_n) + c\,2^{-n}$. Then $G = \bigcup_1^\infty G_n \supset A$, and you have

$$\mathrm{m}(A) \leq \mathrm{m}(G) \leq \mathrm{m}\Big(G \setminus \bigcup_1^n G_j\Big) + \mathrm{m}\Big(\bigcup_1^n G_j\Big)$$

$$< \varepsilon + \sum_1^n \mathrm{m}(G_j) \quad \text{for small } \varepsilon \text{ and large } n \text{ by Carathéodory's lemma}$$

$$< \varepsilon + \sum_1^n \mathrm{m}(A_j) + c$$

$$< \varepsilon + \sum_1^n \mathrm{m}(K_j) + 2c$$

$$< \varepsilon + \mathrm{m}\Big(\bigcup_1^n K_j\Big) + 2c$$

$$< \varepsilon + \mathrm{m}(A) + 2c,$$

line 5 being justified by the fact that Ks do not abut, being compact and disjoint by inclusion in the As – in short, ε being small and c arbitrary, you have

$$\mathrm{m}(A) \leq \sup_{K \subset A} \mathrm{m}(K) \text{ by line 5 and } \mathrm{m}(A) \geq \inf_{G \supset A} \mathrm{m}(G) \text{ by line 6,}$$

which is to say $A \in \mathfrak{F}''$, and you have also a bonus from the third line:

(3) $\mathrm{m}(A) = \sum_1^\infty \mathrm{m}(A_n)$.

Discussion. In the parlance adopted here, (1) and (2) state $\mathfrak{F}''$ is a "field", unlike $\mathfrak{F}'$, meaning that it is closed under complements and both countable unions and countable intersections. (3) states that m is "countably additive" on $\mathfrak{F}''$; equivalently, if $A_1 \supset A_2 \supset \cdots$ and $\bigcap_1^\infty A_n = A$ then $\mathrm{m}(A_n) \downarrow \mathrm{m}(A)$, generalizing Carathéodory's lemma from $\mathfrak{F}'$ to $\mathfrak{F}''$. Below, I dispense with $\mathfrak{F}''$ and write $\mathfrak{F}$, plain, in place of $\mathfrak{F}''$. The extension of all this from $[0,1]$ to the whole line $\mathbb{R}$ will be obvious, with one proviso: $\mathrm{m}(A_n) \downarrow \mathrm{m}(A)$ as above if $\mathrm{m}(A_1) < \infty$, but not, for example, if $A_n = [n, \infty)$.

1.1.3 The integral

The function $f\colon [0,1] \to \mathbb{R}$ is "measurable" if $(x : a \leq f(x) \leq b)$ belongs to $\mathfrak{F}$ for every $a < b$, this measurability being assumed for all functions mentioned below. Then if $f \geq 0$, the sums

$$I_n = \sum_1^\infty (k-1)2^{-n}\, \mathrm{m}(0 \leq x \leq 1 : (k-1)2^{-n} \leq f(x) < k2^{-n})$$

increase with n, and you declare the integral $I = \int_0^1 f(x)\,\mathrm{d}x$ to be $\lim_{n\uparrow\infty} I_n \leq \infty$; f is "summable" if $I < +\infty$. The integral for $f = f^+ - f^-$ of indefinite signature is $\int f = \int f^+ - \int f^-$ provided either $\int f^+ < \infty$ or $\int f^- < \infty$.

The integral for $f \geq 0$ has a pleasing geometrical interpretation. The 1-dimensional measure $\mathrm{m} = \mathrm{m}_1$ on $\mathbb{R}$ may be imitated by the 2-dimensional measure m_2 on $\mathbb{R}^2$. The plan is the same, only now you start with rectangles $I \times J$ with measure (area) $\mathrm{m}_2(I \times J) = |I| \times |J|$. Then the figure

$$F_n = \bigcup_1^\infty (0 \leq x \leq 1 : (k-1)2^{-n} \leq f(x) < k2^{-n}) \times [0, (k-1)2^{-n}) \subset \mathbb{R}^2$$

increases to the region "under the graph of f", meaning the 2-dimensional figure

$$F = [(x,y) : 0 \leq x \leq 1 \ \text{ and } \ 0 \leq y < f(x)],$$

and the sum $I_n = \mathrm{m}_2(F_n)$ increases to $I = \mathrm{m}_2(F) = \int_0^1 f(x)\,\mathrm{d}x$. From this point of view, Lebesgue's theorem of monotone convergence is obvious: if $0 \leq f_n \uparrow f$, then $\int f_n \uparrow \int f$, which is nothing but the fact that $F(f_n) \uparrow F(f)$ implies $\mathrm{m}_2\, F(f_n) \uparrow \mathrm{m}_2\, F(f)$. The rest of Lebesgue's theorems follow: if $\lim f_n(x) = f(x)$ except perhaps on a set of measure 0, then $\lim \int f_n = \int f$ provided either:

(1) $f_n \downarrow f$ and $\int f_1 < \infty$; or
(2) f_n is dominated by a summable function h, i.e. $|f_n| \leq h$ and $\int h < \infty$; or, more simply,
(3) $|f_n|$ is bounded, independently of n.

This stuff extends to $\mathbb{R}$, but once more look out: (2) is fine, but (3) may be insufficient because $\mathrm{m}(\mathbb{R}) = +\infty$. A useful variant is Fatou's lemma: if $f_n \geq 0$, then $\int \liminf f_n \leq \liminf \int f_n$ without exception.

Exercise 1.1.2 Prove it.

The story ends with Fubini's theorem: if $0 \le f(x,y)$ is measurable on $\mathbb{R}^2$, then

$$\iint f \, dx \, dy = \int f \, dm_2 = \int dx \left[\int f(x,y) \, dy \right] = \int dy \left[\int f(x,y) \, dx \right]$$

without exception: either all these integrals are finite and they agree or else they are all $+\infty$, with a self-evident extension to signed functions $f(x,y)$ if either f^+ or f^- is summable.

Exercise 1.1.3 Try to prove it if you want.
Hint: $m_2(I \times J) = |I| \times |J|$ is the simplest instance. Take it from there.

1.2 Probabilities and expectations

The mathematical set-up imitates Lebesgue measure on the unit interval – only the language changes. There are three parts to it:

(1) A *sample space* **X** of 1, 2, 3, or even countably many isolated points **x**, representing, e.g. the possible outcomes of some experiment; or as it may be $[0,1]$ itself; or any (even ∞-dimensional) geometrical figure you want with a nice (countable) topology on it.

(2) A *field* $\mathfrak{F}$ of *events* A, B, C, etc. $\subset$ **X** which is closed under complements $A' = \mathbf{X} \backslash A$, (countable) unions $A_1 \cup A_2 \cup$ etc., and so also under (countable) intersections $B_1 \cap B_2 \cap$ etc., and containing every open $G \subset \mathbf{X}$ and so also every compact K.

(3) *Probabilities* $\mathbb{P} \colon \mathfrak{F} \to [0,1]$, i.e. a non-negative, countably additive measure on $\mathfrak{F}$ of total mass $\mathbb{P}(\mathbf{X}) = 1$, such that

$$\mathbb{P}(A) = \inf_{G \supset A} \mathbb{P}(G) = \sup_{K \subset A} \mathbb{P}(K) \quad \text{for any } A \in \mathfrak{F}.$$

1.2.1 Reduction to Lebesgue measure

This whole scheme looks very general but is not. The context can be quite elaborate, but if that be ignored, then (in principle) you know all about $(\mathbf{X}, \mathfrak{F}, \mathbb{P})$ already. I explain, following Halmos–von Neumann [1942:335–336]. The present account is simplified. To begin with, there may be "lumps" of mass, i.e. "indivisible" events A with $\mathbb{P}(A) > 0$ such that $\mathbb{P}(B) = \mathbb{P}(A)$ or 0 for any $B \subset A$. Obviously, these are countable in number, and if they do not exhaust **X**, put them aside and look at $\mathbb{P}$ on the rest of **X**, renormalized to make $\mathbb{P}(\mathbf{X}) = 1$ again. Now suppose, what is natural, that **X** has a fixed countable family of open sets $G_n : n \ge 1$

such that each point $\mathbf{x} \in \mathbf{X}$ is the intersection of a sub-collection of these. Then there is a isomorphism between $\mathbf{X}$ and $[0,1]$, mapping $\mathbb{P}$ to the Lebesgue measure m. This means what it has to mean, as will appear presently.

Discussion.   Let e_n be the indicator of G_n and map

$$\mathbf{x} \in \mathbf{X} \to \mathbf{y} = \mathbf{e}_1/2 + \mathbf{e}_2/4 + \mathbf{e}_3/8 + \cdots \to \mathbf{z} = F(\mathbf{y}),$$

in which $\mathbf{e}_n = e_n(\mathbf{x})$ and $F(y) = \mathbb{P}(\mathbf{y} \le y)$ for $0 \le y \le 1$. The first map $\mathbf{x} \to \mathbf{y}$ of $\mathbf{X}$ into $[0,1]$ is $1:1$ – in fact, any ambiguity such as $\mathbf{y} = \frac{1}{4} + \frac{1}{8} + \cdots = \frac{1}{2}$ cannot appear since $\bigcap_{n \ge 1} G_n$ is then void. (Why?). Besides, F is jump-free: any jump would come from a single point $\mathbf{x}$ and $\mathbb{P}(\mathbf{x}) = 0$ since no lumps are left. Now the second map $\mathbf{x} \to \mathbf{z}$ of $\mathbf{X}$ into $[0,1]$ is not $1:1$ if F has flat places, but who cares? The totality of these come from a null set inside $\mathbf{X}$ and this may be removed. Then the map inverse to $\mathbf{x} \to \mathbf{z}$ induces a faithful pairing, not of points, but of events, placing the event $(\mathbf{z} \in A)$ in correspondence with $B = (\mathbf{x} : \mathbf{y} \in F^{-1}(A))$ via the inverse function $F^{-1}(z) = \max\{y : F(y) \le z\}$, modulo null sets fore and aft. Now

$$\mathbb{P}(\mathbf{z} \le z) = \mathbb{P}(\mathbf{y} \le F^{-1}(z)) = F(F^{-1}(z)) = z,$$

so that $\mathbf{z}$ is distributed by the Lebesgue measure m and $\mathbb{P}(B) = \mathrm{m}(A)$. In this way, $\mathbf{X}$ and $\mathbb{P}$ may be identified with $[0,1]$ and m – in short, the Lebesgue space is all you need to think about from a merely technical point of view.

But not so fast. The technical aspect is one thing, the context is another, and context is what it's all about, really, so this is *not* the way to compute anything of practical interest.

1.2.2 Expectation

A function $f : \mathbf{X} \to \mathbb{R}$, measurable over $\mathfrak{F}$, i.e. with $(\mathbf{x} : a \le f(\mathbf{x}) < b) \in \mathfrak{F}$ for any $a < b$ is now a *random variable*, so-called, and the integral $\mathbb{E}(f) = \int_X f \, d\mathbb{P}$ is its mean value or expectation, as in

$$\mathbb{E}(f) = \lim_{n \uparrow \infty} \sum_{k=1}^{\infty} \frac{k}{2^n} \mathbb{P}\left(\mathbf{x} : \frac{k-1}{2^n} \le f(\mathbf{x}) < \frac{k}{2^n}\right) \le \infty \quad \text{if } f \ge 0.$$

If $f = f^+ - f^-$ is capable of both signatures, then its expectation $\mathbb{E}(f) = \mathbb{E}(f^+) - \mathbb{E}(f^-)$ is defined only if either $\mathbb{E}(f^+) < \infty$ or $\mathbb{E}(f^-) < \infty$. Obviously, the whole of §1.1 now applies unchanged in view of the reduction just explained.

1.2.3 Independence

This notion is peculiar to probability. Its importance for all our business cannot be over-emphasized. Events A and B are *independent* if $\mathbb{P}(A \cap B) = \mathbb{P}(A) \times \mathbb{P}(B)$. Then A and B' are also independent, as per

$$\mathbb{P}(A \cap B') = \mathbb{P}(A) - \mathbb{P}(A \cap B) = \mathbb{P}(A)\big(1 - \mathbb{P}(B)\big) = \mathbb{P}(A)\,\mathbb{P}(B'),$$

and likewise A' and B, and A' and B' – in other words, any event from the little field $\mathfrak{A} = [\varnothing, A, A', \mathbf{X}]$ is independent of any event from $\mathfrak{B} = [\varnothing, B, B', \mathbf{X}]$. For the independence of three events A, B, C, you ask that $\mathbb{P}(A \cap B \cap C) = \mathbb{P}(A)\,\mathbb{P}(B)\,\mathbb{P}(C)$ and likewise for any triple from the associated little fields. More generally, two fields $\mathfrak{A}$ and $\mathfrak{B}$ are independent if $\mathbb{P}(A \cap B) = \mathbb{P}(A)\,\mathbb{P}(B)$ for any $A \in \mathfrak{A}$ and $B \in \mathfrak{B}$, and similarly for the independence of three fields or more. Likewise for random variables f: associate to f the smallest field containing all the events $(\mathbf{x} : a \le f(\mathbf{x}) < b)$. Then two or more variables are independent if their fields are such. This leads to an important principle: namely, $\mathbb{E}(f_1 f_2) = \mathbb{E}(f_1) \times \mathbb{E}(f_2)$ if f_1 and f_2 are independent, assuming the expectations make sense.

Exercise 1.2.1 Check this rule from scratch when the functions are non-negative.

Caution! In general, the independence of three events taken two at a time is *not* the same as their full independence as per the next exercise.

Exercise 1.2.2 Take independent variables $\mathbf{e}_1$, $\mathbf{e}_2$, $\mathbf{e}_3$ with values $= 0$ or 1 and probabilities $\mathbb{P}(\mathbf{e} = 1) = \mathbb{P}(\mathbf{e} = 0) = \frac{1}{2}$. Check that the events $A = (\mathbf{e}_1 + \mathbf{e}_2 \text{ even})$, $B = (\mathbf{e}_2 + \mathbf{e}_3 \text{ even})$, $C = (\mathbf{e}_3 + \mathbf{e}_1 \text{ odd})$ are independent taken two at a time but that $0 = \mathbb{P}(A \cap B \cap C) \ne \mathbb{P}(A)\,\mathbb{P}(B)\,\mathbb{P}(C) = \frac{1}{8}$.

Example 1.2.3 $\mathbf{X}$ is comprised of the 2^n strings $\mathbf{x} = (\mathbf{e}_1, \mathbf{e}_2, \ldots, \mathbf{e}_n)$ of 0s or 1s representing n tosses of a coin (1 for heads, 0 for tails). $\mathfrak{F}$ is the class of all subsets of $\mathbf{X}$, and the $\mathbf{e}$s have the common distribution $\mathbb{P}(\mathbf{e} = 1) = p$ and $\mathbb{P}(\mathbf{e} = 0) = 1 - p = q$ with $0 < p < 1$. The coin is honest if $p = 1/2$. Here, it is natural to suppose that the individual $\mathbf{e}$s are statistically independent so that probabilities multiply as per $\mathbb{P}(\mathbf{x}) = p^{\#} q^{n - \#}$, $\#$ being the number of successes ($\mathbf{e} = 1$) in the string $\mathbf{x} = (\mathbf{e}_1, \ldots, \mathbf{e}_n)$.

Then $\mathbb{P}(\# = k) = \binom{n}{k}p^k q^{n-k}$, in which $\binom{n}{k} = n!/k!(n-k)!$ – pronounced "n choose k" – is the number of ways of selecting $0 \le k \le n$ objects out of n like objects. These are the Bernoulli trials, so-called, to which Chapter 2 is devoted, with special attention to their behavior for $n \uparrow \infty$.

Exercise 1.2.4   If the interpretation "n choose k" is not quite clear in your mind, think it over. Justify $\binom{2n}{n} = \sum_{k=0}^n \binom{n}{k}^2$ as a check that you've understood. Don't compute. Just think what it says.

Exercise 1.2.5   Compute $\mathbb{E}(\#) = np$ and $\mathbb{E}(\# - np)^2 = npq$ using the binomial theorem.

Example 1.2.6 X is now the unit interval $0 \le \mathbf{x} \le 1$, $\mathfrak{F}$ is the familiar field including all subintervals, $\mathbb{P}$ is the standard Lebesgue (or should I say Borel) measure defined on it. You may expand $2\mathbf{x}-1$ as $\mathbf{e}_1/2+\mathbf{e}_2/4+\mathbf{e}_3/8 + \cdots$. The present $\mathbf{e}$ imitates the old $2\mathbf{e} - 1$ of Example 1.2.3 for $p = 1/2$. They are independent, with common distribution $\mathbb{P}(\mathbf{e} = \pm 1) = 1/2$. Now replace $\mathbf{x} \in [0,1]$ by the equivalent *sample path* $\mathbf{x}: n \ge 0 \to \mathbb{Z}$ defined by $\mathbf{x}(0) = 0$ and $\mathbf{x}(n) = \mathbf{e}_1 + \cdots + \mathbf{e}_n$, $n \ge 1$, stepping right or left with equal probabilities at times $n = 1, 2, 3, \ldots$. This is the standard random walk RW(1). It occupies Chapter 3.

Example 1.2.7 X is now $C[0, \infty) =$ the space of (continuous) sample paths $\mathbf{x}: t \in [0, \infty) \to \mathbf{x}(t) \in \mathbb{R}$ with $\mathbf{x}(0) = 0$, say; $\mathfrak{F}$ is the smallest field containing all events

$$Z = \left[\mathbf{x} : \bigcap_{i=1}^n (a_i \le \mathbf{x}(t_i) < b_i)\right]$$

for any $n \ge 1$, any $0 < t_1 < \cdots < t_n$, and any as and bs; finally,

$$\mathbb{P}(Z) = \int_{\bigcap_{i=1}^n (a_i \le x_i \le b_i)} \frac{e^{-x_1^2/2t_1}}{\sqrt{2\pi t_1}} \frac{e^{-(x_2-x_1)^2/2(t_2-t_1)}}{\sqrt{2\pi(t_2 - t_1)}} \cdots \frac{e^{-(x_n-x_{n-1})^2/2(t_n-t_{n-1})}}{\sqrt{2\pi(t_n - t_{n-1})}} \, \mathrm{d}^n x.$$

This set-up is the standard Brownian motion BM(1) to be introduced in Chapter 6, not so much for itself, though it is endlessly fascinating, but as an aid to proving limit theorems. BM(1) is what comes out when RW(1) is speeded up (jump time $1/n$) and scaled back (jump size $1/\sqrt{n}$) and n is taken to $+\infty$, as will be proved in §6.4. By the way, it is not obvious that the probabilities displayed above can be extended to the whole of $\mathfrak{F}$ in a countably additive way: the continuity of the sample path $\mathbf{x}$ is descriptive of an *uncountable* number of observations, and $\mathfrak{F}$, with its countable additivity only, does not permit you to speak about such things. Wiener [1923] adapted Carathéodory's lemma to overcome this technical difficulty; see §6.2 for this and for P. Lévy's [1948] beautiful, more economical method.

Exercise 1.2.8 Check that, for any $c > 0$, $c\mathbf{x}(t/c^2)$ is a copy of BM(1), and likewise $t\mathbf{x}(1/t)$. This means that BM(1) has lots of internal "symmetries", of which much more later on.

1.3 Conditional probabilities and expectations

The naïve version of conditional probabilities is the familiar $\mathbb{P}(A|B) = \mathbb{P}(A \cap B)/\mathbb{P}(B)$, signifying the probability of A when B is known to have occurred. I need a more subtle and flexible version of this idea.

1.3.1 Signed measures

Let Q be the difference $Q^+ - Q^-$ of two non-negative, countably additive measures $Q^\pm$ on a field $\mathfrak{F}$ of subsets A, B, etc. of a space $\mathbf{X}$ and suppose $Q^\pm(\mathbf{X}) < \infty$. Then it is possible to refine the splitting so that Q^+ and Q^- live on disjoint parts of $\mathbf{X}$. This is obvious if $Q(A) = \int_A f \, d\mathbb{P}$ for some summable function f on $\mathbf{X}$ and some probabilities $\mathbb{P}$ on $\mathfrak{F}$: just take $Q^+(A) = \int_A f^+ \, d\mathbb{P}$ with the positive part f^+ of f. More generally, define

$$Q^+(A) = \sup_{B \subset A} Q(B) \quad \text{and} \quad Q^-(A) = -\inf_{B \subset A} Q(B).$$

Claim 1 Q^+ and Q^- are countably additive.

Proof. Take $A_i \cap A_j = \varnothing$ if $i \neq j$ and $B \subset A = \bigcup_1^\infty A_n$ so that $Q(B) > Q^+(A) - \varepsilon$, and so also

$$\sum_{n=1}^\infty Q^+(A_n) \geq \sum_{n=1}^\infty Q(B \cap A_n) = Q(B) > Q^+(A) - \varepsilon.$$

Similarly, if $B_n \subset A_n$ is taken so that $Q(B_n) > Q^+(A_n) - \varepsilon \, 2^{-n}$, then with $B = \bigcup_1^\infty B_n$, you find

$$\sum_{n=1}^\infty Q^+(A_n) - \varepsilon < \sum_{n=1}^\infty Q(B_n) = Q(B) \leq Q^+(A),$$

and so on. The same applies to Q^-.

Claim 2 $Q^+ - Q^- = Q$.

Proof.   You have

$$Q(A) + Q^-(A) = Q(A) - \inf_{B \subset A} Q(A)$$

$$= \sup_{B \subset A} [Q(A) - Q(B)]$$

$$= \sup_{B' \subset A} Q(B') \qquad \text{with } B' = A \setminus B \text{ for the moment}$$

$$= Q^+(A).$$

Claim 3 Q^+ and Q^- *live on disjoint parts of* $\mathbf{X}$.

Proof. $Q^-(B') + Q^+(B) = Q^-(\mathbf{X}) + Q(B)$ by Claim 2, so $Q^-(B'_n) + Q^+(B_n) < 2^{-n}$ by choice of B_n in view of $Q^-(\mathbf{X}) = -\inf Q(B)$. Then with

$$C = \bigcap_{n=1}^{\infty} \bigcup_{k=n}^{\infty} B_k,$$

you have

$$Q^-(C') + Q^+(C) = Q^- \left(\bigcup_{n=1}^{\infty} \bigcap_{k=n}^{\infty} B'_k \right) + Q^+ \left(\bigcap_{n=1}^{\infty} \bigcup_{k=n}^{\infty} B_k \right)$$

$$\leq Q^- \left(\bigcup_{k=n}^{\infty} B'_k \right) + Q^+ \left(\bigcup_{k=n}^{\infty} B_k \right) \qquad \text{for any } n \geq 1$$

$$\leq \sum_{k=n}^{\infty} [Q^-(B'_k) + Q^+(B_k)]$$

$$< 2^{-n} \downarrow 0 \quad \text{as } n \uparrow \infty,$$

i.e. Q^- lives on C and Q^+ on its complement C'.

1.3.2 Radon–Nikodym

Let Q and P be non-negative, countably additive set functions on $\mathfrak{F}$, of finite total mass, and let $Q(A)$ vanish whenever $P(A) = 0$. Then there is a non-negative function f, measurable over the field $\mathfrak{F}$, such that $Q(A) = \int_A f \, \mathrm{d}P$ for any $A \in \mathfrak{F}$. This function is the so-called *Radon–Nikodym derivative* of Q with respect to P: symbolically, $f = \mathrm{d}Q/\mathrm{d}P$.

Proof.   Fix $n \geq 1$ and let $k \geq 1$ vary. $Q - k \, 2^{-n}P$ is negative on some set A_k and positive on its complement A'_k, as per the decomposition described above; $(k-1)2^{-n}P \le Q \le k\,2^{-n}P$ on $A_k \setminus A_{k-1} = B_k$ for every $k \ge 1$; and with $f_n = (k-1)2^{-n}$ on B_k, you have

$$\int_A f_n \, \mathrm{d}P \le Q(A) \le \int_A f_n \, \mathrm{d}P + 2^{-n}P(\mathbf{X}) \quad \text{for any } A \in \mathfrak{F}$$

provided $B = \bigcup_{k=1}^{\infty} B_k$ exhausts $\mathbf{X}$. This is automatic in the sense that $Q(B') = 0$: indeed, if $Q(B') > 0$, then also $P(B') > 0$, and (what is impossible) $Q(A) \ge k\,2^{-n}P(A)$ for any $A \subset B'$ and $k \ge 1$. The proof is finished by noting that f_n increases with n to some measurable function f for which $Q(A) = \int_A f \, \mathrm{d}P$. Obviously, there is only one such function, up to trivialities.

1.3.3 Conditional probabilities and expectations

Let $f \ge 0$ be a summable function on the probability space $(\mathbf{X}, \mathfrak{F}, \mathbb{P})$ and let $\mathfrak{F}_0$ be any subfield of $\mathfrak{F}$. Then the conditional expectation $\mathbb{E}(f|\mathfrak{F}_0)$ of f relative to $\mathfrak{F}_0$ is *the* function f_0 subject to

(1) f_0 is measurable over $\mathfrak{F}_0$ and

(2) $\int_A f \, \mathrm{d}\mathbb{P} = \int_A f_0 \, \mathrm{d}\mathbb{P}$ for any $A \in \mathfrak{F}_0$.

The existence of f_0 is easy: it is the Radon–Nikodym derivative of $Q(A) \equiv \int_A f \, \mathrm{d}\mathbb{P}$, restricted to the little field $\mathfrak{F}_0$, with respect to $\mathbb{P}(A)$ similarly restricted. Both (1) and (2) are satisfied. Obviously, f_0 is the only function of this kind. The extension to signed functions f is self-evident: $\mathbb{E}(f|\mathfrak{F}_0) = \mathbb{E}(f^+|\mathfrak{F}_0) - \mathbb{E}(f^-|\mathfrak{F}_0)$ with the usual proviso, that the expectations make sense. The conditional probability of $A \in \mathfrak{F}$ relative to $\mathfrak{F}_0$ is just the conditional expectation of its indicator function: $\mathbb{P}(A|\mathfrak{F}_0) = \mathbb{E}(1_A|\mathfrak{F}_0)$. Notice that these conditional objects are *not* numbers any more: they are functions on $\mathbf{X}$, i.e. random variables.

Example 1.3.1 Take $B \in \mathfrak{F}$ with $0 < \mathbb{P}(B) < 1$ and let $\mathfrak{F}_0$ be the little field $[\varnothing, B, B', \mathbf{X}]$. Because $\mathbb{P}(B|\mathfrak{F}_0)$ is measurable over $\mathfrak{F}_0$, it is constant on B and likewise on B'. Then (2) implies

$$\mathbb{P}(A \cap B) = \int_B 1_A \, \mathrm{d}\mathbb{P} = \int_B \mathbb{P}(A|\mathfrak{F}_0) \, \mathrm{d}\mathbb{P} = \mathbb{P}(A|\mathfrak{F}_0)\,\mathbb{P}(B) \quad \text{on } B$$

and

$$\mathbb{P}(A \cap B') = \mathbb{P}(A|\mathfrak{F}_0)\,\mathbb{P}(B') \quad \text{on } B',$$

i.e.

$$\mathbb{P}(A|\mathfrak{F}_0) = \frac{\mathbb{P}(A \cap B)}{\mathbb{P}(B)} \quad \text{on } B,$$

$$= \frac{\mathbb{P}(A \cap B')}{\mathbb{P}(B')} \quad \text{on } B',$$

in which you see a reduction to the conventional conditional probabilities.

Example 1.3.2   Let $(\mathbf{X}, \mathfrak{F}, \mathbb{P})$ be the usual Lebesgue space with $\mathbf{X} = [0,1]$ and $\mathfrak{F}_0$ the field produced by the intervals $[(k-1)2^{-n}, k2^{-n}] : 1 \leq k \leq 2^n$. Then for any summable function f,

$$\mathbb{E}(f|\mathfrak{F}_0) = 2^n \int_{(k-1)/2^n}^{k/2^n} f(x)\,\mathrm{d}x \quad \text{on} \quad [(k-1)/2^n, k/2^n)],$$

by inspection, following Example 1.3.1.

Example 1.3.3 $(\mathbf{X}, \mathfrak{F}, \mathbb{P})$ is now the unit square equipped with the 2-dimensional Lebesgue measure $\mathrm{d}x\,\mathrm{d}y$ (area). I take for $\mathfrak{F}_0$ the "horizontal" field produced by rectangles $[a,b) \times [0,1]$. By (1), $f_0 = \mathbb{E}(f|\mathfrak{F}_0)$ is horizontally measurable, i.e. it is a function of x alone. Then by (2),

$$\int_{[a,b) \times [0,1]} f\,\mathrm{d}x\,\mathrm{d}y = \int_a^b \mathrm{d}x \int_0^1 f(x,y)\,\mathrm{d}y = \int_{[a,b) \times [0,1]} f_0\,\mathrm{d}x\,\mathrm{d}y$$

$$= \int_a^b f_0\,\mathrm{d}x,$$

i.e. $\mathbb{E}(f|\mathfrak{F}_0) = \int_0^1 f(x,y)\,\mathrm{d}y$ for $0 \leq x \leq 1$. This exemplifies the correct attitude towards conditional expectation: *Freeze all the information about f expressible over the little field $\mathfrak{F}_0$ and integrate out the rest.*

Rules

Conditional probabilities and expectations obey all the rules of ordinary measure and integral, with one precaution: $\mathbb{E}(f|\mathfrak{F}_0)$ is a measurable function, so when you say, for example, $\lim \mathbb{E}(f_n|\mathfrak{F}_0) = \mathbb{E}(f_\infty|\mathfrak{F}_0)$, as is the case if $0 \leq f_n \uparrow f_\infty$, then you must keep in mind that the statement is meant up to sets of measure 0.

Exercise 1.3.4   Check this rule.

There are special rules, too.

Rule 1. $\mathbb{E}\big(\mathbb{E}(f|\mathfrak{F}_2)|\mathfrak{F}_1\big) = \mathbb{E}(f|\mathfrak{F}_1)$ if $\mathfrak{F}_2 \supset \mathfrak{F}_1$; in particular, $\mathbb{E}\big(\mathbb{E}(f|\mathfrak{F}_0)\big) = \mathbb{E}(f)$.

Rule 2. $\mathbb{E}(ff_0|\mathfrak{F}_0) = \mathbb{E}(f|\mathfrak{F}_0) \times f_0$ if f_0 is measurable over $\mathfrak{F}_0$.

Rule 3. $\mathbb{E}(f|\mathfrak{F}_0) = \mathbb{E}(f)$ if f is independent of $\mathfrak{F}_0$.

Rule 4'. $\lim_{n\uparrow\infty} \mathbb{E}(f|\mathfrak{F}_n) = \mathbb{E}(f|\mathfrak{F}_\infty)$ if $\mathfrak{F}_1 \supset \mathfrak{F}_2 \supset \cdots$ and $\mathfrak{F}_\infty = \bigcap_1^\infty \mathfrak{F}_n$.

Rule 4''. $\lim_{n\uparrow\infty} \mathbb{E}(f|\mathfrak{F}_n) = \mathbb{E}(f|\mathfrak{F}_\infty)$ if $\mathfrak{F}_1 \subset \mathfrak{F}_2 \subset \cdots$ and $\mathfrak{F}_\infty$ is the smallest field containing every $\mathfrak{F}_n$: $n \geq 1$.

Exercise 1.3.5 Prove Rules 1, 2, and 3.

Exercise 1.3.6 Comparing Rules 2 and 3, you might hope that $\mathbb{E}(ff_0|\mathfrak{F}_0) = \mathbb{E}(f|\mathfrak{F}_0) \times \mathbb{E}(f_0)$ if f_0 is independent of $\mathfrak{F}_0$. Show that this is not always so.
Hint: see Exercise 1.2.2.

Rule 4 is due to P. Lévy [1937]. Its modern proof requires a little machinery, simple to explain and of wide versatility. I postpone it for a little in favor of a few more instructive examples and exercises.

Example 1.3.7 The horizontal field $\mathfrak{F}_0$ of Example 1.3.3 is independent of the analogous vertical field $\mathfrak{F}_1$ and $\mathfrak{F} = \mathfrak{F}_0 \times \mathfrak{F}_1$ in an obvious sense. This made the computation of $\mathbb{E}(f|\mathfrak{F}_0)$ particularly simple, and it would be nice, for the implementation of the rule – *freeze $\mathfrak{F}_0$ and integrate out the rest* – if the big field $\mathfrak{F}$ could always be expressed in this way as $\mathfrak{F}_0 \times \mathfrak{F}_1$ with an independent field $\mathfrak{F}_1$. Take, for example, $\mathfrak{F}_0 =$ the 45° field and $\mathfrak{F}_1 =$ the 135° field. You will see what I mean from the picture. $\mathfrak{F}_0$ measures $x + y$, $\mathfrak{F}_1$ measures $x - y$, and while $\mathfrak{F}_0 \times \mathfrak{F}_1 = \mathfrak{F}$

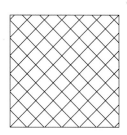

Figure 1.3.1

as before, it is not true that they are independent: for instance,

$$\mathbb{E}\big((x+y)^2 \times (x-y)^2\big) = \mathbb{E}(x^4 - 2x^2y^2 + y^4) = \frac{1}{5} - \frac{2}{9} + \frac{1}{5} = \frac{8}{45}$$

but

$$\mathbb{E}(x+y)^2 \times \mathbb{E}(x-y)^2 = \mathbb{E}(x^2 + 2xy + y^2)\,\mathbb{E}(x^2 - 2xy + y^2)$$
$$= \left(\frac{2}{3} + \frac{1}{2}\right)\left(\frac{2}{3} - \frac{1}{2}\right) = \frac{7}{36}$$

is off.

Exercise 1.3.8 Actually, this lack of independence is geometrically obvious from Figure 1.3.1. Explain.

Exercise 1.3.9 This defect can, however, be cured by taking $\mathfrak{F}_1$ to be the field of *all* events, independent of $\mathfrak{F}_0$. To see more easily what is going on, rotate Figure 1.3.1 counter-clockwise by $45°$ and blow it by a factor $\sqrt{2}$, as in Figure 1.3.2 where the old $45°/135°$ fields appear as the horizontal/vertical fields. I ask you to check that the variables

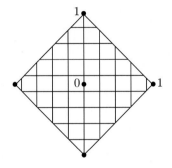

Figure 1.3.2

$\mathbf{x}' = \mathbf{x}$ and $\mathbf{y}' = \mathbf{y}$ over $1 - |\mathbf{x}|$ are independent and to observe that since they jointly determine both $\mathbf{x}$ and $\mathbf{y}$, so the product $\mathfrak{F}_0' \times \mathfrak{F}_1'$ of the corresponding fields is the whole field $\mathfrak{F}$. Good.

Exercise 1.3.10 But maybe that was luck. Take now variables $\mathbf{e}_1$ and $\mathbf{e}_2$ taking values 0 and 1, and let $\mathfrak{F}_0$ be the field of $\mathbf{f} = \mathbf{e}_1 + \mathbf{e}_2$. What now is the field $\mathfrak{F}_1$ of all events, independent of $\mathfrak{F}_0$? *Answer*: It's the trivial field, so there is no way to express the big field $\mathfrak{F}$ as $\mathfrak{F}_0 \times \mathfrak{F}_1$. Too bad.

Example 1.3.11[1] Better luck this time. Figure 1.3.3 shows the square once more, distinguishing the "base" $[0,1] \times 0$ where $\mathbf{x}$ lives from the "fibers" $x \times [0,1]$ where $\mathbf{y}$ lives, and I suppose that $\mathbf{x}$ and $\mathbf{y}$ have a

---

[1] I owe these remarks to S.R.S. Varadhan.

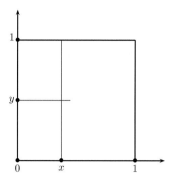

Figure 1.3.3

nice, positive joint density $p(x, y)$. For fixed x in the base, you may use the conditional density $p_x(y) = p(x, y)/\int_0^1 p(x, y)\, dy$ to change the coordinate in the fiber, from y to y', in such a way that $\mathbf{y}'$ is uniformly distributed in $[0, 1]$: in detail, if $F_x(y)$ is the conditional distribution function of $\mathbf{y}$ and if $\mathbf{y}' = F_x(\mathbf{y})$, then

$$\mathbb{P}\left[\mathbf{y}' \le c \,|\, \mathbf{x} = x\right] = \mathbb{P}[\mathbf{y} \le F_x^{-1}(c)] = F_x \circ F_x^{-1}(c) = c.$$

Now the variables $\mathbf{x}$ and $\mathbf{y}' = y'(\mathbf{x}, \mathbf{y})$ are independent as per

$$\mathbb{E}_0[f_0(\mathbf{x})f_1(\mathbf{y}')] = \int_0^1 f_0(x)\, dx \int_0^1 f_1(y')\, dy' = \mathbb{E}[f_0(\mathbf{x})] \times \mathbb{E}[f_1(\mathbf{y}')],$$

and since x and y' together determine also y, so ($\mathfrak{F}_0 =$ the field of $\mathbf{x}$) $\times$ ($\mathfrak{F}_1 =$ the field of $\mathbf{y}'$) is the big field $\mathfrak{F}$.

You see what's going on. Here, the fibers all *look alike*, so you can rescale fiber-wise to make the conditional distribution of $\mathbf{y}$ the same in each of them. The case of Exercise 1.3.10 is different: $f = 0, 1, 2$ is the base and the fibers are *not* the same. For $f = 0$ or 2, the fiber has only *one* point; for $f = 1$, it has *two*. Think it over and think, too of how the present scheme could be made more general still.

1.3.4 Games, fair and otherwise

The context is gambling, just to have a picture to keep in mind. The increasing fields $\mathfrak{F}_0 \subset \mathfrak{F}_1 \subset \mathfrak{F}_2 \cdots \subset \mathfrak{F}_n$ describe the play up to the nth trial. The variables $\mathbf{z}_0, \mathbf{z}_1, \mathbf{z}_2, \ldots, \mathbf{z}_n$ record your capital; these are summable and successively measurable over the $\mathfrak{F}$s. The game is *fair* if

$\mathbb{E}(\mathbf{z}_j|\mathfrak{F}_i) = \mathbf{z}_i$ for $j > i$; more generally, it is *favorable* if $\mathbb{E}(\mathbf{z}_j|\mathfrak{F}_i) \geq \mathbf{z}_i$ for $j > i$. If, in addition, the game goes on indefinitely and $\sup \mathbb{E}(\mathbf{z}_n^+)$ is finite, then $\mathbf{z}_\infty = \lim_{n\uparrow\infty} \mathbf{z}_n$ exists with probability 1. The fact is due to Doob [1953] as is the stylish proof presented below. P. Lévy's Rule 4 follows easily, as will be seen.

The upcrossing lemma

This controls the fluctuations of a favorable game. Fix $-\infty < a < b < \infty$ and let $A_1 < B_1 < A_2 < B_2$ etc. be the successive passage times of $\mathbf{z}$ from a or below to b or above, i.e.

$$A_1 = \min(n \geq 0 : \mathbf{z}_n \leq a)$$
$$B_1 = \min(n > A_1 : \mathbf{z}_n \geq b)$$
$$A_2 = \min(n > B_1 : \mathbf{z}_n \leq a)$$
$$B_2 = \min(n > A_2 : \mathbf{z}_n \geq b)$$

and so on.

The indices $A \leq j < B$ mark the *upcrossings*, starting at a or below when $j = A$, continuing below b while $j < B$, completed when $j = B$ at b or above. Figure 1.3.4 shows a complete upcrossing, starting at $j = 1$,

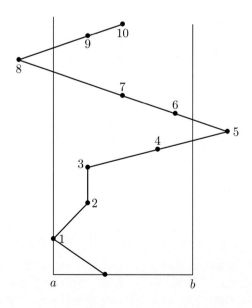

Figure 1.3.4

continuing for $j = 2, 3, 4$, completed at $j = 5$, and also an uncompleted upcrossing, starting at $j = 8$, continuing for $j = 9$ and 10. It is desired to control the number of completed upcrossings by $\mathbf{z}_0, \mathbf{z}_1, \ldots, \mathbf{z}_n$. Look at $\mathbf{z}_j' = [(\mathbf{z}_j - a)/(b - a)]^+$ for $j \le n$. This, too, is a favorable game, and as its upcrossings, from 0 to 1 or above, recapitulate those of $\mathbf{z}_0, \ldots, \mathbf{z}_n$ from a or below to b or above, you may reduce to the case $\mathbf{z} \ge 0$, $a = 0$, $b = 1$. Now any event $(A \le j < B)$ that $\mathbf{z}_j$ is in the act of crossing up is measurable over $\mathfrak{F}_j$, as you will check. Let $\mathbf{e}_j$ be its indicator. Obviously, each upcrossing completed by time n contributes 1 or more to the sum $S = \sum_0^{n-1}(\mathbf{z}_{j+1}' - \mathbf{z}_j')\mathbf{e}_j$, while any final, uncompleted upcrossing contributes something ≥ 0, so that the number $(\#)$ of upcrossings completed by time n is not more than S. Now take expectations:

$$\mathbb{E}(\#) \le \mathbb{E}(S)$$

$$= \sum_0^{n-1} \mathbb{E}[(\mathbf{z}_{j+1}' - \mathbf{z}_j')\,\mathbf{e}_j]$$

$$= \sum_0^{n-1} \mathbb{E}\left[(\mathbb{E}(\mathbf{z}_{j+1}'|\mathfrak{F}_j) - \mathbf{z}_j')\,\mathbf{e}_j\right] \qquad \text{by Rule 1,} \atop \mathbf{e}_j \text{ being measurable over } \mathfrak{F}_j$$

$$\le \sum_0^{n-1} \mathbb{E}\left[\mathbb{E}(\mathbf{z}_{j+1}'|\mathfrak{F}_j) - \mathbf{z}_j'\right] \qquad \text{since } \mathbb{E}(\mathbf{z}_{j+1}'|\mathfrak{F}_j) \ge \mathbf{z}_j'$$

$$= \sum_0^{n-1} \mathbb{E}(\mathbf{z}_{j+1}' - \mathbf{z}_j') \qquad \text{by Rule 1 again}$$

$$= \mathbb{E}(\mathbf{z}_n' - \mathbf{z}_0')$$

$$\le \mathbb{E}\left(\frac{\mathbf{z}_n - a}{b - a}\right)^+ \le \frac{c + (-a)^+}{b - a} \qquad \text{with } c = \mathbb{E}(\mathbf{z}_n^+).$$

That's the upcrossing lemma, where $\#$ is identified as the number of completed upcrossings by $\mathbf{z}_0, \mathbf{z}_1, \ldots, \mathbf{z}_n$ from a or below to b or above.

Evidently, the number of such upcrossings can only increase with n and if $\sup \exp(\mathbf{z}_n^+)$ is finite then $\mathbb{E}(\#) < \infty$ where now $\#$ is the total number of (completed) upcrossings by the whole string $\mathbf{z}_n : n \ge 0$. In particular, $\mathbb{P}(\# < \infty) = 1$, guaranteeing the existence of $\mathbf{z}_\infty = \lim_{n\uparrow\infty} \mathbf{z}_n$ since $\liminf < \limsup$ requires an infinite number of upcrossings of some, e.g. rational, interval ab with $\liminf < a < b < \limsup$.

P. Lévy's Rules

Rule $4''$ with its increasing fields may now be proved. Take $f \geq 0$ (it does not matter). Then $\mathbf{z}_n = \mathbb{E}(f|\mathfrak{F}_n)$ is non-negative and fair and $\mathbb{E}(\mathbf{z}_n^+) \leq \mathbb{E}(f^+) < \infty$, so $\mathbf{z}_\infty$ exists, and

$$\int_A \mathbf{z}_\infty \, d\mathbb{P} \leq \liminf_{n\uparrow\infty} \int_A \mathbf{z}_n \, d\mathbb{P} \qquad \text{by Fatou's lemma}$$

$$= \liminf_{n\uparrow\infty} \int_A \mathbb{E}(f|\mathfrak{F}_n) \, d\mathbb{P}$$

$$= \int_A f \, d\mathbb{P} \qquad \text{if } A \in \mathfrak{F}_m \text{ for some fixed } m$$

$$= \int_A \mathbb{E}(f|\mathfrak{F}_\infty) \, d\mathbb{P} \qquad \text{by Rule 1,}$$

with the implication, m being arbitrary, that the same is true for any event $A \in \mathfrak{F}_\infty$; in short $\mathbf{z}_\infty \leq \mathbb{E}(f|\mathfrak{F}_\infty)$. But also, for fixed $0 \leq N < \infty$ and $\mathbf{z}_n' = \mathbb{E}(f \wedge N|\mathfrak{F}_n)$, you have $\mathbf{z}_\infty \geq \mathbf{z}_\infty' = \mathbb{E}(f \wedge N|\mathfrak{F}_\infty)$ by inspection of the display just above, and $\mathbf{z}_\infty \geq \mathbb{E}(f|\mathfrak{F}_\infty)$ is seen upon making $N \uparrow \infty$. That does the trick.

But what about Rule $4'$ with its *decreasing* fields? $\mathbf{z}^\uparrow = \mathbf{z}_0\mathbf{z}_1\mathbf{z}_2 \ldots \mathbf{z}_n$ is *not* fair anymore, but its reversal $\mathbf{z}^\downarrow = \mathbf{z}_n \ldots \mathbf{z}_2\mathbf{z}_1\mathbf{z}_0$ *is*, and the *up*crossings of $\mathbf{z}^\downarrow$ control the *down*crossings of $\mathbf{z}^\uparrow$, so $\mathbf{z}_\infty$ still exists and its identification as $\mathbb{E}(f|\mathfrak{F}_\infty)$ proceeds as before in view of $\mathbb{E}(\mathbf{z}_0^+) \leq \mathbb{E}(f^+) < \infty$, only now more simply since the events A may be taken from the fixed field $\mathfrak{F}_\infty$.

Example 1.3.12 (Example 1.3.2 continued) f is now summable on $[0,1]$ relative to the standard Lebesgue measure, and

$$\mathbb{E}(f|\mathfrak{F}_n) = 2^n \int_{(k-1)/2^n}^{k/2^n} f(x) \, dx \quad \text{on the interval } \left[\frac{k-1}{2^n}, \frac{k}{2^n}\right],$$

$\mathfrak{F}_n$ being generated by these for $0 < k \leq 2^n$. Then $\mathfrak{F}_\infty$ is the field of all Borel (or should I say Lebesgue) measurable sets in $[0,1]$ and Rule $4''$ applies, with the outcome:

$$\lim_{n\uparrow\infty} 2^n \int_{(k-1)/2^n}^{k/2^n} f(x') \, dx' = f(x) \quad \text{almost everywhere,}$$

this being a variant of Lebesgue's version of the fundamental theorem of calculus.

Historical Note

Conventionally, a fair game is called "martingale", a favorable game a "submartingale", but I prefer my present usage – after all, names should suggest what is named, and who now knows that a martingale is part of a farm horse's tack or an old, unprofitable gambling system: *Double your bets until you win?* The Oxford English Dictionary appends to the second meaning a line of Thackeray (1815): "You have not played as yet? Do not do so; above all avoid a martingale if you do." I explain. Let's play heads and tails with an honest coin, doubling the bet until heads comes up at the Nth toss. Then stop and take your winnings, $\mathbb{P}(N = n) = 2^{-n}$, so $\mathbb{P}(N < \infty) = 1$. Take $\mathbf{z}_0 = 0$ and bet \$1. The first play leaves you with $\mathbf{z}_1 = +1$ or -1 according as $N = 1$ or not; if the game is not over, the second play leaves you with $\mathbf{z}_2 = -1 + 2 = 1$ or $-1 - 2$ according as $N = 2$ or not; and if the game is *still* not over, $\mathbf{z}_3 = -1 - 2 + 4 = 1$ or $-1 - 2 - 4$ according as $N = 3$ or not. You see how it goes: either $\mathbf{z}_n = 1$ or else $\mathbf{z}_n = -1 - 2 - 4 - \cdots - 2^{n-1} = -2^n + 1$ according as $N = n$ or not, so with the natural fields you have

$$
\begin{aligned}
&\mathbb{E}(\mathbf{z}_n | \mathfrak{F}_{n-1}) \\
&= \mathbb{E}(\mathbf{z}_n, N \leq n | \mathfrak{F}_{n-1}) + \mathbb{E}(\mathbf{z}_n, N > n | \mathfrak{F}_{n-1}) \\
&= \mathbb{P}(N \leq n | \mathfrak{F}_{n-1}) + (1 - 2^n) \, \mathbb{P}(N > n | \mathfrak{F}_{n-1}) \\
&= 1 - 2^n \, \mathbb{P}(N > n | \mathfrak{F}_{n-1}),
\end{aligned}
$$

and as the events $n < N$ and $n \geq N$ are both measurable over $\mathfrak{F}_{n-1}$, this reduces to 1 or to $1 - \mathbb{P}(N > n | N \geq n) = 1 - 2^{n-1}$ according as $N < n$ or not – in short, $\mathbb{E}(\mathbf{z}_n | \mathfrak{F}_{n-1}) = \mathbf{z}_{n-1}$, i.e. $\mathbf{z}$ is a fair game. But look: it costs $2^n - 1$ to play to the end, up to time $N = n$, so the expected cost of play is $\sum_1^\infty (2^n - 1) 2^{-n} = +\infty$, and, at best, you win only $\mathbf{z}_\infty = 1$, *to wit*, your original bet. Why bother? Besides, if you have only a limited capital K, the probability of going broke before you win is $\mathbb{P}(2^N - 1 > K) \simeq K^{-1}$ for large K, so you had better listen to Thackeray.

1.4 Rademacher functions and Wiener's trick

The reduction to Lebesgue measure of §1.2.1 means that you can build anything you want over $\mathbf{X} = [0, 1]$, $\mathfrak{F} = $ the smallest field containing all intervals, and $\mathbb{P} = $ the standard Lebesgue measure m. Wiener [1948], coming before, had realized this intuitively. The idea that anything more elaborate is needed is just wrong. Here, I want you to see how this is implemented concretely.

The variable $0 \leq \mathbf{x} \leq 1$ itself, with its uniform distribution $\mathbb{P}(\mathbf{x} \leq \lambda) = \lambda$, $0 \leq \lambda \leq 1$, is the basic object. Expand $\mathbf{x}$ as in $.\mathbf{e}_1\mathbf{e}_2\mathbf{e}_3 \ldots$ with $\mathbf{e} = 1$ or 0, meaning $\mathbf{x} = \mathbf{e}_1/2 + \mathbf{e}_2/4 + \mathbf{e}_3/8$ and so on. The es are the Rademacher functions, so-called, and are interesting enough in themselves. The first three are seen in the pictures.

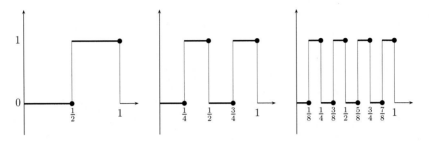

Figure 1.4.1 Rademacher functions: $\mathbf{e}_1$, $\mathbf{e}_2$, $\mathbf{e}_3$

They have the common distribution $\mathbb{P}(\mathbf{e}_n = 1) = \mathbb{P}(\mathbf{e}_n = 0)$ since $\mathbf{e}_n = 1$ on 2^{n-1} intervals, each of length 2^{-n}. What is most important is that *they are independent*. I spell it out: specification of $\mathbf{e}_1$, $\mathbf{e}_2$, up to $\mathbf{e}_n$ cuts out a particular interval of length 2^{-n}, this number being the product of the individual probabilities $1/2 = \mathbb{P}(\mathbf{e}_1 = 0 \text{ or } 1)$ etc. But what if there are gaps in the specification, as in $\mathbf{e}_1 = 1$, $\mathbf{e}_2 = 0$, $\mathbf{e}_3 = 1$, $\mathbf{e}_6 = 1$, $\mathbf{e}_8 = 0$? Then $\mathbb{P}(\mathbf{e}_1 = 1, \mathbf{e}_2 = 0, \mathbf{e}_3 = 1, \mathbf{e}_6 = 1, \mathbf{e}_8 = 0) = 2^{-5}$, as it should be: indeed, specifying the missing $\mathbf{e}_4$, $\mathbf{e}_5$, $\mathbf{e}_7$ reduces the probability to 2^{-8} and there are 8 mutually exclusive and equally likely ways of doing that, so the original string has probability $8 \times 2^{-8} = 2^{-5}$, as I said.

OK. Now suppose you want an infinite number of independent statistically identical copies of $\mathbf{x}$. Here they are:

$$\mathbf{x}_1 = .\mathbf{e}_1\mathbf{e}_3\mathbf{e}_6\mathbf{e}_{10}\mathbf{e}_{15} \text{ etc.}$$

$$\mathbf{x}_2 = .\mathbf{e}_2\mathbf{e}_5\mathbf{e}_9\mathbf{e}_{14} \text{ etc.}$$

$$\mathbf{x}_3 = .\mathbf{e}_4\mathbf{e}_8\mathbf{e}_{13} \text{ etc.}$$

$$\mathbf{x}_4 = .\mathbf{e}_7\mathbf{e}_{12} \text{ etc.}$$

$$\mathbf{x}_5 = .\mathbf{e}_{11} \text{ etc.}$$

and so on. You see the pattern. The rest is obvious. But maybe you like Gaussian better. If so, you have only to apply the inverse function of

$G(x) = \int_{-\infty}^{x} \frac{e^{-y^2/2}}{\sqrt{2\pi}}\, dy$: in fact, the variables $\mathbf{y}_n = G^{-1}(\mathbf{x}_n)$ are independent, with common distribution

$$\mathbb{P}(\mathbf{y}_n \le \lambda) = \mathbb{P}(\mathbf{x}_n \le G(\lambda)) = G(\lambda).$$

This device is due to Wiener [1948: 23–24].

Exercise 1.4.1 Fix $p \in (0,1)$ not equal to $1/2$. How do you make independent variables $\mathbf{e} = 1$ or 0 with common distribution $\mathbb{P}(\mathbf{e}=1) = p$ and $\mathbb{P}(\mathbf{e}=0) = 1-p$?

Exercise 1.4.2 Take $f \in C[0,1]$ and let $\mathfrak{F}_n$ be the field of the first n Rademacher functions. What is $\mathbb{E}(f|\mathfrak{F}_n)$? What does it do as $n \uparrow \infty$? Compare Example 1.3.2.

Exercise 1.4.3 Check that the "shift" $\mathbf{x} \to \mathbf{x}^+ = 2\mathbf{x}$ if $\mathbf{x} \le 1/2$ and $2(\mathbf{x}-1/2)$ if $\mathbf{x} \ge 1/2$ preserves probabilities. What does it do to the **e**s? What (measurable) functions of $\mathbf{x}$ are shift-invariant?

Exercise 1.4.4 Recall the 2-dimensional Lebesgue measure m_2 on the square $[0,1] \times [0,1]$ constructed just as in dimension 1, but starting from little squares instead of intervals as in §1.1. Check that everything you know about measures and integrals for $d = 1$ lifts automatically to $d = 2$ via the map from $\mathbf{x} = .e_1 e_2 e_3 e_4 e_5 e_6 \ldots$ to the pair $\mathbf{x}_1 = .e_1 e_3 e_5 \ldots$ and $\mathbf{x}_2 = .e_2 e_4 e_6 \ldots$ which converts the 1-dimensional measure m_1 into m_2. Obviously, the same can be done in any dimension $d \ge 2$.

Exercise 1.4.5 after Kac [1959b:1–12]. $2\mathbf{x} - 1 = \sum \mathbf{e}'_n/2^n$ with $\mathbf{e}'_n = 2\mathbf{e}_n - 1 = \pm 1$ so

$$\frac{\sin\theta}{\theta} = \int_0^1 e^{\sqrt{-1}\,\theta(2x-1)}\, dx = \mathbb{E}\left[\prod_{n=1}^{\infty} e^{\sqrt{-1}\,\theta \mathbf{e}'_n/2^n}\right]$$

$$= \prod_{n=1}^{\infty} \cos(\theta/2^n) \quad \text{by independence;}$$

in particular, if $\theta = \pi/2$, you find

$$\frac{2}{\pi} = \frac{\sqrt{2}}{2} \times \frac{\sqrt{2+\sqrt{2}}}{2} \times \frac{\sqrt{2+\sqrt{2+\sqrt{2}}}}{2} \times \cdots$$

Curious. These little formulae are due to Vieta (1600) with a different proof:

$$\sin x = 2\sin\frac{x}{2}\cos\frac{x}{2} = 4\cos\frac{x}{2}\sin\frac{x}{4}\cos\frac{x}{4} = 8\cos\frac{x}{2}\cos\frac{x}{4}\sin\frac{x}{8}\cos\frac{x}{8}$$

and so on. See it?

1.5 Stirling's approximation

The Gaussian approximation of de Moivre (1718) and Laplace (1812) for independent Bernoulli trials, to be derived in §2.2, requires a simple but accurate appraisal of $n!$ for large n. This is Stirling's approximation:

$$n! = n^{n+\frac{1}{2}} e^{-n} \sqrt{2\pi} \times [1 + o(1)].$$

The corrective factor $e^{-1/12n}$, not derived here, shows that the percentage error is pretty small: it is 8% for $n = 1$, $\leq 2\%$ for $n \geq 5$, $\leq 1\%$ for $n \geq 9$, and so on. The presence of $e = 2.718+$ is not surprising since it is roughly $(1 + 1/n)^n$, but $\sqrt{2\pi}$ looks odd (no circles here); actually, it is the same number Z as in de Moivre–Laplace, required to make the total probability $Z^{-1} \int_{-\infty}^{\infty} e^{-x^2/2} \, dx$ be 1.

1.5.1 The poor man's Stirling

I mean the approximation without identification of the correct constant $C = \sqrt{2\pi}$. Write $r_n = n!/n^{n+1/2}e^{-n}$. Then $r_n/r_{n-1} = (1 - 1/n)^{n-1/2}e$ as you will check, and

$$\begin{aligned}
\log(r_n/r_{n-1}) &= (n - 1/2)\log(1 - 1/n) + 1 \\
&= -(n - 1/2)(1/n + 1/2n^2 + 1/3n^3 + \cdots) + 1 \\
&= -(1 - 1/2n + 1/2n - 1/4n^2 + 1/3n^2 - 1/6n^3 + \cdots) + 1 \\
&\simeq -1/12n^2.
\end{aligned}$$

Now $\sum_1^\infty 1/n^2 < \infty$, so with $r_0 = 1$, it is obvious that

$$\sum_1^\infty \log(r_n/r_{n-1}) = \lim_{n\uparrow\infty} \log r_n = \log(\lim_{n\uparrow\infty} r_n) = \log r_\infty$$

exists, i.e.

$$n! \simeq n^{n+1/2}e^{-n} \times C \quad \text{with some positive number } C = r_\infty.$$

Example 1.5.1   Here is a cute application for which the actual value of C is not needed. N people are met together. Their birth dates are independently distributed over the 365 days of the year, the chance of any particular date being 1/365. What is the probability that two of them were born the same day? The probability this does *not* happen is

$$365 \times 364 \times \cdots \times (365 - N + 1) \times (365)^{-N}$$

$$= \frac{(365)!}{(365 - N)!} \times (365)^{-N}$$

$$\simeq \frac{(365)^{365+1/2}}{(365 - N)^{365-N+1/2}} \times \frac{e^{-365}}{e^{-365+N}} (365)^{-N} \quad \text{by Stirling}$$

$$= e^{-N}(1 - N/365)^{-365+N-1/2}$$

$$\simeq e^{-N(N-1/2)/365}$$

if N is small compared to 365. Now adjust N to make the probability $1/3$ more or less, which is to say

$$N^2 \simeq N(N - 1/2) \simeq 365 \times (\log 3 = 1.0986) \simeq 400,$$

i.e., for $N \simeq 20$, the probability of a match is about $2/3$. A little surprising, no? And what would you do without Stirling?

But what is C? The obvious identity $1 = \sum_0^n \binom{n}{k} 2^{-k}$ supplies an easy answer:

$$\sum_{k=0}^{n} \binom{n}{k} x^k = (1 + x)^n,$$

from which it is a simple exercise to derive

$$\sum_{k=0}^{n} \binom{n}{k} 2^{-n} (k - n/2)^2 = n/4.$$

Then

$$\sum_{|k-n/2|>n^{2/3}} \binom{n}{k} 2^{-n} < \sum_{k=0}^{n} \binom{n}{k} 2^{-n} \frac{(k - n/2)^2}{n^{4/3}} < n^{-1/3},$$

so the bulk of the sum $\sum_0^n \binom{n}{k} 2^{-n} = 1$ comes from the restricted range $|k - n/2| \leq n^{2/3}$ where both k and $n - k$ are big and the poor man's Stirling's formula may be used with confidence to justify

$$1 \simeq \frac{1}{C} \sum_{|k-n/2|\leq n^{2/3}} \left[\left(\frac{k}{n/2}\right)^{k+1/2} \left(\frac{n - k}{n/2}\right)^{n-k+1/2} \right] \frac{2}{\sqrt{n}},$$

as you will check. Now write $k = n/2 + x\sqrt{n}/2$, with $|x| < 2n^{1/6}$ and $\Delta x = 2/\sqrt{n}$ in response to $\Delta k = 1$, and transcribe the previous display in this format as

$$C \simeq \sum_{|x| \le 2n^{1/6}} \left[\left(1 + \frac{x}{\sqrt{n}}\right)^{n/2 + x\sqrt{n}/2 + 1/2} \left(1 - \frac{x}{\sqrt{n}}\right)^{n/2 - x\sqrt{n}/2 + 1/2} \right]^{-1} \Delta x$$

$$\simeq \sum_{|x| \le 2n^{1/6}} \left[\left(1 - \frac{x^2}{n}\right)^{n/2} \left(1 + \frac{x}{\sqrt{n}}\right)^{x\sqrt{n}/2} \left(1 - \frac{x}{\sqrt{n}}\right)^{-x\sqrt{n}/2} \right]^{-1} \Delta x$$

neglecting the exponent $1/2$ since $x/\sqrt{n}$ is small

$$\simeq \sum_{|x| \le 2n^{1/6}} \left[e^{-x^2/2} \times e^{x^2/2} \times e^{x^2/2} \right]^{-1} \Delta x$$

$$= \sum_{|x| \le 2n^{1/6}} e^{-x^2/2} \Delta x$$

$$\simeq \int_{-\infty}^{+\infty} e^{-x^2/2} \, \mathrm{d}x$$

$$= \sqrt{2\pi}$$

as promised. This type of computation will be seen again in §2.1. Incidentally, do you know the trick used here? It's this: $\left(\int_{-\infty}^{+\infty} e^{-x^2/2} \, \mathrm{d}x\right)^2 = \int_0^{2\pi} \mathrm{d}\theta \int_0^{\infty} r \mathrm{d}r \, e^{-r^2/2} = 2\pi$. Right? See also Exercise 3.5.2 for a swift, more sophisticated route to Stirling's approximation, $C = \sqrt{2\pi}$ and all.

1.5.2 Kelvin's method

The method of "steepest descent" of Lord Kelvin (1870) is neater; it has also a very general scope, illustrated here only in the simplest case. You have

$$n! = \int_0^{\infty} e^{-x} x^n \, \mathrm{d}x = \int_0^{\infty} e^{-x + n \log x} \, \mathrm{d}x,$$

and as integrals are a lot easier than arithmetic, this may have some advantage. The idea is to locate the main contribution to the integral, centered at the place where the exponent $f(x) = -x + n \log x$ is biggest, namely at $x = n$, as $f'(x) = -1 + n/x$ shows. Now $f''(x) = -n/x^2 = -1/n$ at the peak, and you may expand about $x = n$ so:

$$f(x) = f(n) + f'(n)(x - n) + \frac{1}{2} f''(c)(x - n)^2 = -n + n \log n - \frac{n}{2c^2}(x - n)^2$$

with some number c between n and x. Now for $|x - n| \le n^{2/3}$, the ratio of $1/n$ to n/c^2 lies between the numbers $(1 \pm n^{-1/3})^2 \simeq 1$, so

$$\int_{|x-n| \le n^{2/3}} e^{-x} x^n \, dx \simeq \int_{|x-n| \le n^{2/3}} e^{-n + n \log n - (x-n)^2 / 2n} \, dx$$

$$= n^n e^{-n} \int_{|x| \le n^{2/3}} e^{-x^2 / 2n} \, dx$$

$$= n^{n+1/2} e^{-n} \int_{|x| \le n^{1/6}} e^{-x^2 / 2} \, dx$$

$$\simeq n^{n+1/2} e^{-n} \times \sqrt{2\pi}.$$

Naturally, this would be silly if either of the tails coming from $x > n + n^{2/3}$ or from $x < n - n^{2/3}$ were comparable to $n!$, but it is not so: for example, with $\vartheta = n^{-1/3}$, you have $n + n^{2/3} = (1 + \vartheta)n$, so

$$\int_{n+n^{2/3}}^{\infty} e^{-x} x^n \, dx = \int_n^{\infty} e^{-(1+\vartheta)x} (1+\vartheta)^n x^n \, dx \times (1+\vartheta)$$

$$< 2 \int_n^{\infty} e^{-x} x^n \, dx \times [e^{-\vartheta}(1+\vartheta)]^n,$$

in which the integral is not more than $n!$ and

$$e^{-\vartheta}(1+\vartheta) < (1 - \vartheta + \vartheta^2/2)(1+\vartheta) = 1 - \vartheta^2/2 + \vartheta^3/2 < 1 - \vartheta^2/3, \quad \text{say,}$$

so that $[e^{-\vartheta}(1+\vartheta)]^n < \exp(-1/3 \, n^{1/3})$ is tiny. The proof is finished by a similar appraisal of the other tail where $x < n - n^{2/3}$.

Exercise 1.5.2 Check that tail for me.

Exercise 1.5.3 Use Stirling's approximation to confirm the estimate of Laplace (1826):

$$\lim_{x \uparrow \infty} e^{-x} \sum_{x + a\sqrt{x} \le n \le x + b\sqrt{x}} \frac{x^n}{n!} = \int_a^b \frac{e^{-y^2/2}}{\sqrt{2\pi}} \, dy.$$

Hint: With $n = x + y\sqrt{x}$, you have $x/n \times e^{1-x/n} \simeq 1 - y^2/2n$.

1.6 Fast Fourier: series and integrals

I explain in a somewhat idiosyncratic but efficient way only what is needed here. Seeley [1966] is just right if you want more.

1.6.1 Fourier series

Think of the complex d-dimensional space $\mathbb{C}^d$ as functions $f(n) : 0 \le n < d$ extended periodically so that $f(n + d) = f(n)$ for any $n \in \mathbb{Z}$. Let $\omega = e^{2\pi\sqrt{-1}/d}$ be the primitive dth root of unity and introduce the periodic functions $e_k(n) = \omega^{kn}$, one such for each $0 \le k < d$. The inner product[2]

$$(e_j, e_k) = \sum_{n=0}^{d-1} e_j(n)\overline{e_k(n)} = \sum_{n=0}^{d-1} \omega^{n(j-k)} = d \text{ if } j = k;$$

otherwise, $r = \omega^{j-k}$ is a non-trivial dth root of unity and the sum vanishes: $\sum_0^{d-1} r^n = (r^d - 1)/(r - 1) = 0$. This means that $d^{-1/2}e_k :$ $0 \le k < d$ is a unit perpendicular basis of $\mathbb{C}^d$, permitting the expression of any vector $f = [f(n) : 0 \le n < d]$ as

$$f = \sum_{k=0}^{d-1} d^{-1/2}e_k(f, d^{-1/2}e_k);$$

in extenso,

$$f(n) = \sum_{k=0}^{d-1} e^{2\pi\sqrt{-1}\,kn/d} \times \frac{1}{d}\sum_{m=0}^{d-1} f(m)e^{-2\pi\sqrt{-1}\,km/d},$$

the two factors $d^{-1/2}$ having been combined. This is the Fourier series for the additive group of integers modulo d. Note for future use that if Δ is the symmetric second difference, $\Delta f(n) = f(n+1) - 2f(n) + f(n-1)$, then $\Delta e_k = -2[1-\cos(2\pi k/d)]e_k$. Now let the vector f be formed out of a smooth function $f(x) : 0 \le x < 1$ of period 1, as in $[f(n/d) : 0 \le n < d]$, and write, as you may,

$$f\left(\frac{n}{d}\right) = \sum_{-d/2 \le k < d/2} e^{2\pi\sqrt{-1}\,kn/d} \times \frac{1}{d}\sum_{m=0}^{d-1} f\left(\frac{m}{d}\right)e^{-2\pi\sqrt{-1}\,km/d},$$

the outer sum being the same as before but re-ordered, e.g. for $d = 3$, you have $k = -1, 0, +1$ and $e_{-1} = e_2$. Here, I make d increase to ∞ and n/d approximate a fixed number $0 \le x < 1$. Then $f(x)$ appears to the left and

$$\sum_{k \in \mathbb{Z}} e^{2\pi\sqrt{-1}\,kx} \times \left[\hat{f}(k) \equiv \int_0^1 f(y)e^{-2\pi\sqrt{-1}\,ky}\,dy\right]$$

---

[2] The bar signifies the complex conjugate.

to the right, the integral being well approximated by the inner (Riemann) sum in the display just before. This is the Fourier series for such a smooth function f. Naturally, the exchange of the outer sum and the limit $d \uparrow \infty$ of the inner sum needs to be justified. That is where the smoothness of f comes in. The individual inner sum is fine and the over-estimate

$$\left| \frac{1}{d} \sum_{m=0}^{d-1} f\left(\frac{m}{d}\right) e^{-2\pi\sqrt{-1}\,km/d} \right| \leq \max_{0 \leq x < 1} |f''(x)| \times \frac{1}{8k^2}$$

takes care of the tails. The details are simple: the sum is the inner product (f, e_k) of $f(n/d) : 0 \leq n < d$ and $e_k(n) = \omega^{kn} : 0 \leq n < d$, multiplied by $1/d$. Taking f real for simplicity, three applications of the mean-value theorem show that

$$\left| \Delta f\left(\frac{n}{d}\right) \right| = \left| \left[f\left(\frac{n+1}{d}\right) - f\left(\frac{n}{d}\right) \right] - \left[f\left(\frac{n}{d}\right) - f\left(\frac{n-1}{d}\right) \right] \right|$$

$$= \left| f'(x_+)\frac{1}{d} - f'(x_-)\frac{1}{d} \right|$$

$$= |f''(x_0)| \times \frac{x_+ - x_-}{d}$$

$$\leq \max_{0 \leq x < 1} |f''(x)| \times \frac{2}{d^2}$$

with $\frac{n-1}{d} < x_- < \frac{n}{d} < x_+ < \frac{n+1}{d}$ and x_0 between x_- and x_+. Also

$$1 - \cos 2\theta = 2\sin^2\theta \geq 2\left(\frac{\theta}{\pi/2}\right)^2 = \frac{8\theta^2}{\pi^2} \quad \text{for } |\theta| \leq \pi/2,$$

as is plain from a picture, so from

$$-2[1 - \cos(2\pi k/d)] \times (f, e_k) = (f, \Delta e_k) = (\Delta f, e_k),$$

and $-d/2 \leq k < d/2$ follows

$$\left| \frac{1}{d}(f, e_k) \right| = \frac{\left| \frac{1}{d}(f, \Delta e_k) \right|}{2[1 - \cos(2\pi k/d)]}$$

$$\leq \max_{0 \leq x < 1} |f''(x)| \times \frac{2}{d^2} \times \frac{1}{2 \cdot 8(\pi k/d)^2/\pi^2}$$

$$= \max_{0 \leq x < 1} |f''(x)| \times \frac{1}{8k^2},$$

independently of d, as promised. The same estimate justifies the Pythagorean rule:

$$\int_0^1 |f(x)|^2 \, dx = \sum_{k \in \mathbb{Z}} |\hat{f}(k)|^2.$$

1.6.2 Fourier integral

This is even easier and more symmetrical. Take a smooth function f on $\mathbb{R}$ with rapid decay at $\pm\infty$, such as $f(x) = e^{-x^2/2}$ or what you will. It is a simple exercise to adapt Fourier's series from period 1 to any period p as in

$$f(x) = \sum_{k \in \mathbb{Z}} e^{2\pi\sqrt{-1}\,kx/p} \times \frac{1}{p} \int_0^p f(y) e^{-2\pi\sqrt{-1}\,ky/p} \, dy.$$

This is applied to the function f in hand after making it periodic by formation of the sum $f_0(x) = \sum_{n \in \mathbb{Z}} f(x + np)$. Then the inner integral is

$$\sum_{n \in \mathbb{Z}} \int_{(n-1)p}^{np} f(y) e^{-2\pi\sqrt{-1}\,ky/p} \, dy = \int_{-\infty}^{+\infty} f(y) e^{-2\pi\sqrt{-1}\,ky/p} \, dy \equiv \hat{f}(k/p),$$

and you have

$$\sum_{n \in \mathbb{Z}} f(x + np) = \sum_{n \in \mathbb{Z}} e^{2\pi\sqrt{-1}\,k/p} \times \hat{f}(k/p) \times 1/p.$$

Now make $p \uparrow \infty$ for fixed x. Figure 1.6.1 shows that the left-hand sum

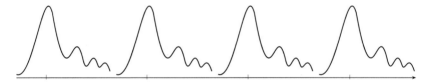

Figure 1.6.1

reduces to $f(x)$, plain, because of the rapid decay of f. To the right, you see once more a Riemann sum, reducing to

$$\int_{-\infty}^{\infty} e^{2\pi\sqrt{-1}\,ky} \hat{f}(k) \, dk$$

with

$$\hat{f}(k) = \int_{-\infty}^{\infty} f(y) e^{-2\pi\sqrt{-1}\,ky} \, dy \quad \text{for arbitrary } k \in \mathbb{R};$$

in short, with a self-evident notation,

$$f(x) = \int_{-\infty}^{\infty} e^{2\pi\sqrt{-1}\,kx} \hat{f}(k) \, dk \equiv (\hat{f})^{\vee}(x).$$

This is Fourier's integral on the line. The justification is much as before, assuming that $f \in C^2(\mathbb{R})$ and that f' and f'' decay nicely, too. Obviously, the Riemann sum reduces to the integral in any moderate range $|k/p| \leq L$, and the easily verified identity $(f'')^\wedge(k) = -4\pi^2 k^2 \hat{f}(k)$ takes care of the tails: $|\hat{f}(k/p)|$ is not more than

$$\int_{-\infty}^{\infty} |f''(x)|\, \mathrm{d}x \times [4\pi^2 (k/p)^2]^{-1}$$

and

$$\sum_{|k/p|>L} \frac{p^2}{k^2} \frac{1}{p} \leq \int_L^{\infty} \frac{\mathrm{d}x}{x^2}$$

is small for large L, independently of p. The estimate also justifies the newly symmetrical Pythagorean rule:

$$\int_{-\infty}^{\infty} |f(x)|^2\, \mathrm{d}x = \int_{-\infty}^{\infty} |\hat{f}(k)|^2\, \mathrm{d}k.$$

Exercise 1.6.1 Let $C_\downarrow^\infty(\mathbb{R})$ be the class of smooth, rapidly vanishing functions f on $\mathbb{R}$, meaning that[3] $x^p D^q f(x)$ vanishes at $x = \pm\infty$ for every choice of p and $q \geq 0$. Prove that $\wedge$ (and so also the inverse transform $\vee$) maps $C_\downarrow^\infty(\mathbb{R})$ $1\!:\!1$ onto itself.

Warning. $\hat{f}(k)$ is often taken to mean $\int e^{-ikx} f(x)\, \mathrm{d}x$ without the 2π in the exponential. Then $f(x) = \frac{1}{2\pi} \int e^{+ikx} \hat{f}(k)\, \mathrm{d}k$ with a nuisance factor $\frac{1}{2\pi}$ in front. Check it out and keep it in mind.

1.6.3 Poisson summation

Take $f \in C_\downarrow^\infty(\mathbb{R})$ and form the sum $f_0(x) = \sum_{n\in\mathbb{Z}} f(x+n)$. This function is smooth and of period 1 and so may be expanded into a Fourier series with coefficients

$$\hat{f_0}(n) = \int_0^1 f_0(x) e^{-2\pi\sqrt{-1}\,nx}\, \mathrm{d}x = \int_{-\infty}^{\infty} f(x) e^{-2\pi\sqrt{-1}\,nx}\, \mathrm{d}x = \hat{f}(n).$$

Now write out

$$\sum_{n\in\mathbb{Z}} f(x+n) = f_0(x) = \sum_{n\in\mathbb{Z}} e^{2\pi\sqrt{-1}\,nx} \hat{f_0}(n) = \sum_{n\in\mathbb{Z}} e^{2\pi\sqrt{-1}\,nx} \hat{f}(n)$$

---

[3] D is the derivative $\mathrm{d}/\mathrm{d}x$.

and note the pleasing special case $x = 0$:

$$\sum_{n \in \mathbb{Z}} f(n) = \sum_{n \in \mathbb{Z}} \hat{f}(n).$$

That's Poisson summation.

Exercise 1.6.2   Jacobi's (1820) identity

$$\sum_{n \in \mathbb{Z}} \frac{e^{-(x-n)^2/4t}}{\sqrt{4\pi t}} = \sum_{n \in \mathbb{Z}} e^{2\pi\sqrt{-1}\,nx} e^{-4\pi^2 n^2 t}$$

is produced by choice of f as $p = (4\pi t)^{-1/2} e^{-x^2/4t}$. Jacobi's identity will come into play from time to time. The case $x = 0$ with $t/4\pi$ in place of t, namely

$$\sum_{n \in \mathbb{Z}} e^{-\pi n^2/t} = \sqrt{t} \sum_{n \in \mathbb{Z}} e^{-\pi n^2 t},$$

is specially striking. It was known to Gauss before (1800).
Hint: $\partial p/\partial t = \partial^2 p/\partial x^2$ as you will check by hand.

Exercise 1.6.3   Note that Jacobi's identity provides a simple justification of Fourier series: namely, if $f(x+1) = f(x)$, then

$$\sum_{n \in \mathbb{Z}} e^{2\pi\sqrt{-1}\,nx} e^{-4\pi^2 n^2 t} \hat{f}(n) = \int_{-\infty}^{\infty} \frac{e^{-(x-y)^2/4t}}{\sqrt{4\pi t}} f(y)\, dy$$

with

$$\hat{f}(n) = \int_0^1 f(x) e^{-2\pi\sqrt{-1}\,nx}\, dx.$$

Now make $t \downarrow 0$.

★Exercise 1.6.4   Here is a second proof of Jacobi's identity if you know a little complex function theory. Look at

$$p(x) = \sum_{n \in \mathbb{Z}} \frac{e^{-(x-n)^2/4t}}{\sqrt{4\pi t}} \quad \text{and} \quad q(x) = \sum_{n \in \mathbb{Z}} e^{2\pi\sqrt{-1}\,nx} e^{-4\pi^2 n^2 t}$$

for *complex* x with fixed $t > 0$. Obviously the sums are just fine. Now p and q are unchanged by the shift $x \to x + 1$, while the substitution $x \to x + (\omega = 4\pi\sqrt{-1}\,t)$ multiplies them both by the common factor $e^{-2\pi\sqrt{-1}\,x} e^{4\pi^2 t}$, as you will check. Next, look at the "fundamental cell" seen in Figure 1.6.2 and compute
(1) $\oint q'/q = 2\pi\sqrt{-1}$ and
(2) $\oint p/q = 0$,

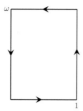

Figure 1.6.2

the integrals being taken about the boundary. (1) says that q has just one (simple) root in the cell; (2) says that p/q has no pole; p/q is then bounded in the whole plane and so must be constant (see why?). Now check that the constant is 1.

1.6.4 Several dimensions

Fourier series and integrals for dimension 1 are easily adapted to dimensions 2 or more. For example, if $f \in C^m((\mathbb{R}/\mathbb{Z})^d)$, you find that

$$f(x) = \sum_{n \in \mathbb{Z}^d} e^{2\pi\sqrt{-1}\, n \cdot x} \hat{f}(n)$$

with

$$\hat{f}(n) = \int_{(\mathbb{R}/\mathbb{Z})^d} e^{-2\pi\sqrt{-1}\, n \cdot x} f(x)\, dx$$

and absolute convergence of the sum if $m > d/2$: in detail, $\sum |\hat{f}(n)|^2 (1 + n^2)^m$ is finite, as you will check, so

$$\left(\sum_{n \in \mathbb{Z}^d} |\hat{f}(n)| \right)^2 \leq \sum_{n \in \mathbb{Z}^d} |\hat{f}(n)|^2 (1 + n^2)^m \times \sum_{n \in \mathbb{Z}^d} (1 + n^2)^{-m},$$

in which the final sum is comparable to $\int_1^\infty r^{-2m} r^{d-1}\, dr < \infty$ if $m > d/2$; also

$$\sum_{n \in \mathbb{Z}^d} |\hat{f}(n)|^2 = \int_{(\mathbb{R}/\mathbb{Z})^d} |f(x)|^2\, dx,$$

as in dimension 1. The extension to $f \in C_\downarrow^\infty(\mathbb{R}^d)$ is just as simple, only more symmetrical:

$$f(x) = \int_{\mathbb{R}^d} e^{2\pi\sqrt{-1}\, k \cdot x} \hat{f}(k)\, dk \quad \text{with} \quad \hat{f}(k) = \int_{\mathbb{R}^d} e^{-2\pi\sqrt{-1}\, k \cdot x} f(x)\, dx$$

and

$$\int |\hat{f}(k)|^2 \, dk = \int |f(x)|^2 \, dx.$$

The proofs are much the same.

The more technical aspects of Fourier's series and integrals are now easy to elicit, for which see, e.g. Dym–McKean [1972], but the present machinery is about all you need here. You see how simple it really is. Why don't we always do it this way?

1.7 Distribution functions and densities

Let $-\infty < \mathbf{x} < \infty$ be a measurable function on $(X, \mathfrak{F}, \mathbb{P})$. Its distribution function is $F(x) = \mathbb{P}(\mathbf{x} \leq x)$. Obviously,

  (1) $F(x) \leq F(y)$ for any $x \leq y$,

  (2) $F(x+) = \lim_{y \downarrow x} F(y) = F(x)$, and also

  (3) $\lim_{x \downarrow -\infty} F(x) = 0$ and $\lim_{x \uparrow \infty} F(x) = 1$.

F may have a (countable) number of jumps and/or a (countable number of flat places, as in Figure 1.7.1 (left). F has a density f if it can be written $F(x) = \int_{-\infty}^{x} f(y) \, dy$ with (naturally) $f \geq 0$ and $\int_{-\infty}^{\infty} f = 1$.

1.7.1 Compactness

Distribution functions have a certain compactness, most easily seen by replacing $F(x)$ by $G(x) =$ the distance from $x \in \mathbb{R}$ to the graph of F, measured along $135°$ lines in Figure 1.7.1 (left). G is also a distribution

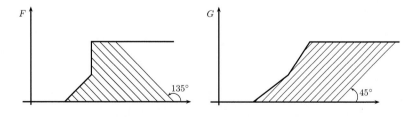

Figure 1.7.1

function (apart from a scaling by $\sqrt{2}$) but has no jumps; indeed

  (4) $0 \leq G(x+h) - G(x) \leq \sqrt{2}h$ for $h > 0$,

as the picture shows. F is recovered from G by a similar recipe using the

$45°$ lines in Figure 1.7.1 (right). Now the Gs are (almost) compact: by (4) and Exercise 1.7.2 below, such a family G_n, $n \geq 1$, may be weeded out so as to converge, locally uniformly on the whole line, to some G_∞, which is to say that $F_n(x)$ approaches some (possibly defective) distribution function $F_\infty(x)$ except, perhaps, at jumps of the latter, as the picture shows. $F_n \rightsquigarrow F_\infty$ indicates this mode of convergence. The only difficulty into which you may fall is that mass may escape to $\pm\infty$, i.e. you may have $\lim_{x\downarrow-\infty} F_\infty(x) > 0$ or, as it may be, $\lim_{x\uparrow\infty} F_\infty(x) < 1$, in which case you may think of F_∞ as having a jump at $-\infty$ or at $+\infty$. That's what I mean by "defective". This is prevented if you have control of tails by, e.g. $-\infty < \int_{-\infty}^{0} x \, dF(x) < LF(-L)$ and $+\infty > \int_{0}^{\infty} x \, dF(x) \geq L(1 - F(L))$ for $L > 0$: indeed, with such or some similar control, the Fs are compact relative to the distance $d(F_1, F_2) = \max_{x\in\mathbb{R}} |G_1(x) - G_2(x)|$. This nice picture of things is due to P. Lévy [1937].

Exercise 1.7.1 Let F_1, F_2 etc. and F_∞ be true distribution functions subject to (1), (2), and (3). Then $F_n \rightsquigarrow F_\infty$ is the same as saying $\lim_{n\uparrow\infty} \int \varphi \, dF_n = \int \varphi \, dF_\infty$ for every $\varphi \in C(\mathbb{R})$.

Exercise 1.7.2 The type of compactness employed above may be illustrated in the space $C[0,1]$ equipped with its natural distance $d(f_1, f_2) = \max_{0\leq x\leq1} |f_1(x) - f_2(x)|$. *Question*: When is a family of such functions compact in the sense that any sequence $f_n : n \geq 1$ from the family can be weeded out, so as to converge to some function f_∞? *Answer*: If and only if $\max_{0\leq x\leq1} |f(x)|$ is controlled independently of f, and likewise the "modulus of continuity" $\omega(h) = \max_{0\leq x\leq1-h} |f(x+h) - f(x)|$, meaning that $\omega(h) \leq \overline{\omega}(h)$ with a fixed function $\overline{\omega}(h) \downarrow 0$ as $h \downarrow 0$. If this is unfamiliar, think it over.

1.7.2 Convolution

If $\mathbf{x}_1$ and $\mathbf{x}_2$ are independent, with distribution functions F_1 and F_2, then their sum $\mathbf{x}_1 + \mathbf{x}_2$ is distributed by the convolution

$$F_1 * F_2(x) = \int_{-\infty}^{\infty} F_1(x - y) \, dF_2(y).$$

Here is the simple proof. Obviously,

$$\left(\mathbf{x}_1 \le x - \frac{k}{n}\right) \cap \left(\frac{k-1}{n} < \mathbf{x}_2 \le \frac{k}{n}\right)$$

$$\subset (\mathbf{x}_1 + \mathbf{x}_2 \le x) \cap \left(\frac{k-1}{n} < \mathbf{x}_2 \le \frac{k}{n}\right)$$

$$\subset \left(\mathbf{x}_1 \le x - \frac{k-1}{n}\right) \cap \left(\frac{k-1}{n} < \mathbf{x}_2 \le \frac{k}{n}\right).$$

Now take probabilities and sum over $k \in \mathbb{Z}$ to obtain

$$F_1 * F_2(x) \le \mathbb{P}(\mathbf{x}_1 + \mathbf{x}_2 \le x) \le F_1 * F_2(x + 1/n)$$

and make $n \uparrow \infty$.

The rule reduces to $f_1 * f_2(x) = \int_{-\infty}^{\infty} f_1(x - y) f_2(y) \, dy$ if F_1 and F_2 have densities. Obviously, convolution commutes by its very meaning; no proof is needed.

1.7.3 Fourier transforms

In connection with sums of independent variables and their distribution functions, the transform[4]

$$\hat{F}(k) = \mathbb{E}(e^{\sqrt{-1}\,k\mathbf{x}}) = \int e^{\sqrt{-1}\,kx} \, dF(x)$$

is often useful, for three reasons:

(1) $(F_1 * F_2)^\wedge = \mathbb{E}\left[e^{\sqrt{-1}\,k(\mathbf{x}_1 + \mathbf{x}_2)}\right] = \mathbb{E}(e^{\sqrt{-1}\,k\mathbf{x}_1})\,\mathbb{E}(e^{\sqrt{-1}\,k\mathbf{x}_2}) = \hat{F}_1\hat{F}_2$ if $\mathbf{x}_1$ and $\mathbf{x}_2$ are independent, converting the complicated convolution $F_1 * F_2$ into the simpler, ordinary pointwise product $\hat{F}_1 \times \hat{F}_2$.

(2) F may always be recovered from $\hat{F}$: indeed, if $\varphi \in C_{\downarrow}^{\infty}(\mathbb{R})$, then the (inverse) transform $\varphi^\vee(k) = (2\pi)^{-1} \int e^{-\sqrt{-1}\,kx} \varphi(x) \, dx$ is of the same type and $\varphi(x) = \int e^{\sqrt{-1}\,kx} \varphi^\vee(k) \, dk$ as explained in §1.6. Then

$$\int \varphi \, dF = \int \left[\int e^{\sqrt{-1}\,kx} \varphi^\vee(k) \, dk\right] dF(x) = \int \varphi^\vee \hat{F} \, dk$$

does the trick.

(3) By the same reason, $\lim_{n\uparrow\infty} F_n =$ the true distribution F_∞ if and only if $\lim_{n\uparrow\infty} \hat{F}_n = \hat{F}_\infty$ pointwise; compare Exercise 1.7.1 above.

[4] Look, first, at §1.6 if this is unfamiliar ground.

1.7.4 Laplace transform

This is more appropriate for non-negative variables $0 \le \mathbf{x} < \infty$. The transform is $\hat{F}(k) = \int_0^\infty e^{-kx}\, dF(x)$ for $k \ge 0$. Now the convolution reads $F_1 * F_2(x) = \int_0^x F_1(x-y)\, dF_2(y)$ for $x \ge 0$; (1) is as before; and (2) and (3) likewise since the functions $e^{-nx} : n \ge 0$ span $C[0, \infty]$, as will be proved in §2.1.3.

Complaint

It is the most natural thing in the world to decompose an oscillating electrical signal into sines and cosines of various frequencies via the Fourier transform – but probabilities, no. Basically, these are positive numbers adding up to 1, and what have sines and cosines to do with that? Indeed, in many applications done first by Fourier, a simpler, more understandable proof may emerge upon taking Fourier away. Still, the Fourier transform is a very effective, sometimes indispensable *technical* tool in much of our business here, as the next article illustrates.

1.7.5 Some special densities

Three types of special densities are needed below.

Gauss. The density is $p_t(x) = (2\pi t)^{-1/2} e^{-x^2/2t}$ for $x \in \mathbb{R}$, with parameter $t > 0$. $\partial p/\partial t = \frac{1}{2}\partial^2 p/\partial x^2$, as you will check by hand if you didn't do it already. This explains the parameter $t = \mathbb{E}(\mathbf{x}^2)$ since $\frac{d}{dt}\mathbb{E}(\mathbf{x}^2) = \int x^2 \frac{1}{2} p'' = \int p = 1$. It also tells you what the Fourier transform has to be: $\partial \hat{p}/\partial t = \int e^{\sqrt{-1}\,kx} \frac{1}{2} p'' = -(k^2/2)\hat{p}$ and $\hat{p}(t = 0+, k) = 1$, so $\hat{p} = e^{-tk^2/2}$, from which you also see, by (1) and (2) above that $p_{t_1} * p_{t_2} = p_t$ with $t = t_1 + t_2$, i.e. the sum of independent Gaussian variables is again Gaussian with addition of mean-squares: $\mathbb{E}(\mathbf{x}_1 + \mathbf{x}_2)^2 = \mathbb{E}(\mathbf{x}_1^2) + \mathbb{E}(\mathbf{x}_2^2)$.

Exercise 1.7.3 Check that directly, without the aid of transforms. Just write the convolution integral and complete the square upstairs.

Cauchy. Now the density is $p_c(x) = \frac{c}{\pi}(x^2+c^2)^{-1}$ on $\mathbb{R}$, with parameter $c > 0$, the meaning of which is not so clear any more. Here, $\partial^2 p/\partial c^2 = -\partial^2 p/\partial x^2$, producing $\partial^2 \hat{p}/\partial c^2 = k^2 \hat{p}$, so $\hat{p} = e^{-c|k|}$, and $p_a * p_b = p_c$ with $c = a + b$.

Exercise 1.7.4 That's a little harder to check directly. Try it.

1-sided stable (exponent 1/2). Now it's $p_x(t) = (2\pi t^3)^{-1/2} x e^{-x^2/2t}$ on $[0, \infty)$, with parameter $x > 0$. The function is $-\partial/\partial x\, (2\pi t)^{-1/2} e^{-x^2/2t}$, so $\partial p/\partial t = \frac{1}{2}\partial^2 p/\partial x^2$ once more, and the Laplace transform $\hat{p} =$

$\int_0^\infty e^{-kt} p \, dt$ solves $\frac{1}{2} \partial^2 \hat{p} / \partial x^2 \equiv k\hat{p}$, so $\hat{p} = e^{-\sqrt{2k}\,x}$ and $p_x * p_y = p_z$ with $z = x + y$. The same pattern, right?

★Exercise 1.7.5   Try this directly, too. It's hard.

Exercise 1.7.6 $\displaystyle\int_0^\infty \frac{e^{-x^2/2t}}{\sqrt{2\pi t}} \frac{c e^{-c^2/2t}}{\sqrt{2\pi t^3}} \, dt = \frac{c}{\pi} \frac{1}{x^2 + c^2}$.   Curious, no? More later in §4.5 and elsewhere.

Exercise 1.7.7   Consider the sum $\mathbf{z} = \mathbf{x}_1 + \cdots + \mathbf{x}_n$ of independent variables with parameter 1. For Gauss, $\mathbf{z}$ looks like $\sqrt{n}\,\mathbf{x}_1$; for Cauchy, like $n\mathbf{x}_1$; for 1-sided stable, like $n^2\mathbf{x}_1$; or, to put the matter differently, the distribution of the arithmetic mean $\mathbf{z}/n$ sharpens for Gauss, does nothing for Cauchy, spreads for 1-sided stable.

Remark   These three types of distribution belong to the class of stable laws, so-called. The others are seldom seen, and as I have no call for them, I leave the matter there. Feller [1966] tells the whole story.

1.8 More probability: some useful tools illustrated

These devices come in all time and must be well digested.

1.8.1 Chebyshev's Inequality

Take any variable $\mathbf{x}$ and a positive number λ. Then

$$\mathbb{P}(\mathbf{x} \geq \lambda) = \int_{\mathbf{x} \geq \lambda} d\mathbb{P} \leq \int_{\mathbf{x} \geq \lambda} \frac{\mathbf{x}}{\lambda} \, d\mathbb{P} \leq \frac{1}{\lambda} \int \mathbf{x}^+ \, d\mathbb{P} = \frac{1}{\lambda} \mathbb{E}(\mathbf{x}^+).$$

This seemingly innocuous remark is used continually in various alternative forms: for example, $\mathbb{P}(|\mathbf{x}| \geq \lambda) = \mathbb{P}(\mathbf{x}^2 \geq \lambda^2) \leq \lambda^{-2} \mathbb{E}(\mathbf{x}^2)$, or again, for positive α and β, $\mathbb{P}(\mathbf{x} \geq \beta) = \mathbb{P}(e^{\alpha \mathbf{x}} \geq e^{\alpha\beta}) \leq e^{-\alpha\beta} \mathbb{E}(e^{\alpha \mathbf{x}})$, and here the right-hand side could still be minimized in respect to α.

Exercise 1.8.1   To illustrate the second variant, let $\mathbf{x}$ be Gaussian with density $(2\pi)^{-1/2} e^{-x^2/2}$. Then $\mathbb{E}(e^{\alpha \mathbf{x}}) = e^{\alpha^2/2}$ as you will check. Conclude that $\mathbb{P}(\mathbf{x} > \beta) \leq e^{-\beta^2/2}$, the moral being that though Chebyshev's inequality is crude, it may be not far from the truth if used intelligently.

1.8.2 Mills's Ratio

But what, in fact, is the real truth of Exercise 1.8.1? You have

$$\int_\lambda^\infty \frac{e^{-x^2/2}}{\sqrt{2\pi}} < \int_\lambda^\infty \frac{x}{\lambda} \frac{e^{-x^2/2}}{\sqrt{2\pi}}\, dx = \frac{1}{\lambda\sqrt{2\pi}} e^{-\lambda^2/2},$$

which is a shade better.

Exercise 1.8.2 Prove that the ratio of Mills's over-estimate to the left-hand side tends to 1 as $\lambda \uparrow \infty$, so *that's* the truth.

1.8.3 Doob's Inequality

Doob's inequality [1953] is the mother of all Chebyshev-type inequalities. It says that if $\mathbf{z}_0, \mathbf{z}_1, \dots, \mathbf{z}_n$ is a favorable game over the fields $\mathfrak{F}_0 \subset \mathfrak{F}_1 \subset \cdots \subset \mathfrak{F}_n$, as in §1.3, then

$$\mathbb{P}\left[\max_{k\le n} \mathbf{z}_k > \lambda\right] \le \frac{1}{\lambda} \mathbb{E}(\mathbf{z}_n^+) \quad \text{for any } \lambda > 0,$$

controlling the entire play up to the nth trial by the single number $\mathbb{E}(\mathbf{z}_n^+)$.

Proof. Let $N = \min(k \le n : z_k > \lambda)$ or $+\infty$ according as $\max z_k > \lambda$ or no. Evidently,

$$\mathbb{P}\left[\max_{k\le n} \mathbf{z}_k > \lambda\right] = \mathbb{P}(N \le n)$$

$$\le \sum_0^n \mathbb{E}\left[\frac{\mathbf{z}_k}{\lambda}, N = k\right]$$

$$\le \sum_0^n \frac{1}{\lambda} \mathbb{E}\left[(\mathbf{z}_n|\mathfrak{F}_k), N = k\right]$$

$$= \frac{1}{\lambda} \sum_0^n \mathbb{E}(\mathbf{z}_n, N = k) \quad \text{since } (N = k) \in \mathfrak{F}_k$$

$$\le \frac{1}{\lambda} \mathbb{E}(\mathbf{z}_n^+).$$

Easy enough.

Exercise 1.8.3 Let f be a convex function so that $pf(a)+(1-p)f(b) \ge f(pa + (1 - p)b)$ for $a \le b$ and $0 \le p \le 1$. Deduce $\mathbb{E}[f(\mathbf{x})] \ge f[\mathbb{E}(\mathbf{x})]$ for any variable $\mathbf{x}$ provided the expectations make sense. That's Jensen inequality.
Hint: When $f \in C^2(\mathbb{R})$: $f(x) = f(m) + f'(m)(x - m) + \frac{1}{2}f''(y)(x - m)^2$ for some y between x and m. Take it from there.

Exercise 1.8.4 Conclude that $f(\mathbf{z}_0), \dots, f(\mathbf{z}_n)$ is favorable if either $\mathbf{z}_0, \dots, \mathbf{z}_n$ is fair and f is convex or if it is favorable and f is not only convex but increasing as well.

Exercise 1.8.5 Let the independent variables $\mathbf{x}_0, \dots, \mathbf{x}_n$ have mean $\mathbb{E}(\mathbf{x}) = 0$ and $\mathbb{E}(\mathbf{x}^2) < \infty$. Then with the natural fields, $\mathbf{z}_k = \mathbf{x}_0 + \mathbf{x}_1 + \dots + \mathbf{x}_k$ $(k \le n)$ is a fair game. Apply Exercise 1.8.4 with $f(x) = x^2$ to obtain Kolmogorov's inequality:

$$\mathbb{P}\left[\max_{k \le n} \mathbf{x}_0 + \mathbf{x}_1 + \dots + \mathbf{x}_k > \lambda\right] \le \frac{1}{\lambda^2} \sum_0^n \mathbb{E}(\mathbf{x}_k^2) \text{ for } \lambda > 0.$$

Exercise 1.8.6 Let $\mathbf{x}_1$, $\mathbf{x}_2$, etc. be independent copies of a variable $\mathbf{x}$ with $\mathbb{E}(\mathbf{x}) = 0$ and $\mathbb{E}(\mathbf{x}^2) < \infty$ and take $y = (y_1, y_2, \dots) \in \mathbb{R}^\infty$. Prove that $\mathbf{x} \cdot y = \mathbf{x}_1 y_1 + \mathbf{x}_2 y_2 + \dots$ converges with probability 1 if $y^2 = \sum y_n^2 < \infty$. The simplest example is $\sum(\pm 1/n)$ with independent, equally likely ± 1s. Think about it. It's not so obvious beforehand.
Hint: The event of divergence may be expressed in terms of the partial sums $\mathbf{z}_n = \mathbf{x}_1 y_1 + \dots + \mathbf{x}_n y_n$ by

$$\bigcup_{m=1}^\infty \bigcap_{n=1}^\infty \bigcup_{i>n} \left(\sup_{j>i}|\mathbf{z}_j - \mathbf{z}_i| > \frac{1}{m}\right),$$

in which the inner union is contained in the event $\sup_{k>n}|\mathbf{z}_k - \mathbf{z}_n| > \frac{1}{2n}$. Take it from there.

1.8.4 Kolmogorov's 01 Law

Let $\mathbf{x}_1$, $\mathbf{x}_2$, etc. be independent and let $\mathfrak{F}_n$ be the smallest field over which the variables $\mathbf{x}_n$, $\mathbf{x}_{n+1}$, etc. are measurable. Kolmogorov's 01 law [1933b] states that $\mathbb{P}(A) = 0$ or 1 for any event $A \in \mathfrak{F}_\infty =$ the "tail field" $\bigcap_1^\infty \mathfrak{F}_n$. The point is that $\mathfrak{F}_1$ is the smallest field containing all the "tame" events involving only a finite number of the $\mathbf{x}$s, so that $A \in \mathfrak{F}_\infty$ may be well approximated by a tame event B in the sense that $\mathbb{P}(A' \cap B) + \mathbb{P}(A \cap B')$ is small. But $A \in \mathfrak{F}_\infty$ is also independent of any such tame B, so $\mathbb{P}(A) \simeq \mathbb{P}(A \cap B) = \mathbb{P}(A)\mathbb{P}(B) \simeq [\mathbb{P}(A)]^2$, i.e. $\mathbb{P}(A) = [\mathbb{P}(A)]^2$ of which the only solution is $\mathbb{P}(A) = 0$ or 1.

Exercise 1.8.7 Kolmogorov's 01 law is refined to that of Hewitt–Savage [1955]. $\mathfrak{F}_\infty$ is now the field of events A invariant under any finite permutation of the $\mathbf{x}$s, and the conclusion is the same: $\mathbb{P}(A) = 0$ or 1. Prove it. The scope is now wider: for example, the event that $\mathbf{x}_1 + \dots + \mathbf{x}_n$ is ≥ 0 infinitely often as $n \uparrow \infty$ belongs to the present $\mathfrak{F}_\infty$ but not to Kolmogorov's tail field.

Exercise 1.8.8 This completes Exercise 1.8.6 by showing that $\mathbf{x} \cdot y$ diverges with probability 1 if $y^2 = +\infty$.

Hints: (1) $\mathbf{x} \cdot y$ converges with probability 0 or 1, by Kolmogorov's law. (2) The function $\varphi(k) = \mathbb{E}(e^{\sqrt{-1}\,k\mathbf{x}})$ is of class $C^2(\mathbb{R})$ with $\varphi(0) = 1$, $\varphi'(0) = 0$, and $\varphi''(0) = -k^2$ so that $\varphi(k) \simeq 1 - k^2/2$ for small k.

1.8.5 Borel–Cantelli Lemmas

Let A_1, A_2, etc. be any events and look at

$$B = \bigcap_{n=1}^{\infty} \bigcup_{m \geq n} A_m$$

comprising the points of $\mathbf{X}$ which belong to infinitely many of the As. B is written (A_n i.o.), in which i.o. stands for "infinitely often." There are two Borel–Cantelli lemmas.

First Lemma $\mathbb{P}(B) = 0$ *if* $\sum_1^{\infty} \mathbb{P}(A_n) < \infty$.

Proof. $\mathbb{P}(B) \leq \sum_{m \geq n} \mathbb{P}(A_m)$ for any n, and this decreases to 0 as $n \uparrow \infty$.

Second Lemma $\mathbb{P}(B) = 1$ *if* $\sum_1^{\infty} \mathbb{P}(A_n) = +\infty$ *and the* A_n *are independent.*

Proof. $1 - \mathbb{P}(B) = \mathbb{P}\left[\bigcup_1^{\infty} \bigcap_{m \geq n} A'_m\right]$ and

$$\mathbb{P}\left[\bigcap_{m \geq n} A'_m\right] = \text{ the product over } m \geq n \text{ of } 1 - \mathbb{P}(A_m)$$

$$< \exp\left[-\sum_{m \geq n} \mathbb{P}(A_m)\right] \quad \text{since } 1 - x < e^{-x}$$

$$= 0.$$

Cheap but wonderfully effective if varied intelligently, as the next exercises illustrate.

Exercise 1.8.9 Let $\mathbf{x}_1$, $\mathbf{x}_2$, etc. be independent with common density $(2\pi)^{-1/2}e^{-x^2/2}$. Prove that

$$\mathbb{P}\left[\max_{k \leq n} \mathbf{x}_k > \sqrt{(2-)\log n} \text{ as } n \uparrow \infty\right] = 1.$$

Here $(2-)$ signifies any fixed number a shade less than 2.

Hints: $\mathbb{P}(\max \leq \lambda) = [1 - \mathbb{P}(\mathbf{x} > \lambda)]^n$. For the rest, use Mills's ratio for large λ and the first Borel–Cantelli lemma.

Exercise 1.8.10   Use Mills's ratio and the first Borel–Cantelli lemma again to prove that

$$\mathbb{P}\Big[\max_{k\le n}\mathbf{x}_k < (2+)\sqrt{\log n} \text{ as } n\uparrow\infty\Big]=1$$

with a number $(2+)$ a shade more than 2.
Hint: $(1-x)^n > 1-nx$ for $0<x<1$.

Exercise 1.8.11   Refine Exercise 1.8.10 using the fact that Mills's ratio is an overestimate of the Gaussian integral to obtain

$$\mathbb{P}\left[\frac{\max}{\sqrt{2\log n}}\ge\sqrt{1+\frac{\gamma}{\log n}}\right]\le 1-\left(1-\frac{e^{-\gamma}}{n}\right)^n < e^{-\gamma}$$

for any $\gamma>0$ and $n\ge 3$. Take $\alpha>1$ and $\gamma=\alpha\log n$. Then $e^{-\gamma}=n^{-\alpha}$ and Borel–Cantelli tells you that $\limsup\big(\max/\sqrt{2\log n}\big)\le\sqrt{2}$ by choice of $\alpha\downarrow 1$. This is no improvement. The way forward is to take small $\alpha>0$, a whole number β so as to make $\alpha\beta>1$, and replace n by n^β. See what Borel–Cantelli says then. Now fill in between n^β and $n=2,3,4$ etc. to get the true picture of the matter: $\mathbb{P}\big(\lim_{n\uparrow\infty}\max/\sqrt{2\log n}=1\big)=1$.

★Exercise 1.8.12   Check that

$$\lim_{n\uparrow\infty}\mathbb{P}\Big[\max_{k\le n}\mathbf{x}_k\le\sqrt{2\log n}-\log(2\log n)-\log 2\pi+2x\Big]=\exp(-e^{-x}),$$

i.e. $\max/\sqrt{2\log n}$ likes to be near 1 or even a bit below.

★Exercise 1.8.13   Gauss (1800) made an extensive table of the prime numbers $p=2,3,5$, etc. and conjectured that the number of primes $\le n$ should be $n/\log n$, more or less. This was proved much later by Hadamard and de la Vallée Poussin (1890). H. Cramér [1936] proposed the following amusement. Declare the whole number $n\ge 2$ to be "prime" with probability $1/\log n$, independently, with the convention $\log 2=1$. It is to be proved that Gauss's conjecture holds with probability 1. Let $\#(n)$ be the number of such "primes" $\le n$. Then $m(n)=\mathbb{E}(\#)\simeq n/\log n$, $\mathbb{E}(\#-m(n))^2\le m(n)$, and $\mathbb{P}\big[|\#-m(n)|>n^{2/3}\big]<n^{-1/3}\log n$: not good enough for Borel–Cantelli. To improve it, replace n by n^4. Then Borel–Cantelli works and $\#(k)$ may be estimated for $(n-1)^4\le k<n^4$, from above by $m(k)+5k^{3/4}$, and similarly from below by $m(k)-5k^{3/4}$. In short, changing k to n, you have

$$\mathbb{P}\Big[\#(n)=m(n)+\text{an absolute error}<6n^{3/4}\text{ as }n\uparrow\infty\Big]=1,$$

just as Gauss would have it and more.

★**Exercise 1.8.14**   Exercise 1.8.13 employs Borel–Cantelli only in a crude form. Try this: with $\beta = \sqrt{(2+)m \log m}$, $\alpha = \beta/m$ and large n,

$$\mathbb{P}(\# > m + \beta) \le \mathbb{E}\big[e^{\alpha(\#-m)}\big]e^{-\alpha\beta}$$
$$< \exp\big[m(e^\alpha - 1 - \alpha) - \alpha\beta\big]$$
$$\le e^{-\beta^2/(2+)m} = e^{-(2+)\log m} = m^{-(2+)},$$

the last being the general term of a convergent sum. A similar estimate of $m - \#$ produces the final result:

$$\mathbb{P}\bigg[\#(n) = m(n) + \text{an absolute error} < \sqrt{(2+)m \log m} \text{ as } n \uparrow \infty\bigg] = 1,$$

by the cheap first Borel–Cantelli lemma.

Hint: $e^\alpha < 1 + \alpha + \vartheta\alpha^2/2$ for small α and any fixed $\vartheta > 1$. *Note*: The same estimate for the true primes with $\int_2^n \frac{dx}{\log x}$ in place of $m(n) = \sum_2^n \frac{1}{\log k}$ and $n^{(1/2)+}$ in place of $\sqrt{m \log m}$ has been commonly believed since Riemann (1860) but is still not proven.

2

Bernoulli Trials

Fix $0 < p < 1$ and let $\mathbf{e}_n : n \geq 1$ be independent copies of a variable $\mathbf{e} = 0$ or 1, with common distribution $\mathbb{P}(\mathbf{e} = 1) = p$ and $\mathbb{P}(\mathbf{e} = 0) = 1 - p = q$. These are the Bernoulli trials, with success probability, p of Example 1.2.3. It seems best to prove the main limit theorems of classical probability, first in this primitive setting, using only the simplest machinery. That is the route followed in the present chapter, with extensive generalizations to follow. You may think of the sums $\mathbf{s}_0 = 0$, $\mathbf{s}_n = \mathbf{e}_1 + \cdots + \mathbf{e}_n$ ($n \geq 1$) as the number of heads in n tosses of a coin with $\mathbb{P}(\text{heads}) = p$ and $\mathbb{P}(\text{tails}) = q$. It is the behavior of $\mathbf{s}_n$ for $n \uparrow \infty$ that is to be understood.

The distribution of $\mathbf{s}_n$ is binomial:

$$\mathbb{P}(\mathbf{s}_n = k) = \binom{n}{k} p^k q^{n-k} \quad (0 \leq k \leq n),$$

in which $\binom{n}{k}$ is the usual symbol "n choose k"$= n!/k!(n-k)!$.

Exercise 2.0.15 Check that $\binom{n}{k} p^k q^{n-k}$ rises for $k < np$ and falls for $k > np$ – in particular, it peaks within one unit of the mean $np = \mathbb{E}(\mathbf{s}_n)$. What is its approximate value at the peak for large n?

Of course, these Bernoulli trials can be built over the standard Lebesgue space $([0,1], \mathfrak{F}, \mathrm{m})$ by Wiener's trick, but I prefer to think, more in context, that $\mathbf{X}$ is the (sample) space of points $\mathbf{x} = (\mathbf{e}_1, \mathbf{e}_2, \mathbf{e}_3, \dots)$, $\mathfrak{F}$ the smallest field containing all the events $(\mathbf{e}_n = 1)$ and $(\mathbf{e}_n = 0)$, and $\mathbb{P}$ signifies the probabilities induced on $\mathbf{X}$ by the Lebesgue measure m.

2.1 The law of large numbers (LLN)

Write $\mathbf{s}_0 = 0$ and $\mathbf{s}_n = \mathbf{e}_1 + \cdots + \mathbf{e}_n$ as before. The law of large numbers states that $\mathbf{s}_n/n \simeq p$ for large n, which is to say that the empirical frequency, $\mathbf{s}_n/n = \frac{1}{n}\#(k \leq n : \mathbf{e}_k = 1)$ of success in a large number of trials recapitulates the a priori probability of success in a single trial – in short, what you *see* is the same as what you *could have figured out beforehand*. The law comes in two versions:

2.1.1 The weak law

The weak law   states that $\lim\limits_{n \uparrow \infty} \mathbb{P}[p - \varepsilon \leq \mathbf{s}_n/n < p + \varepsilon] = 1$ for any $\varepsilon > 0$, which is to say that the law holds with overwhelming probability.

Proof of the weak law.   This is a simple application of Chebyshev's inequality. Write $\mathbf{e}' = \mathbf{e} - p$ so that $\mathbb{E}(\mathbf{e}') = 0$ and $\mathbb{E}(\mathbf{e}')^2 = pq$, and let $\mathbf{s}'_n = \mathbf{e}'_1 + \cdots + \mathbf{e}'_n$. Then

$$\mathbb{P}(|\mathbf{s}'_n| > n\varepsilon) \leq \frac{\mathbb{E}(\mathbf{s}'_n)^2}{\varepsilon^2 n^2} = \frac{1}{\varepsilon^2 n^2} \sum_{1 \leq i,j \leq n} \mathbb{E}(\mathbf{e}'_i \mathbf{e}'_j).$$

Here, $\mathbb{E}(\mathbf{e}'_i \mathbf{e}'_j) = 0$ if $i \neq j$, by independence of the $\mathbf{e}$'s, so the sum reduces to npq, and $\mathbb{P}(|\mathbf{s}'_n| > n\varepsilon) \leq pq/\varepsilon n$ which decreases to 0 as $n \uparrow \infty$. That does it.

2.1.2 The individual or strong law

This is the sharper version. The statement is now $\mathbb{P}[\lim\limits_{n \uparrow \infty} \mathbf{s}_n/n = p] = 1$.

The weak law does not prohibit the events $Z_n = (p - \varepsilon \leq \mathbf{s}_n/n < p + \varepsilon)$ from wandering about in the sample space so that any individual $\mathbf{x} = (\mathbf{e}_1, \mathbf{e}_2, \mathbf{e}_3, \dots)$ may fall in Z_n for most, but not for all, even very large n. The content of the strong law is that, with probability 1, $\mathbf{x} \in Z_n$ without exception from some large n on, in which it is to be understood that what is "large" can (and will) depend upon the particular $\mathbf{x}$ in hand.

Proof of the strong law.   The estimate employed for the weak law is not good enough to permit the first Borel–Cantelli lemma to be used. But wait. Replace, in the last line of §2.1.1, n by n^2 and ε by $n^{-1/3}$ to obtain

$$\mathbb{P}\left[|\mathbf{s}'_{n^2}| > n^{5/3}\right] \leq pq n^{-4/3}.$$

Now Borel–Cantelli works, showing that

$$\mathbb{P}\left[|\mathbf{s}'_{n^2}| \le n^{5/3} \text{ for } n \uparrow \infty\right] = 1.$$

Next take $n^2 \le m < (n+1)^2$ and make m (and so also n) increase to ∞. Obviously,

$$\begin{aligned} |\mathbf{s}'_m| &\le |\mathbf{s}'_{n^2}| + (n+1)^2 - n^2 \quad \text{since } |\mathbf{e}'| \le 1 \\ &\le n^{5/3} + 2n + 1 \\ &< 2m^{5/6}, \end{aligned}$$

i.e.

$$\mathbb{P}\left[|\mathbf{s}'_m/m| \le 2m^{-1/6} \text{ as } m \uparrow \infty\right] = 1,$$

proving the strong law and somewhat more.

2.1.3 Application of the weak law: polynomial approximation

It is nice to see so simple an observation in action, producing an easy proof of a much less obvious fact. I refer to S. Bernstein's pretty proof (1912) of Weierstraß's theorem (1840) that any function $F \in \mathrm{C}[0,1]$ can be approximated by a polynomial G so as to make $\|F - G\| = \max_{0 \le x \le 1} |F(x) - G(x)|$ as small as you like.

Aside. This illustrates an important point; that it is always advantageous to have two or more different pictures of a mathematical question so that you may take the easier route to the answer. Here, the matter may be transparently expressed by means of Bernoulli trials – contrariwise, large factorials are best understood, not combinatorially, but by Stirling's formula.

Proof. Take Bernoulli trials with success probability $0 < x < 1$, introduce the polynomial

$$G(x) = \mathbb{E}\left[F\left(\frac{\mathbf{s}_n}{n}\right)\right] = \sum_{k=0}^{n} \binom{n}{k} x^k (1-x)^{n-k} F\left(\frac{k}{n}\right),$$

and write

$$G(x) - F(x) = \mathbb{E}\Big[F\Big(\frac{\mathbf{S}_n}{n}\Big) - F(x)\Big]$$
$$= \mathbb{E}\Big[F\Big(\frac{\mathbf{S}_n}{n}\Big) - F(x), \Big|\frac{\mathbf{S}_n}{n} - x\Big| \le \varepsilon\Big] \qquad = (1)$$
$$+ \mathbb{E}\Big[F\Big(\frac{\mathbf{S}_n}{n}\Big) - F(x), \Big|\frac{\mathbf{S}_n}{n} - x\Big| > \varepsilon\Big] \qquad = (2).$$

Let $\omega(h)$ be the modulus of continuity of F: *to wit,* $\max\big(|F(b) - F(a)| :$ $|b - a| \le h\big)$. Then the absolute value of (1) cannot exceed $\omega(\varepsilon)$, nor can that of (2) exceed

$$2\|F\| \times \mathbb{P}(|\mathbf{s}_n'| > n\varepsilon) \le 2\|F\| \frac{x(1-x)}{\varepsilon^2 n},$$

and both can be made small by choice of small $\varepsilon > 0$ and large n, in that order.

Exercise 2.1.1 $\|F - G\| \le \frac{3}{2}\|F\|^{1/3}\|F'\|^{2/3} \times n^{-1/3}$ if F is piecewise smooth.

2.1.4 Application of the strong law: empirical distributions

These provide a nice illustration of the individual law of large numbers. Let $\mathbf{x}_1, \mathbf{x}_2, \ldots$ be independent copies of some random variable $\mathbf{x}$ with common distribution $F(\lambda) = \mathbb{P}(\mathbf{x} \le \lambda)$. Evidently, the empirical distribution

$$F_n(\lambda) = \frac{1}{n}\#(k \le n : \mathbf{x}_k \le \lambda)$$
$$= \frac{1}{n}\sum_{k=1}^{n} \mathbf{1}_{(-\infty,\lambda]}(\mathbf{x}_k)$$
$$\simeq \mathbb{E}[\mathbf{1}_{(-\infty,\lambda]}(\mathbf{x})]$$
$$= \mathbb{P}(\mathbf{x} \le \lambda) = F(\lambda)$$

for each fixed λ, separately. The reason is obvious: the variables in the second line are just independent Bernoulli trials with success probability $p = F(\lambda)$. I will come back to this subject in §§5.2 and 6.8 for a deeper look.

Runs

The individual law states that the empirical frequency, $\lim_{n\uparrow\infty}\frac{1}{n}\#(k\leq n : \mathbf{e}_k = 1)$ is nothing but $\mathbb{P}(\mathbf{e}_1 = 1) = p$. Let's elaborate the question and ask: what is the empirical frequency of runs such as $\mathbf{e}_n = 0$ followed by $\mathbf{e}_{n+1} = 1$? A run is now indicated by $(1 - \mathbf{e}_n)\mathbf{e}_{n+1} = 1$, so the empirical frequency $\lim_{n\uparrow\infty}\frac{1}{n}\sum_{k=1}^{n}(1 - \mathbf{e}_k)\mathbf{e}_{k+1}$ is or ought to be qp since $(1 - \mathbf{e}_n)\mathbf{e}_{n+1}$ can only be 1 or 0 with success probability $\mathbb{P}(\mathbf{e}_1 = 0, \mathbf{e}_2 = 1) = qp$. I want to apply LLN to these Bernoulli-like variables, but they are not independent: for example, $(1 - \mathbf{e}_1)\mathbf{e}_2 \times (1 - \mathbf{e}_2)\mathbf{e}_3$ is always 0. This can be remedied by splitting the sum in two according as $k = 0$ or $1 \mod 2$. Then, e.g. $(1-\mathbf{e}_1)\mathbf{e}_2$ and $(1-\mathbf{e}_3)\mathbf{e}_4$ are independent, LLN applies each sum separately, and

$$\lim_{n\uparrow\infty}\frac{1}{n}\sum_{k=1}^{n}(1-\mathbf{e}_k)\mathbf{e}_{k+1} = \lim_{n\uparrow\infty}\frac{1}{n}\sum_{\substack{1\leq k\leq n \\ k \text{ odd}}} + \lim_{n\uparrow\infty}\frac{1}{n}\sum_{\substack{1\leq k\leq n \\ k \text{ even}}} = \frac{1}{2}qp+\frac{1}{2}qp = qp,$$

as you would like. The same applies to any pattern such as 10010: $\mathbf{e}_n(1-\mathbf{e}_{n+1})(1-\mathbf{e}_{n+2})\mathbf{e}_{n+3}(1-\mathbf{e}_{n+4}) = 1$ now indicates a run, and splitting the sum for the empirical frequency into 5 pieces according as $k = 0, 1, 2, 3$, or $4 \mod 5$ reestablishes independence, whereupon LLN produces the empirical frequency $\mathbb{P}(\mathbf{e}_1 = 1, \mathbf{e}_2 = 0, \mathbf{e}_3 = 0, \mathbf{e}_4 = 1, \mathbf{e}_5 = 0) = p^2q^3$ you would hope to see, the moral being that any such pattern or run of which $\mathbf{e}_1, \mathbf{e}_2, \mathbf{e}_3, \ldots$ is capable is seen, not just once, but over and over with the "correct" empirical frequency.

Improved estimates

The crude bound employed above, namely

(1) $\mathbb{P}(|\mathbf{s}_n'| > n\varepsilon) \leq pq/\varepsilon^2 n$

can be vastly improved. For example,

$$\mathbb{E}(\mathbf{s}_n')^4 = \mathbb{E}\left[\sum_{1\leq i,j,k,l\leq n}\mathbf{e}_i'\mathbf{e}_j'\mathbf{e}_k'\mathbf{e}_l'\right]$$

$$= \text{the } n \text{ ways of picking 4 equal indices} \times \mathbb{E}(\mathbf{e}')^4 = pq^4 + qp^4$$

$$+ \text{ the } 3n(n - 1) \text{ ways of picking 2 unequal pairs} \times [\mathbb{E}(\mathbf{e}')^2]^2$$

$$= (pq)^2,$$

so

(2) $\mathbb{P}(|\mathbf{s}_n'| > n\varepsilon) \leq \mathbb{E}(\mathbf{s}_n')^4/\varepsilon^4 n^4$ is comparable to $1/\varepsilon^4 n^2$.

Exercise 2.1.2 Continue in this way, using higher powers of s_n' to verify that $\mathbb{P}(|s_n'| > n\varepsilon)$ is comparable

to $1/\varepsilon^6 n^3$,

to $1/\varepsilon^8 n^4$,

and so on. This will lead to $\mathbb{P}(|s_n'| < n^{(1/2)+}$ as $n \uparrow \infty) = 1$. Check it out.

Exercise 2.1.3 Use Exercise 2.1.2 to improve Exercise 2.1.1 to $\|F - G\| \le$ a constant depending upon $F \times n^{(-1/2)+}$. This is pretty sharp as the next exercise shows.

Exercise 2.1.4 For the function

$$F(x) = \begin{cases} x & x \le 1/2 \\ (1-x) & x > 1/2 \end{cases}$$

and odd n, the error $F(1/2) - G(1/2) = 2 \times \sum_{k < n/2} \binom{n}{k} 2^{-n}(1/2 - k/n)$ is approximately

$$\frac{2}{\sqrt{n}} \times \left(\int_0^\infty \frac{e^{-x^2/2}}{\sqrt{2\pi}} x \, dx = 1 \right).$$

Hint: Argue as in the first computation of $C = \sqrt{2\pi}$ in §1.5.

But why fool with powers? Let's go to the exponential-type Chebyshev inequality:

$$\mathbb{P}(s_n' > \beta) \le e^{-\alpha\beta} \, \mathbb{E}(e^{\alpha s_n'}), \quad \text{with } \alpha \text{ and } \beta > 0.$$

Here, you are free to decide how β should depend upon n – something like $\gamma\sqrt{n}$ looks right if γ would increase to ∞ slowly – and then to adjust α to get the best bound of this kind.

Now

$$\mathbb{E}(e^{\alpha s_n'}) = \left[\mathbb{E}(e^{\alpha e'}) \right]^n = [pe^{\alpha q} + qe^{-\alpha p}]^n \quad \text{by independence,}$$

and α wants to be small like $\beta/npq = \gamma/\sqrt{n} \, pq$: in fact, at the minimum with respect to α of $e^{-\alpha\beta}[pe^{\alpha q} + qe^{-\alpha p}]^n$,

$$\gamma\sqrt{n} = \beta = npq(e^{\alpha q} - e^{-\alpha p}) \times (pe^{\alpha q} + qe^{-\alpha p})^{-1} \simeq \alpha npq$$

if α is small. Take that at face value even if it's not perfect, i.e. let $\alpha = \gamma/\sqrt{n}\,pq$. Then

$$e^{-\alpha\beta}[pe^{\alpha q} + qe^{-\alpha p}]^n$$

$$< e^{-\gamma^2/pq} \times \left[p\left(1 + \alpha q + \frac{\alpha^2}{(2-)}q^2\right) + q\left(1 - \alpha p + \frac{\alpha^2 p^2}{(2-)}\right)\right]^n$$

$$< e^{-\gamma^2/pq} \times \left[\frac{1 + \alpha^2 pq}{(2-)}\right]^n$$

$$= e^{-\gamma^2/pq} \times \left[1 + \frac{\gamma^2}{(2-)npq}\right]^n$$

$$< e^{-\gamma^2/(2+)pq},$$

with numbers 2− a little less than 2 and 2+ a little more. Then choose $\gamma = \sqrt{(2+)pq\log n}$ with a new 2+ a little bigger than the old to produce $\mathbb{P}\left(s_n' > \sqrt{(2+)pq\,n\log n}\right) \le e^{-(1+)\log n} = n^{-(1+)}$ and conclude from the first Borel–Cantelli lemma that $\mathbb{P}\left[s_n' \le \sqrt{(2+)pq\,n\log n} \text{ as } n \uparrow \infty\right] = 1$.

Exercise 2.1.5 Make a similar estimate of s_n' from below to obtain the individual law in the sharpened form

$$\mathbb{P}\left[|s_n - np| \le \sqrt{(2+)pq\,n\log n} \text{ as } n \uparrow \infty\right] = 1.$$

Some History

The estimate of Example 2.1.5 is due to Hardy–Littlewood (1914). Khinchine (1924) improved the error to constant $\times \sqrt{n\log\log n}$, in which loglog means log(log), whence the name of the definitive "law of the iterated logarithm":

$$\mathbb{P}\left[\limsup_{n\uparrow\infty} \frac{s_n - np}{\sqrt{2pq\,n\log\log n}} = +1\right] = 1$$

and

$$\mathbb{P}\left[\liminf_{n\uparrow\infty} \frac{s_n - np}{\sqrt{2pq\,n\log\log n}} = -1\right] = 1$$

also due to him. I come to it shortly in §2.3.

2.2 The Gaussian approximation (CLT)

This is the "central limit theorem" going back to de Moivre (1718) for $p = 1/2$ and to Laplace (1812) for general $0 < p < 1$. It refines upon the law of large numbers as follows:

$$\lim_{n\uparrow\infty} \mathbb{P}\left[a \leq \frac{\mathbf{s}_n - np}{\sqrt{npq}} < b\right] = \int_a^b \frac{e^{-x^2/2}}{\sqrt{2\pi}}\, dx \quad \text{for any } a < b$$

or, as you may say,

$\mathbf{s}_n = $ a bulk effect $np + $ Gaussian fluctuations on the scale $\sqrt{npq}$.

It is amusing to know that de Moivre was a bookie working in London, taking bets on the horses, prize fights, what you will – a practical appraiser of risk and, evidently, an accomplished mathematician as well.

This kind of bulk/fluctuation picture is at the heart of classical probability and its practical applications, of which much more below. I emphasize that unlike the law of large numbers, the Gaussian refinement is a weak law, only: the ultimate behavior of $(\mathbf{s}_n - np)/\sqrt{npq}$ is subject to Kolmogorov's 01 law, so there are just two possibilities: with probability 1, either $\lim_{n\uparrow\infty}(\mathbf{s}_n - np)/\sqrt{npq}$ fails to exist or else it *does* exist and is constant $(= 0)$, contradicting de Moivre–Laplace.

Exercise 2.2.1 Check the second alternative: that

$$\lim_{n\uparrow\infty}(\mathbf{s}_n - np)/\sqrt{npq},$$

if it exists, can only be 0.
Hint: $\mathbf{s}_{2n}$ is the sum of two independent copies of $\mathbf{s}_n$.

Proof of CLT. The computation proceeds (by hand) much as in the verification of $C = \sqrt{2\pi}$ in §1.5, but it seems best to spell it out anyhow. Fix $-\infty < a < b < \infty$ and look at

$$\mathbb{P}\left[a \leq \frac{\mathbf{s}_n - np}{\sqrt{npq}} < b\right] = \sum_{np+a\sqrt{npq}\leq k<np+b\sqrt{npq}} \binom{n}{k}p^k q^{n-k}.$$

Here, both k and $n - k$ are comparable to n, so all the factorials may be well approximated by Stirling's formula, i.e. the general term of the sum may be replaced by

$$\frac{n^{n+1/2}p^k q^{n-k}}{k^{k+1/2}(n-k)^{n-k+1/2}} \times \frac{1}{\sqrt{2\pi}}.$$

Now write $k = np + x\sqrt{npq}$ with $a \leq x < b$ and $\Delta x = 1/\sqrt{npq}$ in response to $\Delta k = 1$. In this format, $\sqrt{2\pi} \times$ the general summand is

$$\frac{n^{n+1}p^{k+1/2}q^{n-k+1/2}}{k^{k+1/2}(n-k)^{n-k+1/2}} \times \frac{1}{\sqrt{npq}}$$

$$= \left[\left(\frac{k}{np}\right)^{k+1/2}\left(\frac{n-k}{nq}\right)^{n-k+1/2}\right]^{-1} \times \frac{1}{\sqrt{npq}}$$

$$= \left[\left(1+\frac{x}{\sqrt{n}}\frac{\sqrt{q}}{\sqrt{p}}\right)^{np+x\sqrt{npq}+1/2}\left(1-\frac{x}{\sqrt{n}}\frac{\sqrt{p}}{\sqrt{q}}\right)^{nq-x\sqrt{npq}+1/2}\right]^{-1} \times \Delta x$$

in which the $1/2$ in the two exponents can be neglected since $x/\sqrt{n}$ is small. Take the logarithm of what's inside the bracket and compute up to (but not including) terms in $n^{-1/2}$. You find

$$(np + x\sqrt{npq})\log\left(1+\frac{x}{\sqrt{n}}\frac{\sqrt{q}}{\sqrt{p}}\right) + (nq - x\sqrt{npq})\log\left(1-\frac{x}{\sqrt{n}}\frac{\sqrt{p}}{\sqrt{q}}\right)$$

$$= (np + x\sqrt{npq})\left(\frac{x}{\sqrt{n}}\frac{\sqrt{q}}{\sqrt{p}} - \frac{x^2}{2n}\frac{q}{p}\right)$$

$$+ (nq - x\sqrt{npq})\left(-\frac{x}{\sqrt{n}}\frac{\sqrt{p}}{\sqrt{q}} - \frac{x^2}{2n}\frac{p}{q}\right)$$

$$= \left(x\sqrt{npq} - \frac{x^2}{2}q + x^2 q\right) + \left(-x\sqrt{npq} - \frac{x^2}{2}p + x^2 p\right)$$

$$= \frac{x^2}{2},$$

so

$$\mathbb{P}\left[a \leq \frac{\mathbf{s}_n - np}{\sqrt{npq}} < b\right] \simeq \sum_{a \leq x < b}\frac{e^{-x^2/2}}{\sqrt{2\pi}}\Delta x \simeq \int_a^b \frac{e^{-x^2/2}}{\sqrt{2\pi}}\,dx.$$

That's fine for finite $a < b$, and if you want it for $a = -\infty$ and/or $b = +\infty$ you have only to notice that

$$\mathbb{P}\left[\left|\frac{\mathbf{s}_n - np}{\sqrt{npq}}\right| \geq c\right] \leq \mathbb{E}\left[\frac{\mathbf{s}_n - np}{\sqrt{npq}}\right]^2 \times \frac{1}{c^2} = \frac{1}{c^2}$$

is small for large c. Note that for $p = 1/2$,

$$\mathbb{P}\left[\frac{\mathbf{s}_n - np}{\sqrt{npq}} = 0\right] = \begin{cases} 0 & \text{if } n \text{ is odd,} \\ \binom{n}{n/2}2^{-n} \simeq \sqrt{\frac{2}{\pi n}} & \text{if } n \text{ is even,} \end{cases}$$

so this is the kind of error you may expect.

Important Note

I want you to know that *beyond* LLN and CLT, there is *no*, so to say, *philosophically correct refinement*. You might hope that $\mathbf{s}_n = np +$ Gauss $\times \sqrt{npq}+$ corrections on more delicate scales $n^0, n^{-1/2}$, etc., but that is simply *wrong*. I do not have a suitable example for you yet, but see §11.6.3 for a convincing illustration of this mantra.

2.3 The law of iterated logarithm

The Hardy–Littlewood estimate of §2.1 states that $\mathbf{s}'_n \le \sqrt{(2+)pqn \log n}$ for $n \uparrow \infty$. On the other hand, $\limsup n^{-1/2}\mathbf{s}'_n = +\infty$: indeed, it is measurable over Kolmogorov's tail field and so constant $(= c \le +\infty)$, the same for almost every path. But if $c < \infty$ and $f(x) = 1 - \exp(-(x-c)^+)$, then by CLT and Fatou,

$$0 < \int_0^\infty f(x)\,\frac{e^{-x^2/2}}{\sqrt{2\pi}}\,dx = \lim_{n\uparrow\infty} \mathbb{E}\Big[f\Big(\frac{\mathbf{s}'_n}{\sqrt{npq}}\Big)\Big] \le \mathbb{E}\Big[\limsup_{n\uparrow\infty}\Big] = f(c) = 0,$$

and that won't do. A. Ya. Khinchine (1924) discovered the truth of the matter: *to wit* his law of the iterated logarithm, stated in §2.1:

$$\mathbb{P}\left[\limsup \frac{\mathbf{s}'_n}{\sqrt{2pqn\,\mathrm{loglog}\,n}} = 1\right] = 1$$

and similarly

$$\mathbb{P}\left[\liminf \frac{\mathbf{s}'_n}{\sqrt{2pqn\,\mathrm{loglog}\,n}} = -1\right] = 1.$$

I deal with the first of these. Then the second follows by exchange of $\mathbf{e}$ and $\mathbf{e}' = 1 - \mathbf{e}$.

Proof of the cheap half: $\mathbb{P}(\limsup \le 1) = 1$.

Step 1 provides control of $\max_{k\le n} \mathbf{s}'_k$ in terms of $\mathbf{s}'_n$ itself, as in

$$\mathbb{P}\left[\max_{k\le n} \mathbf{s}'_k \ge m\right] \le \text{a constant multiple of } \mathbb{P}\left[\mathbf{s}'_n \ge m\right].$$

The argument is much like the proof of Doob's inequality in §1.8.3, but I spell it out anyhow. Fix $m > 0$ and introduce the passage time $T = \min(k : \mathbf{s}'_k \ge m)$. $\mathbb{P}(\mathbf{s}'_n \ge 0)$ is positive and tends to $1/2$ as $n\uparrow\infty$ by

CLT, so with $C^{-1} = \min \mathbb{P}(\mathbf{s}'_n \geq 0)$, you have

$$\mathbb{P}\left[\max_{k \leq n} \mathbf{s}'_k \geq m\right]$$

$$= \mathbb{P}(T \leq n)$$

$$= \sum_{k=0}^{n} \mathbb{P}(T = k)$$

$$\leq \sum_{k=0}^{n} \mathbb{P}(T = k) C \; \mathbb{P}(\mathbf{s}'_n - \mathbf{s}'_k \geq 0) \qquad \begin{array}{l} \text{since the event } (T = k) = \\ (\mathbf{s}'_1, \ldots, \mathbf{s}'_{k-1} < m, \mathbf{s}'_k \geq m) \\ \text{is independent of } \mathbf{s}'_n - \mathbf{s}'_k \end{array}$$

$$= C \sum_{k=0}^{n} \mathbb{P}(T = k, \mathbf{s}'_n - \mathbf{s}'_k \geq 0)$$

$$\leq C \sum_{k=0}^{n} \mathbb{P}(T = k, \mathbf{s}'_n \geq m)$$

$$= C \, \mathbb{P}(\mathbf{s}'_n \geq m) \qquad \qquad .$$

This will be used in the next step in the form

$$\mathbb{P}\left[\max_{k \leq n} \mathbf{s}'_k \geq m\right] \leq e^{-m^2/(2+)npq} \;\; \text{for} \;\; \frac{m}{\sqrt{n}} \uparrow \infty,$$

which is immediate from the final "improved estimate" $\mathbb{P}(\mathbf{s}'_n > \gamma\sqrt{n}) \leq e^{-\gamma^2/(2+)pq}$ of §2.1.

Step 2. Take a number ϑ a shade bigger than 1 and a number β just a little bigger than the $(2+) \equiv \alpha$ just before, chosen so that $\beta/\alpha\vartheta = 1$. Then

$$\mathbb{P}\left[\mathbf{s}'_k \geq \sqrt{\beta pq \, k \log\log k} \text{ for some } k \text{ between } \vartheta^{n-1} \text{ and } \vartheta^n\right]$$

$$\leq \mathbb{P}\left[\max_{k \leq \vartheta^n} \mathbf{s}'_k \geq \sqrt{\beta pq \, \vartheta^{n-1} \log \vartheta^{n-1}}\right]$$

$$\leq \exp\left[-\beta pq \, \vartheta^{n-1} \log\log \vartheta^{n-1}/\alpha pq \, \vartheta^n\right]$$

$$\leq \exp\left[-(1+) \log n\right]$$

$$= n^{-(1+)},$$

by choice of $\vartheta > 1$, with a new $(1+)$ just a shade smaller than $\beta/\alpha\vartheta$. This is the general term of a convergent sum, so the first Borel–Cantelli lemma does the rest.

Proof of the second half: $\mathbb{P}(\limsup \geq 1) = 1$.

Now let ϑ be a large whole number and look at the independent events

$$B_n : \mathbf{s}'_{\vartheta^n} - \mathbf{s}'_{\vartheta^{n-1}} \geq \sqrt{(2-)\vartheta^n \log\log \vartheta^n},$$

pretending, by abuse of notation, that ϑ^n is a whole number.

Step 1. If $\sum \mathbb{P}(B_n)$ were to diverge, then by the second Borel–Cantelli lemma and the cheap half of the proof, it would happen, infinitely often, that

$$\mathbf{s}'_{\vartheta^n} \geq \sqrt{(2-)pq\,\vartheta^n \log\log \vartheta^n} - \sqrt{(2+)pq\,\vartheta^{n-1} \log\log \vartheta^{n-1}}$$

$$\geq \sqrt{(2-)pq\,\vartheta^n \log\log \vartheta^n} \times \left[1 - \sqrt{\frac{(2+)}{(2-)}\frac{1}{\vartheta}}\right]$$

$$> \sqrt{(2-)pq\,\vartheta^n \log\log \vartheta^n}$$

with a new $(2-)$ a bit smaller than before since ϑ can be taken as big as you want.

Step 2 validates Step 1 by underestimating $\mathbb{P}(B_n)$ by the general term of a divergent sum. Write $m = \vartheta^n - \vartheta^{n-1}$ and note that $\vartheta^n > m = \vartheta^n(1 - 1/\vartheta) > (1-)\vartheta^n$ if ϑ is big. Now

$$\mathbb{P}(B_n) = \mathbb{P}\left[\mathbf{s}'_m \geq \sqrt{(2-)pq\,m \log\log m}\right]$$

$$= \sum_{k \geq \sqrt{(2-)pq\,m \log\log m}} \binom{m}{k} p^k q^{m-k}.$$

Cutting off the sum at $k = m^{2/3}$ makes both m and $m - k$ large. Then Stirling works fine, and the sum may be underestimated as in §2.2 by a multiple of

$$\int_{\sqrt{(2-)\log\log m}}^{\sqrt[6]{m}} \frac{e^{-x^2/2}}{\sqrt{2\pi}}\,dx$$

$$\simeq \frac{1}{\sqrt{2\pi}}\frac{1}{\sqrt{(2-)\log\log m}} e^{-(1-)\log\log m} \qquad \text{by Mills's ratio}$$

$$> n^{-(1-)} \qquad \text{with a new } (1-) \text{ a shade smaller than before.}$$

That is the general term of a divergent sum, and all is well.

Comment The cheap half of the proof requires tight control of $\mathbf{s}'_k$ for $\vartheta^{n-1} < k \leq \vartheta^n$. This is why $\vartheta > 1$ must be small. Contrariwise, for the second half, it is necessary that $\mathbf{s}'_{\vartheta^n} - \mathbf{s}'_{\vartheta^{n-1}}$ should dominate $\mathbf{s}'_{\vartheta^{n-1}}$. That is why ϑ must be big.

★**Finer details**   V. Strassen [1964] proved the following remarkable addendum to Khinchine's law. Let $\mathbf{f}_n(x) : 0 \le x \le 1$ be the broken line connecting the points $(k/n, \mathbf{s}'_k/\sqrt{npq}) : 0 \le k \le n$, and let K be the class of functions $f \in C[0,1]$ with $f(0) = 0$ and $f(x) = \int_0^x f'$, f' being Lebesgue measurable with $\int_0^1 (f')^2 \le 1$. Then

$$\mathbb{P}\left[\liminf_{n\uparrow\infty} \max_{0\le x\le 1} |\mathbf{f}_n(x) - f(x)| = 0 \text{ for } f \in K \text{ and not otherwise}\right] = 1.$$

It is an (unobvious) corollary that

$$\mathbb{P}\left[\limsup_{n\uparrow\infty} \frac{1}{n}\#(k \le n : \mathbf{s}'_k \ge c\sqrt{2pq\,k\,\mathrm{loglog}\,k}) = 1 - \mathrm{e}^{-4(1-1/c^2)}\right.$$

$$\left. \text{for } 0 \le c \le 1\right] = 1,$$

with the following extraordinary consequence: $\mathrm{e}^{4(1-1/c^2)} = \mathrm{e}^{-12}$ for $c = 1/2$ and this number is smaller than 10^{-5} but larger than 10^{-6} so that $\frac{1}{n}\#(k \le n : \mathbf{s}'_k > \frac{1}{2}\sqrt{2pqk\,\mathrm{loglog}\,k})$ is larger than .99999 for infinitely many $n \uparrow \infty$ but smaller than .999999 for all sufficiently large n ! Who would have thought it? I mean to say that the same type of thing happens to $\mathbf{s}'_k$ in comparison to $-\frac{1}{2}\sqrt{2pqk\,\mathrm{loglog}\,k}$, showing that $\mathbf{s}'_n$ is subject to unbelievably wild fluctuations. But more of this below in §5.8.

2.4 Large deviations

This topic represents a remarkable refinement of the law of large numbers, initiated by Cramér [1938]. Here, I will explain it in its simplest form, for Bernoulli trials with success probability $0 < p < 1$. You know already about the bulk effect, described by the law of large numbers $\mathbb{P}[\lim_{n\uparrow\infty} \mathbf{s}_n/n = p] = 1$, and about *small* deviations from that typical behavior, on the scale $\sqrt{n}$, described by CLT. The idea behind *large* deviations is exemplified by the next estimate, presented as a warm-up.

2.4.1 Cramér's estimate

Warning. The old $q = 1-p$ is now abandoned. Then for fixed $q \in (p, 1)$

$$\lim_{n\uparrow\infty} \frac{1}{n} \log \mathbb{P}(\mathbf{s}_n > nq) = -I(q)$$

with

$$I(q) = q\log\frac{q}{p} + (1-q)\frac{1-q}{1-p},$$

which is to say $\mathbb{P}(\mathbf{s}_n > nq) \simeq \mathrm{e}^{-nI}$, i.e. what is *not* typical is exponentially rare.

Detour Note the form of the number $-I$. It is a "relative entropy" of the general shape $\int \mathrm{d}Q \log(\mathrm{d}P/\mathrm{d}Q) = -\int \mathrm{d}Q/\mathrm{d}P \log(\mathrm{d}Q/\mathrm{d}P) \mathrm{d}P$. I say no more now, but do please take a look at §9.1.4 to see what "entropy" is all about.

Exercise 2.4.1 Use Gibbs's lemma, $x \log x - x + 1 > 0$ for $0 \le x \ne 1$, to verify that $I(q) > 0$ unless $q = p$.

Exercise 2.4.2 Check also that $I(q)$ increases with $q \ge p$, as will be wanted in a moment.

Proof of Cramér's estimate. $\mathbb{P}(\mathbf{s}_n > nq) = \sum_{k>nq} \binom{n}{k} p^k (1-p)^{n-k}$ is approximated with the help of Stirling's formula in its crude form: $\log n! = n \log n - n + \mathrm{O}(\log n)$ for $n \ge 2$, from which it is immediate that

$$\frac{1}{n} \log \binom{n}{k} = -\frac{k}{n} \log \frac{k}{n} - \left(1 - \frac{k}{n}\right) \log\left(1 - \frac{k}{n}\right) + \mathrm{O}\left(\frac{\log n}{n}\right),$$

independently of $0 \le k \le n$. Now very crudely, the sum for $\mathbb{P}(\mathbf{s}_n > nq)$ lies between its biggest term and n times that number, and if all you want is $\frac{1}{n} \log \mathbb{P}(\mathbf{s}_n > nq)$, then the two are indistinguishable for large n so that

$$\frac{1}{n} \log \mathbb{P}(\mathbf{s}_n > nq)$$

$$\simeq -\min_{k>nq} \left[\frac{k}{n} \log \frac{k}{n} + \left(1 - \frac{k}{n}\right) \log\left(1 - \frac{k}{n}\right) - \frac{k}{n} \log p \right.$$

$$\left. - \left(1 - \frac{k}{n}\right) \log(1 - p) \right]$$

$$= -\min_{m>q} \left[m \log \frac{m}{p} + (1 - m) \log \frac{1 - m}{1 - p} \right] \quad \text{with } m = \frac{k}{n}$$

$$\simeq -I(q) \quad \text{by Exercise 2.4.2.}$$

The complementary estimate $\mathbb{P}(\mathbf{s}_n < nq) \simeq \mathrm{e}^{-nI}$ for $q < p$ is proved in the same way.

2.4.2 Empirical distributions

Cramér's estimate is promoted to a second level in a more elaborate context. Replace Bernoulli's $\mathbf{e}_1$, $\mathbf{e}_2$, etc. by new independent variables $\mathbf{x}_1$, $\mathbf{x}_2$, etc. capable of $d = 3$ or more distinct values $x = x_1, \ldots, x_d$, with common distribution $\mathbb{P}(\mathbf{x} = x) = p_x > 0$, and introduce the empirical distribution

$$\mathbf{f}_n(x) \equiv \frac{1}{n}\#(k \le n : \mathbf{x}_k = x).$$

For fixed x, $\mathbf{f}_n(x)$ is just the empirical frequency of success in n Bernoulli trials with success probability p_x, so the law of large numbers, applied for each x separately, tells you that $\mathbb{P}\big[\lim_{n\uparrow\infty} \mathbf{f}_n = p\big] = 1$ in which p is now the vector $(p_x : x = x_1, \dots, x_d)$. That's typical behavior.

Now for what's *atypical*. Let Q be the compact simplex of all probability distributions $q = (q_1, \dots, q_d)$ and ∂Q its boundary where one or more of the individual probabilities vanish; let $A \subset Q$ be an open neighborhood of $q = p$; and let's estimate $\mathbb{P}(\mathbf{f}_n \notin A)$ in Cramér's style: with $B = $ the (compact) complement of A, large n, and a temporary re-indexing of the ps, you have

$$\frac{1}{n}\log \mathbb{P}(\mathbf{f}_n \notin A)$$

$$= \frac{1}{n}\log\left[\sum_{\substack{n_1+\cdots+n_d=n \\ (\frac{n_1}{n},\dots,\frac{n_d}{n})\in B}} \frac{n!}{n_1!\dots n_d!} p_1^{n_1} \cdots p_d^{n_d} \right]$$

$$\simeq - \min_{(\frac{n_1}{n},\dots,\frac{n_d}{n})\in B} \sum_{i=1}^{d} \left[\frac{n_i}{n}\log\frac{n_i}{n} - \log p_i \right] \qquad \begin{array}{l}\text{as you will check by the}\\ \text{same reasoning as before}\end{array}$$

$$\simeq - \min_{q\in B} I(q) \quad \text{with } I(q) = \sum_{i=1}^{d} q_i \log\frac{q_i}{p_i},$$

or, what is more or less the same,

$$\mathbb{P}(\mathbf{f}_n \notin A) \simeq \exp\left[-n \min_{q\in B} I(q) \right].$$

Exercise 2.4.3 $-I$ is a relative entropy, as before. Check that $I(q) > 0$ unless $q = p$ and also $I(q) \simeq \frac{1}{2}\sum_{i=1}^{d}(q_i - p_i)^2/p_i$ near $q = p$, exhibiting $I(q)$ as a kind of distance from q to p. Note that A could be taken as the open region where $I(q) < h$, in which case $\mathbb{P}(\mathbf{f}_n \notin A) \simeq e^{-nh}$.

2.4.3 The descent

Now it's not hard to "descend" from $\mathbf{f}_n$ to the sums $\mathbf{s}_n = \mathbf{x}_1 + \cdots + \mathbf{x}_n$ and *their* large deviations from the bulk estimate

$$\frac{\mathbf{s}_n}{n} = \frac{1}{n}\sum_{k=1}^{n} \mathbf{x}_k = \sum_x x\mathbf{f}_n(x) \simeq \sum_x x p_x = \mathbb{E}(\mathbf{x}).$$

In fact,

$$\mathbb{P}\left(\frac{\mathbf{S}_n}{n} \notin A\right) \simeq \exp\left[-n \min_{q \in B} I(q)\right]$$

where A is now an open neighborhood of the number $\mathbb{E}(\mathbf{x})$, $B = Q \cap (q : \sum_x x q_x \notin A)$, and q has been re-indexed by $x = x_1, \ldots, x_d$ for uniformity of notation. The minimum of I over B is explicated in two stages.

Stage 1 is to minimize I over B, subject to the natural constraints

(1) $q_x \geq 0$,

(2) $\sum q_x = 1$, plus the additional condition

(3) $\sum x q_x = m$ for fixed m strictly between $\min x$ and $\max x$.

Obviously, the minimum is taken on, B being compact and $I \in C(B)$. (3) implies that the minimum is not on ∂Q. Then, ignoring the first constraint, Lagrange's recipe says that at this place $\partial I/\partial q_x = \log(q_x/p_x) + 1 = a + bx$, i.e. $q_x = p_x e^{bx}/Z$ with normalizer $Z = \sum p_x e^{bx}$ to satisfy (2). (3) now fixes b: in fact, $\sum x q_x = \partial \log Z/\partial b$ approximates $\max x$ or $\min x$ as $b \uparrow \infty$ or $\downarrow -\infty$, so you can hit any m between these two extremes, and indeed just once in view of $\partial^2 \log Z/\partial b^2 > 0$, which you will check please. Notice that the q so produced is necessarily positive so (1) was superfluous anyhow.

The upshot of this first stage is then

$$\min\left(I : q \in B \text{ and } \sum x q_x = m\right) = \sum \frac{q_x}{Z} e^{bx}(bx - \log Z)$$
$$= bm - \log Z$$
$$\equiv J(m).$$

Stage 2 is to determine $K = \min J(m)$ for $m \notin A$. Now $m = \partial \log Z/\partial b$ and $\partial m/\partial b = \partial^2 \log Z/\partial b^2$ is positive, so you may think of $J(m)$ as a function of b. Then

$$\partial J/\partial b = m + b \,\partial m/\partial b - \partial \log Z/\partial b = b \,\partial m/\partial b$$

shows that $\partial J/\partial b$ vanishes only if $b = 0$, in which case $m = \mathbb{E}(\mathbf{x}) \in A$, but that isn't wanted. It follows that $\partial J/\partial b > 0$ to the right of A since m tends to $\max x$ at $b = +\infty$, and similarly, $\partial J/\partial b < 0$ to the left of A since m tends to $\min x$ at $b = -\infty$ – in short, K is the smaller of the two values of J at ∂A, and you have $\mathbb{P}\left(\frac{\mathbf{S}_n}{n} \notin A\right) \simeq e^{-nK}$, much as for simple Bernoulli trials.

The function $J(m)$ may be expressed as $\max_{c \in \mathbb{R}}[cm - \log Z(c)]$: indeed, for $c \uparrow \infty$, $\log Z \simeq c \max x$ so that $cm - \log Z(c)$ is large negative, and similarly for $c \downarrow -\infty$, so the minimum really exists. There $m = \partial \log Z / \partial c$, and this happens only at $c = b$, as you know already – in short, $J(m)$ is the Legendre transform or dual of $\log Z$. I explain.

2.4.4 Legendre–Fenchel duality

The idea goes back to Legendre (1820). Fenchel [1949] perfected it. I present it here in its simplest form. Let $F \colon \mathbb{R} \to \mathbb{R}$ be smooth and strictly convex, i.e. $F''(x) > 0$, and define its Legendre dual by the rule $G(y) = \max_{x \in \mathbb{R}} [xy - F(y)]$ for each fixed y separately. Obviously, the maximum really exists if, as I now suppose, $F'(-\infty) = -\infty$ and $F'(+\infty) = \infty$, and $y = F'(x)$ at that place. Now F' is strictly increasing, so it has a nice inverse function $(F')^{-1}$, and $G(y) = y(F')^{-1}(y) - F((F')^{-1}(y))$. Here $G'(y) = (F')^{-1}(y)$, as you will check, which is to say $G'(F')$ is the identity, and from $G''(F')F'' = 1$, you see that G is a function of the same type as F. Now repeat. I mean compute $\max_{y \in \mathbb{R}} [xy - G(y)]$: at the maximum $x = G'(y)$, which is to say $y = F'(x)$ and $xy - G(y) = F(x)$, i.e. with the notation $G = \hat{F}$, you have $\hat{\hat{F}} = F$. That is the duality.

The conditions imposed are more than is really called for, as the next two exercises illustrate.

Exercise 2.4.4   Take $F(x) = e^x$. Then for $y > 0$, $G(y) = \max[xy - F(x)] = y \log y - y$ and $\max_{y>0}[xy - G(y)] = e^x$.

Exercise 2.4.5   Take $F(x) = x \log x + (1 - x) \log(1 - x)$ for $0 < x < 1$ and $G(y) = \max_{0<x<1}[xy - F(x)]$ for unrestricted y. Then $\max_{-\infty<y<\infty}[xy - G(y)] = F(x)$, as it should be.

$\star$**Exercise 2.4.6**   The duality extends to the second level prior to the descent. $I(q)$ is convex in Q, as you will check; $\max_{q \in Q}[v \cdot q - I(q)] = \log Z(v)$ for fixed $v \in \mathbb{R}^d$ where now $Z(v) = \sum_x p_x e^{v_x}$; and $\max_{v \in \mathbb{R}^d}[v \cdot q - \log Z(v)] = I(q)$ for fixed $q \in Q$. Note that the first max might have been assumed on ∂Q. That must be ruled out.

2.4.5 General variables

Now let $\mathbf{x}_1$, $\mathbf{x}_2$, etc. be independent copies of any variable $\mathbf{x}$, with common distribution function $P(x)$, subject to $Z(b) = \int e^{bx} \, dP < \infty$ for any $b \in \mathbb{R}$. The empirical distribution function $\mathbf{f}_n(x) = \frac{1}{n}(k \leq n : \mathbf{x}_k \leq x)$

tends to $P(x)$, as in §2.4.2, and you have an estimate of the now familiar type:

$$\mathbb{P}(\mathbf{f}_n \notin A) \simeq \exp\left[-n \min_{Q \in B} I(Q)\right],$$

in which A is an (open) neighborhood of P, B its complement in the space of distribution functions Q with $\int e^{bx} \, dQ < \infty$, and $-I(Q)$ is the relative entropy $\int dQ \log(dP/dQ)$. For the descent, you want the law of large numbers – $\mathbb{P}[\lim \mathbf{s}_n/n = \mathbb{E}(\mathbf{x})] = 1$ – which is easily proved as in §2.1: with $\mathbf{s}'_n = \mathbf{s}_n - n\,\mathbb{E}(\mathbf{x})$, you have $\mathbb{E}(\mathbf{s}'_n)^4 = O(n^{-2})$ and so on. That's typical behavior. Contrariwise, if A is an open neighborhood of $\mathbb{E}(\mathbf{x})$ and J the Legendre dual of $\log Z$, then

$$\mathbb{P}\left(\frac{\mathbf{s}_n}{n} \notin A\right) \simeq \exp\left[-n \min_{m \notin A} J(m)\right]$$

just as before. The proof does not offer any real difficulty.

Exercise 2.4.7 Compute $J(m)$ for Gaussian variables with mean 0 and mean-square 1. *Answer*: $m^2/2$, whence $\mathbb{P}(|\mathbf{s}_n/n| > c) \simeq e^{-nc^2/2}$, as you already know by Mill's ratio. Check it out.

The idea of large deviations à la Cramér should now be plain, and natural to your mind. More is to come. Donsker and Varadhan have raised it to a general principle in the statistical picture of things. You can see it all in miniature in §7.7 where I have spelled it out for Markov chains. For more information, Kac's charming [1980] introduction is specially recommended; see also Varadhan [1984] for the best general account.

3
The Standard Random Walk

The format for Bernoulli trials is modified a little: the basic object is now the "sample path" $\mathbf{x} : n \geq 0 \to \mathbf{x}(n) = \mathbf{e}_1 + \cdots + \mathbf{e}_n$ with $\mathbf{x}(0) = 0$ by convention, and independent $\mathbf{e}$'s $= \pm 1$ with common distribution $\mathbb{P}(\mathbf{e} = +1) = p$ and $\mathbb{P}(\mathbf{e} = -1) = 1 - p = q$. You may think of $\mathbf{x}$ as a record of the winnings and/or losses of a gambler in a game of heads and tails with a possibly biased coin $(p \neq q)$, or what I like better, as a "random walk" in which the traveller steps right $(\mathbf{e} = +1)$ or left $(\mathbf{e} = -1)$ with probabilities p and q, at each time n, independently of the history of the walk before. Here, $\mathbb{E}(\mathbf{e}) = p - q \equiv m$, $\mathbb{E}(\mathbf{e} - m)^2 = 4pq$ and

$$\mathbf{x}(n) = nm \qquad\qquad \text{by LLN}$$

$$+ \text{ a Gaussian correction on the scale } 2\sqrt{npq} \qquad \text{by CLT;}$$

in particular, $\mathbb{P}\left[\mathbf{x}(n) \text{ tends to } +\infty \text{ as } n \uparrow \infty\right] = 1$ if $p > q$. The most interesting case is when the coin is honest so that the walker steps right/left with equal probabilities $p = q = 1/2$. That is the standard random walk $RW(1)$ of Example 1.2.6 to which the present chapter is devoted, with occasional glances at the general case. Notice that the expression of the binomial distribution is a little changed. In the context of the standard walk: $\mathbf{x}(n) = k$ is the difference $n_+ - n_-$ of n_+/n_- positive/negative steps, while $n_- + n_+ = n$ itself, so that $n_\pm = \frac{1}{2}(n \pm k)$, n and k have the same parity, and

$$\mathbb{P}\left[\mathbf{x}(n) = k\right] = \binom{n}{n_+} 2^{-n} \quad \text{for } |k| \leq n, \ k \equiv n \mod 2.$$

As for Bernoulli trials, $\mathfrak{F}_n$ is the field descriptive of the walk up to time n and $\mathfrak{F}$ is the smallest field containing all of these.

LLN, CLT, and the law of the iterated logarithm may be adapted from Chapter 2:

$$\text{LLN:} \quad \mathbb{P}\left[\lim_{n\uparrow\infty} \frac{\mathbf{x}(n)}{n} = 0\right] = 1,$$

$$\text{CLT:} \quad \lim_{n\uparrow\infty} \mathbb{P}\left[a \leq \frac{\mathbf{x}(n)}{\sqrt{n}} < b\right] = \int_a^b \frac{e^{-x^2/2}}{\sqrt{2\pi}},$$

and

$$\mathbb{P}\left[\lim_{\substack{\text{sup}\\\text{inf}}} \frac{\mathbf{x}(n)}{\sqrt{2n\log\log n}} = \pm 1\right] = 1,$$

to which may be added the delicate refinement of Erdős [1942]: for positive increasing $h(n)$,

$$\mathbb{P}\left[\mathbf{x}(n) < \sqrt{n}\,h(n) \text{ as } n \uparrow \infty\right] = 1 \text{ or } 0$$

according as $\sum_1^\infty n^{-1} h(n) e^{-h^2(n)/2} < +\infty$ or not; compare Feller [1968(1): 210–211].

3.1 The Markov property

I do this for the standard walk only; its extension to the general case will be obvious. The path space is extended to permit any starting point $\mathbf{x}(0) = x \in \mathbb{Z}$; $\mathfrak{F}$ is extended accordingly, and to each $x \in \mathbb{Z}$ is assigned probabilities $\mathbb{P}_x(A) = \mathbb{P}_0(x + \mathbf{x} \in A)$, pronounced "the probability of A for paths starting at $\mathbf{x}(0) = x$", in which $\mathbb{P}_0$ denotes the old probabilities for $\mathbf{x}(0) = 0$; in other words, the walk starting at x is just the old walk starting at 0 moved bodily over by addition of $\mathbf{x}(0) = x$.

3.1.1 The simple Markov property

The simple Markov property describes how the probabilities $\mathbb{P}_x : x \in \mathbb{Z}$ are knit together. You know it all already, but I spell it out in the present language anyhow. Fix $x \in \mathbb{Z}$, $n > 0$, and look at the picture.

It will be clear that the "future"

$$\mathbf{x}^+ : \mathbb{Z}^+ \to \mathbf{x}(n), \mathbf{x}(n+1), \mathbf{x}(n+2), \mathbf{x}(n+3) \text{ etc.}$$

is just a copy of the walk starting at the "present" place $\mathbf{x}(n) = y$, involving, besides $\mathbf{x}(n)$, only $\mathbf{e}_k : k > n$, and so independent of how it got to y, in the sense that it is otherwise independent of the past

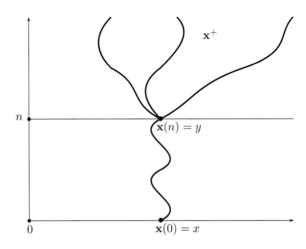

Figure 3.1.1

$\mathbf{x}^- = [\mathbf{x}(k) : k \leq n]$. The latter is described by the field $\mathfrak{F}_n$ of $\mathbf{e}_k : k \leq n$, so you may write

$$\mathbb{P}_x(\mathbf{x}^+ \in B | \mathbf{x}(k) : k \leq n) = \mathbb{P}_x(\mathbf{x}^+ \in B | \mathfrak{F}_n) = \mathbb{P}_y(B) \quad \text{with } y = \mathbf{x}(n);$$

less formally, you may say that the walk has *no memory*, starting *from scratch* at time n at the place $y = \mathbf{x}(n)$.

Exercise 3.1.1   Really, there's nothing to prove, but think it over carefully. It could not be more important.

3.1.2 The strict Markov property

The strict Markov property permits a more flexible and effective use of the Markovian character of the walk. It states that the walk starts from scratch, not only at each *fixed* time $T = n$, but also at certain random "stopping times" $T : \mathbf{x} \to \mathbb{Z}^+$. As before, T is a stopping time if the event $(T = n)$ belongs to $\mathfrak{F}_n$ for every $n \geq 0$; equivalently, if you want to decide if $T = n$, you have only to look at the past $\mathbf{x}(k) : k \leq n$.

Example 3.1.2 $T = \min(n \geq 0 : \mathbf{x}(n) = 1)$ is a stopping time, *aka* the "passage time to 1", of which much more below.

Example 3.1.3 $T = \min(n \geq 1 : \mathbf{x}(n) = 0)$ is another; for paths starting at 0, it is the "loop time".

Example 3.1.4 $T = \max(n \le 10 : \mathbf{x}(n) = 0)$ is *not* a stopping time: e.g. to decide if $T = 6$, you must look into the future, $\mathbf{x}(7)$ up to $\mathbf{x}(10)$, as well as at $\mathbf{x}(0)$ up to $\mathbf{x}(6)$. Besides, if $T = 6$, the future is abnormal in that you're starting at 0, but cannot return in less than 5 steps.

The statement of the Markov property is nearly the same as before: for any $x \in \mathbb{Z}$, $B \in \mathfrak{F}$, and stopping time T,

$$\mathbb{P}_x(\mathbf{x}^+ \in B | \mathfrak{F}_T) = \mathbb{P}_y(B) \ \text{ with } y = \mathbf{x}(T)$$

where now $\mathbf{x}^+$ is the future, $\mathbf{x}(n) : n \ge T$, and $\mathfrak{F}_T$ is the field $(A \in \mathfrak{F} : A \cap (T = n) \in \mathfrak{F}_n$ for any $n \ge 0)$.

Exercise 3.1.5 Check that $\mathfrak{F}_T$ is indeed a field and that it measures exactly the "stopped path" $\mathbf{x}^- : n \to \mathbf{x}(n)$ if $n < T$ or else $\mathbf{x}(T)$ if $n \ge T$. Does it measure T itself? Of course.

Proof. Here's another way to say the same thing, more convenient for the proof itself:

$$\mathbb{P}_x(\mathbf{x}^- \in A, \, \mathbf{x}(T) = y, \, \mathbf{x}^+ \in B) = \mathbb{P}_x(\mathbf{x}^- \in A, \, \mathbf{x}(T) = y) \times \mathbb{P}_y(B).$$

With a self-evident notation, the left-hand side is

$$\sum_{n=0}^{\infty} \mathbb{P}_x(T = n, \, \mathbf{x}^{-n} \in A, \, \mathbf{x}(n) = y, \, \mathbf{x}^{+n} \in B),$$

and since $(T = n) \in \mathfrak{F}_n$, you can split the probabilities by the simple Markov property, as in

$$\sum_{n=0}^{\infty} \mathbb{P}_x(T = n, \, \mathbf{x}^{-n} \in A, \, \mathbf{x}(n) = y) \times \mathbb{P}_y(B).$$

Now re-assemble the sum into $\mathbb{P}(\mathbf{x}^- \in A, \, \mathbf{x}(T) = y)$. That's all there is to it.

Note 3.1.6 It is implicit that $\mathbb{P}(T < \infty) = 1$. If it is not so, you must insert the event $(T < \infty)$ at the left. Then the proof goes as before, with the result that $\mathbb{P}_x(\mathbf{x}^+ \in B | \mathfrak{F}_T) = \mathbb{P}_y(B)$ with $y = \mathbf{x}(T)$ on the set where $T < \infty$. Nothing is said on the complement where $T = \infty$.

Note 3.1.7 The Markov property can be expressed in this way: if $T < \infty$, then past and future are independent, conditional upon the knowledge of the present. Check it out.

Note 3.1.8 In the present context, to call T a "stopping time" seems wrong-headed; it is, rather, a "starting time", but never mind. §3.2 will explain the usage. Anyhow, it's too late to change it now.

3.2 Passage times

A simple application of stopping times and the strict Markov property is to passage times such as $T_1 = \min(n : \mathbf{x}(n) = 1)$, $\mathbf{x}(0) = 0$ being understood. This is indeed a stopping time since the event $(T_1 = n)$ depends upon $\mathbf{e}_1, \ldots, \mathbf{e}_n$ only, as noted in Example 3.1.2. I drop the subscript and write T plain for a little while. Let's check that $\mathbb{P}_0(T < \infty) = 1$ as a warm-up. This is self-evident by the iterated log, but let's do it more simply.

First proof. Suppose $\mathbb{P}_0(T < \infty) = p < 1$ and let R be the time of first return: $\min(n \geq 1 : \mathbf{x}(n) = 0)$. R is a stopping time, identical in law to $1 +$ a copy of T, so $\mathbb{P}_0(R < \infty) = p$. Now the walk starts from scratch at each return (if any), its future being fully independent of its past since $\mathbf{x}(R) = 0$ is known, so the probability of n returns (no more, no less) is $p^n(1 - p)$, and the expected number of returns is $\sum_1^\infty np^n(1 - p) = p/(1-p)$. But the number of returns is also the number of visits of $\mathbf{x}(2n)$ to the origin for $n \geq 1$, and *that* has mean $\sum_1^\infty \mathbb{P}_0[\mathbf{x}(2n) = 0] = +\infty$ in view of $\mathbb{P}_0[\mathbf{x}(2n) = 0] = \binom{2n}{n} 2^{-2n} \simeq 1/\sqrt{\pi n}$, contradicting $p < 1$.

Second proof. Here's another, simpler way: if $\mathbf{e}_1 = 1$, then $T = 1$, and that costs you $1/2$. If $\mathbf{e}_1 = -1$, then you must still go from -1 back to 0 and from 0 to $+1$, in which case T looks like $1 + 2$ independent copies of itself, costing $\frac{1}{2} \times p \times p$ for the whole passage time to be finite in this case, i.e. $p = \mathbb{P}_0(T < \infty) = \frac{1}{2} + \frac{1}{2}p^2$, of which the only root is $p = 1$. That's better now.

Warning. $T = T_1$ may be finite, but it can be very large: for example, if $\mathbf{e}_1 = -1$, you could spend days wandering about to the left of -1 before even coming back to 0. The following is indicative:

$$
\begin{aligned}
\mathbb{E}_0(T_1) &= \mathbb{E}_0(T_1, \mathbf{e}_1 = +1) + \mathbb{E}_0(T_1, \mathbf{e}_1 = -1) \\
&= \frac{1}{2} + \frac{1}{2}\,\mathbb{E}_{-1}(1 + T_1) \\
&= 1 + \frac{1}{2}\,\mathbb{E}_0(T_2) \quad \text{with a self-evident notation} \\
&= 1 + \frac{1}{2} \times 2\,\mathbb{E}_0(T_1),
\end{aligned}
$$

by use of the strict Markov property in line 4 and the fact that T_2 looks like the sum of two independent copies of T_1. In short, $\mathbb{E}_0(T_1) = +\infty$. But more of this below.

Distribution of T_1

I drop the subscript again and introduce the "generating function":

$$f(\gamma) = \mathbb{E}_0(\gamma^T) = \sum_{n=0}^{\infty} \gamma^n \, \mathbb{P}_0(T = n) \quad \text{for } 0 < \gamma < 1.$$

This is a handy package containing all the probabilities $\mathbb{P}_0(T = n)$ at once. You will see in a moment what an effective tool it is. The computation of f is divided into two cases, according to whether $\mathbf{e}_1 = +1$ or -1, as in the second proof of $\mathbb{P}_0(T < \infty) = 1$.

Case 1. $\mathbf{e}_1 = +1$. Then $T = 1$ and that's all there is to it: $\mathbb{P}_0(T = 1) = 1/2$ and $\mathbb{E}_0[\gamma^T, \mathbf{e}_1 = +1] = \gamma/2$.

Case 2. $\mathbf{e}_1 = -1$. Now $T = 1$ *plus* the passage time from -1 to $+1$ which is identical in law to the sum of two independent copies of T, as above, so that

$$\mathbb{E}_0[\gamma^T, \mathbf{e}_1 = -1] = \frac{1}{2}\gamma\, \mathbb{E}_{-1}(\gamma^{T_1}) = \frac{1}{2}\gamma[\mathbb{E}_0(\gamma^T)]^2 = \frac{1}{2}\gamma[f(\gamma)]^2.$$

The upshot is

$$f = \frac{\gamma}{2} + \frac{\gamma}{2}\,f^2, \quad \text{which is to say} \quad f^2 - \frac{2}{\gamma}f + \frac{1}{\gamma^2} = \frac{1}{\gamma^2} - 1.$$

Now solve to obtain

$$f(\gamma) = \frac{1}{\gamma} \pm \sqrt{\frac{1}{\gamma^2} - 1} = \frac{1 \pm \sqrt{1 - \gamma^2}}{\gamma}.$$

But which signature is to be used? That's clear: $0 < \gamma < 1$, so $f(\gamma) < 1$ so you better take the minus sign. The rest comes from the power series

$$\sqrt{1 - x} = 1 - \sum_{n=1}^{\infty} \frac{(2n-2)!}{n!(n-1)!}\, 2^{-2n+1} x^n \quad \text{for } 0 \leq x < 1.$$

Exercise 3.2.1 Check the power series. Deduce

$$\mathbb{P}_0(T = 2n - 1) = \binom{2n}{n} 2^{-2n} \frac{1}{2n - 1} \quad \text{for } n \geq 1.$$

Exercise 3.2.2 $\mathbb{P}_0(T < \infty) = 1$ comes from $f(1-) = 1$ and $\mathbb{E}_0(T) = \infty$ from $f'(1-) = \infty$. The very long tail of the distribution: $\mathbb{P}_0(T = 2n - 1) \simeq 1/2\sqrt{\pi}n^{3/2}$ shows clearly why T likes to be large.

3.2.1 A better way: the reflection principle of D. André.

More attention to the actual sample path gives a better proof, as is often the case. $T = \min(n : \mathbf{x}(n) = 1)$ is a stopping time, so the future $\mathbf{x}^+ : n \to \mathbf{x}(n+T)$ is a copy of RW(1), starting at $\mathbf{x}(T) = 1$ and just as likely to go right or left, as in Figure 3.2.1. This is the picture behind

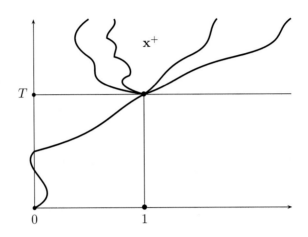

Figure 3.2.1

the "reflection principle" of D. André (1887) leading at once to:

$$\mathbb{P}_0[\mathbf{x}(2n) > 0]$$

$$= \mathbb{P}_0[T \le 2n - 1, \mathbf{x}(2n) > 0] \qquad\qquad T \text{ being odd}$$

$$= \sum_{k=1}^{n} \mathbb{P}_0[T = 2k - 1, \mathbf{x}(2n) - \mathbf{x}(2k-1) > 0] \qquad \begin{array}{l}\text{since } 2n - (2k-1) \\ \text{is odd}\end{array}$$

$$= \sum_{k=1}^{n} \mathbb{P}_0[T = 2k - 1, \mathbf{x}(2n) - \mathbf{x}(2k-1) < 0] \qquad \begin{array}{l}\text{by the reflection} \\ \mathbf{x} \to -\mathbf{x}\end{array}$$

$$= \frac{1}{2} \sum_{k=1}^{n} \mathbb{P}_0(T = 2k - 1) \qquad\qquad \text{see why?}$$

$$= \frac{1}{2} \mathbb{P}_0(T \le 2n - 1),$$

or, what is the same,

$$\mathbb{P}_0(T \le 2n - 1) = 2 \, \mathbb{P}_0[\mathbf{x}(2n) > 0] = 1 - \mathbb{P}_0[\mathbf{x}(2n) = 0],$$

whence

$$\mathbb{P}_0(T = 2n - 1) = \mathbb{P}_0(T \leq 2n - 1) - \mathbb{P}_0(T \leq 2n - 3)$$
$$= \mathbb{P}_0[\mathbf{x}(2n - 2) = 0] - \mathbb{P}_0[\mathbf{x}(2n) = 0]$$
$$= \binom{2n - 2}{n - 1} 2^{-2n+2} - \binom{2n}{n} 2^{-2n}$$
$$= \binom{2n}{n} 2^{-2n} \frac{1}{2n - 1}$$

when you work it out.

Exercise 3.2.3 Here are some variants on D. André's theme for you to verify: with $x \in \mathbb{Z}^+$ and $T_x = \min(n : \mathbf{x}(n) = x)$, you have

$$2\,\mathbb{P}_0[\mathbf{x}(2n) > x] = \mathbb{P}_0\Big[\max_{k \leq 2n-1} \mathbf{x}(k) \geq x\Big] = \mathbb{P}_0(T_x \leq 2n - 1) \text{ if } x \text{ is odd,}$$

$$2\,\mathbb{P}_0[\mathbf{x}(2n + 1) > x] = \mathbb{P}_0\Big[\max_{k \leq 2n} \mathbf{x}(k) \geq x\Big] = \mathbb{P}_0(T_x \leq 2n) \quad \text{if } x \text{ is even,}$$

and

$$\mathbb{P}_0(T_x \leq n) = 2\,\mathbb{P}_0[\mathbf{x}(n) > x] + \mathbb{P}_0[\mathbf{x}(n) = x] \qquad\qquad \text{in general.}$$

Exercise 3.2.4 Check the formula

$$\mathbb{P}_0\Big[\mathbf{x}(n) = a \text{ and } \max_{k \leq n} \mathbf{x}(k) = b\Big] = \mathbb{P}_0[\mathbf{x}(n) = 2b - a]$$

for $b \geq 0$ and $-\infty < a \leq b$ by looking at Figure 3.2.2 and noting its self-evident symmetries.

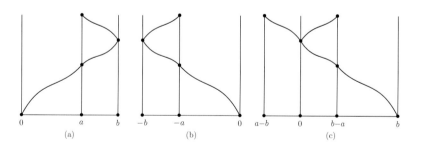

Figure 3.2.2

3.2.2 Yet another take: stopping

The formula $\mathbb{E}_0(\gamma^T) = \gamma^{-1}(1 - \sqrt{1 - \gamma^2})$ can also be obtained with the help of a fair game, for which see §1.3. The idea is useful, so I want you to see it now in this simple case.

With fixed α and $\beta = \cosh \alpha$, $\mathbf{z}(n) = \beta^{-1} \exp\left[\alpha \mathbf{x}(n)\right]$ is a fair game, as you will check. In particular, $\mathbb{E}[\mathbf{z}(n)] = \mathbb{E}[\mathbf{z}(0)] = 1$, and it would be nice if this would be true for stopping times as well, e.g. with the passage time T in place of n, for then you would have $\mathbb{E}_0(\gamma^T) = \frac{1}{\gamma}(1 - \sqrt{1 - \gamma^2})$ with $\gamma = 1/\beta$. Take that as an exercise. But be warned that it is not always so: for example, the walk itself is fair, but $\mathbb{E}_0[\mathbf{x}(T)] = 1$ is not the same as $\mathbb{E}_0[\mathbf{x}(0)] = 0$. So what does it take? A simple sufficient condition is that $|\mathbf{z}(n)|$ should be bounded up to the stopping time, as it is here if $\alpha \geq 0$ since $0 < \mathbf{z}(n) \leq e^\alpha$ while $\mathbf{x}(n)$ is ≤ 1, or more generally, if $\max_{n \leq T}|\mathbf{z}(n)|$ is summable. The proof is easy: with a fixed number N,

$$\mathbb{E}_0[\mathbf{z}(T \wedge N)]$$

$$= \mathbb{E}_0[\mathbf{z}(N), T > N] + \sum_0^N \mathbb{E}_0[\mathbf{z}(n), T = n]$$

$$= \mathbb{E}_0[\mathbf{z}(N), T > N] + \sum_0^N \mathbb{E}_0\left[\mathbb{E}_0(\mathbf{z}(N)|\mathfrak{F}_n), T = n\right]$$

$$= \mathbb{E}_0[\mathbf{z}(N), T > N] + \sum_0^N \mathbb{E}_0[\mathbf{z}(N), T = n] \quad \text{since } (T = n) \in \mathfrak{F}_n$$

$$= \mathbb{E}_0[\mathbf{z}(N)]$$

$$= \mathbb{E}_0[\mathbf{z}(0)] = 1.$$

Now make $N \uparrow \infty$. That's all there is to it.

Notice that the "stopped game" $\mathbf{z}(n \wedge T)$ is still fair. That's the origin of the name "stopping time".

A number of further illustrations of this general principle can be found in §3.2.4.

3.2.3 Passage to a distance

The fact that $\mathbb{P}_0(T_1 < \infty) = 1$, together with the strict Markov property, makes it obvious that the walk visits every place in $\mathbb{Z}$ infinitely often as $n \uparrow \infty$. But how long does it really take to come to some distant site $n > 0$? The passage time $T_n = \min(k : \mathbf{x}(k) = n)$ is the sum of n independent copies of T_1, so you might think it would be comparable to

n. This is wildly wrong. The correct scale is n^2, as you see from

$$\lim_{n\uparrow\infty} \mathbb{P}_0\big(T_n/n^2 \le c\big) = \int_0^c \frac{e^{-1/2t}}{\sqrt{2\pi t^3}}\, dt.$$

The distribution to the right is the 1-sided stable law of exponent $1/2$ from §1.7. Notice its long tail comparable to $t^{-3/2}$, imitating the $n^{-3/2}$ of Exercise 3.2.2 above.

Proof. Replace the γ in $\mathbb{E}_0(e^{\gamma T_1}) = \gamma^{-1}(1 - \sqrt{1 - \gamma^2})$ by $e^{-\beta}$ with $\beta > 0$. Then

$$\mathbb{E}_0(e^{-\beta T_n}) = \big[\mathbb{E}_0(e^{-\beta T_1})\big]^n = \big[e^\beta(1 - \sqrt{1 - e^{-2\beta}})\big]^n \simeq (1 - \sqrt{2\beta})^n$$

for small β: in fact, with fixed α and $\beta = \alpha/n^2$,

$$\lim_{n\uparrow\infty} \mathbb{E}_0(e^{-\alpha T_n/n^2}) = e^{-\sqrt{2\alpha}} = \int_0^\infty e^{-\alpha t}\, \frac{e^{-1/2t}}{\sqrt{2\pi t^3}}\, dt,$$

as per §1.7. Now take a nice function $F(t)$ of class $C[0,\infty]$, think of it as a function of $s = e^{-t}$, and approximate it by a polynomial G in that variable, as in §2.1. Then

$$\lim_{n\uparrow\infty} \mathbb{E}\big[F(T_n/n^2)\big] = \int_0^\infty F(t)\, \frac{e^{-1/2t}}{\sqrt{2\pi t^3}}\, dt,$$

first for G and then for F itself. The rest will be obvious.

Exercise 3.2.5   Convert the law of distant passage into

$$\lim_{n\uparrow\infty} \mathbb{P}_0\left[n^{-1/2} \max_{k\le n} \mathbf{x}(k) \ge c\right] = 2\int_c^\infty \frac{e^{-x^2/2}}{\sqrt{2\pi}}\, dx \quad \text{for } c \ge 0$$

and confirm by the reflection principle of D. André and CLT. This is simpler, but I wanted you to see it both ways.

3.2.4 Two-sided passage: the Gambler's Ruin

Fix $a < b$ and look at the exit time $T = \min(n : \mathbf{x}(n) = a \text{ or } b) = T_a \wedge T_b$ for paths starting at x between a and b, as in Figure 3.2.3. Evidently, $f(x) = \mathbb{P}_x(\mathbf{x}(T) = a)$ satisfies $f(x) = \frac{1}{2}f(x+1) + \frac{1}{2}f(x-1)$ for $a < x < b$ with $f(a) = 1$ and $f(b) = 0$, from which you find the law for the Gambler's Ruin, so-called:

$$\mathbb{P}_x(T_a < T_b) = \frac{b - x}{b - a}.$$

Exercise 3.2.6   The walk is fair and may be stopped at time T. Do it that way.

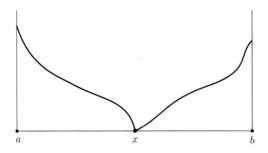

Figure 3.2.3

Exercise 3.2.7 Now back to the 1-sided passage time $T = \min(n : \mathbf{x}(n) = 1)$. $\mathbb{E}_0[\mathbf{x}(n)] = 0$ for any $n \geq 0$, but $\mathbb{E}_0[\mathbf{x}(T)] = 1$. This could not be so if $\mathbb{E}_0[\min_{n<T} \mathbf{x}(n)] > -\infty$ – in fact, for any whole number N, $\mathbb{E}_0[\mathbf{x}(T \wedge N)] = 0$, as you will check, producing a contradiction when $N \uparrow \infty$ if $\mathbb{E}(\min) > -\infty$. Now justify this:

$$- \mathbb{E}_0\left[\min_{n<T} \mathbf{x}(n)\right] = \sum_{k=1}^{\infty} \mathbb{P}_0\left[\min_{n<T} \mathbf{x}(k) \leq -k\right] = \sum_{k=1}^{\infty} \frac{1}{1+k} = +\infty.$$

Exercise 3.2.8 That's all pretty obvious, but how long does it take to come to a or b? I claim that $\mathbb{E}_x(T) = (b-x)(x-a)/(b-a)$ for $a \leq x \leq b$. To prove it try this: $\mathbf{x}^2(n) - n$ is a fair game, and while it is not bounded for $n \leq T$, it may be stopped at $T \wedge N$ for any whole number N. Do that and make $N \uparrow \infty$.

More information is obtained by stopping the fair game

$$\beta^{-n} \cosh \alpha \left[\mathbf{x}(n) - \frac{a+b}{2}\right]$$

in which $\beta = \cosh \alpha$ as before, the result being

$$\mathbb{E}_x(\beta^{-T}) = \frac{\cosh \alpha (x - \frac{a+b}{2})}{\cosh \alpha \left(\frac{b-a}{2}\right)},$$

by inspection.

Exercise 3.2.9 How big can γ be and still keep $\mathbb{E}_0(\gamma^T) < \infty$ in the case $a = -b < 0 < b$? *Answer*: It must be smaller than the reciprocal of $\cos[\pi/(b-a)]$.

Now take $-a = b = n$ and $x = 0$, so that $\mathbb{E}_0(\beta^{-T}) = 1/\cosh(\alpha n)$, replace $\cosh \alpha$ by e^{α/n^2} with fixed $\alpha > 0$ and make $n \uparrow \infty$ to produce

$$\lim_{n \uparrow \infty} \mathbb{E}_0[e^{-\alpha T/n^2}] = \frac{1}{\cosh \sqrt{2\alpha}}$$

$$= 2 \times [e^{-\sqrt{2\alpha}} - e^{-3\sqrt{2\alpha}} + e^{-5\sqrt{2\alpha}} - \text{etc.}],$$

or, what is the same,

$$\lim_{n \uparrow \infty} \mathbb{P}_0(T/n^2 \le c) =$$

$$2 \times \left[\int_0^c \frac{e^{-1/2t}}{\sqrt{2\pi t}} - 3 \int_0^c \frac{e^{-9/2t}}{\sqrt{2\pi t^3}} + 5 \int_0^c \frac{e^{-25/2t}}{\sqrt{2\pi t^5}} - \text{etc.} \right],$$

as per §1.7. This does not even *look* positive but it must be so, as will now be verified in a different style. The density

$$\vartheta(t) = 2 \times (2\pi t)^{-3/2} \sum_0^\infty (-1)^n (2n+1) e^{-(2n+1)^2/2t}$$

figuring in the previous display is an alternating sum of diminishing terms and so must be positive if $t < 4/\log 3$, as you will please check. It can also be expressed as

$$\vartheta(t) = 2 \times \sum_{\mathbb{Z}} f(n) \text{ with } f(x) = (2\pi t)^{-3/2} e^{\sqrt{-1}\,\pi x}(2x+1) e^{-(2x+1)^2/2t}$$

to which the Poisson summation formula of §1.6.3 may be applied.

Exercise 3.2.10 Do this to obtain $\vartheta(2t/\pi) =$ a positive multiple of $\vartheta(2/\pi t)$ and finish the proof that way.

3.3 Loops

Now the walk starts at $x = 0$ and is divided into "loops", beginning at 0 and ending at the next return to that place (which is bound to happen). The first return time $R = \min(n \ge 1 : \mathbf{x}(n) = 0)$ is a stopping time and $\mathbf{x}(R) = 0$ is known, so everything starts over, independently of what went before, i.e. the successive loops are independent copies of the first loop $\mathbf{x}(n) : 0 < n \le R$. This nice picture is the idea of W. Döblin (1937). I will use it often.

3.3.1 Equidistribution

The question is: How often does the walk visit a particular place $x \in \mathbb{Z}$? Let $\#(N, x) = \#(n \le N : \mathbf{x}(n) = x)$ be the number of visits in the first N steps and let R_n be the duration of the first n loops. $\#(R_n, 0) = n$ by definition; also, $\#(R_n, x)$ is the sum of n independent copies of $\#(R_1, x)$;

and if you knew that $\mathbb{E}\#(R_1, x) = 1$ (which looks odd, but is quite correct and will be proved in a moment) then the general law of large numbers of §5.1 would tell you that

$$\mathbb{P}\left[\lim_{n\uparrow\infty} \frac{1}{n}\#(R_n, x) = 1\right] = 1 \quad \text{for any } x \in \mathbb{Z},$$

and indeed, for every x simultaneously. More generally, if $R_n \leq N < R_{n+1}$ and $n \uparrow \infty$, then $\#(R_n, x) \leq \#(N, x) \leq \#(R_{n+1}, x)$ implies $\lim_{N\uparrow\infty} \frac{1}{N}\#(N, x) = 1$, but as $N \simeq R_n$ is the sum of n independent copies of $1 + T_1$ and so comparable to n^2, you cannot expect a law of large numbers like $\lim_{N\uparrow\infty} N^{-1}\#(N, x) = $ some constant, but rather a weak CLT-type law for $N^{-1/2}\#(N, x)$, as will be found in the next article. What you *do* have already is the equidistribution of visits to different places expressed by

$$\mathbb{P}_0\left[\lim_{N\uparrow\infty} \frac{\#(N, x)}{\#(N, y)} = 1 \text{ for every } x \text{ and } y \in \mathbb{Z}\right] = 1.$$

Proof of $\mathbb{E}_0[\#(R, x)] = 1$. That $\#(R, 0) = 1$ is automatic, so let's try $\#(R, 1)$. Obviously, $\#(R, 1) = 0$ if the walk goes left. If it goes right, then $\#(R, 1) = 1 +$ the number of loops from $x = 1$ and back that don't hit 0. But the probability of n such loops (and no more) is 2^{-n} (step left and it's over), so

$$\mathbb{E}[\#(R, 1)] = \left(\frac{1}{2} \text{ for going right}\right) \times \left(1 + \frac{1}{2} + \frac{1}{4} + \cdots\right) = \frac{1}{2} \times 2 = 1.$$

$\mathbb{E}[\#(R, 2)] = 1$ can be proved similarly, but let's take a better way for general $x \geq 2$. Here it is:

$\mathbb{E}_0[\#(R, x)]$

$$= \sum_{n=1}^{\infty} \mathbb{P}_0\left[\mathbf{x}(1), \ldots, \mathbf{x}(n-1) > 0, \mathbf{x}(n) = x, n \leq R\right]$$

$$= \sum_{n=1}^{\infty} \mathbb{P}_0\left[\mathbf{x}(1), \ldots, \mathbf{x}(n-1) > 0, \mathbf{x}(n) = x\right] \text{ since } n \leq R \text{ is redundant}$$

$$= \sum_{n=1}^{\infty} \mathbb{P}_0\left[\mathbf{x}(n) - \mathbf{x}(1), \ldots, \mathbf{x}(n) - \mathbf{x}(n-1) < x, \mathbf{x}(n) - \mathbf{x}(0) = x\right]$$

$$= \sum_{n=1}^{\infty} \mathbb{P}_0\left[\mathbf{x}(1), \ldots, \mathbf{x}(n-1) < x, \mathbf{x}(n) = x\right] \quad \text{see it?}$$

$$= \mathbb{P}_0(T_x < \infty)$$

$$= 1.$$

A surprising fact I think.

Exercise 3.3.1   Do it again this neater way: $f(x) = \mathbb{E}_0[\#(R, x)]$ is a solution of $\frac{1}{2}f(x-1) + \frac{1}{2}f(x+1) = f(x)$ with $f(0) = 1$; in particular, $0 \le f(x) < \infty$ for any x. Check it out and conclude $f(x) = f(0) = 1$. *Hint*: Think about the number of visits to 0 for a loop from x back to x.

Exercise 3.3.2   Here's yet a third way:

$$f = \lim_{n\uparrow\infty} \frac{1}{n}\#(R_n, x) = \lim_{N\uparrow\infty} \frac{\#(N, x)}{\#(N, 0)}$$

is the empirical mean number of visits to x per loop, so

$$\frac{1}{f} = \lim_{N\uparrow\infty} \#(N, 0)/\#(N, -x),$$

by symmetry. Deduce $1/f = f$, i.e. $f = 1$.

Exercise 3.3.3   One more time: Use the Gambler's Ruin from §3.2.4 to show that the probability of exactly $\# = n$ visits to $x > 0$ per loop is

$$\frac{1}{2} \times \frac{1}{x} \times \sum_{\ell=0}^{n-1} \binom{n-1}{\ell} \left(\frac{1}{2}\left(1 - \frac{1}{x}\right)\right)^{\ell} \left(\frac{1}{2}\right)^{n-1-\ell} \times \frac{1}{2} \times \frac{1}{x},$$

explaining what each term means. Then check $\mathbb{E}(\#) = 1$.

3.3.2 The actual number of visits

This is all well and good, but how many times does the walk really visit a particular site? In this matter, all sites are the same in the long run, so it is enough to look at $\#(N, 0)$ for paths starting at $x = 0$. Then you find

$$\lim_{n\uparrow\infty} \mathbb{P}_0\left[\frac{\#(N)}{\sqrt{N}} \ge x\right] = \sqrt{\frac{2}{\pi}} \int_x^\infty e^{-y^2/2} \quad \text{for } x \ge 0.$$

Proof.   Note three things:

(1) that each loop contributes just one more visit to 0;

(2) that the duration R_n of n loops is $n +$ a copy of the passage time T_n;

(3) that the event $\#(N, 0) \ge n$ is the same as $N \ge R_n$.

Then, with fixed $x > 0$ and the pretense that $x\sqrt{N}$ is a whole number m, you have

$$\mathbb{P}_0\left[\frac{\#(N)}{\sqrt{N}} \ge x\right] = \mathbb{P}_0[N \ge m + T_m] \simeq \mathbb{P}_0\left[\frac{1}{x^2} \ge \frac{T_m}{m^2}\right] \quad \text{with } \frac{N}{m^2} = \frac{1}{x^2},$$

so

$$\lim_{n\uparrow\infty} \mathbb{P}_0\left[\frac{\#(N)}{\sqrt{N}} \geq x\right] = \int_0^{1/x^2} \frac{e^{-1/2t}}{\sqrt{2\pi t^3}}\, dt = \sqrt{\frac{2}{\pi}} \int_x^{\infty} e^{-y^2/2}\, dy$$

by the law of distant passage of §3.2 and the substitution $t = 1/y^2$.

Exercise 3.3.4   Here, there cannot be an individual law. Why?

Exercise 3.3.5   Check directly that $\mathbb{E}_0[\#(2n,0)] = (2n+1)\binom{2n}{n}2^{-2n} \simeq 2\sqrt{n/\pi}$. Does this agree?

3.3.3 Long runs

R_1, R_2, etc. are now the durations of the individual loops, distributed the same as $1 +$ the passage time from 0 to 1.

Exercise 3.3.6   Check that

$$\lim_{n\uparrow\infty} \mathbb{P}_0\left[\frac{1}{n^2} \max_{k\leq n} R_k \leq c\right] = e^{-1/\sqrt{\pi c}}.$$

You will want $\mathbb{P}_0(R > n) \simeq \sqrt{2/\pi n}$, for which see Exercise 3.2.2.

Now by the law of distant passage of §3.2,

$$\lim_{n\uparrow\infty} \mathbb{P}_0\left[\frac{1}{n^2}(R_1 + \cdots + R_n) \leq c\right] = \int_0^c \frac{e^{-1/2t}}{\sqrt{2\pi t^3}}\, dt,$$

the suggestion being that $\max_{k\leq n} R_k$ is comparable to $R_1 + \cdots + R_n$, i.e. if you play for a long time $N = R_1 + \cdots + R_n$, very likely you will see a run of luck (good or bad) comparable to N itself. Notice how different this is from independent standard Gaussian variables: there $\mathbf{x}_1 + \cdots + \mathbf{x}_n$ is comparable to $\sqrt{n}$ but $\max_{k\leq n} \mathbf{x}_k$ is like $\sqrt{\log n}$ only, as per Exercise 1.8.11. Now I ask you, what is the probability that a single loop is longer than the total duration of all the others. I claim it's about $2/3$ if the number of loops is large. Remarkable, no? This phenomenon is closely allied to the arcsine law coming next.

Proof.   The events C_k – that the kth loop is longer than the duration of all the rest – are disjoint and equally likely, so

$$\mathbb{P}_0\left(\bigcup_{k=1}^{n} C_k\right) = n\,\mathbb{P}_0(R_1 > R_2 + \cdots + R_n)$$

$$\simeq \sqrt{\frac{2}{\pi}}\,n\,\mathbb{E}_0(R_2 + \cdots + R_n)^{-1/2} \text{ since } \mathbb{P}_0(R > n) \simeq \sqrt{2/\pi n}$$

$$= \sqrt{\frac{2}{\pi}}\,\mathbb{E}_0\left[\frac{R_2 + \cdots + R_n}{n^2}\right]^{-1/2}$$

$$\simeq \sqrt{\frac{2}{\pi}}\int_0^\infty \frac{1}{\sqrt{t}}\,\frac{e^{-1/2t}}{\sqrt{2\pi t^3}}\,dt \qquad \text{by the law of distant passage}$$

$$= \frac{2}{\pi} = .636 + .$$

Exercise 3.3.7 The $2/\pi$ is quite correct, but line 4 needs to be justified from line 3. Try it this way:

$$\frac{1}{\pi}\int_0^\infty e^{-Rx}\,\frac{dx}{\sqrt{x}} = \frac{1}{\sqrt{\pi R}},$$

so line 3 can be written

$$\frac{\sqrt{2}}{\pi}\int_0^\infty \mathbb{E}_0\left[e^{-x(R_1 + \cdots + R_n)/n^2}\right]\frac{dx}{\sqrt{x}} = \frac{\sqrt{2}}{\pi}\int_0^\infty\left(1 - \sqrt{1 - e^{-2x/n^2}}\right)^n\frac{dx}{\sqrt{x}}.$$

To proceed, you need a domination such as $1 - \sqrt{1 - e^{-x^2}} \le e^{-x}$. Check it. Then

$$\frac{\sqrt{2}}{\pi}\int_0^\infty e^{-\sqrt{2x}}\,\frac{dx}{\sqrt{x}} = \frac{2}{\pi}$$

comes out.

★**Exercise 3.3.8** Now let C_k be the event that the kth loop is longer than some positive number $\theta \times$ the total duration of the loops $1 \le j \ne k \le n$. The same method produces

$$\lim_{n\uparrow\infty}\mathbb{P}_0\left(\bigcup_{k=1}^{n} C_k\right) = \frac{\sqrt{2}}{\pi}\,\frac{1}{\sqrt{\theta}} \quad \text{if } \theta \ge 1,$$

but if $\theta < 1$, the events C are not disjoint any more. If still $\theta \ge 1/2$, the intersection of any three of them is void, so by inclusion–exclusion,

$$\mathbb{P}_0\left(\bigcup_{k=1}^{n} C_k\right)$$

$$= n\,\mathbb{P}_0\big[R_1 > \theta(R_2 + \cdots + R_n)\big]$$

$$- \binom{n}{2}\mathbb{P}_0\big[R_1 - \theta R_2 \text{ and } R_2 - \theta R_1 \text{ are both } > \theta(R_3 + \cdots + R_n)\big]$$

$$= (1) - (2).$$

As $n \uparrow \infty$, (1) produces $2/\pi\sqrt{\theta}$, as before. (2) is more complicated. *Hints*: Begin by estimating the sum of $\mathbb{P}(R_1 = 2i) \times \mathbb{P}(R_2 = 2j)$ over i and $j \geq 1$ subject to $i - \theta j$ and $j - \theta i > \theta n$. Here, both i and j must be large, so you may replace $\mathbb{P}(R = 2k)$ by $1/\sqrt{2\pi k^3}$ and also the approximate sum by a double integral to produce $\frac{2}{\pi n}\frac{1-\theta}{\theta}\frac{1-\sqrt{\theta}}{1+\sqrt{\theta}}$. Now go back to $(2) = \frac{2}{\pi}\frac{1-\theta}{\theta}\frac{1-\sqrt{\theta}}{1+\sqrt{\theta}} \times \binom{n}{2}\mathbb{E}_0(R_3 + \cdots + R_n)^{-1}$ more or less, and make $n \uparrow \infty$ to see what comes out. *Answer*: $(1) - (2) = \frac{2}{\pi}\frac{1}{\sqrt{\theta}} - \frac{1}{\pi}\frac{1-\theta}{\theta}\frac{1-\sqrt{\theta}}{1+\sqrt{\theta}}$, reducing to (1) as it should when $\theta = 1$. Now compute for $\theta = 1/2$:

$$\lim_{n\uparrow\infty} \mathbb{P}_0\left(\bigcup_{k=1}^{n} C_k\right) = \frac{2\sqrt{2}}{\pi} - \frac{1}{\pi}\frac{\sqrt{2}-1}{\sqrt{2}+1} = .845 + .$$

For $\theta < 1/2$, it's messier, so let's stop here.

3.4 The arcsine law

It is desired to find the distribution of $\#(n) =$ the number of times $k \leq n$ when the walker is at the right of $\mathbf{x}(0) = 0$. This could mean $\mathbf{x}(k) > 0$ or $\mathbf{x}(k) \geq 0$, but it is more symmetrical and simplifies the proof to count the number of positive "legs". You will see what I mean from Figure 3.4.1 in which $\#(10) = 6$ while the number of non-negative, respectively positive $\mathbf{x}$s is 7, respectively 4. P. Lévy [1939:303–304] indicated the answer:

$$\lim_{n\uparrow\infty} \mathbb{P}_0\left[\frac{\#(n)}{n} \leq x\right] = \frac{1}{\pi}\int_0^x \frac{dy}{\sqrt{y(1-y)}} = \frac{2}{\pi}\arcsin\sqrt{x},$$

whence the name "arcsine". The fact is striking. The conventional wisdom would have it that $\#(n)/n$ should be near $1/2$, i.e. that its "empirical density" should be symmetrical about $x = \frac{1}{2}$ and sharpening as $n \uparrow \infty$ as in Figure 3.4.2 (left), so that $\#(n)/n \simeq 1/2$ for large n. What you see in Figure 3.4.2 (right) depicting the arcsine density is just the opposite! Here are some numbers taken from Feller [1968:80] to illustrate how far from the conventional wisdom the arcsine law really is: for large

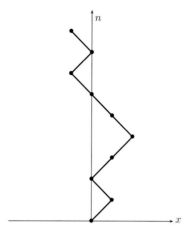

Figure 3.4.1

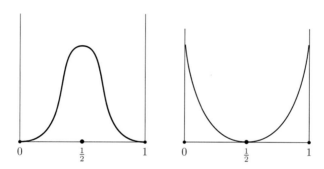

Figure 3.4.2

n, the probability that the *unlucky* player is ahead of the game for not more than a small fraction ϑ of the total time of play is

$$\frac{1}{\pi}\int_0^{\vartheta} \frac{dx}{\sqrt{x(1-x)}} + \frac{1}{\pi}\int_{1-\vartheta}^1 \frac{dx}{\sqrt{x(1-x)}} = \frac{4}{\pi}\sin^{-1}\sqrt{\vartheta}.$$

This number is about:

$$\begin{array}{ll} 1/2 & \text{if } \vartheta = .049 \\ 1/5 & \text{if } \vartheta = .025 \\ 1/10 & \text{if } \vartheta = .006 \\ 1/20 & \text{if } \vartheta = .0015, \end{array}$$

and $1/20$ is not all *that* rare! The arcsine phenomenon is one aspect of the fact that in (honest) coin tossing, runs of luck, both good or bad, can and will be surprisingly long, as in §3.3 above, where it was proved that, out of a large number of loops, the longest loop time will be still longer than the total duration of all the other loops with probability about $2/3$!

The proof of the arcsine law is based on the explicit distribution of $\#(2n)$ to be verified presently. This number must be even ($= 2m$), and you have

$$\mathbb{P}\left[\#(2n) = 2m\right] = \mathbb{P}\left[\mathbf{x}(2m) = 0\right] \mathbb{P}\left[\mathbf{x}(2n - 2m) = 0\right]$$

$$= \binom{2m}{m} 2^{-2m} \times \binom{2n - 2m}{n - m} 2^{-2n+2m}$$

$$\simeq \frac{1}{\sqrt{\pi m}} \times \frac{1}{\sqrt{\pi(n - m)}} \quad \text{by Stirling,}$$

leading at once to

$$\mathbb{P}\left[a \le \frac{\#(2n)}{2n} < b\right] \simeq \sum_{a \le \frac{m}{n} < b} \frac{1}{\pi} \frac{1}{\sqrt{\frac{m}{n}\left(1 - \frac{m}{n}\right)}} \frac{1}{n} \simeq \frac{1}{\pi} \int_a^b \frac{dx}{\sqrt{x(1 - x)}}$$

provided $a > 0$ and $b < 1$ so that both n and m are big for the reliability of Stirling's formula.

Exercise 3.4.1   Remove the restrictions on a and b.
Hint: It is enough to show that all is well for $a = 0$ and $b = 1/2$.

Exercise 3.4.2   Remove the restriction that the number of trials $(2n)$ be even.

First proof.   The evaluation of $\mathbb{P}\left[\#(2n) = 2m\right]$ is done by induction, starting with $m = 0$. Obviously,

$$\mathbb{P}\left[\#(2n) = 0\right] = \mathbb{P}\left[\mathbf{x}(1), \ldots, \mathbf{x}(2n) \le 0\right] = \mathbb{P}(T_1 \ge 2n + 1);$$

also

$$2\mathbb{P}\left[\mathbf{x}(2n) > 0\right] = \sum_{k=1}^n \mathbb{P}\left[T_1 = 2k - 1, \mathbf{x}(2n) > 0\right] \times 2$$

$$= \sum_{k=1}^n \mathbb{P}\left[T_1 = 2k - 1\right] \mathbb{P}\left[\mathbf{x}(2n) - \mathbf{x}(2k - 1) > 0\right] \times 2 \quad \begin{array}{l} 2n - (2k - 1) \\ \text{being odd,} \end{array}$$

$$= \mathbb{P}(T_1 \le 2n - 1),$$

as in §3.2, so by symmetry,

$$\mathbb{P}\left[\#(2n) = 0\right] = 1 - \mathbb{P}(T_1 \le 2n - 1) = 1 - 2\mathbb{P}\left[\mathbf{x}(2n) > 0\right] = \mathbb{P}\left[\mathbf{x}(2n) = 0\right].$$

Now for the induction. Clearly, $\mathbb{P}(\# = 2m)$ and $\mathbb{P}(\# = 2n - 2m)$ are the same, by reflection $\mathbf{x} \to -\mathbf{x}$, so just look at the event $\# = 2m$ for $0 < m < n$. Then the return time $R = \min(2k \geq 2 : \mathbf{x}(2k) = 0)$ is also $\leq 2n - 2$, permitting an induction based upon the number $\#^-(2k)$, respectively $\#^+(2n - 2k)$, of positive legs produced before, respectively after, R. Here it is:

$$\mathbb{P}\left[\#(2n) = 2m\right]$$
$$= \sum_{k=1}^{n-1} \mathbb{P}\left[\mathbf{x}(1) = -1, R = 2k, \#^-(2k) = 0, \#^+(2n - 2k) = 2m\right]$$
$$+ \sum_{k=1}^{n-1} \mathbb{P}\left[\mathbf{x}(1) = +1, R = 2k, \#^-(2k) = 2k, \#^+(2n - 2k) = 2m - 2k\right]$$
$$= \sum_{k=1}^{n-m} \mathbb{P}\left[\mathbf{x}(1) = -1, R = 2k\right] \mathbb{P}\left[\#(2n - 2k) = 2m\right]$$

since $\#^+(2n - 2k)$ is a copy of $\#(2n - 2k)$, independent of what went before

$$+ \sum_{k=1}^{m} \mathbb{P}\left[\mathbf{x}(1) = +1, R = 2k\right] \mathbb{P}\left[\#(2n - 2k) = 2m - 2k\right]$$

for the same reason

$$= \frac{1}{2} \sum_{k=1}^{n-m} \mathbb{P}(R = 2k)\mathbb{P}\left[\mathbf{x}(2n - 2m - 2k) = 0\right] \mathbb{P}\left[\mathbf{x}(2m) = 0\right]$$
$$+ \frac{1}{2} \sum_{k=1}^{m} \mathbb{P}(R = 2k)\mathbb{P}\left[\mathbf{x}(2m - 2k) = 0\right] \mathbb{P}\left[\mathbf{x}(2n - 2m) = 0\right]$$

by induction on n

$$= \frac{1}{2}\mathbb{P}\left[\mathbf{x}(2n - 2m) = 0\right] \mathbb{P}\left[\mathbf{x}(2m) = 0\right]$$
$$+ \frac{1}{2}\mathbb{P}\left[\mathbf{x}(2m) = 0\right] \mathbb{P}\left[\mathbf{x}(2n - 2m) = 0\right] \qquad \text{see why?}$$
$$= \mathbb{P}\left[\mathbf{x}(2m) = 0\right] \times \mathbb{P}\left[\mathbf{x}(2n - 2m) = 0\right],$$

as advertised – though I have always thought there should be a simpler proof just by "looking".

Second proof. Provide each loop with a variable $\mathbf{e}$ to indicate if it goes right ($\mathbf{e} = +1$) or left ($\mathbf{e} = -1$). These are Bernoulli variables, with success probability $1/2$ for $\mathbf{e} = +1$, independent of the Rs.

Exercise 3.4.3   Use $\mathbb{E}(\gamma^R) = 1 - \sqrt{1-\gamma^2}$ to evaluate

$$\sum_0^\infty \gamma^{2n}\mathbb{P}\left[\mathbf{x}(2n)=0\right] \quad \text{as} \quad (1-\gamma^2)^{-1/2}.$$

Exercise 3.4.4 $\#(2n) = 2m$ may be expressed as the intersection, for some $\ell \geq 0$, of the events

(1) $$R_1 + \cdots + R_\ell \leq 2n < R_1 + \cdots + R_\ell + R$$

with an extra (independent) loop time R, and

(2) $$\mathbf{e}_1 R_1 + \cdots + \mathbf{e}_\ell R_\ell + \mathbf{e}(2n - R_1 - \cdots - R_\ell) = 2m$$

with an extra $\mathbf{e}$ assigned to the final (possibly uncompleted) loop. Use this picture to evaluate

$$\sum_{0\leq m\leq n} \alpha^{2n}\beta^{2m}\mathbb{P}\left[\#(2n)=2m\right] \quad \text{as} \quad (1-\alpha^2)^{-1/2}(1-\alpha^2\beta^2)^{-1/2}.$$

Then compare Exercise 3.4.3. *Hint*: Sum first over m and n, in that order, and integrate out the $\mathbf{e}$s.

Third proof from P. Lévy [1939:303–304].   This is a bit sketchy, but that is Lévy's way, and it's so clever that I could not leave it out. You have, or so you may hope,

$$\frac{\#(2n)}{2n} \simeq \frac{\mathbf{e}_1 R_1 + \cdots + \mathbf{e}_\ell R_\ell + \mathbf{e}(2n - R_1 - \cdots - R_\ell)}{R_1 + \cdots + R_\ell + (2n - R_1 - \cdots - R_\ell)}$$

$$\simeq \frac{\mathbf{e}_1 R_1 + \cdots + \mathbf{e}_\ell R_\ell}{\mathbf{e}_1 R_1 + \cdots + \mathbf{e}_\ell R_\ell + (1-\mathbf{e}_1)R_1 + \cdots + (1-\mathbf{e}_\ell)R_\ell},$$

if $2n \simeq R_1 + \cdots + R_\ell$. Now for each choice of the $\mathbf{e}$s, $\mathbf{e}_1 R_1 + \cdots + \mathbf{e}_\ell R_\ell$ is the sum of $\mathbf{e}_1 + \cdots + \mathbf{e}_\ell \equiv m$ independent copies of R_1 while $(1-\mathbf{e}_1)R_1 + \cdots + (1 - \mathbf{e}_\ell)R_\ell$ is the sum of $\ell - m$ *different* copies of R_1, so that the two sums are independent. Then, still hopeful, you may imagine that

$$\lim_{n\uparrow\infty} \mathbb{P}\left[\frac{\#(2n)}{2n} \leq x\right]$$

$$= \lim_{n\uparrow\infty} \mathbb{P}\left[\frac{\mathbf{e}\cdot R}{\mathbf{e}\cdot R + (1-\mathbf{e})\cdot R} \leq x\right] \quad \text{with a self-evident notation}$$

$$= \lim_{n\uparrow\infty} \sum_{m=0}^\ell \binom{\ell}{m} 2^{-\ell}\mathbb{P}\left[\frac{R_1' + \cdots + R_m'}{R_1' + \cdots + R_m' + R_1'' + \cdots + R_{\ell-m}''} \leq x\right]$$

with independent R's and R''s. Here, the binomial distribution is concentrating about $m = \ell/2$ by the law of large numbers, so you ought to have

$$\lim_{n\uparrow\infty} \mathbb{P}\left[\frac{\#(2n)}{2n} \le x\right] = \mathbb{P}\left[\frac{T'}{T'+T''} \le x\right] = \mathbb{P}\left[T'' \ge \frac{1-x}{x}T'\right]$$

with independent T' and T'' distributed alike, as per the law of distant passage of §3.2.3, repeated for convenience:

$$\mathbb{P}(T \le x) = \lim_{n\uparrow\infty} \mathbb{P}\left[\frac{1}{n^2}(R_1 + \cdots + R_n) \le x\right] = \int_0^x \frac{e^{-1/2t}}{\sqrt{2\pi t^3}}\, dt,$$

which is to say

$$\lim_{n\uparrow\infty} \mathbb{P}\left[\frac{\#(2n)}{2n} \le x\right] = \int_0^\infty \frac{e^{-1/2t}}{\sqrt{2\pi t^3}}\, dt \int_{\frac{1-x}{x}t}^\infty \frac{e^{-1/2s}}{\sqrt{2\pi s^3}}\, ds \equiv D(x).$$

Exercise 3.4.5 Finish the proof by showing the density D' of this distribution is nothing but $\frac{2}{\pi}\frac{1}{\sqrt{x(1-x)}}$. Remarkable, no?

But wait. There is some sleight of hand here. The number ℓ is chosen so that $R_1 + \cdots + R_\ell \le 2n < R_1 + \cdots + R_{\ell+1}$. Then n tends to $+\infty$ and so does ℓ in response. Now be clear: n goes $1, 2, 3$, etc. but ℓ is random. Lévy reverses this: for him, it is ℓ that goes $1, 2, 3$, etc. and n that is random. Evidently, it makes no difference since what comes out is correct, *but how did he know?* Like any skillful magician, he's too quick to be easily detected, but that's OK, too. He's always right!

3.4.1 The conditional arcsine law

The conditional arcsine law is perfectly flat:

$$\mathbb{P}[\#(2n) = 2m|\mathbf{x}(2n) = 0] = \frac{1}{n+1}, \quad 0 \le m \le n;$$

a little surprising in view of the unconditional law, but that's typical too. Fluctuations in coin tossing look pretty odd the first time you see them.

Exercise 3.4.6 Use the format of Exercise 3.4.4 to evaluate

$$\sum \mathbb{P}[\#(2n) = 2m, \mathbf{x}(2n) = 0]\, \alpha^{2n}\beta^{2m},$$

noting that $\mathbf{x}(2n) = 0$ is the same as saying $R_1 + \cdots + R_\ell = 2n$ for some $\ell \ge 0$. The sum is a little easier now. *Answer:* $\left[\sqrt{1-\alpha^2} + \sqrt{1-\alpha^2\beta^2}\right]^{-1}$.

Exercise 3.4.7 Rewrite this answer as $\sqrt{1 - \alpha^2\beta^2} - \sqrt{1 - \alpha^2}$ over $\alpha^2(1 - \beta^2)$. Check that the coefficient of α^{2n} is proportional to $\sum_0^n \beta^{2m}$. Read off the conditional law from that.

Exercise 3.4.8 Use the conditional arcsine law to re-derive the passage time distribution of Exercise 3.2.1: $\mathbb{P}_0(T = 2n - 1) = \binom{2n}{n}2^{-2n}\frac{1}{2n-1}$, rewritten here so as to give a hint. A nice picture will show what to do.

3.4.2 Drift and the conventional wisdom

Now let's put in a drift and see what happens. For example, take the 1-step probabilities in Figure 3.4.3. I need to know that $\mathbb{E}_1(T)$ is finite,

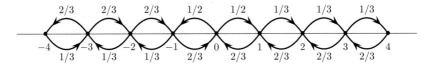

Figure 3.4.3

T being the passage time $\min(n : \mathbf{x}(n) = 0)$. Take $0 < \gamma < 1$ and look at $f(x) = \mathbb{E}_x(\gamma^T)$. For $x > 0$. You have $f(x) = \frac{2}{3}\gamma f(x-1) + \frac{1}{3}\gamma f(x+1)$, and with the guess, $f(x) = \text{a constant} \times \beta^x$, you find two possibilities:

$$\beta_+ = \frac{1}{2\gamma}\left(3 + \sqrt{9 - 8\gamma^2}\right) > \frac{3}{2}$$

and

$$\beta_- = \frac{1}{2\gamma}\left(3 - \sqrt{9 - 8\gamma^2}\right) < 1.$$

β_+ is out since $f < 1$, so $f(x) = \beta_-^x$ taking $f(0) = 1$ into account, and you find $\mathbb{E}_1(T_0) = d\beta_-/d\gamma$ taken at $\gamma = 1$ is 3. Then $\mathbb{E}_0(R) = 1 + \mathbb{E}_1(T) = 4$, by symmetry, R being the loop time $\min(n \geq 1 : \mathbf{x}(n) = 0)$. Now introduce Bernoulli variables $\mathbf{e} = 1$ or 0, indicating whether the loop goes right ($\mathbf{e} = 1$) or left ($\mathbf{e} = 0$), with success probability $1/2$. Then for $n \uparrow \infty$ and $R_1 + \cdots + R_\ell \leq n < R_1 + \cdots + R_{\ell+1}$, you have

$$\frac{\mathbf{e}_1 R_1 + \cdots + \mathbf{e}_\ell R_\ell}{R_1 + \cdots + R_{\ell+1}} \leq \frac{\#(n)}{n} \leq \frac{\mathbf{e}_1 R_1 + \cdots + \mathbf{e}_{\ell+1} R_{\ell+1}}{R_1 + \cdots + R_\ell},$$

to which the general law of large numbers of §5.1 may be applied to produce what the conventional wisdom predicts:

$$\mathbb{P}\left[\lim_{n \uparrow \infty} \frac{\#(n)}{n} = \frac{1}{2}\right] = 1.$$

Obviously, the numbers 1/3, 2/3 are not important – any restoring drift that makes $\mathbb{E}_1(T_0) < \infty$ will do.

Exercise 3.4.9 Check it out, replacing $p = 2/3$ by any $p > 1/2$ and 1/3 by $q = 1 - p$. Will that do?

3.5* Volume

I refer to the number of different places the walk visits up to time n, to wit

$$V(n) = \max_{k \leq n} \mathbf{x}(k) - \min_{k \leq n} \mathbf{x}(k)$$

for the walk starting at, e.g. $\mathbf{x}(0) = 0$. The name "volume" is better associated with the 3-dimensional walk, but no matter. I will come to that in §4.8. The mean volume is easy:

$$\mathbb{E}[V(n)] = \sum_{k=0}^{n} \mathbb{P}\left[\mathbf{x}(k) \neq \mathbf{x}(k-1), \ldots, \mathbf{x}(1), \mathbf{x}(0)\right]$$

$$= \sum_{k=0}^{n} \mathbb{P}\left[\mathbf{x}(1), \mathbf{x}(2), \ldots, \mathbf{x}(k) \neq 0\right]$$

$$= \sum_{k=0}^{n} \mathbb{P}(R > k)$$

$$\simeq \sum_{k=0}^{n} \sqrt{2/\pi k} \quad \text{as in Exercise 3.4.5}$$

$$\simeq 2^{3/2}\sqrt{n/\pi} \quad \text{for large } n,$$

i.e.

$$\lim_{n \uparrow \infty} \mathbb{E}(V/\sqrt{n}) = 2^{3/2}/\sqrt{\pi},$$

in which the scaling by $\sqrt{n}$ could have been foreseen. More sketchily,

$$\mathbb{E}_0[V(n)] = 2\,\mathbb{E}_0\left[\max_{k \leq n} \mathbf{x}(k)\right]$$

$$\simeq 4\,\mathbb{E}_0[\mathbf{x}^+(n)] \qquad \text{by the reflection principle of}$$
$$\qquad\qquad\qquad\qquad\qquad\qquad\quad \text{D. André}$$

$$\simeq 4\sqrt{n}\int_0^\infty x\frac{e^{-x^2/2}}{\sqrt{2\pi}}\,dx \qquad \text{by CLT}$$

$$= 2^{3/2}\sqrt{n/\pi}.$$

The goal is now to find $\lim_{n\uparrow\infty}\mathbb{P}(V/\sqrt{n}\leq c)$. Nothing better can be had, i.e. there is no individual law since

$$\lim_{n\uparrow\infty}\frac{1}{\sqrt{n}}\max_{k\leq n}\mathbf{x}(k)=+\infty$$

by the law of the iterated logarithm. I explain the plan of the computation. The technicalities are skimped as being neither difficult nor interesting.

Fix positive numbers a and b and look at

$$\mathbb{P}\left[\frac{1}{\sqrt{n}}\min_{k\leq nt}\mathbf{x}(k)\leq -a \ \text{ and }\ \frac{1}{\sqrt{n}}\max_{k\leq nt}\mathbf{x}(k)\geq b\right]$$

$$=\mathbb{P}\left[\frac{1}{n}T_{-a\sqrt{n}}\leq t \ \text{ and }\ \frac{1}{n}T_{b\sqrt{n}}\leq t\right]$$

$$=\mathbb{P}(A\cap B)=\mathbb{P}(A)+\mathbb{P}(B)-\mathbb{P}(A\cup B),$$

in which A is the event $(T_{-a\sqrt{n}}\leq nt)$, B the event $(T_{b\sqrt{n}}\leq nt)$, and a and b are to be adjusted as $n\uparrow\infty$ so that $a\sqrt{n}$ and $b\sqrt{n}$ are whole numbers. This is harmless: the changes required are tiny and cannot disturb the outcome. The estimation of $\mathbb{P}(A)$ and $\mathbb{P}(B)$ is easy, by the law of distant passage of §3.2 or by D. André, as above; it is also unnecessary, as you will see, so it's only $\mathbb{P}(A\cup B)$ that's wanted. Here, the 2-sided passage of Exercise 3.2.6 comes into play: with $T=$ the smaller of T_a and T_b, you have

$$\mathbb{E}_x[(\cosh\alpha)^{-T}]=\frac{\cosh\alpha\left(x-\frac{a+b}{2}\right)}{\cosh\alpha\left(\frac{b-a}{2}\right)}\quad\text{for }a<x<b.$$

Replace x by 0, a by $-a\sqrt{n}$, b by $+b\sqrt{n}$, and α by $\beta=\operatorname{arccosh}(\mathrm{e}^{\alpha/n})\simeq\sqrt{2\alpha/n}$ for fixed α to obtain

$$\int_0^\infty \mathrm{e}^{-\alpha t}\,\mathrm{d}\,\mathbb{P}(A\cup B)=\mathbb{E}\left[\exp\left(-\frac{\alpha}{n}T_{-a\sqrt{n}}\wedge T_{b\sqrt{n}}\right)\right]$$

$$=\frac{\cosh\beta\sqrt{n}\,(b-a)/2}{\cosh\beta\sqrt{n}\,(b+a)/2}$$

$$\simeq\frac{\cosh\sqrt{2\alpha}\,(b-a)/2}{\cosh\sqrt{2\alpha}\,(b+a)/2}\quad\text{as }n\uparrow\infty$$

$$\equiv-\hat{p}(a,b).$$

Exercise 3.5.1 Convince yourself that $\lim_{n\uparrow\infty}\mathbb{P}(V/\sqrt{n}\leq c)$ may be computed by inverting the transform $\iint_{a+b\leq c}\partial^2\hat{p}/\partial a\partial b$. Think it over.

Note that $\partial^2/\partial a \partial b$ kills any contribution by $\mathbb{P}(A)$ or $\mathbb{P}(B)$, as suggested before.

Now for the computation:

$$\frac{\partial \hat{p}}{\partial a} = \frac{1}{2}\sqrt{2\alpha}\,\frac{\sinh\sqrt{2\alpha}\,b}{\cosh^2\sqrt{2\alpha}\,(b+a)/2}\,,$$

so

$$\iint_{a+b\leq c} \partial^2 \hat{p}/\partial a \partial b = \int_0^c \left(\frac{\partial \hat{p}}{\partial a}(a,b)\Big|_{b=0}^{b=c-a}\right) da$$

$$= \int_0^c \frac{1}{2}\,\frac{\sqrt{2\alpha}\sinh(\sqrt{2\alpha}\,a)}{\cosh^2(\sqrt{2\alpha}\,c/2)}\,da$$

$$= \frac{1}{2}\,\frac{\cosh(\sqrt{2\alpha}\,c)-1}{\cosh^2(\sqrt{2\alpha}\,c/2)}$$

$$= \frac{\sinh^2(\sqrt{2\alpha}\,c/2)}{\cosh^2(\sqrt{2\alpha}\,c/2)}$$

$$= \left(\frac{1-x^2}{1+x^2}\right)^2 \quad \text{with } x = e^{-\sqrt{2\alpha}\,c/2}$$

$$= 1 + (2x) \times \text{the derivative of } \frac{1}{1+x^2}$$

$$= 1 + \sum_{n=1}^{\infty} (-1)^n 4n x^{2n}.$$

Finally, put back $e^{-\sqrt{2\alpha}\,c/2}$ in place of x, invert with the help of §1.7.5, taking $t=1$ in the formula there to obtain

$$\lim_{n\uparrow\infty} \mathbb{P}(V/\sqrt{n} \leq c) = 1 + \sum_{n=1}^{\infty} (-1)^n 4n \int_c^1 nc\,\frac{e^{-n^2 c^2/2t}}{\sqrt{4\pi t^3}}\,dt$$

$$= 1 + \sum_{n=1}^{\infty} (-1)^n 4n \sqrt{\frac{2}{\pi}} \int_{nc}^{\infty} e^{-x^2/2}\,dx$$

and the corresponding density:

$$D(c) = \sqrt{\frac{2}{\pi}} \sum_{n=1}^{\infty} (-1)^{n-1} 4n^2 e^{-n^2 c^2/2}.$$

Not very attractive, nor obviously positive, but there it is; see §6.5 for a variant of the proof.

Exercise 3.5.2 The sum is alternating, with diminishing terms if $x \geq 1$; as such, it is positive and does not exceed $8(2\pi)^{-1/2}e^{-x^2/2}$. Jacobi's identity from §1.6.3 may be used to confirm that, actually, it *is* positive for any $x > 0$ despite appearances: The function

$$p(t, x) = \sum_{n \in \mathbb{Z}} \frac{e^{-(x-n)^2/4t}}{\sqrt{4\pi t}} = \sum_{n \in \mathbb{Z}} e^{2\pi\sqrt{-1}\,nx} e^{-4\pi^2 n^2 t}$$

is the elementary solution of the heat equation $\partial w/\partial t = \partial^2 w/\partial x^2$ on the circle of perimeter 1: it is the circular Gauss density, so to say. Its maximum lies over $x = 0$, its minimum over $x = \frac{1}{2}$, and as time passes, the maximum falls and the minimum rises until the temperature is flat ($p \equiv 1$) at $t = +\infty$. This picture does the trick: at $x = \frac{1}{2}$, $\partial p/\partial t$ is just

$$\sum_{n \in \mathbb{Z}} (-1)^{n-1} 4\pi^2 n^2 e^{-n^2 x^2/2} \quad \text{where now } 8\pi^2 t \text{ is replaced by } x^2,$$

this being $\sqrt{2} \times \pi^{5/2} \times$ the density. Think it over.

Exercise 3.5.3 Use Jacobi's identity to check that $\int_0^\infty D = 1$, as it should be, and also that $\int_0^\infty xD = 2^{3/2}/\sqrt{\pi}$, in conformity with the value of $\lim_{n \uparrow \infty} \mathbb{E}(V/\sqrt{n})$ found before.

4

The Standard Random Walk in Higher Dimensions

Let $\mathbb{Z}^d$ be the d-dimensional lattice $\mathbb{Z} \times \cdots \times \mathbb{Z}$ (d-fold) and let e denote any one of the $2d$ nearest neighbors of the origin. For example, in dimension 3, $e = (\pm 1, 0, 0), (0, \pm 1, 0), (0, 0, \pm 1)$. The standard d-dimensional random walk $\mathrm{RW}(d)$, is just as in dimension 1 except that the independent steps $\mathbf{e}_n : n \geq 1$ have now these $2d$ equally likely possibilities. It will be obvious that the triple comprised of:

(1) $\mathbf{X} = $ the space of paths $\mathbf{x} \colon n \to \mathbf{x}(n) = x \in \mathbb{Z}^d + \mathbf{e}_1 + \cdots + \mathbf{e}_n$ ($n \geq 0$),
(2) $\mathfrak{F} = $ the natural field descriptive of such paths, and
(3) the system of probabilities $\mathbb{P}_x(A) = \mathbb{P}_0(x + \mathbf{x} \in A)$, one such for each $x \in \mathbb{Z}^d$,

constitutes a strict Markov process, just as in dimension 1, i.e.

$$\mathbb{P}_a(\mathbf{x}^+ \in B | \mathfrak{F}_T) = \mathbb{P}_b(B) \text{ with } b = \mathbf{x}(T)$$

for any stopping time T, with all the former meanings of stopping time, $\mathbf{x}^+ \colon n \to \mathbf{x}(T + n)$, $n \geq 0$, and $\mathfrak{F}_T = (A \in \mathfrak{F} : A \cap (T \leq n) \in \mathfrak{F}_n$ for $n \geq 0)$. Notice that the d-dimensional walk is very nearly the joint motion of d independent copies of $\mathrm{RW}(1)$, run at speed $1/d$, in response to the protocol: that if one copy of $\mathrm{RW}(1)$ moves, the others must wait. I speak mostly about $d = 2$ and 3; after $d = 3$, nothing changes much.

4.1 What $\mathrm{RW}(2)$ and $\mathrm{RW}(3)$ do as $n \uparrow \infty$

In dimension 1, the walk visits every site in $\mathbb{Z}$ infinitely many times. It is the same in dimension 2, but $\mathbb{Z}^3$ is *much* bigger and the walk drifts out to ∞, i.e. $\mathbb{P}_0[\lim_{n \uparrow \infty} |\mathbf{x}(n)| = \infty] = 1$, and this also for $d \geq 4$, automatically, since, for example, RW(4) is (more or less) RW(3) run at speed $3/4$ in combination with a copy of RW(1) run at speed $1/4$. Think of it this way: a drunk wandering in $\mathbb{Z}^2$ is sure to get home, though it may take him a very long time, and he may not know he's there, but for $d = 3$, $\mathbb{P}_0[\mathbf{x}(2n) = 0$ for some $n > 0]$ is only a little more than $1/3$, as will be seen, so with probability about $2/3$, the walker gets hopelessly lost.

Proof for $d = 2$.   This imitates the first proof of $\mathbb{P}_0(T_1) = 1$ in §3.2. The event $\mathbf{x}(2n) = 0$ requires $2i$ horizontal steps, i to the right and an equal number to the left, and $2j = 2n - 2i$ vertical steps, j up and an equal number down, i.e.

$$
\begin{aligned}
\mathbb{P}_0[\mathbf{x}(2n) = 0] &= \sum_{i+j=n} \binom{2n}{2i}\binom{2i}{i}\binom{2j}{j} 4^{-2n} \\
&= \binom{2n}{n} \sum_{k=0}^{n} \binom{n}{k}^2 4^{-2n} \\
&= \left[\binom{2n}{n} 2^{-2n} \right]^2 \qquad \text{by Exercise 1.2.4} \\
&\simeq \frac{1}{\pi n} \qquad\qquad\qquad \text{by Stirling,}
\end{aligned}
$$

so $\sum_0^\infty \mathbb{P}_0[\mathbf{x}(2n) = 0] =$ the expected number of visits to the origin diverges to $+\infty$. Let R be the time of first return (if any), namely $R = \min(n \geq 2 : \mathbf{x}(n) = 0) \leq \infty$. This is a stopping time, so the walk starts over again from scratch if $R < \infty$. The excursion $\mathbf{x}(n) : 0 < n \leq R$ is then a "loop", and it is obvious that successive loops are independent copies of the first loop, just as in dimension 1. Now let $p = \mathbb{P}_0(R < \infty)$ be < 1. Then the probability that there be exactly n loops is $p^n(1 - p)$, and the mean number of visits $= 1 +$ the mean number of loops is

$$
1 + \sum_{n=0}^{\infty} np^n(1 - p) = 1 + \frac{p}{1 - p} = \frac{1}{1 - p} < \infty.
$$

But this is contradictory, i.e. $p = 1$, which is to say there are infinitely many loops. Now fix a lattice point $x \neq 0$. Then the events B_n that x is visited by the nth loop may be regarded as independent Bernoulli trials with success probability $p = \mathbb{P}_0(\text{loop no. 1 visits } x)$, so by the law of large numbers, the walk visits x about np times during n loops. In short, the walk goes everywhere, over and over.

Proof for $d = 3$.   Now $\sum_0^\infty \mathbb{P}_0[\mathbf{x}(2n) = 0] < \infty$, i.e. $p < 1$, as will be proved in a moment. Grant me that and the rest is easy. The actual number of visits to the origin is finite with probability 1, and it is the same for any lattice point x: if the walk ever gets there, it starts from scratch and returns only a finite number of times. Now draw a big sphere of radius N. It contains $\frac{4}{3}\pi N^3$ lattice points more or less so, by and by, the path must come outside and *stay* outside, i.e. $\mathbb{P}_0[|\mathbf{x}(n)| > N$ for $n \uparrow \infty] = 1$. Now make $N \uparrow \infty$. It remains only to overestimate $\mathbb{P}_0[\mathbf{x}(2n) = 0]$ by the general term of a convergent sum.

Proof (after Feller [1966(1):36]). $\mathbf{x}(2n) = 0$ requires an equal number of steps eastward or the reverse ($= i$), northward or the reverse ($= j$), up or down ($= k$), so

$$\mathbb{P}_0[\mathbf{x}(2n) = 0] = \sum_{i+j+k=n} \frac{(2n)!}{(i!)^2(j!)^2(k!)^2} 6^{-2n}$$

$$= \binom{2n}{n} 2^{-2n} \sum_{i+j+k=n} \left(\frac{n!}{i!j!k!} 3^{-n}\right)^2,$$

in which you see the trinomial distribution $(n!/i!j!k!)3^{-n}$ squared, and since these numbers add up to 1, the sum is overestimated by

$$\max_{i+j+k=n} \frac{n!}{i!j!k!} \times 3^{-n}.$$

Now for large n, $n!/i!j!k!$ is comparatively small unless all the numbers i, j, k are also large, as you will check. This permits the application of Stirling's formula, after which you have only to minimize $i^{i+\frac{1}{2}}j^{j+\frac{1}{2}}k^{k+\frac{1}{2}}$ for large i, j, k summing to n. Thinking of i, j, k not as whole numbers but as real variables, a short calculation shows that $i = j = k = n/3$ is best, whereupon the maximum is seen to be not more than a constant multiple of $1/n$, and $\mathbb{P}_0[\mathbf{x}(2n) = 0]$ not more than a multiple of $n^{-3/2}$.

Exercise 4.1.1 Check it out. Then do the same in dimension $d > 3$ to obtain $\mathbb{P}_0[\mathbf{x}(2n) = 0] < C(d)n^{-d/2}$ for $n \uparrow \infty$.

Proof (after Pólya [1921]). This is elegant and has other uses. Take $\theta = (\theta_1, \theta_2, \theta_3)$ in the 3-dimensional torus $[0, 2\pi) \times [0, 2\pi) \times [0, 2\pi)$. You have

$$\sum_{x \in \mathbb{Z}^3} e^{\sqrt{-1}\,\theta \cdot x} \mathbb{P}[\mathbf{x}(n) = x] = \mathbb{E}_0\left[e^{\sqrt{-1}\,\theta \cdot \mathbf{x}(n)}\right]$$

$$= \mathbb{E}_0\left[e^{\sqrt{-1}\,\theta \cdot (\mathbf{e}_1 + \cdots + \mathbf{e}_n)}\right]$$

$$= \left[\mathbb{E}_0(e^{\sqrt{-1}\,\theta \cdot \mathbf{e}_1})\right]^n \text{ by independence of steps}$$

$$= \left[\frac{1}{3}(\cos\theta_1 + \cos\theta_2 + \cos\theta_3)\right]^n,$$

so

$$\mathbb{P}_0[\mathbf{x}(n) = x]$$
$$= \frac{1}{(2\pi)^3} \int_{-\pi}^{\pi} \int_{-\pi}^{\pi} \int_{-\pi}^{\pi} e^{\sqrt{-1}\,\theta \cdot x} \left[\frac{1}{3}(\cos\theta_1 + \cos\theta_2 + \cos\theta_3)\right]^n d^3\theta;$$

in particular,

$$\mathbb{P}_0[\mathbf{x}(2n) = 0] = \frac{1}{(2\pi)^3} \int_{-\pi}^{\pi} \int_{-\pi}^{\pi} \int_{-\pi}^{\pi} \left[\frac{1}{3}(\cos\theta_1 + \cos\theta_2 + \cos\theta_3)\right]^{2n} d^3\theta.$$

Now $\frac{1}{3}(\cos\theta_1 + \text{etc.})$ peaks at $\theta = 0$; nearby, it looks like $1 - \frac{1}{6}(\theta_1^2 + \theta_2^2 + \theta_3^2)$, for large n, with the implication that

$$\mathbb{P}_0[\mathbf{x}(2n) = 0] \simeq \frac{1}{(2\pi)^3} \times (4\pi = \text{the surface of the unit sphere})$$
$$\times \int_0^1 \left(1 - \frac{r^2}{6}\right)^{2n} r^2 \, dr$$
$$< \int_0^1 e^{-nr^2/3} r^2 \, dr$$
$$\simeq \int_0^\infty e^{-r^2/3} r^2 \, dr \times n^{-3/2}$$
$$= O(n^{-3/2})$$

as promised.

Exercise 4.1.2   That's a little sketchy. Fill it in. Do the same for $d \geq 4$ to obtain $\mathbb{P}_0[\mathbf{x}(2n) = 0] \simeq C(d) n^{-d/2}$.

Exercise 4.1.3 (from B. Rider)   In any dimension ≥ 3, the last leaving time $L = \max(n : \mathbf{x}(n) = 0)$ is finite with probability 1. Check that $\mathbb{P}_0(L = 2n) = \mathbb{P}_0[\mathbf{x}(2n) = 0] \times (1 - p)$ with $p = \mathbb{P}_0(R < \infty)$ as before, and so reprove $\sum \mathbb{P}_0[\mathbf{x}(2n) = 0] = 1/(1 - p)$.

★ Watson's integral

The expected number of returns for $d = 3$ was evaluated by G.N. Watson [1939]. It is

$$\frac{1}{1 - p} = \sum_{n=0}^{\infty} \mathbb{P}[\mathbf{x}(2n) = 0]$$
$$= \frac{1}{(2\pi)^3} \int_{-\pi}^{\pi} \int_{-\pi}^{\pi} \int_{-\pi}^{\pi} \left[\frac{1}{3}(\cos\theta_1 + \cos\theta_2 + \cos\theta_3)\right]^{-1} d^3\theta$$
$$= \frac{12}{\pi^2}(18 + 12\sqrt{2} - 10\sqrt{3} - 7\sqrt{6})\,\mathrm{K}^2(k)$$

in which $\mathrm{K}(k)$ is Jacobi's complete elliptic integral

$$\mathrm{K}(k) = \int_0^1 (1 - x^2)^{-1/2}(1 - k^2 x^2)^{-1/2}\, \mathrm{d}x$$

with "modulus" $k = \sqrt{(2 - \sqrt{3})(\sqrt{3} - \sqrt{2})}$. Just the number that springs to mind, yes? By numerical calculation, $p \simeq .344$ or a little more than $1/3$, so the expected number of visits is about $3/2$. The alternative expression, using gamma functions,[1] is $(\sqrt{6}/32\pi^2)\Gamma(\frac{1}{24})\Gamma(\frac{5}{24})\Gamma(\frac{7}{24})\Gamma(\frac{11}{24})$ is from Glasser-Zucker [1977]. For more information, see Doyle-Snell [1984].

4.2 How RW(3) escapes to ∞

4.2.1 Speed

Well, if RW(3) runs off to ∞, how fast does it go? It looks (more or less) like three independent copies of RW(1), run at speed $1/3$, so you cannot expect anything much better than $|\mathbf{x}(n)| \simeq \sqrt{2(n/3)\log\log n}$ now and then. On the other hand, $\mathbb{P}_0[\mathbf{x}(2n) = x]$ or $\mathbb{P}_0[\mathbf{x}(2n+1) = x]$ is not more than $\mathbb{P}_0[\mathbf{x}(2n) = 0] = O(n^{-3/2})$, by Pólya's formula, so

$$\mathbb{P}_0[|\mathbf{x}(n)| \le c] \lesssim \frac{4}{3}\pi c^3 \times O(n^{-3/2})$$

is the general term of a convergent sum if $c = n^{(1/6)-}$, whence

$$\mathbb{P}_0[|\mathbf{x}(n)| \ge n^{(1/6)-} \text{ for } n \uparrow \infty] = 1$$

by Borel–Cantelli. This is obviously crude. The real truth is much nearer to $|\mathbf{x}(n)| \ge \sqrt{n}$, showing more plainly how big $\mathbb{Z}^3$ really is.

Proof.   Fix a small number $\alpha > 0$ and estimate as above to obtain

$$\mathbb{P}[|\mathbf{x}(n)| \le n^{\frac{1}{2}-\alpha}] \le \frac{4}{3}\pi n^{\frac{3}{2}-3\alpha} \times O(n^{-\frac{3}{2}}) = O(n^{-3\alpha}).$$

Then take β so as to make $\alpha\beta > 1/3$, and ignoring the fact that $m = n^\beta$ may not be a whole number, conclude that $\mathbb{P}_0[\mathbf{x}(m) > m^{\frac{1}{2}-\alpha}$ as $m \uparrow \infty] = 1$. The plan for the rest is to make $k \uparrow \infty$ with $m = n^\beta \le k < (n+1)^\beta$ and to estimate $|\mathbf{x}(k)| \ge |\mathbf{x}(m)| - |\mathbf{x}(k) - \mathbf{x}(m)|$ from below, hoping that $|\mathbf{x}(m)|$ may dominate since $k - m < \ell = (n+1)^\beta - n^\beta \simeq$

---

[1] The gamma function is defined as $\Gamma(s) = \int_0^\infty x^{s-1}e^{-x}\, \mathrm{d}x/x$.

$\beta n^{\beta-1}$ is small compared to $m = n^\beta$. This works if $\alpha\beta < 1/2$: indeed, with a small number γ,

$$\mathbb{P}_0\left[\max_{m\leq k<m+\ell}|\mathbf{x}(k)-\mathbf{x}(m)| > \gamma\,m^{\frac{1}{2}-\alpha}\right]$$

$$= \mathbb{P}_0\left[\max_{k<\ell}|\mathbf{x}(k)| > \gamma\,m^{\frac{1}{2}-\alpha}\right]$$

$$\leq 3\,\mathbb{P}_0\left[\max_{k<\ell}|\mathbf{x}_3(k)| > \frac{\gamma}{\sqrt{3}}\,m^{\frac{1}{2}-\alpha}\right] \quad \text{in which } \mathbf{x}_3 \text{ is the vertical part of the 3-dimensional walk}$$

$$\leq 12\,\mathbb{P}_0\left[\mathbf{x}_3(\ell) > \frac{\gamma}{\sqrt{3}}\,m^{\frac{1}{2}-\alpha}\right] \quad \text{by André's reflection principle}$$

$$< e^{-x^2/(2+)\ell} \quad \text{with } x = \frac{\gamma}{\sqrt{3}}m^{\frac{1}{2}-\alpha}, \text{ by the final "improved estimate" of §2.1}$$

$$< e^{-cn^{1-2\alpha\beta}} \quad \text{with a constant } c > 0,$$

and since $\alpha\beta < 1/2$, Borel–Cantelli guarantees that

$$|\mathbf{x}(k)| \geq |\mathbf{x}(m)| - |\mathbf{x}(k)-\mathbf{x}(m)| \geq (1-\gamma)m^{\frac{1}{2}-\alpha};$$

in short,

$$\mathbb{P}_0\left[\mathbf{x}(n) \geq n^{(1/2)-} \text{ as } n\uparrow\infty\right] = 1.$$

This could still be improved, but I leave the matter there; see Dvoretsky–Erdős [1950:363–367] for the whole truth, *to wit*: for $0 < h(n)$ decreasing with n, $\mathbb{P}_0[|\mathbf{x}(n)| > \sqrt{n}\,h(n) \text{ as } n\uparrow\infty] = 1$ or 0 according as $\sum_1^\infty n^{-1}h(n) < \infty$ or not – for example, $h(n) = 1/\log n$ is borderline.

I must also mention Doyle–Snell [1984] for the beautiful connection between resistance in electrical circuits and the tendency to infinity of RW(3) and more general walks.

4.2.2 Direction

Now you know how fast $\mathbf{x}(n)$ runs off to infinity; but what direction does it take? Obviously, *no* particular direction. To spell it out, write $\mathbf{x} = |\mathbf{x}| \times \mathbf{e}$ in spherical polar coordinates, and let C be any spherical cap: $e \cdot o > \cos\varphi$ with fixed $0 < \varphi < \pi/2$, fixed direction o, and variable direction e, as in Figure 4.2.1. Then $\mathbf{e}(n) \in C$ is the same as to say that $\mathbf{x}(n)/\sqrt{n}$ belongs to the cone $K = (x : x \cdot o > |x|\cos\varphi)$. Now recall that RW(3) looks like three independent copies of RW(1) with the protocol that when one of them moves – with probability $1/3$ – the other two wait. Then, roughly speaking, RW(3) looks like three independent copies of

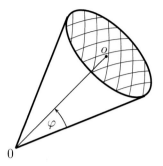

Figure 4.2.1

RW(1) run at $1/3$ speed, and $\mathbf{x}(n)/\sqrt{n}$ approximates a 3-dimensional Gaussian vector with density $(2\pi n/3)^{-3/2}e^{-3x^2/2n}$, so that

$$\mathbb{P}_0[\mathbf{e}(n) \in C] \simeq \int_K \frac{e^{-3x^2/2n}}{(2\pi n/3)^{3/2}} \, d^3x$$

$$= \frac{1}{4\pi} \times \text{the spherical area of the cap}$$

when you work it out. In short, for large n, $\mathbf{e}(n)$ is nearly equidistributed over the spherical surface $|\mathbf{e}| = 1$.

Exercise 4.2.1 That's pretty rough, but perfectly correct; §5.9 explains the 3-dimensional version of CLT which takes care of everything. Meanwhile, try to reproduce the approximation of $\mathbb{P}_0[\mathbf{e}(n) \in C]$ by formal application of Pólya's formula in dimension 3 from §4.1.
Hint: $\frac{1}{3}(\cos\theta + \text{etc.})$ from §4.1 looks like $e^{-\theta^2/6}$ in the vicinity of $\theta = 0$, producing the bulk of Pólya's integral. Take this literally, extend the integral to $\mathbb{R}^3$, and see what comes out.

Exercise 4.2.2 Check that $\mathbf{e}(n)$ enters every spherical cap i.o. as $n \uparrow \infty$.
Hint: What 01 law operates here? Tell me that. Then look at $\mathbb{P}_0[\mathbf{e}(n) \notin C]$ for large n.

The discussion to date may make you think that a naive equidistribution of angles takes place, as per

$$\mathbb{P}_0\left[\lim_{n\uparrow\infty} \frac{1}{n}\#(k \le n : \mathbf{e}(k) \in C = \frac{\text{area}}{4\pi}\right] = 1,$$

but this is wildly wrong. The reason is that when $|\mathbf{x}(n)|$ is large, as it soon is being comparable to $\sqrt{n}$, then $\mathbf{e}(n)$ moves slowly, with a clock comparable, not to n, but to $\sum_1^n |\mathbf{x}(k)|^{-2} \simeq \log n$, in which case the correct statement is, or ought to be,

$$\mathbb{P}_0\left[\lim_{n\uparrow\infty} \frac{1}{\log n} \#(k \le n : \mathbf{e}(k) \in C = \frac{\text{area}}{4\pi}\right] = 1,$$

I do not know how to prove that, but pretty convincing evidence will be found in §6.10 where the analogous question for the 3-dimensional Brownian motion, BM(3), is taken up.

4.3* Gauss–Landen, Pólya and RW(2)

Exercise 4.3.1   RW(2), viewed in the oblique coordinates of Figure 4.3.1, is just a pair of independent copies of RW(1) with step-length $1/\sqrt{2}$. Check it out. Could this trick work in dimension 3 or more? Here

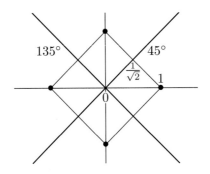

Figure 4.3.1 Oblique coordinates

you see the explanation of the formula of Exercise 1.2.4 employed in §4.1:

$\mathbb{P}_0[\mathbf{x}(2n) = 0]$ in dimension $2 = \mathbb{P}_0[\mathbf{x}(2n) = 0]$ in dimension 1, squared.

This fact, combined with Pólya's formulas from §4.1, has a beautiful consequence. Landen (1775) discovered a remarkable feature of Jacobi's complete elliptic integral:

$$K(k) = \int_0^1 \frac{dx}{\sqrt{(1 - x^2)(1 - k^2 x^2)}} = \int_0^{\pi/2} \frac{d\theta}{\sqrt{1 - k^2 \sin^2 \theta}}, \quad 0 < k < 1,$$

namely

$$(1 + k)\,K(k) = K\big(\sqrt{2}\,k/(1 + k)\big).$$

Gauss [1827] noticed a more intelligible variant: that the allied integral

$$\int_0^{\pi/2} \frac{\mathrm{d}\theta}{\sqrt{a^2 \sin^2 \theta + b^2 \cos^2 \theta}} \quad (\text{with } 0 < b < a) \;=\; \frac{1}{a}\, \mathrm{K}\!\left(\sqrt{1 - \frac{b^2}{a^2}}\, \right)$$

is unchanged if a is replaced by the arithmetic mean $a' = \frac{1}{2}(a+b)$ and b by the geometric mean $b' = \sqrt{ab}$. This will now be proved with the aid of Exercise 4.3.1.

Proof. Take $0 < \gamma < 1$ and let R be the first return to the origin of RW(1). R looks like $1 +$ the passage time from 0 to 1, so $\mathbb{E}_0(\gamma^R) = 1 - \sqrt{1 - \gamma^2}$, as per §3.2, and

$$
\begin{aligned}
G(\gamma^2) &\equiv \sum_{n=0}^{\infty} \mathbb{P}_0[\mathbf{x}(2n) = 0]\gamma^{2n} \\
&= 1 + \sum_{n=1}^{\infty} \mathbb{E}_0(\gamma^{R_1 + \cdots + R_n}) \\
&= \frac{1}{1 - \mathbb{E}_0(\gamma^R)} \\
&= \frac{1}{\sqrt{1 - \gamma^2}}
\end{aligned}
$$

in which R_1, R_2 etc. are the (independent) durations of the successive loops. Now put $\gamma^2 = k e^{\sqrt{-1}\,\theta}$ with $0 < k < 1$. Keeping Exercise 4.3.1 in mind, you see that for RW(2),

$$
\begin{aligned}
\sum_{n=0}^{\infty} \mathbb{P}_0[\mathbf{x}(2n) = 0]\, k^{2n} &= \frac{1}{2\pi} \int_0^{2\pi} |G(k e^{\sqrt{-1}\,\theta})|^2 \, \mathrm{d}\theta \\
&= \frac{1}{2\pi} \int_0^{2\pi} \frac{\mathrm{d}\theta}{|1 - k e^{\sqrt{-1}\,\theta}|} \\
&= \frac{1}{2\pi} \int_0^{2\pi} \frac{\mathrm{d}\theta}{\sqrt{1 - 2k \cos\theta + k^2}} \\
&= \frac{2}{\pi} \int_0^{\pi/2} \frac{\mathrm{d}\theta}{\sqrt{(1+k)^2 \sin^2 \theta + (1-k)^2 \cos^2 \theta}}
\end{aligned}
$$

after the substitution $\theta \to 2\theta$ in line 3 and inspection of Figure 4.3.2. Now my point is that this sum can be computed another way, à la Pólya:

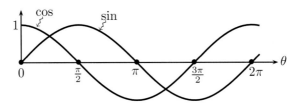

Figure 4.3.2

$$\sum_{n=0}^{\infty} \mathbb{P}_0[\mathbf{x}(2n) = 0]k^{2n}$$

$$= \sum_{n=0}^{\infty} \mathbb{P}_0[\mathbf{x}(n) = 0]k^n \quad \text{since } \mathbf{x}(n) = 0 \text{ only for even } n$$

$$= \frac{1}{(2\pi)^2} \int_0^{2\pi} \int_0^{2\pi} \frac{\mathrm{d}^2\theta}{1 - \frac{k}{2}(\cos\theta_1 + \cos\theta_2)}$$

$$= \frac{1}{(2\pi)^2} \int_0^{2\pi} \int_0^{2\pi} \frac{\mathrm{d}^2\theta}{1 - k\cos\theta_1 \cos\theta_2}$$

$$= \frac{1}{2\pi} \int_0^{2\pi} \frac{\mathrm{d}\theta}{\sqrt{1 - k^2 \cos^2\theta}}$$

$$= \frac{2}{\pi} \mathrm{K}(k).$$

Here, the identity $\cos\theta_1 + \cos\theta_2 = 2\cos\frac{1}{2}(\theta_1 + \theta_2)\cos\frac{1}{2}(\theta_1 - \theta_2)$ is used in line 3, followed by the substitution $\theta_2 \to \theta_1 + 2\theta_2$ and a subsequent integration over θ_1 to obtain line 4; then line 5 comes from

$$\frac{1}{2\pi} \int_0^{2\pi} \frac{\mathrm{d}\theta}{1 - k\cos\theta} = \frac{1}{\sqrt{1 - k^2}}$$

which can be found in any nice table of integrals or derived by Pólya's formula for RW(1), as in

$$\frac{1}{\sqrt{1 - k^2}} = \sum_{n=0}^{\infty} \mathbb{P}_0[\mathbf{x}(n) = 0]k^n = \sum_{n=0}^{\infty} \frac{1}{2\pi} \int_0^{2\pi} \cos^n\theta \, \mathrm{d}\theta \times k^n$$

$$= \frac{1}{2\pi} \int_0^{2\pi} \frac{\mathrm{d}\theta}{1 - k\cos\theta}.$$

Anyhow, the upshot is

$$\int_0^{\pi/2} \frac{d\theta}{\sqrt{1 - k^2 \cos^2 \theta}} = \int_0^{\pi/2} \frac{d\theta}{\sqrt{\sin^2 \theta + (1 - k^2) \cos^2 \theta}}$$

$$= \int_0^{\pi/2} \frac{d\theta}{\sqrt{(1 + k)^2 \sin^2 \theta + (1 - k)^2 \cos^2 \theta}},$$

and that is Gauss's rule when $a = 1 + k$ and $b = 1 - k$ for which $\frac{1}{2}(a + b) = 1$ and $\sqrt{ab} = \sqrt{1 - k^2}$.

McKean–Moll [1997:66] can be consulted for more information about this curious subject. There you will see that Gauss's proof is not really different from mine; it is only differently expressed and a good deal less transparent.

4.4 RW(2): loops and occupation numbers

The equidistribution of visits to the several sites of $\mathbb{Z}^2$ repeats itself as for RW(1). Take $\mathbf{x}(0) = 0$ and let R be the loop time $\min(n \geq 2 : \mathbf{x}(n) = 0)$. Then the mean number of visits per loop to $x \in \mathbb{Z}^2$ is $\mathbb{E}_0(\#) = \mathbb{P}_0(T_x < \infty) = 1$, just as before, in §3.3, and the rest is the same; in particular,

$$\mathbb{P}_0 \left[\lim_{N \uparrow \infty} \frac{\#(N, x)}{\#(N, 0)} = 1 \text{ for every } x \in \mathbb{Z}^2 \right] = 1.$$

What is now more complicated is the distribution of the occupation number for an individual site such as $x = 0$, it makes no difference which. The result is due to Kallianpur–Robbins [1953]:

$$\lim_{N \uparrow \infty} \mathbb{P}_0 \left[\frac{\#(N, 0)}{\log N} \geq \frac{x}{\pi} \right] = e^{-x}.$$

I prove it by A.A. Markov's (1912) "method of moments" – this for variety and also as a preview of future applications of Markov's method in, for example, §12.3.

Proof. I abbreviate $\#(N, 0)$ by $\#$ plain. Now $\mathbb{P}_0[\mathbf{x}(2n) = 0] \simeq 1/\pi n$, by Stirling's formula, so

$$\mathbb{E}_0(\#) \simeq \sum_{2n \leq N} \frac{1}{\pi n} \simeq \frac{1}{\pi} \log N,$$

and similarly,

$$\mathbb{E}(\#^2) \simeq \frac{2}{\pi^2} \sum_{1 \le i < j \le N/2} \frac{1}{i} \frac{1}{j-i}$$

$$= \frac{2}{\pi^2} \sum_{1 \le i < j \le N/2} \left(\frac{1}{i} + \frac{1}{j-i}\right) \frac{1}{j}$$

$$\simeq \frac{2}{\pi^2} \sum_{1 < j \le N/2} 2 \log j \times \frac{1}{j} \qquad \text{by summation on } i < j$$

$$\simeq \frac{2}{\pi^2} \int_1^{x/2} 2 \frac{\log x}{x} \, \mathrm{d}x$$

$$\simeq \frac{2}{\pi^2} (\log N)^2;$$

$$\mathbb{E}(\#^3) \simeq \frac{(6 = 3!)}{\pi^3} \sum_{1 \le i < j < k \le N/2} \frac{1}{i} \frac{1}{j-i} \frac{1}{k-i}$$

$$\simeq \frac{6}{\pi^3} \sum_{1 \le i \le N/2} \frac{1}{i} \left[\log\left(\frac{N}{2} - i\right)\right]^2 \qquad \begin{matrix} \text{by summation on } j \text{ and } k \\ \text{and the previous estimate} \end{matrix}$$

$$\simeq \frac{6}{\pi^3} \int_1^{N/2} \frac{1}{x} \left[\log\left(\frac{N}{2} - x\right)\right]^2 \, \mathrm{d}x$$

$$= \frac{6}{\pi^3} \int_{2/N}^1 \frac{1}{x} \left[\log \frac{N}{2} + \log(1-x)\right]^2 \, \mathrm{d}x \qquad \begin{matrix} \text{by the substitution} \\ x \to \frac{N}{2}x \end{matrix}$$

$$\simeq \frac{6}{\pi^3} (\log N)^3,$$

and so on. You see the pattern: with $\mathbf{x} = \pi \times \#(N)/\log N$,

$$\lim_{N \uparrow \infty} \mathbb{E}_0(\mathbf{x}^n) = n! = \int_0^\infty x^n \mathrm{e}^{-x} \, \mathrm{d}x \quad \text{for all } n \ge 0.$$

That's hopeful, and if you knew that

(1) $\displaystyle \lim_{N \uparrow \infty} \mathbb{E}_0(\mathrm{e}^{-\alpha \mathbf{x}}) = \int_0^\infty \mathrm{e}^{-\alpha x} \mathrm{e}^{-x} \, \mathrm{d}x$ for any $\alpha \ge 0$,

you could finish the proof as in §3.2. Now the distribution of $\mathbf{x}$ has all its moments under control, independently of N, so any limiting distribution F, of which there might be several, must have the same moments: $\int_0^\infty x^n \, \mathrm{d}F = n!$. Then

$$\int_0^\infty \mathrm{e}^{\alpha x} \, \mathrm{d}F = \sum_{n=0}^\infty \frac{\alpha^n}{n!} \int_0^\infty x^n \, \mathrm{d}F = \sum_{n=0}^\infty \alpha^n = \frac{1}{1-\alpha} < \infty \quad \text{if } \alpha < 1,$$

so you can use $e^{\alpha x}$ to dominate the partial sums $\sum_{k=0}^{n}(-1)^k\alpha^k/k!$ for $e^{-\alpha x}$ to obtain

(2) $\displaystyle\int_0^\infty e^{-\alpha x}\,dF = \int_0^\infty e^{-\alpha x}e^{-x}\,dx$ with the same restriction to $\alpha < 1$.

But also, differentiating (2) by α, you see that

$$\int_0^\infty x^n e^{-\alpha x}\,dF = \int_0^\infty x^n e^{-\alpha x}e^{-x}\,dx \quad \text{for any } n\geq 0, \text{ still with } \alpha < 1,$$

so you can repeat the argument to obtain (2) for $\alpha < 2, 3, 4$ etc. You see how it goes – in short, (1) holds.

Aside. There *do* exist variables $\mathbf{x}\geq 0$, with moments $m_n = \mathbb{E}(\mathbf{x}^n) < \infty$ for all $n\geq 0$ but not fully specified by these. Fortunately, they are seldom seen in practice. Carleman's [1922] test – $\sum_{n=1}^\infty m_n^{-1/n} < \infty$ – decides if *that* is the case or not. Dym–McKean [1972:57–58] explains.

Exercise 4.4.1 Markov's method may be awkward, but sometimes it is the only elementary, hands-on tool available; see §12.2 for a case in point. Just for practice, do CLT for RW(1) Markov's way. Of course, the smart way is via

$$\mathbb{E}\left[e^{\sqrt{-1}\,k\mathbf{x}(n)/\sqrt{n}}\right] = \left(\cos\frac{k}{\sqrt{n}}\right)^n \simeq \left(1-\frac{k^2}{2n}\right)^n \simeq e^{-k^2/2}:$$

two lines only. Markov requires

$$\mathbb{E}(\mathbf{x}(n)/\sqrt{n})^{2p} \simeq (2p)!\,2^{-p}/p! = \int x^{2p}(2\pi)^{-1/2}e^{-x^2/2}\,dx,$$

plus control of the left-hand side, independently of $n\uparrow\infty$ if you can get it. Think the whole thing through, starting from $\mathbb{E}(\mathbf{x}(n)/\sqrt{n})^{2p}\leq (2p)!\,2^{-p}/p!$, to be proved by induction on n.

4.4.1 Duration of a large number of loops

Obviously, the loop time is now much longer than in dimension 1 as it is the first time that two independent copies of RW(1) come back to the origin *simultaneously*, as per Exercise 4.3.1. For a more informative picture, let R_n be the duration of the first n loops and take $R_{n-1}\leq N < R_n$. Then $\#(N)/\log N = n/\log N \simeq n/\log R_n$ if $\log R_n$ is not much different from $\log R_{n-1}$, suggesting that

$$\lim_{n\uparrow\infty}\mathbb{P}_0\left(\sqrt[n]{R_n}\leq x\right) = \lim_{n\uparrow\infty}\mathbb{P}_0\left[n/\log R_n \geq \frac{1}{\log x}\right] = e^{-\pi/\log x} \quad \text{for } x\geq 1.$$

I do not know how to prove this seemingly obvious result in this style, but here's a better way.

Proof.   The identity

$$\sum_{n=0}^{\infty} \mathbb{P}_0[\mathbf{x}(2n) = 0] \, k^{2n} = \frac{2}{\pi} \int_0^{\pi/2} \frac{d\theta}{\sqrt{1 - k^2 \sin^2 \theta}} = \frac{2}{\pi} \, \mathrm{K}(k)$$

from §4.3 is remodeled by noting that $\mathbf{x}(2n) = 0$ is the same as saying $2n = R_l$ for some $l \geq 0$, with the convention $R_0 = 0$. This leads at once to

$$\frac{2}{\pi} \, \mathrm{K}(\gamma) = \sum_{l=0}^{\infty} \mathbb{E}_0(\gamma^{R_l}) = \sum_{l=0}^{\infty} [\mathbb{E}_0(\gamma^R)]^l = \frac{1}{1 - \mathbb{E}_0(\gamma^R)}$$

or, what is the same,

$$\mathbb{E}_0(\gamma^R) = 1 - \frac{\pi/2}{\mathrm{K}(\gamma)} \, ,$$

since R_l is the sum of l independent copies of the first loop time $R_1 = R$.

Exercise 4.4.2   I will soon need to know that $\mathrm{K}(k) \simeq -\log \sqrt{1 - k^2}$ for $k \uparrow 1$. Please check this.

Now fix $x > 1$ and look at

$$\mathbb{E}_0\left[e^{-\alpha R_n/x^n}\right] = \left[\mathbb{E}_0(e^{-\alpha R/x^n})\right]^n$$

$$= \left[1 - \frac{\pi/2}{\mathrm{K}(k)}\right]^n \quad \text{with } k = e^{-\alpha/x^n}.$$

Here, $1 - k^2 \simeq 2\alpha/x^n$ is small, so $\mathrm{K}(k) \simeq \frac{1}{2} n \log x - \log \sqrt{2\alpha}$, and

$$\mathbb{E}_0\left[e^{-\alpha R_n/x^n}\right] \simeq \left[1 - \frac{\pi}{n \log x}\right]^n \simeq e^{-\pi/\log x}.$$

Exercise 4.4.3   Take it from there, noting that $\mathbb{P}_0(R_n/x^n \leq y) = \mathbb{P}_0(\sqrt[n]{R_n} \leq x \sqrt[n]{y})$ in which $\sqrt[n]{y}$ tends to 1 for any $y > 0$.

Comment   For RW(1), R_n looks like n^2. Now, for RW(2), it wants to imitate the nth power of a variable $\mathbf{r}$ with $\mathbb{P}(\mathbf{r} \leq x) = e^{-\pi/\log x}$, $1 \leq x < \infty$: more evidence that loops are *much* longer in dimension 2. Note the curious fact that the nth root of the largest of n independent copies of $\mathbf{r}$ is distributed the same as $\mathbf{r}$ itself.

4.4.2 Long runs

OK, long loops, but how much longer? For example, what is the probability that one loop out of n is longer than the total duration of all the others. *Answer*: It's nearly 1 if n is big, unlike dimension 1 where the probability found in §3.3 was about 2/3.

Proof. I need an estimate of $\mathbb{P}_0(R > L)$ for a single loop and $L \uparrow \infty$. To begin with,

$$\mathbb{E}_0(e^{-\alpha R}) = \int_0^\infty e^{-\alpha x}\, d[1 - \mathbb{P}(R > x)] = 1 - \alpha \int_0^\infty e^{-\alpha x}\, \mathbb{P}_0(R > x)\, dx,$$

as you will check, so

$$\alpha \int_0^\infty e^{-\alpha x}\, \mathbb{P}_0(R > x)\, dx = \frac{\pi/2}{\mathrm{K}(k)}$$

with $k = e^{-\alpha} \simeq \pi/|\log \alpha|$ for $\alpha \downarrow 0$, by Exercise 4.4.2.

Exercise 4.4.4 $\mathbb{P}_0(R > L) \simeq \pi/\log L$ is obtained from this. Do it.

Now to the real business: the events (that a particular loop is longer than the total duration of the others) do not overlap, so the probability that's wanted is $n\,\mathbb{P}_0(R > R_{n-1})$ in which R is the duration of a single loop, independent of R_{n-1}. Then with n replaced by $n+1$, the estimate $\mathbb{P}_0(R > L) \simeq \pi/\log L$ shows that

$$(n+1)\,\mathbb{P}_0(R > R_n) \gtrsim \mathbb{E}_0(\pi n/\log R_n)$$
$$= \int_1^\infty \frac{\pi}{\log x}\, d\mathbb{P}_0(\sqrt[n]{R_n} \le x)$$
$$= \int_1^\infty \mathbb{P}_0(\sqrt[n]{R_n} \le x)\, \frac{\pi}{(\log x)^2}\, \frac{dx}{x},$$

to which Fatou's lemma is applied to obtain

$$\liminf_{n\uparrow\infty} n\,\mathbb{P}_0(R > R_{n-1}) \ge \int_1^\infty e^{-\pi/\log x}\, \frac{\pi}{(\log x)^2}\, \frac{dx}{x} = 1.$$

But $n\,\mathbb{P}_0(R > R_{n-1})$, being a probability, cannot exceed 1. That does the trick.

4.5 RW(2): a hitting distribution

Take $\mathbf{x}(0)$ at height n on the vertical line $0 \times \mathbb{Z}$, let T be the passage time to the horizontal line $\mathbb{Z} \times 0$, and let $\mathbf{x}(T)$ be the hitting place, as in Figure 4.5.1.

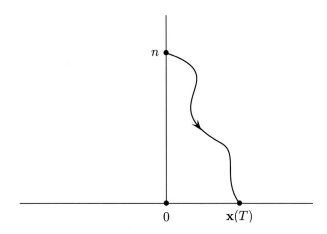

Figure 4.5.1

I will make two proofs of the fact that

$$\lim_{n\uparrow\infty} \mathbb{P}_{(0,n)}\Big[a \le \frac{\mathbf{x}(T)}{n} < b\Big] = \frac{1}{\pi}\int_a^b \frac{\mathrm{d}x}{1+x^2}.$$

First proof.   Split up the expectation

$$F_n = \mathbb{E}\big[e^{\sqrt{-1}\,k\mathbf{x}(T)}\big]$$

according to the outcome of the first step:

$$F_n = \frac{1}{4}F_n e^{-\sqrt{-1}\,k} + \frac{1}{4}F_n e^{+\sqrt{-1}\,k} + \frac{1}{4}F_{n+1} + \frac{1}{4}F_{n-1}$$
$$\qquad\quad right \qquad\qquad\quad left \qquad\qquad up \qquad\quad down$$

Then with F_n in the form G^n, you have $1 - \frac{1}{2}\cos k = \frac{1}{4}G + \frac{1}{4}G^{-1}$, so $G = 2 - \cos k \pm \sqrt{(2 - \cos k)^2 - 1}$, and you better take the lower signature to keep $|F| \le 1$. Replace k by k/n and make $n \uparrow \infty$ to obtain $G \simeq 1 - |k|/n$ and $F_n \simeq e^{-|k|} = \big[\frac{1}{\pi}(1 + x^2)^{-1}\big]^{\wedge}$. The rest follows by general rules, but let's do it by hand, so you see how simple it really is. Take $f(x) = \frac{1}{2\pi}\int_{-\infty}^{\infty} e^{\sqrt{-1}\,kx}\hat{f}(k)\,\mathrm{d}k$ of class $C_{\downarrow}^{\infty}(\mathbb{R})$ where $\hat{f}(k)$ is now $\int e^{\sqrt{-1}\,kx} f(x)\,\mathrm{d}x$. Obviously,

$$\mathbb{E}[f(\mathbf{x}(T)/n)] = \frac{1}{2\pi}\int_{-\infty}^{\infty} G^n \hat{f}\,\mathrm{d}k \simeq \frac{1}{2\pi}\int_{-\infty}^{\infty} e^{-|k|}\hat{f}(k)\,\mathrm{d}k$$
$$= \int_{-\infty}^{\infty} f(x)\,\frac{1}{\pi}\frac{\mathrm{d}x}{1+x^2},$$

so, for fixed $a < b$ and $f^\pm$ as in Figure 4.5.2, $\mathbb{P}\big[a \le \frac{1}{n}\mathbf{x}(T) < b\big]$ is

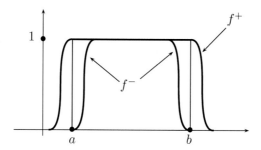

Figure 4.5.2

pinched between $\mathbb{E}[f^\pm(\mathbf{x}(T)/n)]$ and has no choice but to approximate $\frac{1}{\pi}\int_a^b(1+x^2)^{-1}\,\mathrm{d}x$ since $\int f^+$ and $\int f^-$ can be made as close to each other as you want.

Aside on stable laws. Recall CLT for, e.g., RW(1):

$$(1)\ \lim_{n\uparrow\infty}\mathbb{P}\left[\frac{\mathbf{e}_1+\cdots+\mathbf{e}_n}{\sqrt{n}}\le\lambda\right]=\int_{-\infty}^{\lambda}\frac{\mathrm{e}^{-x^2/2}}{\sqrt{2\pi}}\,\mathrm{d}x,$$

the **e**s being independent, unbiased (±1)s. This may be compared to the present hitting distribution: the horizontal displacements $\mathbf{e}_1$, $\mathbf{e}_2$, etc. as the walk passes down, level by level, from height n to height 0, are independent copies of $\mathbf{e}_1$, so you may write

$$(2)\ \lim_{n\uparrow\infty}\mathbb{P}\left[\frac{\mathbf{e}_1+\cdots+\mathbf{e}_n}{n}\le\lambda\right]=\int_{-\infty}^{\lambda}\frac{1}{\pi}\frac{\mathrm{d}x}{1+x^2}.$$

It may also be compared to the law of distant passage for RW(1): with independent copies of the passage time $T = \min(n : \mathbf{x}(n) = 1)$,

$$(3)\ \lim_{n\uparrow\infty}\mathbb{P}\left[\frac{T_1+\cdots+T_n}{n^2}\le\lambda\right]=\int_{-\infty}^{\lambda}\frac{\mathrm{e}^{-1/2t}}{\sqrt{2\pi t^3}}\,\mathrm{d}t.$$

The densities

$(1')$ $(2\pi)^{-1/2}\mathrm{e}^{-x^2/2}$,

$(2')$ $\frac{1}{\pi}(1+x^2)^{-1}$, and

$(3')$ $(2\pi t^3)^{-1/2}\mathrm{e}^{-1/2t}$

were already seen in §1.7. They are "stable" meaning that they arise from (scaled) sums $\mathbf{x}_1+\cdots+\mathbf{x}_n$ of independent copies of a single variable $\mathbf{x}$. You may ask what other stable densities there are. *Answer*: Not too many and anyhow, in my experience, it is mostly only $(1')$, $(2')$, $(3')$ and their variants

(1″) $(2\pi t)^{-1/2} e^{-x^2/2t}$,

(2″) $\frac{c}{\pi}(c^2 + x^2)^{-1}$, and

(3″) $(2\pi t^3)^{-1/2} x e^{-x^2/2t}$

that are seen in practice. (1) holds for sums $\mathbf{x}_1 + \cdots + \mathbf{x}_n$ as soon as $\mathbb{E}(\mathbf{x}_1) = 0$ and $\mathbb{E}(\mathbf{x}_1^2) = 1$. That is CLT, to be proved in such generality in §5.2. The criteria for (2) and (3) are more delicate, and depend upon the tails of the distribution of $\mathbf{x}$: if the tails match pretty well, then the law holds. I leave the matter there, but see Feller [1966(2):167–172; 574–581] for the whole story.

Second proof.   Mark Kač liked to say: "A demonstration is to convince a reasonable man, a proof is to convince an *un*reasonable man". Here's a demonstration to convince you, reasonable as I hope you are, as to why $1/\pi(1 + x^2)$ comes out. Roughly, RW(2) is a pair of independent copies of RW(1) run at half speed, so the time T it takes to come from height n to $\mathbb{Z} \times 0$ is about twice the 1-dimensional passage time T_n. By CLT, the hitting place on $\mathbb{Z} \times 0$ is well approximated by a unit Gaussian variable $G \times \sqrt{\frac{1}{2} \times 2T_n}$, reducing to $G\sqrt{T_n/n^2}$ upon scaling the hit by $1/n$. Then the density for the scaled hit is or ought to be

$$\int_0^\infty \frac{e^{-1/2t}}{\sqrt{2\pi t^3}} \frac{e^{-x^2/2t}}{\sqrt{2\pi t}} \, dt = \frac{1}{2\pi} \int_0^\infty e^{-\frac{1}{2}(1+x^2)/t} \frac{dt}{t^2} = \frac{1}{\pi} \frac{1}{1 + x^2} \,.$$

Right? Think it over. Anyhow, you know this formula already from Exercise 1.7.6.

4.6 RW(2): volume

The result is simpler than in dimension 1 though the proof is not. $\mathbb{Z}^2$ is so much bigger than $\mathbb{Z}^1$ that the "redundancy" is down, the Gaussian fluctuations seen in §3.7 are suppressed, and an individual law is found for $V(n) = $ the number of distinct places visited up to time n, *to wit*

$$\mathbb{P}_0\left[\lim_{n\uparrow\infty} \frac{V(n)}{\pi n/\log n} = 1\right] = 1;$$

see Erdős–Dvoretsky [1950:360–363] as corrected by Jain–Pruitt [1970]. The full proof is lengthy and intricate, so I do not reproduce it here. Only the mean value is really easy: $V(n)$ is the number of sites $x \in \mathbb{Z}^2$

such that $T_x \le n$, so

$$\mathbb{E}_0[V(n)] = \sum_{x\in\mathbb{Z}^2}\sum_{k=0}^{n}\mathbb{P}_0[\mathbf{x}(0),\mathbf{x}(1),\ldots,\mathbf{x}(k-1)\ne\mathbf{x}(k)=x]$$

$$= \sum_{k=0}^{n}\mathbb{P}_0[\mathbf{x}(0),\mathbf{x}(1),\ldots,\mathbf{x}(k-1)\ne\mathbf{x}(k)]$$

$$= \sum_{k=0}^{n}\mathbb{P}_0[\mathbf{x}(1),\ldots,\mathbf{x}(k)\ne 0]$$

$$= \sum_{k=0}^{n}\mathbb{P}_0(R>k)$$

$$\simeq \sum_{k=2}^{n}\frac{\pi}{\log k}\qquad\text{as in Exercise 4.4.4}$$

$$\simeq \int_2^n\frac{\pi}{\log x}\,\mathrm{d}x$$

$$\simeq \frac{\pi n}{\log n}.$$

I leave it at that.

4.7 RW(3): hitting probabilities

I follow mostly Itô–McKean [1960]. Take $K\subset\mathbb{R}^3$ with $\#(K)<\infty$. The point x is inside K if all its nearest neighbors also belong to K; otherwise, it is on the boundary ∂K. The "outer" boundary "facing" ∞ is distinguished. Now let T be the hitting time $\min(n:\mathbf{x}(n)\in K)$ and $p(x)$ the probability $\mathbb{P}_x(T<\infty)$ of hitting K for paths starting at $x\in\mathbb{Z}^3$. RW(3) runs off to ∞, so if it hits K at all, it has a last leaving time $N=\max(n:\mathbf{x}(n)\in K)<\infty$. Following Chung [1973], if $y\in K$ then

$$\mathbb{P}_x[N<\infty,\mathbf{x}(N)=y]=\sum_{n=0}^{\infty}\mathbb{P}_x[\mathbf{x}(n)=y,\mathbf{x}(k)\notin K\text{ for }k>n]$$

$$=\sum_{n=0}^{\infty}\mathbb{P}_x[\mathbf{x}(n)=y]\,\mathbb{P}_y[\mathbf{x}(k)\notin K\text{ for }k>0],$$

in which you see the expected number of visits to y, namely

$$G(x,y)=\sum_{n=0}^{\infty}\mathbb{P}_x[\mathbf{x}(n)=y]=\mathbb{P}_x(T_y<\infty)\times\sum_{n=0}^{\infty}\mathbb{P}_0[\mathbf{x}(2n)=0],$$

multiplied by the last-leaving probabilities

$$q(y) = \mathbb{P}_y[\mathbf{x}(k) \notin K \text{ for } k > 1] = \mathbb{P}_y(N = 0),$$

which vanish except on the "outer" boundary of K; in particular, summing over $y \in \partial K$, you find Chung's formula

$$p(x) = \mathbb{P}_x(T < \infty) = \mathbb{P}_x(N < \infty) = \sum_{y \in \partial K} G(x, y)q(y) \equiv Gq(x).$$

This formula has an elegant interpretation in terms of static electricity, by comparison with $\mathbb{R}^3$. The discussion skips back and forth between $\mathbb{R}^3$ and $\mathbb{Z}^3$; see §4.7.3. Gauss [1884] knew everything I am about to tell you about $\mathbb{R}^3$; see also Doyle–Snell [1984] for more about $\mathbb{Z}^3$, and Kellogg [1928] for more about $\mathbb{R}^3$.

4.7.1 The meaning of G

Introduce Laplace's difference operator on $\mathbb{Z}^3$:

$$\Delta: f \to \frac{1}{6} \sum_e f(x + e) - f(x),$$

in which e runs over the six nearest neighbors of $x = 0$. Obviously,

$$\frac{\text{``}\partial\text{''}}{\partial n} \mathbb{P}_x[\mathbf{x}(n) = y] \equiv \mathbb{P}_x[\mathbf{x}(n+1) = y] - \mathbb{P}_x[\mathbf{x}(n) = y]$$

$$= \frac{1}{6} \sum_e \mathbb{P}_{x+e}[\mathbf{x}(n) = y] - \mathbb{P}_x[\mathbf{x}(n) = y]$$

$$= \Delta \mathbb{P}_x[\mathbf{x}(n) = y],$$

so adding up from $n = 0$ on produces $-\Delta G(x, y) = \mathbb{P}_x[\mathbf{x}(0) = y] = 1$ or 0 according as $x = y$ or not, which is to say that G and $-\Delta$ are inverse to each other; in particular, if $q \geq 0$ and $p = Gq < \infty$, then $\Delta p = -q \leq 0$.

Exercise 4.7.1   Just a minute. It is not obvious that $-G\Delta$ is the identity. See to it.

There is also a converse to Exercise 4.7.1, stated here in a restricted version, good enough for now: if $p \geq 0$, if $\Delta p \equiv -q \leq 0$, and if $p(\infty) = 0$, then $p = Gq$. The proof is easy: with $\mathfrak{F}_n =$ the usual field of $\mathbf{x}(k) : k \leq n$,

$$\mathbb{E}[p \circ \mathbf{x}(n+1)|\mathfrak{F}_n] = p \circ \mathbf{x}(n) + \Delta p \circ \mathbf{x}(n) = p \circ \mathbf{x}(n) - q \circ \mathbf{x}(n).$$

Now take the full expectation, replace n by k, and add from $k = 0$ to n to obtain

$$p(x) = \mathbb{E}_x[p \circ \mathbf{x}(n+1)] + \sum_{k=0}^{n} \mathbb{E}_x[q \circ \mathbf{x}(k)] \quad \text{independently of } n$$

$$= \lim_{n \uparrow \infty} \mathbb{E}_x[p \circ \mathbf{x}(n)] + \sum_{k=0}^{\infty} \mathbb{E}_x[q \circ \mathbf{x}(k)]$$

$$= Gq(x) \quad \text{since } p \circ \mathbf{x}(n) \text{ tends to } p(\infty) = 0.$$

Exercise 4.7.2 Try that out for $p(x) = \mathbb{P}_x(T_K < \infty)$, noting, in particular, how the charges $q(x) = \mathbb{P}_x[\mathbf{x}(1), \mathbf{x}(2), \text{etc.} \notin K]$ on ∂K come in.

4.7.2 Comparison with $\mathbb{R}^3$

OK, but what is one to think of all that? Go to $\mathbb{R}^3$ with its "true" Laplacian $\Delta = \partial^2/\partial x_1^2 + \partial^2/\partial x_2^2 + \partial^2/\partial x_3^2$. The analogue of "$\partial/\partial n$" $\mathbb{P}_x[\mathbf{x}(n) = y]$ = what have you, is the heat equation $\partial p/\partial t = \Delta p$ for $p(t, x, y) = (4\pi t)^{-3/2} e^{-|x-y|^2/4t}$ so the analogue of $G(x, y)$ in $\mathbb{R}^3$ must be

$$G(x, y) = \int_0^{\infty} (4\pi t)^{-3/2} e^{-|x-y|^2 4t} \, dt = \frac{1}{|x-y|} \times \int_0^{\infty} (4\pi t)^{-3/2} e^{-1/4t} \, dt$$

$$= \frac{1}{4\pi} \frac{1}{|x-y|},$$

which is Green's function for the problem $\Delta p = -q$ in $\mathbb{R}^3$, i.e. $G \colon q \to \frac{1}{4\pi} \int |x-y|^{-1} q(y) \, dy$ is inverse to $-\Delta$.

Exercise 4.7.3 If this is unfamiliar, check it this way. Take a smooth compact function $q \geq 0$, let $w(t, x) = \int (4\pi t)^{-3/2} e^{-|x-y|^2/4t} q(y) \, dy$ be the solution of $\partial w/\partial t = \Delta w$ with $w(0+, \cdot) = q$, and abbreviate $\int_0^{\infty} w \, dt = Gq$ by p. Then

$$\Delta p(x) = (G\Delta q)(x) = \int_0^{\infty} dt \int \frac{e^{-|x-y|^2/4t}}{(4\pi t)^{3/2}} \Delta q \, dy = \int_0^{\infty} \frac{\partial w}{\partial t} \, dt$$

$$= -w(0+, x) = -q(x).$$

Justify all that. You could also make a proof using Green's formula applied to $\frac{1}{4\pi}|x-y|^{-1}$ and $p(y)$ in a big sphere less a little sphere centered at x. Try it if you like.

4.7.3 Electrostatics

But still, what is the meaning of

$$G(x,y) = \frac{1}{4\pi} \frac{1}{|x-y|} \equiv \frac{1}{4\pi} \frac{1}{r} ?$$

That comes from gravity, or what is preferable here, from static electricity. In convenient units, Coulomb's law states that the repulsive force between like charges q_1 and q_2, placed a distance r apart, is $q_1 q_2 / 4\pi r^2$, or better, with the direction of the force included, $(q_1 q_2 / 4\pi) \times -\operatorname{grad}(1/r)$. It follows that the work ($=$ force $\times$ distance) required to bring the charge q_2, from ∞ to a distance r from q_1, is $1/4\pi r$, independently of the path. Now think of a (smooth, compact) conductor $K \subset \mathbb{R}^3$ loaded up with positive charge described by a nice density $q(y) \geq 0$. Then the work required bring to a test charge $+1$ from ∞ to the place x in face of the repulsive force field so produced is

$$p(x) = \frac{1}{4\pi} \int_K \frac{q(y)\,\mathrm{d}y}{|x-y|} = (Gq)(x),$$

namely the "electrostatic potential", with the interpretation of voltage.

4.7.4 Back to $\mathbb{Z}^3$

You see where this is headed. $G(x,y) = \sum_{n=0}^{\infty} \mathbb{P}_x[\mathbf{x}(n) = y]$ is the analogue of Coulomb's $1/4\pi r$ law, and $p = Gq$ or, what is the same, $\Delta p = -q$ is the relation between charge and voltage; in particular, $p(x) = \mathbb{P}_x(T < \infty)$ is the voltage produced by the charges $q(y) = \mathbb{P}_y(N = 0)$ on the outer boundary of K. To make the comparison still more plain, let's check that $G(x,y) \simeq 3/2\,\pi r$ for large $r = |x-y|$ – off by a factor of 6, but no matter. It's easy to see where it comes from.

Proof. $G(x,y) = G(x-y,0)$, so take $y = 0$. By Pólya's trick of §4.3,

$$G(x,0) = \frac{1}{(2\pi)^3} \int_{-\pi}^{\pi} \int_{-\pi}^{\pi} \int_{-\pi}^{\pi} e^{\sqrt{-1}\,x\cdot\theta} \frac{\mathrm{d}^3\theta}{w(\theta)}$$

with $w(\theta) = 1 - \frac{1}{3}(\cos\theta_1 + \cos\theta_2 + \cos\theta_3)$. Now w is smooth and positive away from $\theta = 0$; near $\theta = 0$, it looks like $\frac{1}{6}(\theta_1^2 + \theta_2^2 + \theta_3^2) \equiv r^2/6$, and $1/w - 6/r^2$ is smooth there. It follows that $1/w$ may be replaced by $6/r^2$, up to an error which vanishes rapidly as $|x| \uparrow \infty$, so switching to spherical polars, you find

$$
\begin{aligned}
G(x,0) &\simeq \frac{1}{(2\pi)^3} \int_0^1 r^2\, dr \int_0^{2\pi} d\theta \int_0^{\pi} \sin\varphi\, d\varphi\, e^{\sqrt{-1}\,|x|r\cos\varphi}\, \frac{6}{r^2} \\
&= \frac{3}{\pi^2} \int_0^1 \sin(|x|\,r)\frac{dr}{r} \\
&\simeq \frac{3}{\pi^2 |x|} \times \left[\int_0^{\infty} \sin r\, \frac{dr}{r} = \frac{\pi}{2} \right] \\
&= \frac{3}{2\pi} \frac{1}{|x|}.
\end{aligned}
$$

Note, in particular, that if Q is the "total charge" = the sum of $q(y)$ over ∂K, then by Chung's formula and $p = Gq$,

$$
\lim_{|x|\uparrow\infty} \mathbb{P}_x[\mathbf{x}(N) = y \mid N < \infty] = q(y)/Q,
$$

clarifying the meaning of the last-leaving probabilities.

4.7.5 Energy and capacity in $\mathbb{R}^3$

A fixed total charge Q is placed on the (smooth, compact) conductor $K \subset \mathbb{R}^3$. In Nature, the little (positive) charges dQ spread out over K in response to their anti-social character, each wanting to be as far away from the others as it can. At the resulting equilibrium, the whole charge is loaded up on the outer boundary of K with a smooth density q relative to surface area, as for a sphere of radius r when the charge is evenly spread out on the surface with density $Q/4\pi r^2$. Then the associated potential $p = Gq = \frac{1}{4\pi} \int_{\partial K} |x - y|^{-1} q(y)\,d$ area is constant throughout K: otherwise, the force field, $-\operatorname{grad} p \neq 0$, would indicate that further adjustments must be made. Kelvin's principle (1870) asserts that this equilibrium takes place when and only when the electrostatic energy

$$
\mathfrak{E} = \frac{1}{\pi} \int_{K \times K} \frac{dQ(x)\, dQ(y)}{|x - y|}
$$

is least. Since $\mathfrak{E}$ is a measure of the reciprocal of mutual distances, that's plausible enough. The total charge Q is now adjusted to make $p \equiv 1$ on K, and *that* charge Q is declared to be the electrostatic capacity $C(K)$ of the conductor – it's the capacity you learned about in school. Gauss [1884] puts it differently: for unrestricted non-negative charges, the equilibrium takes place only when $\mathfrak{E} - 2Q$ is least, in which case $p \equiv 1$ on K, automatically. Then $C(K) = Q$ without further adjustment. I prove these things only for $\mathbb{Z}^3$.

4.7.6 Finale in $\mathbb{Z}^3$

The punch line is that the principles of Kelvin and Gauss, translated to $\mathbb{Z}^3$ in the obvious way, determine one and the same equilibrium charge distribution, namely the last-leaving probabilities $q(x) = \mathbb{P}(N = 0)$ on the outer boundary of K. The total charge Q is then the capacity $C(K)$, by definition.

Proof. Gauss's principle is now more convenient than Kelvin's. Place unrestricted charges $q \geq 0$ on K. Then

$$\mathfrak{E} - 2Q = q \cdot Gq - 2q \cdot 1 \geq Q^2 \min_{(x,y) \in K \times K} G(x,y) - 2Q$$

is bounded below and so has an honest minimum in respect to q. There, you have $\partial \mathfrak{E}/\partial q(x) - 2 = 2p(x) - 2 = 0$ or ≥ 0 according as q charges x or not. Give these charged points a name: $J \subset K$. Then

(1) $p = 1$ on J,

(2) $\Delta p = 0$ elsewhere, and

(3) $p(\infty) = 0$.

Now (2) and (3) imply that the maximum of p is taken on ∂J since $p(x) = \frac{1}{6} \sum_e p(x + e)$ elsewhere by (2), so (1) can be sharpened to

(1′) $p \equiv 1$ on K, because $p \geq 1$ on $K \setminus J$.

Then the difference $h(x)$ of $p(x)$ and $\mathbb{P}_x(T < \infty)$ vanishes on K and at ∞, $\Delta h = 0$ between, and the same maximum principle just employed shows that $h = 0$ everywhere. That does it.

Exercise 4.7.4 Show that Kelvin's principle produces the same charges.

Exercise 4.7.5 The capacity of a single point is the reciprocal of $G(0,0) =$ the probability of no return $\simeq 2/3$, for which see the end of §4.1. Find the capacity of a pair of points $a \neq b$ and also the approximate capacity $2\pi/3 \times$ the radius of the lattice points enclosed in a big sphere. These examples show that capacity is far from additive. Think it over.

Exercise 4.7.6 Prove: (1) $C(A) \leq C(B)$ if $A \subset B$; (2) $C(A \cup B) \leq C(A) + C(B)$; or more precisely, (3) $C(A \cup B) + C(A \cap B) = C(A) + C(B)$. *Hint for* (3): To hit A but not B is less probable than to hit A but not $A \cap B$.

4.7.7 Grounding

The Green's function G does not exist in dimensions 1 and 2 since $\sum_{n=0}^{\infty} \mathbb{P}_x[\mathbf{x}(n) = y] \equiv +\infty$. A simple cure is to put everything inside a grounded surface ∂D.

Take $D \subset \mathbb{Z}^d$ for $d = 1, 2, 3$ or more with $\#(D) < \infty$ and, as before, declare x to be inside D if all its $2d$ nearest neighbors $x + e$ also belong to D; the rest of D is the boundary ∂D. Now with $T_{\partial D}$ = the exit time $\min[n : \mathbf{x}(n) \in \partial D]$ from the interior of D, the "grounded" Green's function

$$G(x, y) = \sum_{n=0}^{\infty} \mathbb{P}_x[\mathbf{x}(n) = y, \, n < T_{\partial D}]$$

is fine: in fact, any component of RW(3) is a copy of RW(1) run at $\frac{1}{3}$ speed and since $\mathbb{E}(T_{\partial D}) < \infty$ in dimension 1 as per the Gambler's Ruin of §3.2, so also in dimension 3. Then the recipes of Chung, Kelvin, and Gauss still work: in particular, if T is the hitting time to $K \subset D$, then $p(x) = \mathbb{P}_x(T < T_{\partial D})$ is the equilibrium potential of K relative to the surface ∂D which is said to be "grounded" to reflect the fact that the potential vanishes there.

Exercise 4.7.7 The grounded Green's function $G(x, y)$ is symmetric in x and y, vanishes if either x or y belongs to ∂D, and solves $\Delta G = -1$ inside D. Check also the statements of Chung, Kelvin, and Gauss if you like.

4.7.8 Harmonic functions

A function h is said to be harmonic in some region if $\Delta h = 0$ there. Such a function $h(x) = \mathbb{E}_x [f \circ \mathbf{x}(T_{\partial D})]$ solves Dirichlet's problem:

(1) $\Delta h = 0$ inside D with

(2) $h = f$ on ∂D.

(2) is self-evident, and likewise (1) since $h(x) = \frac{1}{6} \sum_e h(x + e)$ inside D as a picture shows. In particular, the exit probabilities

$$H(x, y) = \mathbb{P}_x [\mathbf{x}(T_{\partial D}) = y] \quad \text{for } x \in D \text{ and } y \in \partial D$$

describe how the walk leaves the interior of D after the manner of the Gambler's Ruin for RW(1):

$$\mathbb{P}_x(T_a < T_b) = \frac{b - x}{b - a} \quad \text{for } a \le x \le b.$$

$H(x, y)$ is, if you will, a "flux" – the walk starts at $\mathbf{x}(0) = x$ with probability 1, and $H(x, y)$ is the amount of probability leaving the interior via $y \in \partial D$ – and there is a pretty connection between H and G. For simplicity, let D be the cube seen in Figure 4.7.1: x is inside and the route $y - e \rightarrow y$ is the only way to exit via $\mathbf{x}(T_{\partial D}) = y$.

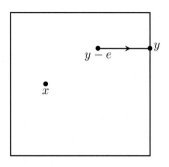

Figure 4.7.1

Then

$$\frac{1}{6}G(x, y - e) = \sum_{0}^{\infty} \mathbb{P}_x \left[\mathbf{x}(n) = y - e, n < T_{\partial D}\right] \times \mathbb{P}_{y-e} \left[\mathbf{x}(1) = y\right]$$

$$= \sum_{0}^{\infty} \mathbb{P}_x \left[T_{\partial D} = n, \mathbf{x}(n) = y\right]$$

$$= H(x, y),$$

which may construed somewhat fancifully as $H(x, y) = 1/6 \times$ the inward-pointing normal derivative at $y \in \partial D$ of the grounded Green's function $G(x, y)$.

Exercise 4.7.8   What is the classical, $\mathbb{R}^3$ version of that? Kellogg [1928] explains with the technical details.

Coda

What you have seen here is, in miniature, a glimpse of the connection between probability and partial differential equations of parabolic and elliptic type, typified by $\partial w / \partial t = \Delta w$, $\Delta p = -q$, and $\Delta h = 0$. It has remarkable, far-reaching ramifications. I cannot say more now, but see, e.g. Itô–McKean [1960], McKean [1969] and/or Feller [1966], and Rogers–Williams [1979, 1987] for lots more information.

4.8 RW(3): volume

$\mathbb{Z}^3$ is *much* bigger than $\mathbb{Z}^2$, so things should be simpler (proof and all) than in §4.6; indeed a bona fide law of large numbers is found:

$$\mathbb{P}_0 \left[\lim_{n \uparrow \infty} \frac{1}{n} V(n) = C\right] = 1,$$

in which C is the capacity of a single point of $\mathbb{Z}^3$, i.e. $C = \mathbb{P}_0(\text{no return}) \simeq 2/3$.

Exercise 4.8.1 $\lim \frac{1}{n} V(n)$ cannot be as large as 1, i.e. there is real redundancy: for example, the number of double points up to time n is at least $n/36$. Why? How about triple points, etc.?

More generally, if $\#(K) < \infty$ and if $V(n)$ is now the volume of the "tube" $\bigcup_{k=0}^n [K + \mathbf{x}(k)]$, then the same law holds with the capacity $C(K)$ in place of $C = C(0)$. This pretty result is due to Spitzer [1964]. I will want both the general law of large numbers (which you will surely believe) and also Birkhoff's theorem which I will explain when I need it. Their proofs come in §§5.1 and 8.4, respectively.

Proof.   Write K_n for $K + \mathbf{x}(n)$. Then

$$V(5) = \#(K_0 \cup \cdots \cup K_5) \leq \#(K_0 \cup K_1 \cup K_2) + \#(K_3 \cup K_4 \cup K_5 - \mathbf{x}(3))$$

since translation of sets does not change the count, i.e. $V(5)$ is not more than the sum of two independent copies of $V(3)$; similarly, $V(nm)$ is not more than the sum of n independent copies of $V(m)$, so that

$$\limsup_{n \uparrow \infty} \frac{1}{nm} V(nm) \leq \frac{1}{m} \mathbb{E}_0[V(m)],$$

by the law of large numbers. Now compute

$$
\begin{aligned}
&\mathbb{E}_0[V(n) - V(n-1)] \\
&\quad = \mathbb{E}_0\big[\#(x : x \in K + \mathbf{x}(k) \text{ for } k = n \text{ but not before})\big] \\
&\quad = \mathbb{E}_0\big[\#(x : x + \mathbf{x}(n) - \mathbf{x}(k) \in K \text{ for } k = n \text{ but not before})\big] \\
&\quad = \mathbb{E}_0\big[\#(x : x + \mathbf{x}(k) \in K \text{ for } k = 0 \text{ but not after})\big] \\
&\quad = \sum_{x \in K} \mathbb{P}_x\big[\mathbf{x}(1), \dots, \mathbf{x}(n) \notin K\big].
\end{aligned}
$$

Recall next from §4.7 that $\mathbb{P}_x[\mathbf{x}(n) \notin K \text{ for } n > 0]$ is the equilibrium charge (if any) at $x \in \partial K$, and conclude that $\mathbb{E}_0[V(n+1) - V(n)] \simeq C(K)$ for large n. Then $\mathbb{E}_0[V(n)] \simeq n\, C(K)$, so for fixed m, $n \uparrow \infty$, and $(n-1)m \leq N < nm$, you find

$$\frac{V(N)}{N} \leq \frac{V(nm)}{n(m-1)} \lesssim \frac{1}{m} \mathbb{E}_0\, V(m) \simeq C(K) \quad \text{if } m \text{ is large},$$

i.e.

$$\mathbb{P}_0\Big[\limsup_{N \uparrow \infty} N^{-1} V(N) \leq C(K)\Big] = 1,$$

which is half the job. For the rest, notice that

$$\begin{aligned}
V(N) &\geq \#(x \in K : x \notin K_n \text{ for } n \geq 1) \\
&\quad + \#(x \in K_n \text{ for some } 1 \leq n < N) \\
&\geq \#(x \in K : x \notin K_n \text{ for } n \geq 1) \\
&\quad + \#(x \in K_1 : x \notin K_n \text{ for } n \geq 2) \\
&\quad + \#(x \in K_2 : x \notin K_n \text{ for } n \geq 3) \\
&\quad + \text{ etc.} \\
&\quad + \#(x \in K_{N-1} : x \notin K_n \text{ for } n \geq N) \\
&= \sum_{n=0}^{N-1} T^n f,
\end{aligned}$$

in which $f = \#(x \in K : x \notin K_n \text{ for } n \geq 1)$ and, by abuse of current notation, T is the probability-preserving "shift" taking $\mathbf{e}_1$, $\mathbf{e}_2$, etc. $\rightarrow \mathbf{e}_2$, $\mathbf{e}_3$, etc. and so also $\mathbf{x}(n) \rightarrow \mathbf{x}(n+1) - \mathbf{x}(1)$. Now Birkhoff's theorem asserts that in such a case,

$$\lim_{N \uparrow \infty} \frac{1}{N} \sum_{n=0}^{N-1} T^n f$$

exists. That's all I need. Obviously, this limit has nothing to do with $\mathbf{e}_1$ or $\mathbf{e}_2$ or, indeed, with any step $\mathbf{e}_n$. Then it must be independent of itself and so constant, and since $f \leq \#(K)$ is bounded, it appears that

$$\begin{aligned}
\liminf_{N \uparrow \infty} \frac{1}{N} V(N) &\geq \lim_{N \uparrow \infty} \frac{1}{N} \sum_{n=0}^{N-1} T^n f \\
&= \mathbb{E}_0 \left[\lim_{N \uparrow \infty} \frac{1}{N} \sum_{n=0}^{N-1} T^n f \right] \\
&= \lim_{N \uparrow \infty} \frac{1}{N} \sum_{n=0}^{N-1} \mathbb{E}_0(T^n f) \\
&= \mathbb{E}_0(f) \\
&= \sum_{x \in K} \mathbb{P}_0 \left[x \notin K + \mathbf{x}(n) \text{ for } n \geq 1 \right] \\
&= \sum_{x \in K} \mathbb{P}_x \left[\mathbf{x}(n) \notin K \text{ for } n \geq 1 \right] \quad \text{by reversal } \mathbf{x} \rightarrow -\mathbf{x} \\
&= \sum_{x \in \partial K} \mathbb{P}_x(N = 0) \\
&= \mathrm{C}(K).
\end{aligned}$$

That's the other half.

4.9 Non-negative harmonic functions

I want to say something more about such functions in low dimensions 1, 2 and 3. These have a deep connection to the behavior of the walk as $n \uparrow \infty$, not only for the standard walk but more widely, for Markovian motion in general. I cannot go into the whole subject, but see, e.g. Rogers–Williams [1987] for a more complete account if the simple examples presented here take your fancy.

4.9.1 Standard walks: first pass

In dimension d, the function $h \colon \mathbb{Z}^d \to \mathbb{R}$ is harmonic if $h(x)$ is the arithmetic mean of its values at the $2d$ nearest neighbors $x + e$ of x. These form a convex cone with compact section. By section, I mean the subclass with $h(0) = 1$. Then $h(x) \leq (2d)^n$ where n is the distance from the origin to x – not the Pythagorean distance (as the crow flies) but the smallest number of steps required to come from the one place to the other (as the walker goes in a rectangular grid of streets). Now the fact is that *there are no non-trivial functions of this style*: $h \equiv 1$ is only possibility. Disappointing? Well, wait – more will be revealed. Let's prove it now.

Dimension 1. Here, harmonicity means $h(x) = \frac{1}{2}h(x-1) + \frac{1}{2}h(x+1)$, $h(x) = ax+b$, and to keep it non-negative you need $a = 0$. Then $h(0) = 1$ does the rest.

Dimension 2 is more subtle, but here's neat proof due to J. Doob [1953]: $h \circ \mathbf{x}(n)$ is a fair game, and as it is non-negative, it must have a limit when $n \uparrow \infty$. But RW(2) visits every place in $\mathbb{Z}^2$ i.o., so you must have $h \equiv 1$.

Dimension 3. Doob's trick doesn't work now since RW(3) runs off to ∞, nor is there any explicit expression of the general function h as in dimension 1, so what to do? The proof is easy enough if h is bounded. Then $h \circ \mathbf{x}(n)$ has a limit as $n \uparrow \infty$, depending only on the starting point $\mathbf{x}(0) = x$ – this by the Hewitt–Savage 01 law of §1.8.4 – and since

$$\lim_{n\uparrow\infty} h \circ \mathbf{x}(n) = \lim_{n'\uparrow\infty} h \circ \big[\mathbf{x}(n) + \mathbf{x}(n+n') - \mathbf{x}(n)\big],$$

the starting point doesn't matter either. In short, $h(x) = \mathbb{E}_x\big[\lim_{n\uparrow\infty} h \circ \mathbf{x}(n)\big]$ is independent of x, and can only be $h(0) = 1$.

But what if h is *unbounded*? I know two proofs: one is geometrical, using the compactness of the section[2]; the other employs a descent from RW(3) to RW(2). They are not so different really, as you will see.

Geometrical proof. H_0 is the whole section determined by $h(0) = 1$. List the neighbors e in some order such as $e_1 = (100)$, $e_2 = (-100)$, $e_3 = (010)$, $e_4 = (0-10)$, $e_5 = (001)$, $e_6 = (00-1)$, and use the compactness of the section to guarantee following subsections are not void:

1 $H_0 \supset H_1$ where $h(e_1)$ is maximal,
2 $H_1 \supset H_2$ where $h(e_2)$ is maximal,

down to

1 $H_5 \supset H_6$ where $h(e_6)$ is maximal.

To proceed, keep in mind that any non-negative function h is *either* everywhere positive *or else* identically 0: if it vanishes at one place, it must vanish at all the neighbors, and that propagates to the whole lattice.

Now fix $h \in H_6$, take $y \in \mathbb{Z}^3$, and look at the function $h(x+y)/h(y)$ as a function of $x \in \mathbb{Z}^3$. Obviously, it is of class H_0, so $h(e_1 + y)/h(y) \leq h(e_1)$, which is to say that $h(e_1)h(y) - h(e_1+y)$, considered as a function of y, is non-negative, harmonic, and so identically 0 since it vanishes at $y = 0$ – in short, $h(e_1 + y) = h(e_1)h(y)$ for every $y \in \mathbb{Z}^3$. In particular, $h(x+y)/h(y)$ is of class H_1 and $h(e_1)h(-e_1) = h(0) = 1$, and the argument may be repeated down the line, to show that $h(e)h(-e) = 1$ for every neighbor e. But then $h(0) = \frac{1}{6}\sum_e h(e)$ cannot balance unless $h(e) = 1$ for every neighbor e, since $h(e) + 1/h(e) \geq 2$, and this propagates to the whole lattice, i.e. $h \equiv 1$.

It follows that $h(e_1) \leq 1$ *for every* $h \in H_6$, and since the neighbors e could have been listed in some other order, so also $h(e) \leq 1$ *for every* e, with the equality by a second application of $h(0) = \frac{1}{6}\sum_e h(e) = 1$. In short, $h(0) = 1$ implies $h(e) \equiv 1$, and that propagates to the whole of $\mathbb{Z}^3$.

Amplification That's a very attractive argument and it's worth a few words to say what's really going on[3].

The simplest interesting convex cone is the positive octant of $\mathbb{R}^3$ where x_1, x_2, and x_3 are all non-negative. The section is then the triangle seen in Figure 4.9.1 where $x_1 + x_2 + x_3 = 1$, with vertices $e_1 = (100)$, $e_2 = (010)$, and $e_3 = (001)$ and every point inside the triangle is a center

---

[2] I follow a private communication of W. Feller (1956).
[3] This was Feller's frequent promise, at the end of a lecture, having run out of

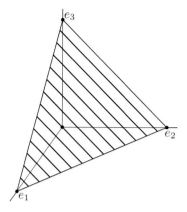

Figure 4.9.1

of gravity of these. Only the vertices are exceptional in this regard not being centers of gravity of anything else. That's a very general fact even in dimension $+\infty$: for any convex cone, with compact section relative to some decent topology, every point of the section is a center of gravity of its vertices – geometrically obvious, but not so easy to prove in wide generality. My point is that H_0 is populated by centers of gravity of vertices, and the proof shows that there is only one of these, namely $h \equiv 1$. Obviously, the present case is too simple to give the real flavor of the thing, but §4.9.4, on space-time walks, deals with a more typical instance with a whole crowd of vertices, indexed, e.g. by $\mathbb{R}$. The general fact is due to G. Choquet [1956]; see also Rogers–Williams [1987] for more than I do here.

Proof by descent. A variant of the geometrical proof consists in maximizing $h(e_1)$ only and using the first part of the argument to show that $h(e_1)h(-e_1) = 1$ and, more generally, that $h(x \pm e_1) = h(x)h(\pm e_1)$, where I revert to the old notation $e_1 = (100)$, $e_2 = (010)$, $e_3 = (001)$. Then, with $h(e_1) = \alpha$, you have

$$h(x) = \frac{1}{6}\left(\alpha + \frac{1}{\alpha}\right) + \frac{2}{3} \times \frac{1}{4}\left[h(x+e_2) + h(x-e_2) + h(x+e_3) + h(x-e_3)\right],$$

which is to say that for fixed x_1, the mean value of h as a function of $(x_2, x_3) \in \mathbb{Z}^2$ is just βh with a necessarily positive number $\beta = \frac{3}{2} \times 1 - \frac{1}{6}\left(\alpha + \frac{1}{\alpha}\right)$. But then, with x_1 still fixed, $\beta^{-n} h \circ \mathbf{x}(n)$ is a fair game

time. But the next week, he would have a new topic and run out of time before he could keep it.

for RW(2), and so has a finite limit as $n \uparrow \infty$. This cannot be correct unless $\alpha = 1$: otherwise, β would be less than 1, and since RW(2) visits the origin i.o., the limit would be $+\infty$. The rest follows.

4.9.2 Standard walks: second pass

So what's interesting at all? Evidently, the whole lattice is not, though even this superficially disappointing fact has its own meaning to which I'll come. For now, let's look at some connected regions $D \subset \mathbb{Z}^d$ to see if it's better there.

Example 4.9.1 $D \subset \mathbb{Z}$ is an interval $a \leq x \leq b$. There, the function $h(x) = (b-x)/(b-a)$ is non-negative and harmonic with values $h(a) = 1$ and $h(b) = 0$, and if you start RW(1) at $\mathbf{x}(0) = x \in D$ and stop it at the exit time $T = T_a \wedge T_b$, you see the rule for the Gambler's Ruin coming out:

$$\frac{x-a}{b-a} = h(x) = \mathbb{E}_x\big[h \circ \mathbf{x}(T)\big] = \mathbb{P}_x(T_a < T_b),$$

just as in §3.2.

Example 4.9.2   Now take a finite connected region $D \subset \mathbb{Z}^2$, distinguishing, as before, the interior, where all the nearest neighbors of a point still lie in D, from the boundary ∂D, and let T be the exit time $\min(n : \mathbf{x}(n) \in \partial D)$. Then, for fixed $y \in \partial D$, the function $h(x) = \mathbb{P}_x\big[\mathbf{x}(T) = y\big]$ is harmonic inside D and reduces, on the boundary, to 1 or 0 according as $x = y$ or no – a sort of two-dimensional Gambler's Ruin if you will. More generally, $h(x) = \mathbb{E}_x\big[f \circ \mathbf{x}(T)\big]$ is *the* harmonic function with $h = f$ on the boundary as you know already.

Exercise 4.9.3   Figure 4.9.2 shows the "disc of radius 2" relative to lattice distance with the exit probabilities indicated, up to a normalizing factor: in part (a) for $\mathbf{x}(0) = (00)$ and, in part (b) for $\mathbf{x}(0) = (10)$. Check these numbers for me.

Example 4.9.4 $D \subset \mathbb{Z}^2$ is now the half-lattice $\mathbb{Z}^{2+}$ where the vertical coordinate is non-negative and there are lots of harmonic functions, some bounded, some not. For example, writing RW(2) as $\mathbf{x}(n) = [\mathbf{x}_1(n), \mathbf{x}_2(n)]$, it is clear that the exit time $T = \min[n : \mathbf{x}_2(n) = 0]$ is finite and that $h(x) = \mathbb{P}_x[\mathbf{x}(T) = x]$ is harmonic. (The Gambler's Ruin again). More generally, $h(x) = \mathbb{E}_x[f \circ \mathbf{x}(T)]$ also fits the bill if the non-negative function f on $\mathbb{Z} \times 0$ is not too big.

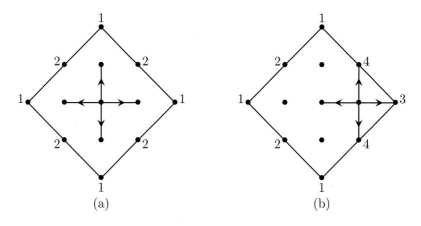

Figure 4.9.2

★**Exercise 4.9.5** The fact is that $\mathbb{P}_x[\mathbf{x}(T) = \ell]$ is comparable to $(1 + \ell^2)^{-1}$ for any x above $\mathbb{Z} \times 0$, so what you need is $\sum_{\mathbb{Z}} (1 + \ell^2)^{-1} f(\ell) < \infty$. The comparison is not so easy to check and is not needed for what follows, so skip it if you like, though it's amusing to do.

Hints for $x = (01)$. By the protocol for RW(2), you may write $\mathbf{x}(n)$ as the pair $\mathbf{x}_1^0(\mathbf{k})$ and $\mathbf{x}_2^0(n - \mathbf{k})$ with two independent copies of RW(1) and the Bernoulli-type clock $\mathbf{k}(n) = $ the number of successes in n honest Bernoulli trials independent of these walks. Let $1 \leq \mathbf{n}_1 < \mathbf{n}_1 + \mathbf{n}_2 < \mathbf{n}_1 + \mathbf{n}_2 + \mathbf{n}_3$ etc. be the jump-times of $\mathbf{k}$, the $\mathbf{n}$s being independent copies of $\mathbf{n}_1$ with common distribution $\mathbb{P}(\mathbf{n} = k) = 2^{-k}$ for $k \geq 1$. Then

$$\mathbb{P}_{01}[\mathbf{x}(T) = \ell]$$
$$= \sum_{m=1}^{\infty} \mathbb{P}_{01}\Big[\mathbb{P}_0\big(\mathbf{x}_2^0\big) = m \text{ and } \mathbf{x}_1^0(\mathbf{n}_1 + \cdots + \mathbf{n}_m - m) = \ell\Big],$$

from which you may compute

$$\mathbb{E}_{01}\Big[e^{\sqrt{-1}\theta\mathbf{x}(T)}\Big] = 2 - \cos\theta - \sqrt{3 - 4\cos\theta + \cos^2\theta} \text{ for } |\theta| \leq \pi.$$

This function is smooth except at the origin $\theta = 0$ where it has a corner like that of $1 - |\theta|$, and you may smooth it out elsewhere, as in Figure 4.9.3 so that

$$\mathbb{P}_{01}[\mathbf{x}(T) = \ell] = \frac{1}{2\pi} \int_{-\pi}^{\pi} \cos\theta\ell\, f(\theta)\, d\theta,$$

up to an error vanishing rapidly for large ℓ. Explain, please. It remains only to estimate the integral by a couple of partial integrations. Note that the result might have been foreseen from the law of distant passage of §4.5.

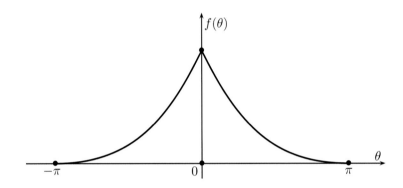

Figure 4.9.3

Now back to business and let's ask if all the non-negative harmonic functions in $\mathbb{Z}^{2+}$ have been accounted for. The answer is *no*. The function $h(x) = x_2$ is exceptional, showing that the general solution $h \geq 0$ of $\Delta h = 0$ in $\mathbb{Z}^{2+}$ is not fully determined by its values on $\mathbb{Z} \times 0$: any non-negative multiple of x_2 may be added to it. I return to the meaning of this function for RW(2) after some instructive variants of RW(1) have been described, but note in passing that its presence cannot be accounted for by saying that $\mathbb{Z}^{2+}$ is too wide: the quadrant where both x_1 and x_2 are non-negative has also such an extra function $x_1 x_2$, vanishing on the whole boundary, and even the strip $[0, \infty) \times [-1, 1]$ has the anomalous function $h(x) = 0$ on the boundary with

$$h(\ell, 0) = \frac{1}{2\sqrt{3}}\left[(2 + \sqrt{3})^\ell - (2 - \sqrt{3})^\ell\right]$$

on $[1, \infty) \times 0$ inside.

4.9.3 Variants of RW(1)

Simple drift

Figure 4.9.4(a), descriptive of RW(1), is modified as in Figure 4.9.4(b) to impart to the walk a drift towards $+\infty$. The mean step is now $\frac{2}{3} - \frac{1}{3} = \frac{1}{3}$, so $\lim_{n \uparrow \infty} \mathbf{x}(n) = \frac{1}{3}$, by LLN. Here harmonicity means $h(x) = \frac{1}{3}h(x-1) + \frac{2}{3}h(x + 1)$ of which there are two basic solutions: $h \equiv 1$ of course, and

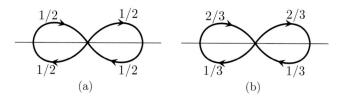

Figure 4.9.4

the new function $h(x) = 2^{-x}$, unbounded to the left. But what is the meaning of that? The short answer is: it has to do with the Gambler's Ruin. In fact, the function

$$h(x) = \frac{2^{-x} - 2^{-b}}{2^{-a} - 2^{-b}}$$

is harmonic between $x = a$ and $x = b > a$, with values $h(a) = 1$ and $h(b) = 0$, so stopping the fair game $h \circ \mathbf{x}(n)$ at the exit time $T = T_a \wedge T_b$ produces

$$\mathbb{P}_x(T_a < T_b) = \frac{2^{-x} - 2^{-b}}{2^{-a} - 2^{-b}} \text{ for } a \leq x \leq b,$$

and, in particular, $\mathbb{P}_x(T_a < \infty) = 2^{a-x}$ for $x \geq a$. Simple enough, but more is here: $h(x) = 2^{-x}$ being harmonic, you may make new n-step probabilities

$$\mathbb{P}_x^h[\mathbf{x}(n) = y] = \frac{1}{h(x)} \mathbb{P}_x[\mathbf{x}(n) = y] h(y),$$

built up from the 1-step probabilities as in, for example,

$$\mathbb{P}_x^h[\mathbf{x}(3) = y] = \frac{1}{h(x)} \mathbb{P}_x[\mathbf{x}(1) = a] h(a) \times \frac{1}{h(a)} \mathbb{P}_a[\mathbf{x}(1) = b] h(b) \times$$
$$\times \frac{1}{h(b)} \mathbb{P}_b[\mathbf{x}(1) = y] h(y), \quad \text{summed over } a \text{ and } b,$$

determining in this way a new walk, and what is that? You have

$$\mathbb{P}_x[\mathbf{x}(1) = x + e] = 2^x \times \frac{2}{3} \text{ or } \frac{1}{3} \times 2^{-x-e} = \frac{1}{3} \text{ or } \frac{2}{3} \text{ according as } e = \pm 1,$$

and this independently of x, so the new walk is nothing but the old with drift reversed, running off, not to $+\infty$, but to $-\infty$ instead. Could it be that this reversed walk is nothing but the old walk *conditioned to do just that*? Of course, it's not clear what such a conditioning means, but let's see what happens if you take it this way: that the walk is conditioned to reach $\ell < 0$ some time and ℓ is taken down to $-\infty$. The computation is easy: for fixed x, y and n, and large $\ell < 0$,

$$\mathbb{P}_x\big[\mathbf{x}(n) = y \mid T_\ell < \infty\big] = \mathbb{P}_x\big[\mathbf{x}(n) = y\big] \times \left[\frac{\mathbb{P}_y(T_\ell < \infty)}{\mathbb{P}_x(T_\ell < \infty)} = 2^{x-y}\right]$$

$$\simeq \mathbb{P}_x^h\big[\mathbf{x}(n) = y\big],$$

and there you have it, the provisional suggestion being that *un*bounded, non-negative harmonic functions have to do with conditioning the walk to behave in some unwonted way. This type of thing, originating with J. Doob [1953], will play an important role later.

A more complicated drift

Now look at Figure 4.9.5 where the drift is always outward, away from

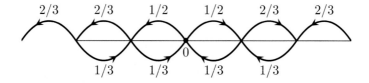

Figure 4.9.5

the origin. Now harmonicity means

$$h(x) = \begin{cases} \dfrac{2}{3}h(x-1) + \dfrac{1}{3}h(x+1) & x < 0, \\[2mm] \dfrac{1}{2}h(-1) + \dfrac{1}{2}h(+1) & \text{when} \quad x = 0, \\[2mm] \dfrac{1}{3}h(x-1) + \dfrac{2}{3}h(x+1) & x > 0, \end{cases}$$

and you may check that there are the two nice, independent solutions h_- and h_+ seen in Figure 4.9.6:

$$h_-(x) = \begin{cases} 2^{-x-1} & x > 0 \\ \tfrac{1}{2} & x = 0 \\ 1 - 2^{x-1} & x < 0 \end{cases} \quad \text{and} \quad h_+(x) = \begin{cases} 2^{x-1} & x < 0 \\ \tfrac{1}{2} & x = 0 \\ 1 - 2^{-x-1} & x > 0 \end{cases}$$

Both of these determine fair games, and a moment's thought provides the proper variant of the Gambler's Ruin:

$$\mathbb{P}_x(T_a < T_b) = \frac{h_+(b) - h_+(x)}{h_+(b) - h_+(a)} \quad \begin{array}{c} \text{or what is the same} \\ \text{in view of} \\ h_- + h_+ = 1 \end{array} \quad \frac{h_-(x) - h_-(b)}{h_-(a) - h_-(b)}.$$

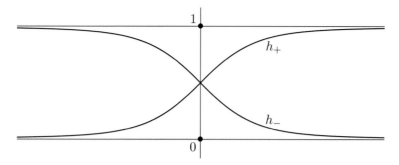

Figure 4.9.6

Notice, further, that $\lim_{n\uparrow\infty}\mathbf{x}(n)$ exists and can only be $+$ or $-\infty$, $h\circ\mathbf{x}(n)$ being a fair game, from which you obtain

$$\mathbb{P}_x\left[\lim_{n\uparrow\infty}\mathbf{x}(n)=\pm\infty\right]=h_\pm(x)$$

by inspection of Figure 4.9.6, with the suggestion that the presence of *two* bounded, non-negative harmonic functions indicates that the tail field descriptive of the ultimate behavior of the walk is of the form $[\emptyset, Z, Z', \mathbf{X}]$ with a single event Z and its complement. Look finally at the walk seen in Figure 4.9.7 where the drift of Figure 4.9.6 is reversed. Obviously, the walk now visits every site in $\mathbb{Z}$ infinitely often, and there

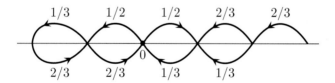

Figure 4.9.7

are no interesting, non-negative harmonic functions, as you may check by explicit computation or by the argument for $\mathbb{Z}^2$: that in such a case, $\lim_{n\uparrow\infty}h\circ\mathbf{x}(n)$ could not exist if h were not constant.

RW(1) on the half-line

On $\mathbb{Z}^+ = (0, 1, 2, 3, \text{etc.})$, the function $h(x) = x$ is surely non-negative and harmonic, and I want to explain *its* meaning. To begin with, the corresponding Doob-type 1-step probabilities

$$\mathbb{P}_x^h\left[\mathbf{x}(1) = x \pm 1\right] = \frac{1}{2} + \frac{1}{2x} \text{ to the right and } \frac{1}{2} - \frac{1}{2x} \text{ to the left}$$

$$= \frac{1}{h(x)} \mathbb{P}_x\left[\mathbf{x}(1) = x \pm 1\right] h(x \pm 1),$$

see Figure 4.9.8, prevent the walk from coming to the origin, imparting to it also a mild drift to the right, the mean step being $\frac{1}{x}$: just a little push,

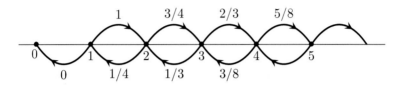

Figure 4.9.8

diminishing far out, but enough to make $\mathbb{P}\left[\lim_{n\uparrow\infty} \mathbf{x}(n) = +\infty\right] = 1$. That's easy to see: $1/\mathbf{x}(n)$ is a fair game for the h-walk, so $\lim_{n\uparrow\infty} \mathbf{x}(n)$ exists and cannot be finite. Yes? OK. But still, what does $h(x) = x$ *mean* for RW(1). Presumably, it has to do with some unwonted behavior of that standard walk, as to which it is an educated guess that the h-walk is nothing but RW(1), started at $\mathbf{x}(0) \geq 1$, *conditioned not to reach the origin*. As before, the meaning of such a conditioning is not so clear, but take it this way: that the free walk, starting at $x \geq 1$, is conditioned not to reach the origin before time N, and N is then taken to $+\infty$. The computation requires the estimate $\mathbb{P}_x(T_0 \geq N) \simeq \text{constant} \times x N^{-1/2}$ for any $x \geq 1$ and $N \uparrow \infty$. The rest is easy: for fixed x and $y \geq 1$, fixed n, and large N,

$$\mathbb{P}_x\left[\mathbf{x}(n) = y | T \geq N\right]$$
$$= \mathbb{P}_x\left[\mathbf{x}(n) = y\right] \times \left[\mathbb{P}_y(T \geq N - n) \text{ over } \mathbb{P}_x(T_x \geq N) \simeq \frac{y}{x}\right],$$

and here you have it. Note that the anomalous function x_2 on $\mathbb{Z}^{2+}$ must play the same role there as x does here: *to wit*, it has to do with RW(2), conditioned not to reach $\mathbb{Z} \times 0$.

Exercise 4.9.6   Check the estimate of $\mathbb{P}_x(T_0 \geq N)$ by induction in respect to $x \geq 1$, starting from the explicit distribution

$$\mathbb{P}_0(T_1 = 2k - 1) = \binom{2k}{k} 2^{-2k} \frac{1}{2k - 1}, \quad k \geq 1$$

determined in §3.2.

Hint: For $x = a + b$, $\mathbb{P}_0(T_x \geq N)$ is the sum of $\mathbb{P}_0(T_a = i) \times \mathbb{P}_0(T_b = j)$ for $i + j \geq N$, and contrary to what you might think, the bulk of the sum comes, not from the middle range where i and j are both comparable to N, but from the tails where i is less than some large number L and j is $\geq N - L$, or the other way around, where $j \leq L$ and $i \geq N - L$.

★Aside You know already that the walk tends to $+\infty$, but how fast does it go?

Exercise 4.9.7 Check that $\mathbb{E}_1[\mathbf{x}^{2p}(n)] \simeq n^p \times (2p + 1)!$ over $2^p p!$ for fixed $p \geq 1$ and $n \uparrow \infty$.

Hints: Compute to leading order only, using the fact that if $\mathbf{x}(n) = x$, the next step **e** has mean $1/x$. Here's a sample:

$$\mathbb{E}_1\big[\mathbf{x}^2(n+1) - \mathbf{x}^2(n)\big] = \mathbb{E}_1\big(2\mathbf{x}(n)\mathbf{e} + 1\big) = 3, \text{ so } \mathbb{E}_1\big[\mathbf{x}^2(n)\big] = 3n + 1.$$

Exercise 4.9.8 Use Markov's method of moments, introduced in §4.4, to show that

$$\lim_{n \uparrow \infty} \mathbb{P}_1\big[n^{-1/2}\mathbf{x}(n) \leq c\big] = 2 \int_0^c x^2 \frac{e^{-x^2/2}}{\sqrt{2\pi}} \, dx.$$

Exercise 4.9.9 Prove also that $\mathbb{P}_1\big[\limsup_{n \uparrow \infty} \frac{\log \mathbf{x}(n)}{\log n} \leq \frac{1}{2}\big] = 1$, with the suggestion that $\mathbf{x}(n)$ goes to ∞ like $\sqrt{n}$.

Exercise 4.9.10 $\mathbb{E}_1\big[\mathbf{x}^{-1}(n)\big] \simeq \sqrt{2/\pi n}$, as you will check, and it would be nice if $\mathbb{E}_1\big[\mathbf{x}^{-3}(n)\big]$ would be comparable to $n^{-3/2}$ and so on, as you would then have $\mathbb{P}_1\big[\liminf_{n \uparrow \infty} \frac{\log \mathbf{x}(n)}{\log n} \leq \frac{1}{2}\big] = 1$. That won't work: already the suggested estimate for $\mathbf{x}^{-3}$ cannot be correct. Explain, please.

Hint for the rest: $\mathbb{E}_1\big[\log \mathbf{x}(n+1)|\mathfrak{F}_n\big]$ is not more than $\log \mathbf{x}(n)$, so $\frac{\log \mathbf{x}(n)}{\log n}$ is an *unfavorable* game. Check the inequality and take it from there to obtain $\mathbb{P}_1\big[\lim_{n \uparrow \infty} \frac{\log \mathbf{x}(n)}{\log n} = \frac{1}{2}\big] = 1$.

Summary

Judging by these variants of RW(1), one may say that *bounded*, non-negative harmonic functions are descriptive of the tail field of the walk, and, at a second level, that any extra *unbounded* functions describe how the walk may be conditioned to *do things it would not ordinarily think of.*

4.9.4 Space-time walks, mostly in dimension 1

The picture is made more clear if you take into account, not only the position $\mathbf{x}(n) \in \mathbb{Z}^d$, but also the time $n = 1, 2, 3 \ldots$ That's the space-time walk

$$n \longrightarrow \text{the pair } n \text{ and } \mathbf{x}(n).$$

Now harmonicity means that

$$h(n, x) = \frac{1}{2d} \sum_e h(n+1, x+e),$$

and a whole new world is revealed. Let's see what happens for RW(1).

Computation of h

Things are simplified if the walk is replaced by the corresponding (honest) Bernoulli trials $\mathbf{s}(n) = \frac{1}{2}\mathbf{x}(n) + \frac{n}{2}$ with equally likely steps $e = 0$ or 1. For paths starting at $\mathbf{s}(0) = 0$, the function $h(n, x)$ is then defined for $0 \le x \le n$ only, and $h(n, x) = \frac{1}{2}h(n+1, x) + \frac{1}{2}h(n+1, x+1)$. To see what such a function looks like, start this way:

$$h(n, x) = \mathbb{E}\big[h(n+m, x+\mathbf{s}(m))\big] \quad \text{for any } m \ge 1$$

$$= \sum_0^m \binom{m}{k} 2^{-m} h(n+m, x+k)$$

$$= \sum_0^m \left[\binom{m}{k} 2^{-m} \Big/ \binom{n+m}{x+k} 2^{-n-m}\right]$$

$$\times \left[h(n+m, x+k)\binom{n+m}{x+k} 2^{-n-m}\right]$$

$$= 2^n \sum_0^m \frac{(k+x)\ldots(k+1) \times (n+m)\ldots(n+m-x-k+1)}{(n+m)\ldots(m+1)}$$

$$\times \left[h(n+m, x+k)\binom{n+m}{x+k} 2^{-n-m}\right]$$

$$\simeq 2^n \sum_0^m \left(\frac{k}{m}\right)^x \left(1 - \frac{k}{m}\right)^{n-x} \pi\left(\frac{k}{m}\right) \quad \text{if } m \text{ is large}$$

where the weights

$$\pi\left(\frac{k}{m}\right) = h(n+m, x+k)\binom{n+m}{x+k} 2^{-n-m}, \quad 0 \le k \le m$$

may be viewed as a non-negative mass distribution on the interval $0 \le \theta \le 1$ of total mass not more than $\mathbb{E}[h(n+m, \mathbf{s}(n+m))] = h(00)$ which I take to be 1. These weights depend, of course, on n and m, but for fixed n, you may make m tend to $+\infty$ in such a way that they converge, in the sense of §1.7.1, to some n-dependent mass distribution $d\pi(\theta)$, with the result that[4]

$$h(n,x) = 2^n \int_0^1 \theta^x (1-\theta)^{n-x} \, d\pi(\theta).$$

Happily, the dependence of π on n can be removed:

$$h(n-1,x) = \frac{1}{2}h(n,x) + \frac{1}{2}h(n,x+1) = \theta^{n-1} \int_0^1 \theta^x (1-\theta)^{n-1-x} \, d\pi,$$

as you will check, which is to say that if π works for $n = n'$, then it also serves for any $n < n'$, whereupon you may send n' to $+\infty$ to produce a *single* mass distribution π that serves for every $n \geq 1$ at once. In short,

$$h(n,x) = 2^n \int_0^1 \theta^x (1-\theta)^{n-x} \, d\pi(\theta) \quad \text{for } 0 \leq x \leq n \text{ and } n \geq 1;$$

and conversely: *every such function with non-negative π of total mass 1 is a space-time harmonic function for honest Bernoulli trials.*[5]

The result is actually an old one, due in a different language to the Italian actuary B. de Finetti (1929). It is the promised second illustration of the general fact that, with a little compactness, every point of the section of a convex cone is the center of gravity of its vertices, those being the functions $h_\theta(n,x) = 2^n \theta^x (1-\theta)^{n-x}$ in the present case.

Now back to the standard walk: $\mathbf{x}(n) = 2\mathbf{s}(n) - n$. Evidently, the general space-time function for that is

$$h(n,x) = 2^n \int_0^1 \theta^{(n+x)/2} (1-\theta)^{(n-x)/2} \, d\pi(\theta)$$

$$\text{for } n \geq 1 \text{ and } -n \leq x \leq n,$$

but this is not a good format any more. Better to write $\sqrt{\theta/(1-\theta)} = e^\alpha$ with $\alpha \in \mathbb{R}$ and to re-write $d\pi(\theta)$ in that language, whereupon you find

$$h(n,x) = \int_{-\infty}^{\infty} \beta^{-n} e^{\alpha x} \, d\pi(\alpha) \text{ with } \beta = \cosh \alpha.$$

The vertices are now the special functions $\beta^{-n} e^{\alpha x}$, *aka* the fair games

---

[4] Some technical proviso is called for here, as to the meaning of θ^0 and $(1-\theta)^0$, but I ignore this for the moment.

[5] As to the ambiguity of θ^0 and $(1-\theta)^0$ cited in the prior footnote, this can be ignored since, e.g. $\pi(0)$ can contribute to h only a constant multiple of the uninteresting function $h_0(n,x) = 2^n$ at $x = 0$, with $h_0(n,x) = 0$ elsewhere.

$(\cosh \alpha)^{-n} \exp[\alpha \mathbf{x}(n)]$ so familiar from Chapter 3 (surprise?) and it's easy to see what they mean for the standard walk: $\alpha = 0$ does nothing; otherwise, the 1-step probabilities

$$\mathbb{P}^h_{nx}\left[\mathbf{x}(n+1) = x + e\right] = \frac{1}{h(n,x)} \mathbb{P}_x\left[\mathbf{x}(1) = x + e\right] h(n+1, x+e)$$
$$= \frac{e^{\alpha e}}{2\beta}$$

do not depend on x and so describe a standard-type walk with drift, typified by Figure 4.9.3 which appears for $\alpha = \sqrt{2}$. That's already interesting – and all the more so when you reflect that $\mathbf{x}(n)$ now tends to ∞ like $n \times \gamma = $ the mean step $\sinh \alpha / \cosh \alpha$, and prove as I do next, that this is nothing but RW(1) *conditioned so as to make* $\lim_{n \uparrow \infty} n^{-1}\mathbf{x}(n) = \gamma$. As usual, it is not clear what such a conditioning should mean but take it this way: that the standard walk should be conditioned to make $|n^{-1}\mathbf{x}(n) - \gamma| < \varepsilon$ for large n, after which n is taken to $+\infty$ and ε is taken down to $0+$. As before, the computation is best laid out in the Bernoulli format: the conditioning requires that $|n^{-1}\mathbf{s}(n) - \frac{1}{2}(\gamma+1)|$ be less than ε and so may be referred to the large deviations of §2.4 since the frequency

$$f = \frac{1}{2}(\gamma+1) = \frac{1}{2}\frac{\sinh \alpha}{\cosh \alpha} + \frac{1}{2} \text{ reduces to } \frac{1}{2} \text{ only if } \alpha = 0.$$

I do only the 1-step probabilities and ask you to accept a simple demonstration – this in the sense of M. Kac: that a *demonstration* serves to convince a reasonable man – only an unreasonable man requires a *proof*.

Demonstration. Take $\alpha > 0$ so the frequency f is more than $1/2$, and ε so small that $1/2 < f - \varepsilon$ as well. Cramér's rule for large deviations of honest Bernoulli trials states that, for $1/2 < a < b \le 1$,

$$\lim_{n \uparrow \infty} \frac{1}{n} \log \mathbb{P}\left[a \le \frac{\mathbf{s}(n)}{n} \le b\right] = -J$$

where J is the minimum of the (positive) function

$$I(q) = q \log q + (1-q) \log(1-q \quad \text{for } a \le q \le b.$$

This value is assumed at the left, at $q = a$, and I take it *literally, without any logarithm*, so that the prior display reads

$$\mathbb{P}\left[a \le \frac{\mathbf{s}(n)}{n} \le b\right] \simeq e^{-nI(a)} \text{ for any } a > \frac{1}{2} \text{ and } 1 \ge b > a.$$

Then you should have

$$\mathbb{P}_0\left[\mathbf{s}(1) = e \ \Big| \ \left|\frac{\mathbf{s}(n)}{n} - f\right| < \varepsilon\right]$$

$$= \mathbb{P}_0\left[\mathbf{s}(1) = e\right] \times \frac{\mathbb{P}_e\left[\left|\frac{\mathbf{s}(n-1)}{n} - f\right| < \varepsilon\right]}{\mathbb{P}_0\left[\left|\frac{\mathbf{s}(n)}{n} - f\right| < \varepsilon\right]}$$

$$= \frac{1}{2} \times \frac{\mathbb{P}_0\left[\frac{n}{n-1}(f-\varepsilon) - \frac{e}{n-1} < \frac{\mathbf{s}(n-1)}{n-1} < \frac{n}{n-1}(f+\varepsilon) - \frac{e}{n-1}\right]}{\mathbb{P}_0\left[f - \varepsilon < \frac{\mathbf{s}(n)}{n} < f + \varepsilon\right]}$$

$$\simeq \exp\left[-(n-1)I\left(\frac{n}{n-1}(f-\varepsilon) - \frac{e}{n-1}\right) + nI(f+\varepsilon) - \log 2\right]$$

in which the exponent reduces to $I(f) - I'(f)(f - e) - \log 2$ when n is sent to $+\infty$ and ε is taken down to 0 *in that order*. In extenso,

$$I(f) - I'(f)(f - e) = \log f \text{ if } e = 1 \text{ or } \log(1 - f) \text{ if } e = 0$$

when you spell it out, with the promised result

$$\mathbb{P}_0\left[\mathbf{s}(1) = e \ \Big| \ \lim_{n\uparrow\infty} \frac{\mathbf{s}(n)}{n} = f\right] = f \text{ or } 1 - f \text{ according as } e = 1 \text{ or } 0.$$

Indeed, the Bernoulli probabilities f and $1-f$ correspond in the language of walks to $\mathbb{P}(0 \to 1) = f$ and $\mathbb{P}(0 \to -1)$, with mean step $2f - 1 = \gamma$, as advertised.

The meaning of π

Now look at the general space-time function in the standard walk format:

$$h(n, x) = \int_{-\infty}^{\infty} \beta^{-n} e^{\alpha x} \, d\pi(\alpha) \text{ with } \beta = \cosh \alpha, \ n \geq 1 \text{ and } -n \leq x \leq n.$$

Evidently, for any event $B \in \mathfrak{F}_n$,

$$\mathbb{P}_0^h(\mathbf{x} \in B) = \int_{-\infty}^{\infty} \mathbb{P}_0\left[\mathbf{x} \in B \ \Big| \ \lim_{n\uparrow\infty} \frac{\mathbf{s}(n)}{n} = \gamma\right] d\pi(\alpha),$$

showing that the h-game is played this way. First, you pick α, determining the honesty $0 \leq f = \frac{1}{2}(\gamma + 1)$ of the coin, according to the probabilities π. Then you play *that* game. The meaning of π is then plain: rewritten in terms of γ, it is the distribution

$$\mathbb{P}_0^h\left[\lim_{n\uparrow\infty} \frac{\mathbf{x}(n)}{n} \in \Gamma\right] \text{ for } \Gamma \subset \mathbb{R}.$$

From this point of view, Figure 4.9.9 is natural, showing the sector $n \geq 0$,

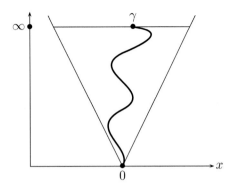

Figure 4.9.9

$|x| \leq n$, where $h(n, x)$ lives, completed at ∞, by adjunction of the line to which the walk tends in the sense that $\lim_{n \uparrow \infty} n^{-1}\mathbf{x}(n) = \gamma$.

Amplification 1. Nothing really changes for dimensions 2 and 3, though the proof is no quite so simple:

$$h(n, x) = \int_{\mathbb{R}^d} \beta^{-n} e^{\alpha \cdot x} \, d\pi(\alpha)$$

with

$$\beta = \begin{array}{ll} \frac{1}{2}(\cosh \alpha_1 + \cosh \alpha_2) & \text{for } d = 2 \\ \frac{1}{3}(\cosh \alpha_1 + \cosh \alpha_2 + \cosh \alpha_3) & \text{for } d = 3. \end{array}$$

Amplification 2. In the Bernoulli format, the sector $n \geq 0, 0 \leq x \leq n$ is compactified at ∞ by the segment $0 \leq f \leq 1$ to which $n^{-1}\mathbf{s}(n)$ tends, as in Figure 4.9.10, suggesting comparison to two well-known classical problems.

The first is the "problem of moments", *viz.* to reconstruct the mass distribution $d\pi(f)$ from its moments $\int_0^1 f^n \, d\pi(f)$. Obviously, you have only to form the function $h(n, x) = 2^n \int_0^1 f^x (1-f)^{n-x} \, d\pi(f)$ and to put

$$\pi(\Theta) = \mathbb{P}_0^h \left[\lim_{n \uparrow \infty} n^{-1}\mathbf{s}(n) \in \Theta \right] \text{ for } \Theta \subset [0, 1].$$

The second problem is entirely different, having to do with heat flow in the unit disc of $\mathbb{R}^2$ with fixed temperatures $f(\theta) : 0 \leq \theta < 2\pi$ on the circle. The temperatures $w(t, x)$ inside are governed by $\partial w / \partial t = \Delta w$ with the "real" Laplacian $\Delta = \partial^2 / \partial x_1^2 + \partial^2 / \partial x_2^2$. These equilibrate with the passing time, the steady state $w(\infty, \cdot) = h$ being *the* harmonic

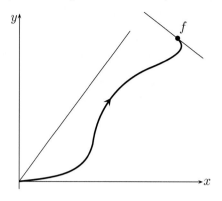

Figure 4.9.10

function with the same values $h = f$ at the boundary, and there is a pretty formula for that, due to Poisson (1820) which bears its name[6]: in polar coordinates $x = (r, \theta)$, it reads

$$h(x) = \frac{1}{2\pi} \int_0^{2\pi} \frac{1 - r^2}{1 - 2r\cos(\theta - \theta') + r^2}\, d\pi(\theta') \text{ with } d\pi(\theta') = f(\theta')\, d\theta'.$$

More generally, any non-negative function, harmonic in the disc, may be expressed in this way with $f(\theta')\, d\theta'$ replaced by the general, non-negative mass distribution $d\pi(\theta')$. These form a convex cone with compact section determined by $h(0) = 1$, *aka* the total mass of π, and the formula shows that every point of that section is the center of gravity of its vertices $\frac{1}{2\pi}(1 - r^2)$ over $(1 - 2r\cos(\theta - \theta') + r^2)$ indexed by $0 \le \theta' < 2\pi$.

I stop here, referring you to Rogers–Williams [1987] for the rest.

---

[6] Seeley [1966:6–23] is a good place to look if this is unfamiliar.

5

LLN, CLT, Iterated Log, and Arcsine in General

Now let's move from Bernoulli trials and RW(1) to a more general view of these things.

5.1 LLN

The statement is that if $\mathbf{x}_1$, $\mathbf{x}_2$, etc. are independent copies of a variable $\mathbf{x}$ with (absolute) mean $\mathbb{E}|\mathbf{x}| < \infty$, then the sums $\mathbf{s}_n = \mathbf{x}_1 + \cdots + \mathbf{x}_n$ obey the law of large numbers:

$$\mathbb{P}\left[\lim_{n \uparrow \infty} \frac{\mathbf{s}_n}{n} = m\right] \quad \text{with } m = \mathbb{E}(\mathbf{x}).$$

One may ask: How far can Bernoulli trials take us? Let $\mathbf{x}$ be capable of several values, not just 0 and 1; three values, $x = 0$, 1 and 2, will make the point. Then, for each x separately, $\#(n, x) = \#(k \leq n : \mathbf{x}_k = x)$ is the number of successes in n independent Bernoulli trials with success probability $\mathbb{P}(\mathbf{x} = x)$, so

$$\lim_{n \uparrow \infty} \frac{\mathbf{s}_n}{n} = \sum_{x=0}^{2} x \lim_{n \uparrow \infty} \frac{1}{n} \#(n, x) = \sum_{x=0}^{2} \mathbb{P}(\mathbf{x} = x) = \mathbb{E}(\mathbf{x}).$$

Simple enough.

Exercise 5.1.1 Deduce that LLN holds if, for example, $0 \leq \mathbf{x} \leq 1$, and also in general, that $\mathbb{P}\left[\liminf_{n \uparrow \infty} \mathbf{s}_n / n \geq m\right] = 1$ if $\mathbf{x} \geq 0$.

That's about as far as you can get this way.

5.1.1 Another way

Here's an easy proof, seen in §2.1 already, if $\mathbb{E}(\mathbf{x}^2) < \infty$. Write $\mathbf{x}'_n = \mathbf{x}_n - m$. Then

$$\mathbb{P}\left[\left|\frac{\mathbf{s}'_n}{n}\right| > \varepsilon\right] \leq \frac{1}{\varepsilon^2 n^2} \mathbb{E}\left[(\mathbf{s}'_n)^2\right] = \frac{1}{\varepsilon^2 n} \mathbb{E}\left[(\mathbf{x}')^2\right].$$

This is not good enough for Borel–Cantelli, but if you replace n by n^2, it works.

Exercise 5.1.2 Finish the proof by looking at $\mathbf{s}_k/k$ for $k \uparrow \infty$ with $(n-1)^2 \leq k < n^2$.
Hint: It is legitimate to take $\mathbf{x} \geq 0$. Yes?

5.1.2 Kolmogorov's 1933 proof

This is stylish, general, and very perfect of its kind, without any extra conditions. Other proofs will be found in the next article, §5.1.3, and in §9.3. Take $\mathbf{x} \geq 0$ as in Exercise 5.1.2 and define $\mathbf{x}'_n = \mathbf{x}_n$ or 0 according as $\mathbf{x}_n < n$ or not. You have

$$\sum_{n=1}^{\infty} \mathbb{P}(\mathbf{x}_n \geq n) = \sum_{n=1}^{\infty} \sum_{k=n}^{\infty} \mathbb{P}(k \leq \mathbf{x} < k+1) = \sum_{k=1}^{\infty} k\,\mathbb{P}(k \leq \mathbf{x} < k+1)$$
$$\leq \mathbb{E}(\mathbf{x}) < \infty,$$

so $\mathbf{x}'_n = \mathbf{x}_n$ with a finite number of exceptions, by Borel–Cantelli, and it is enough to deal with the reduced sums $\mathbf{s}'_n$. Now center $\mathbf{x}'_n$ as in $\mathbf{x}''_n = \mathbf{x}'_n - \mathbb{E}(\mathbf{x}'_n)$ and estimate

$$\sum_{n=1}^{\infty} \frac{1}{n^2} \mathbb{E}(\mathbf{x}''_n)^2 = \sum_{n=1}^{\infty} \frac{1}{n^2} \mathbb{E}(\mathbf{x}'_n)^2$$
$$= \sum_{n=1}^{\infty} \frac{1}{n^2} \sum_{k=1}^{\infty} \mathbb{E}(\mathbf{x}^2, \, k-1 \leq \mathbf{x} < k)$$
$$\leq \sum_{k=1}^{\infty} k\,\mathbb{E}(\mathbf{x}, \, k-1 \leq \mathbf{x} < k) \sum_{n=k}^{\infty} \frac{1}{n^2}$$
$$< \infty,$$

since $\sum_{n=k}^{\infty} n^{-2}$ is comparable to $1/k$. Then $\sum_{n=1}^{\infty}(\mathbf{x}''_n/n)$ converges by Exercise 1.8.5. The rest comes from

Kronecker's lemma $\sum_{n=1}^{\infty} c_n/n$ *converges only if*

$$\frac{1}{n}(c_1 + \cdots + c_n) = o(1).$$

In fact, with $c'_n = \sum_{k=1}^{n} c_k/k$ and $c'_0 = 0$, you have

$$\sum_{k=1}^{n} c_k = \sum_{k=1}^{n} k(c'_k - c'_{k-1}) = \sum_{k=1}^{n} kc'_k - \sum_{k=0}^{n-1}(k+1)c'_k = nc'_n - \sum_{k=0}^{n-1} c'_k,$$

from which the lemma follows.

Now put the pieces together:

$$\frac{\mathbf{s}_n}{n} \simeq \frac{\mathbf{s}'_n}{n} = \frac{\mathbf{s}''_n}{n} + \frac{1}{n}\sum_{k=1}^{n} \mathbb{E}(\mathbf{x}, \mathbf{x} < k) = o(1) + \mathbb{E}(\mathbf{x}) + o(1),$$

as promised.

Exercise 5.1.3   Keeping $\mathbf{x} \geq 0$, prove that $\lim_{n\uparrow\infty}|(\mathbf{s}_n/n) - m| = 0$. *Hint*: $F_n = (\mathbf{s}_n/n) - m = F_n^+ - F_n^-$ with $0 \leq F_n^- \leq m$. Check that $\lim_{n\uparrow\infty} \mathbb{E}(F_n^-) = 0$. Take it from there.

★Exercise 5.1.4   Take independent copies of a variable $\mathbf{x} \geq 0$ with density $\frac{2}{\pi}\frac{1}{1+x^2}$. Obviously, $\lim_{n\uparrow\infty}\frac{1}{n}(\mathbf{x}_1 + \cdots + \mathbf{x}_n) = +\infty$. Check that

(1) $\lim_{n\uparrow\infty} \sqrt[n]{\mathbf{x}_1 \ldots \mathbf{x}_n} = 1$ and

(2) $\lim_{n\uparrow\infty} \frac{1}{n\log n}(\mathbf{x}_1 + \cdots + \mathbf{x}_n) = \frac{2}{\pi}$ in the weak sense.

Hints: LLN covers (1). For (2): $\lim_{n\uparrow\infty} \frac{1}{n\log n} \mathbb{E}\exp\left[-k(\mathbf{x}_1 + \cdots + \mathbf{x}_n)\right] = e^{-(2/\pi)k}$ for $k > 0$ does it. This follows from

$$\mathbb{E}(e^{-k\mathbf{x}}) = \frac{2}{\pi}\int_k^{\infty} \sin(k' - k)\,\frac{dk'}{k'} = 1 + \frac{2}{\pi} k\log k \quad \text{for small } k.$$

5.1.3 Doob's proof

This is really slick. Take $\mathbf{x} \geq 0$ as before, let $\mathfrak{F}_n$ be the field of $\mathbf{s}_n, \mathbf{s}_{n+1}$, etc., and write $\mathbf{z}_n = \mathbb{E}(\mathbf{x}_1|\mathfrak{F}_n)$. The fields are decreasing, so $\mathbf{z}_1, \mathbf{z}_2, \ldots, \mathbf{z}_n$ is not a fair game, but read it backwards: $\mathbf{z}_n, \ldots, \mathbf{z}_2, \mathbf{z}_1$ *is* fair; its *up*-crossings, *aka* the *down*-crossings of $\mathbf{z}_1, \mathbf{z}_2, \ldots, \mathbf{z}_n$, are controlled by $\mathbb{E}(\mathbf{z}_1) = \mathbb{E}(\mathbf{x})$; and as these can only increase in number with n, the limit $\mathbf{z}_\infty$ exists. Now here's the key: $\mathbf{x}_1, \mathbf{x}_2, \ldots, \mathbf{x}_n$ enter symmetrically into the field $\mathfrak{F}_n$ so $\mathbb{E}(\mathbf{x}_1|\mathfrak{F}_n) = \mathbb{E}(\mathbf{x}_2|\mathfrak{F}_n) = \mathbb{E}(\mathbf{x}_3|\mathfrak{F}_n)$ and so on, up to $\mathbb{E}(\mathbf{x}_n|\mathfrak{F}_n)$, with the result that $\mathbf{z}_n = \mathbb{E}(\mathbf{x}_1|\mathfrak{F}_n) = \mathbf{s}_n/n$ itself. But then

$$\lim_{n\uparrow\infty} \frac{\mathbf{s}_n}{n} = \mathbb{E}(\mathbf{x}_1|\mathfrak{F}_\infty) \quad \text{by P. Lévy's rule } (4'') \text{ of §1.3.3,}$$

and since the tail field $\mathfrak{F}_\infty = \bigcap_1^\infty \mathfrak{F}_n$ is trivial, by the Hewitt–Savage 01 law of Exercise 1.8.7, $\mathbb{E}(\mathbf{x}_1|\mathfrak{F}_\infty)$ reduces to $\mathbb{E}(\mathbf{x})$, plain. Can you beat that?

5.2★ Kolmogorov–Smirnov statistics

As noted in §2.1, if the independent variables $\mathbf{x}_1$, $\mathbf{x}_2$, etc. have a common, jump-free distribution F, then the empirical distribution $\mathbf{F}_N(x) = \frac{1}{N}\#[k \leq N : \mathbf{x}_k \leq x]$ obeys the law of large numbers:

$$\lim_{N\uparrow\infty} \mathbf{F}_N(x) = F(x) \quad \text{simultaneously for every } x \in \mathbb{R}.$$

The Kolmogorov–Smirnov statistic $\mathbf{D}_N = \max_{x\in\mathbb{R}}|\mathbf{F}_N(x) - F(x)|$ measures the discrepancy. Actually, if F is jump-free, then $\mathbf{D}_N$ is *universal*, meaning that the old variables $\mathbf{x}_n$ may be replaced by the new variables $\mathbf{x}'_n = F^{-1}(\mathbf{x}_n)$, uniformly distributed on $[0,1]$, reducing the statistic to $\mathbf{D}_N = \max_{0\leq x\leq 1}|\mathbf{F}_N(x) - x|$ but not, in fact, changing it at all. It is this reduction which makes possible the computation of

$$\lim_{N\uparrow\infty} \mathbb{P}\big(\sqrt{N}\,\mathbf{D}_N \leq x\big) = \frac{\sqrt{2\pi}}{x} \sum_{\text{odd } n\geq 1} e^{-n^2\pi^2/8x^2}.$$

Pretty complicated: it's not even clear if the right side *is* a distribution function, but more about that later.

The fact is due to Kolmogorov [1933a] and Smirnov [1939] with a complicated proof. Feller [1948] found an "elementary" proof, but that's not simple either. Then Doob [1949] suggested a very attractive method based upon the Brownian motion, with an acknowledged gap, to be repaired in §7.7 below. Here, I will explain an easy version of Doob's idea (with its own gap). I begin as a warm-up with the simpler result of Feller [1948]: *viz.*, with $\mathbf{D}_N^+ = \max_{0\leq x\leq 1}[\mathbf{F}_N(x) - x]^+$, then

$$\lim_{N\uparrow\infty} \mathbb{P}\left[\sqrt{N}\,\mathbf{D}_N^+ \geq x\right] = e^{-2x^2}.$$

The proof is divided into a number of small steps.

Step 1. $\mathbf{D}_N^+ = \max_{k \leq N} \left[\mathbf{F}_N \left(\frac{k}{N} \right) - \frac{k}{N} \right]$, up to an error not more than $1/N$ which can be ignored: indeed, for $(k-1)/N \leq x < k/N$,

$$\mathbf{F}_N \left(\frac{k-1}{N} \right) - \frac{k-1}{N} - \frac{1}{N} \leq \mathbf{F}_N(x) - x < \mathbf{F}_N \left(\frac{k}{N} \right) - \frac{k}{N} + \frac{1}{N},$$

by inspection.

Step 2.   Now let $\#_n$ be the number of variables $\mathbf{x}_1$, $\mathbf{x}_2$, up to $\mathbf{x}_N$ which fall between $(n-1)/N$ and n/N. They have the joint distribution

$$\mathbb{P}\left[\#_1 = n_1, \#_2 = n_2, \ldots, \#_N = n_N \right]$$
$$= \frac{N!}{n_1! n_2! \ldots n_N!} \left(\frac{1}{N} \right)^{n_1} \left(\frac{1}{N} \right)^{n_2} \cdots \left(\frac{1}{N} \right)^{n_N},$$

which may be interpreted as saying that the variables $\#_k : k \leq N$ are independent copies of $\# = \#_1$, with common (Poisson) distribution $\mathbb{P}(\# = n) = 1/en!$, *conditioned* so that $\#_1 + \cdots + \#_N = N$.

Exercise 5.2.1   Check it out.

Step 3.   Next, introduce the (*unconditioned*) Poisson walk with drift: $\mathbf{x}(n) = (\#_1 - 1) + \cdots + (\#_n - 1)$ $(n \geq 0)$. Then

$$\mathbf{D}_N^+ = \max_{n \leq N} (\#_1 + \cdots + \#_n - n) = \max_{n \leq N} \mathbf{x}(n)$$

and the statement, $\mathbf{D}_N^+ \geq c/N$ with a whole number $c \geq 1$, is the same as saying that the passage time $T = \min(n : \mathbf{x}(n) \geq c)$ is $\leq N$, i.e.

$$\mathbb{P}\left[\mathbf{D}_N^+ \geq c/N \right] = \mathbb{P}\left[T \leq N | \mathbf{x}(N) = 0 \right],$$

in which the conditioning by $\mathbf{x}(N) = 0$ enforces $\#_1 + \cdots + \#_N = N$, as required before.

Step 4.   Now the (unconditional) mean and mean-square of $\# = \#_1$ are

$$\mathbb{E}(\#) = \sum_{n=0}^{\infty} \frac{n}{en!} = 1 \quad \text{and} \quad \mathbb{E}(\#^2) = \sum_{n=0}^{\infty} \frac{n^2}{en!} = 2,$$

so the unit step of the walk has mean 0 and mean-square 1, just like RW(1). Doob's idea, in its present simplified version, is that it does no harm to replace the present walk by RW(1) plain, at least for large N, which is all that matters here.

Step 5.   To implement this, it is necessary to replace N by $2N$ since RW(1) cannot visit 0 at odd times. This will be corrected at the end. T

is now the passage time $\min(n : \mathbf{x}(n) = c)$, and you can use D. André's reflection principle from §3.2 to compute

$$\mathbb{P}_0[T \le 2N, \mathbf{x}(2N) = 0] = \sum_{n=1}^{2N} \mathbb{P}_0(T = n)\, \mathbb{P}_c[\mathbf{x}(2N - n) = 0]$$

$$= \sum_{n=1}^{2N} \mathbb{P}_0(T = n)\, \mathbb{P}_c[\mathbf{x}(2N - n) = 2c]$$

$$= \mathbb{P}_0[\mathbf{x}(2N) = 2c],$$

from which you find

$$\mathbb{P}\big(\mathbf{D}_N^+ \ge c/2N\big) = \frac{\mathbb{P}_0[\mathbf{x}(2N) = 2c]}{\mathbb{P}_0[\mathbf{x}(2N) = 0]} = \frac{(N!)^2}{(N - c)!(N + c)!}\,.$$

Step 6.   Now take $c \simeq x\sqrt{2N}$. Then Stirling's approximation produces

$$\mathbb{P}\big(\sqrt{2N}\,\mathbf{D}_{2N}^+ \ge x\big)$$

$$\simeq \frac{N^{2N+1}}{(N - c)^{N-c+\frac{1}{2}}\,(N + c)^{N+c+\frac{1}{2}}}$$

$$\simeq \left(1 - \frac{2x^2}{N}\right)^{-N} \times \left(1 - \frac{\sqrt{2}\,x}{\sqrt{N}}\right)^{x\sqrt{2N}} \times \left(1 + \frac{\sqrt{2}\,x}{\sqrt{N}}\right)^{-x\sqrt{2N}}$$

$$\simeq \mathrm{e}^{2x^2} \times \mathrm{e}^{-2x^2} \times \mathrm{e}^{-2x^2}$$

$$= \mathrm{e}^{-2x^2},$$

as promised.

Exercise 5.2.2   Check the computation and remove the restriction to even N.

Now look at the original Kolmogorov–Smirnov statistic

$$\mathbf{D}_N = \max_{0 \le x \le 1} |F_N(x) - x|$$

and continue the proof, imitating Steps 1–6: $\mathbf{D}_N$ is replaced by $\frac{1}{N}\max_{k \le N}|\mathbf{x}(k)|$, N is replaced by $2N$ as before, and $\mathbb{P}(\mathbf{D}_{2N} \ge c/2N)$ is identified as $\mathbb{P}_0[T \le 2N | \mathbf{x}(2N) = 0]$, where T is now the two-sided passage time $\min(n : |\mathbf{x}(n)| = c)$.

Step 7 evaluates $\mathbb{P}_0[T \le 2N \mid \mathbf{x}(2N) = 0]$ by a two-sided variant of D. André's reflection principle, known in its present form as Kelvin's "method of images". I spell it out: with $T_{\pm} = \min(n : \mathbf{x}(n) = \pm c)$ and $2N = n$ for temporary brevity, you find

$$\frac{1}{2}\mathbb{P}_0[T \le n, \mathbf{x}(n) = 0]$$

$$= \mathbb{P}_0[T_- > T_+ \le n, \mathbf{x}(n) = 0]$$

$$= \sum_{k=1}^{n} \mathbb{P}_0(T_+ = k < T_-)\,\mathbb{P}_c[\mathbf{x}(n-k) = 0] \qquad \text{by the symmetry } \mathbf{x} \to -\mathbf{x}$$

$$= \sum_{k=1}^{n} \mathbb{P}_0(T_+ = k < T_-)\,\mathbb{P}_c[\mathbf{x}(n-k) = 2c] \qquad \text{by reflection}$$

$$= \mathbb{P}_0[T_+ < T_-, \mathbf{x}(n) = 2c]$$

$$= \mathbb{P}_0[\mathbf{x}(n) = 2c] - \mathbb{P}_0[T_- < T_+, \mathbf{x}(n) = 2c]$$

$$= \mathbb{P}_0[\mathbf{x}(n) = 2c] - \sum_{k=1}^{n} \mathbb{P}_0(T_- = k < T_+)\,\mathbb{P}_{-c}[\mathbf{x}(n-k) = 2c]$$

$$= \mathbb{P}_0[\mathbf{x}(n) = 2c] - \sum_{k=1}^{n} \mathbb{P}_0(T_- = k < T_+)\,\mathbb{P}_{-c}[\mathbf{x}(n-k) = -4c]$$

$$\text{by another reflection}$$

$$= \mathbb{P}_0[\mathbf{x}(n) = 2c] - \mathbb{P}_0[T_- < T_+, \mathbf{x}(n) = -4c]$$

$$= \mathbb{P}_0[\mathbf{x}(n) = 2c] - \mathbb{P}_0[\mathbf{x}(n) = -4c] + \mathbb{P}_0[T_+ < T_-, \mathbf{x}(n) = -4c]$$

$$= \mathbb{P}_0[\mathbf{x}(n) = 2c] - \mathbb{P}_0[\mathbf{x}(n) = -4c]$$
$$\qquad + \mathbb{P}_0[\mathbf{x}(n) = 6c] - \mathbb{P}_0[T_- < T_+, \mathbf{x}(n) = -6c],$$

and so on. The pattern will now be plain, and you will agree that

$$\frac{1}{2}\mathbb{P}_0[T \le 2N, \mathbf{x}(2N) = 0]$$

$$= \sum_{\text{odd } n \ge 1} \mathbb{P}_0[\mathbf{x}(2N) = 2nc] - \sum_{\text{even } n \ge 2} \mathbb{P}_0[\mathbf{x}(2N) = 2nc],$$

which is to say

$$\mathbb{P}\left[\mathbf{D}_{2N} < \frac{c}{2N}\right] = \mathbb{P}_0[T > 2N | \mathbf{x}(2N) = 0] = \sum_{n \in \mathbb{Z}} (-1)^n \frac{\mathbb{P}_0[\mathbf{x}(2N) = 2nc]}{\mathbb{P}_0[\mathbf{x}(2N) = 0]}.$$

Step 8. Now take $c \simeq \sqrt{2N}\,x$. Then as in Step 6 and Exercise 5.2.2, you find

$$\lim_{n \uparrow \infty} \mathbb{P}\left(\sqrt{N}\,\mathbf{D}_N \le x\right) = \sum_{n \in \mathbb{Z}} (-1)^n e^{-2n^2 x^2}$$

without restriction to even N. Naturally, some estimate is needed here, but the final sum of Step 7 is alternating, with diminishing terms to right and left, and tails controlled by the first neglected summand. Then no trouble can arrive.

Step 9. Of course, $\sum_{n\in\mathbb{Z}}(-1)^n e^{-2n^2x^2}$ do not even *look* like a distribution function. Jacobi's identity from §1.7 comes to the rescue, as in Exercises 3.2.9 and 3.5.2:

$$\sum_{n\in\mathbb{Z}} e^{2\pi\sqrt{-1}\,nx} e^{-4\pi^2 n^2 t} = \sum_{n\in\mathbb{Z}} (4\pi t)^{-1/2} e^{-(x-n)^2/4t}$$

is the elementary solution $p(t,x)$ of the heat equation $\partial w/\partial t = \partial^2 w/\partial x^2$ on the circle $0 \le x < 1$ of perimeter 1. At $x = 1/2$, it reduces to

$$\sum_{n\in\mathbb{Z}} (-1)^n e^{-4\pi^2 n^2 t} = 2 \sum_{\text{odd } n\ge 1} (4\pi t)^{-1/2} e^{-n^2/16t},$$

and this is positive, increasing from 0 (at the right) to 1 (at the left) as t runs from 0 to $+\infty$. Now take $t = x^2/2\pi^2$ to confirm that

$$\lim_{N\uparrow\infty} \mathbb{P}\big(\sqrt{N}\,\mathbf{D}_N \le x\big) = \frac{\sqrt{2\pi}}{x} \sum_{\text{odd } n\ge 1} e^{-n^2\pi^2/8x^2}$$

is indeed, a distribution function – not that it really looks so even now.

5.3 CLT in general

Here, $\mathbf{x}_1$, $\mathbf{x}_2$, etc. are independent copies of $\mathbf{x} = \mathbf{x}_1$, subject to the sole condition $\mathbb{E}(\mathbf{x}^2) < \infty$. Write $\mathbb{E}(\mathbf{x}) = m$ and $\mathbb{E}(\mathbf{x} - m)^2 = \sigma^2$. Then, with $\mathbf{s}_n = \mathbf{x}_1 + \cdots + \mathbf{x}_n$ as usual, you have

$$\lim_{n\uparrow\infty} \mathbb{P}\left[a \le \frac{\mathbf{s}_n - nm}{\sigma\sqrt{n}} < b \right] = (\text{what else?}) \int_a^b \frac{e^{-x^2/2}}{\sqrt{2\pi}}\,dx,$$

this being a weak law *only*, as for Bernoulli trials. The fact goes back to A. de Moivre (1667–1754), as noted in §2.2, and was extended to a vague but more inclusive statement by Gauss and Laplace. It was put a somewhat better technical footing by P. L. Chebyshev (1860) and A. A. Markov (1890), but as Poincaré complained, *"Tout le monde y croit (à la loi des erreurs) parce que les mathématiciens s'imaginent que c'est un fait d'observation et les observateurs que c'est un théorème de mathématiques"*. The missing ingredient was an unambiguous concept of independence, supplied by Steinhaus (1930). I will make two proofs.

The conventional proof, by Fourier transform, comes first. But as is not uncommon, if you reflect upon what Fourier is saying, you can get to the bottom of the thing more directly, as will be seen in the second proof.

Aside. LLN and CLT, together, provide a general picture of the sums $\mathbf{s}_n$ as in

$$\mathbf{s}_n = \text{the bulk effect } n \times \mathbb{E}(\mathbf{x}) \text{ on the scale } n$$
$$+ \text{ Gaussian fluctuations on the scale } \sqrt{n},$$

and it is natural to ask if there may not be further corrections, beyond CLT, on the more delicate scales $n^0 = 1$, $n^{-1/2}$, and so on. The answer is *no*. Already, CLT is only a *weak* law, on the level of probabilities, and you cannot expect anything more than *numerical* corrections to that, without, so to say, *any philosophical content*.

5.3.1 Conventional proof

Take $m = 0$ and $\sigma = 1$: a mere rescaling to simplify the writing. You have

$$\mathbb{E}\big[\mathrm{e}^{\sqrt{-1}\,k\mathbf{s}_n/\sqrt{n}}\big] = \mathbb{E}\left[\prod_{j=1}^{n} \mathrm{e}^{\sqrt{-1}\,k\mathbf{x}_j/\sqrt{n}}\right]$$

$$= \big[\mathbb{E}\big(\mathrm{e}^{\sqrt{-1}\,k\mathbf{x}/\sqrt{n}}\big)\big]^n \qquad \text{by independence.}$$

Here, $\mathbb{E}(\mathrm{e}^{\sqrt{-1}\,k\mathbf{x}}) \equiv \varphi(k)$ is of class $C^2(\mathbb{R})$, as you will easily check using $\mathbb{E}(\mathbf{x})^2 < \infty$ to provide the necessary control; also $\varphi(0) = 1$, $\varphi'(0) = 0$, and $\varphi''(0) = 1$, so for fixed k, $\varphi\big(\frac{k}{\sqrt{n}}\big) = 1 - \frac{k^2}{2n} + \mathrm{o}\big(\frac{1}{n}\big)$, and the prior display reduces to

$$\mathbb{E}\big[\mathrm{e}^{\sqrt{-1}\,k\mathbf{s}_n/\sqrt{n}}\big] = \Big[\varphi\Big(\frac{k}{\sqrt{n}}\Big)\Big]^n \simeq \mathrm{e}^{-k^2/2},$$

with convergence under the fixed (absolute) bound ≤ 1. Now take $f \in C_\downarrow^\infty$, written as $\frac{1}{2\pi}\int \mathrm{e}^{-\sqrt{-1}\,kx}\hat{f}(k)\,\mathrm{d}k$ with (automatically)

$$\hat{f}(k) = \int \mathrm{e}^{\sqrt{-1}\,kx} f(x)\,\mathrm{d}x$$

of class $C_\downarrow^\infty$. Then for large n,

$$\mathbb{E}\left[f\Big(\frac{\mathbf{s}_n}{\sqrt{n}}\Big)\right] = \frac{1}{2\pi}\int \mathbb{E}\big(\mathrm{e}^{\sqrt{-1}\,k\mathbf{s}_n/\sqrt{n}}\big)\hat{f}(k)\,\mathrm{d}k \;\simeq\; \frac{1}{2\pi}\int \mathrm{e}^{-k^2/2}\hat{f}(k)\,\mathrm{d}k$$

$$= \int \frac{\mathrm{e}^{-x^2/2}}{\sqrt{2\pi}}\, f(x)\,\mathrm{d}x,$$

which you will justify please, and with such functions, f_- and f_+ as in the Figure 5.3.1, it appears that

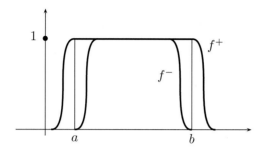

Figure 5.3.1

$$\mathbb{E}\left[f_-\left(\frac{\mathbf{s}_n}{\sqrt{n}}\right)\right] \simeq \int \frac{\mathrm{e}^{-x^2/2}}{\sqrt{2\pi}}\, f_-$$

$$\leq \mathbb{P}\left[a \leq \frac{\mathbf{s}_n}{\sqrt{n}} < b\right]$$

$$\leq \mathbb{E}\left[f_+\left(\frac{\mathbf{s}_n}{\sqrt{n}}\right)\right]$$

$$\simeq \int \frac{\mathrm{e}^{-x^2/2}}{\sqrt{2\pi}}\, f_+.$$

But both $\int \frac{\mathrm{e}^{-x^2/2}}{\sqrt{2\pi}}\, f_\pm$ can be made as close to $\int_a^b \frac{\mathrm{e}^{-x^2/2}}{\sqrt{2\pi}}$ as you like, so $\mathbb{P}\big[a \leq \mathbf{s}_n/\sqrt{n} < b\big]$, caught between $\mathbb{E}(f_-)$ and $\mathbb{E}(f_+)$, has no choice but to do what it should, just as in §4.5. Implicit in the proof is the finiteness of a and b, which is easily removed by the crude estimate $\mathbb{P}(|\mathbf{s}_n/\sqrt{n}| \geq c) \leq 1/c^2$. You see how efficient the Fourier transform really is, despite the complaint of §1.7.4.

5.3.2 Second proof, adapted from H. Trotter [1959]

Coming more directly to the point, what you really need to know is that

$$\lim_{n\uparrow\infty} \mathbb{E}\left[f\left(\frac{\mathbf{s}_n}{\sqrt{n}}\right)\right] = \int_{-\infty}^{+\infty} \frac{\mathrm{e}^{-x^2/2}}{\sqrt{2\pi}}\, f(x)\,\mathrm{d}x$$

for "enough" functions f, e.g. for compact f of class $C^2(\mathbb{R})$. Here's how to do it with no fuss. Fix f, let $\mathbf{y}_n : n \geq 1$ be independent with common density $e^{-x^2/2}/\sqrt{2\pi}$, and let these $\mathbf{y}$s be independent of the $\mathbf{x}$s, too. Now $n^{-1/2}(\mathbf{y}_1 + \cdots + \mathbf{y}_n)$ has the *same* Gaussian density, so

$$\mathbb{E}\left[f\left(\frac{\mathbf{S}_n}{\sqrt{n}}\right)\right] - \int_{-\infty}^{+\infty} \frac{e^{-x^2/2}}{\sqrt{2\pi}} f(x)\,\mathrm{d}x$$

$$= \mathbb{E}\left[f\left(\frac{\mathbf{x}_1 + \cdots + \mathbf{x}_n}{\sqrt{n}}\right)\right] - \mathbb{E}\left[f\left(\frac{\mathbf{y}_1 + \cdots + \mathbf{y}_n}{\sqrt{n}}\right)\right],$$

which may be written as a telescoping sum

$$\sum_{k=1}^{n} \mathbb{E}\left[f\left(\frac{\mathbf{x}_1 + \cdots + \mathbf{x}_k + \mathbf{y}_{k+1} + \cdots + \mathbf{y}_n}{\sqrt{n}}\right)\right.$$

$$\left. - f\left(\frac{\mathbf{x}_1 + \cdots + \mathbf{x}_{k-1} + \mathbf{y}_k + \cdots + \mathbf{y}_n}{\sqrt{n}}\right)\right]$$

$$= \sum_{k=1}^{n} \mathbb{E}\left[f\left(\mathbf{z}_k + \frac{\mathbf{x}_k}{\sqrt{n}}\right) - f\left(\mathbf{z}_k + \frac{\mathbf{y}_k}{\sqrt{n}}\right)\right]$$

with

$$\mathbf{z}_k = n^{-1/2}(\mathbf{x}_1 + \cdots + \mathbf{x}_{k-1} + \mathbf{y}_{k+1} + \cdots + \mathbf{y}_n) \quad \begin{array}{l} \text{independent} \\ \text{of both } \mathbf{x}_k \text{ and } \mathbf{y}_k. \end{array}$$

Here, the typical $f(\mathbf{z} + \mathbf{x}/\sqrt{n})$ may be expressed as

$$f(\mathbf{z}) + f'(\mathbf{z})\,\frac{\mathbf{x}}{\sqrt{n}} + \frac{1}{2}\,f''(\mathbf{z})\,\frac{\mathbf{x}^2}{n} \quad \text{with an (absolute) error} \leq \omega\left(\frac{|\mathbf{x}|}{\sqrt{n}}\right) \times \frac{\mathbf{x}^2}{2n},$$

ω being the modulus of continuity of f''. Now take the expectation, first over $\mathbf{x}$ with the (independent) variable $\mathbf{z}$ fixed, and then over $\mathbf{z}$ itself. This leaves only $\mathbb{E}\left[f(\mathbf{z}) + \frac{1}{2n}f''(\mathbf{z})\right]$ up to an (absolute) error not more than

$$\frac{1}{2n}\mathbb{E}\left[\omega\left(\frac{\mathbf{x}}{\sqrt{n}}\right)\mathbf{x}^2\right] = \mathrm{o}\left(\frac{1}{n}\right) \quad \begin{array}{l} \text{since } \omega \text{ is bounded by } 2\max|f''|, \\ \omega(0+) = 0, \text{ and } \mathbf{x}^2 \text{ dominates.} \end{array}$$

The same is true of the typical $f(\mathbf{z} + \mathbf{y}/\sqrt{n})$, so there is a cancellation in each term of the telescoping sum, leaving after summation only a cumulative error comparable to $n \times \mathrm{o}\left(\frac{1}{n}\right) = \mathrm{o}(1)$. That's all there is to it: I mean that the excursion from f to $\hat{f}$ and back was not needed at all. This type of proof goes back to Khinchine [1933:1–8].

★**Exercise 5.3.1**   CLT may also hold for independent trials with variable distributions. H. Cramér's "primes" of Exercise 1.8.13 provide a simple example. There, $\mathbf{e}_n = 0$ or 1 with $\mathbb{P}(\mathbf{e}_n = 1) = 1/\log n$, and for $\#(n) = $ the number of "primes"$\leq n$, you have

$$m \equiv \mathbb{E}(\#) = \sum_{p \leq n} \frac{1}{\log n} \simeq \frac{n}{\log n} \quad \text{and} \quad \mathbb{E}(\# - m)^2 \simeq m.$$

Prove that

$$\lim_{n \uparrow \infty} \mathbb{P}\left[\frac{\# - m}{\sqrt{m}} \leq x\right] = \int_{-\infty}^{x} \frac{e^{-y^2/2}}{\sqrt{2\pi}}\,dy.$$

Exercise 5.3.2 after Billingsley [1979:370]. Think about $\mathbf{s}_n = \mathbf{x}_1 + \cdots + \mathbf{x}_n$ with independent Poisson-distributed $\mathbf{x}$s: $\mathbb{P}(\mathbf{x} = k) = 1/ek!$. Then $\mathbf{s}_n$ is Poisson-distributed with parameter n, as in $\mathbb{P}(\mathbf{s}_n = k) = n^k e^{-n}/k!$, and $(\mathbf{s}_n - n)/\sqrt{n}$ obeys CLT. Now compute $\mathbb{E}(n - \mathbf{s}_n)^+$ exactly and justify

$$\lim_{n \uparrow \infty} \mathbb{E}\left[\frac{(n - \mathbf{s}_n)^+}{\sqrt{n}}\right] = \int_0^{\infty} x \frac{e^{-x^2/2}}{\sqrt{2\pi}}\,dx = \frac{1}{\sqrt{2\pi}}.$$

You will see Stirling's formula coming out. Cute, no?

Exercise 5.3.3 Let $\mathbb{E}(\mathbf{x}_1^2)$ and $\mathbb{E}(\mathbf{x}_2^2)$ be finite. Show that if $\mathbf{x}_1$ and $\mathbf{x}_2$ are independent and likewise $\mathbf{x}_1 + \mathbf{x}_2$ and $\mathbf{x}_1 - \mathbf{x}_2$, then both $\mathbf{x}_1$ and $\mathbf{x}_2$ have a (common) Gaussian distribution. The fact is due to Kac [1959b]; compare the article on Maxwell's distribution in §5.9.4.
Hint: $2\mathbf{x}_1$ is distributed like two copies of $\mathbf{x}_1$ + copies of $\mathbf{x}_2$ and of $-\mathbf{x}_2$, all copies independent, and similarly for $2\mathbf{x}_2$. Repeat and apply CLT four times.

5.3.3 Errors

The discrepancy between $\mathbb{P}(\mathbf{s}_n/\sqrt{n} \leq x)$ and $(2\pi)^{-1/2} \int_{-\infty}^{x} e^{-y^2/2}\,dy$ depends upon how nice the distribution of $\mathbf{x}$ may be. For the standard random walk,

$$\mathbb{P}_0(\mathbf{s}_{2n} = 0) = \binom{2n}{n} 2^{-2n} \simeq \frac{1}{\sqrt{\pi n}},$$

so

$$\left|\mathbb{P}_0(\mathbf{s}_{2n} \leq 0) - \int_{-\infty}^{0} \frac{e^{-x^2/2}}{\sqrt{2\pi}}\,dx\right| \simeq \frac{1}{2}\frac{1}{\sqrt{\pi n}},$$

as you know from §2.2 already. That's pretty much the worst that can happen: indeed, it is a deep fact, due to Berry [1941] and Esseen [1942], that if $\mathbb{E}(\mathbf{x}) = 0$ and $\mathbb{E}(\mathbf{x}^2) = 1$, then the discrepancy

$$\max_{x \in \mathbb{R}} \left| \mathbb{P}\left[\frac{\mathbf{s}_n}{\sqrt{n}} \leq x \right] - \frac{1}{\sqrt{2\pi}} \int_{-\infty}^{x} e^{-y^2/2} \, dy \right| \quad \text{is not more than} \quad \frac{3}{\sqrt{n}} \mathbb{E}|\mathbf{x}|^3.$$

I do not prove this, but see Feller [1966(2):542–546]. The discrepancy vanishes rapidly as $n \uparrow \infty$ if $\mathbf{x}$ has a density and $\mathbb{E}(\mathbf{x}^{2n}) < \infty$ for every $n \geq 1$, for which also see Feller [1966(2):535].

5.4* The local limit

Take $\mathbb{E}(\mathbf{x}) = 0$ and $\mathbb{E}(\mathbf{x}^2) = 1$, as before. Then $\mathbb{P}(\mathbf{s}_n/\sqrt{n} \leq x)$ approximates $\frac{1}{\sqrt{2\pi}} \int_{-\infty}^{x} e^{-y^2/2} \, dy$ as $n \uparrow \infty$, but what about the corresponding densities $p_n = \sqrt{n}(p * \cdots * p)(\sqrt{n}\, x)$, assuming $\mathbf{x}_1$ *has* a density – that would be $p_1 = p$. Let's suppose that this density is

(1) symmetrical about $x = 0$,
(2) of class $C^1(\mathbb{R})$ with
(3) $\int |p'| < \infty$ and
(4) $\int x^4 p < \infty$,

just to convey the idea easily.

Then for $n \uparrow \infty$,

$$p_n(x) = \frac{e^{-x^2/2}}{\sqrt{2\pi}} \times \left[1 + \frac{c}{n}(x^4 - 6x^2 + 3) \right]$$

$$\text{with } c = \frac{1}{24} \mathbb{E}(\mathbf{x}^4) - \frac{1}{8} + \text{an error } o\left(\frac{1}{n}\right),$$

independently of $x \in \mathbb{R}$. That's the "local limit theorem". Notwithstanding the complaint of §1.7.4, the Fourier transform is indispensable here.

Proof. $(x^4 - 6x^2 + 3)e^{-x^2/2}$ is the fourth derivative of $e^{-x^2/2}$, so the discrepancy between p_n and its stated approximation is

$$\frac{1}{2\pi} \int_{-\infty}^{+\infty} e^{\sqrt{-1}\,kx} \left[\left(\hat{p}\left(\frac{k}{\sqrt{n}}\right) \right)^n - e^{-k^2/2}\left(1 + \frac{c}{n}k^4 \right) \right] dk,$$

and the task is to estimate what's inside the brackets. Fix a small number $h > 0$. By (1),

$$|\hat{p}(k)|^2 = |\mathbb{E}(\cos kx)|^2 \leq \mathbb{E}(\cos^2 kx) < 1 \quad \text{for } k \neq 0;$$

also, by (2) and (3), $\hat{p}(k) = \hat{p}'(k)\sqrt{-1}\,k$ vanishes like $1/k$ at $\pm\infty$, so $|\hat{p}(k)| \leq \vartheta < 1$ for $|k|$ bigger than a fixed number $h > 0$, and

$$\int_{|k| \geq h\sqrt{n}} \left| \hat{p}\left(\frac{k}{\sqrt{n}}\right) \right|^n = \sqrt{n} \int_{|k| \geq h} |\hat{p}(k)|^n \leq \sqrt{n}\,\vartheta^{n-2} \int_{|k| \geq h} |\hat{p}|^2$$

can be neglected, i.e. if it suffices to look at the restricted range $|k| \leq h\sqrt{n}$. There, $k/\sqrt{n}$ is small, and

$$\hat{p}\left(\frac{k}{\sqrt{n}}\right) = \mathbb{E}\left[\cos\left(\frac{k\mathbf{x}}{\sqrt{n}}\right) \right]$$

$$= \mathbb{E}\left[1 - \frac{k^2\mathbf{x}^2}{2n} + \frac{1}{24}\frac{k^4\mathbf{x}^4}{n^2} - \frac{1}{24}\frac{k^4\mathbf{x}^4}{n^2}(1 - \cos\xi) \right]$$

by choice of ξ between 0 and $k\mathbf{x}/\sqrt{n}$

$$= 1 - \frac{k^2}{2n} + \frac{1}{24}\mathbb{E}(\mathbf{x}^4)\frac{k^4}{n^2} \times [1 + o(1)] \qquad \text{by (4)}.$$

Here also $\hat{p}(k/\sqrt{n})$ is real and positive, and

$$\log \hat{p}\left(\frac{k}{\sqrt{n}}\right) = -\frac{k^2}{2n} + \frac{1}{24}\mathbb{E}(\mathbf{x}^4)\frac{k^4}{n^2} - \frac{1}{2}\left(\frac{k^2}{2n}\right)^2 \times [1 + o(1)]$$

$$= -\frac{k^2}{2n} + \left[\frac{1}{24}\mathbb{E}(\mathbf{x}^4) - \frac{1}{8} + o(1) \right]\frac{k^4}{n^2}$$

$$= -\frac{k^2}{2n} + [c + o(1)]\frac{k^4}{n^2}\,,$$

as you will check; in particular, $\log \hat{p}(k/\sqrt{n}) < -k^2/3n$ by choice h, permitting the range $|k| \geq \sqrt{6\log n}$ to be neglected entirely, in comparison to the promised error $o(1/n)$. In the remaining range, $|k|$ is not more than $\sqrt{6\log n}$, so $k/\sqrt{n}$ is really small and

$$\left| \left[\hat{p}\left(\frac{k}{\sqrt{n}}\right) \right]^n - e^{-k^2/2}\left(1 + c\frac{k^4}{n}\right) \right| = e^{-k^2/2}\left| e^{-[c+o(1)]k^4/n} - 1 - c\frac{k^4}{n} \right|$$

$$= o\left(\frac{1}{n}\right) \times k^4 e^{-k^2/2}$$

since k^4/n is also small, i.e. the whole error is comparable to $o(1/n) \times \int k^4 e^{-k^2/2}\,dk$. That does it, but see Feller [1966(2): 533–536] for more detailed information of this type.

5.5 Figures of merit

Let q be a probability density with mean 0 and mean-square 1, say, and let's ask how close it is to the Gaussian density $p(x) = (2\pi)^{-1/2}e^{-x^2/2}$, and by what "figure of merit" is the discrepancy to be measured? Obviously, there are innumerable ways to do this. The figures of merit described here seem to be the best, having a natural statistical meaning and a simple, often computable form.

5.5.1 Gibbs's lemma and entropy

Look at the function $x \log x - x + 1$ seen already in §2.4.1 and now again

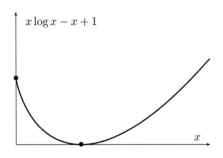

Figure 5.5.1

in Figure 5.5.1: it is convex and also positive, except at $x = 1$ where it vanishes. That's what I call "Gibbs's lemma", and here's the kind of thing Gibbs did with it: For any density q with $\int x^2 q = 1$,

$$0 \le \int \left(\frac{q}{p}\log\frac{q}{p} - \frac{q}{p} + 1\right) \times p = \int q \log \frac{q}{p} \equiv I(q) \le +\infty$$

vanishes only if $q = p$. In short, it is a figure of merit, comparable to the function $I(q)$ of §2.4.1. The present $I(q)$ is now spelled out:

$$+\infty \ge I(q) = \int q \log q + \int p\left(\log \sqrt{2\pi} + \frac{x^2}{2}\right) = \int q \log q - \int p \log p \ge 0,$$

from which three things are apparent: write $H(q) = -\int q \log q$. Then evidently

(1) this integral really exists if the value $-\infty$ be permitted,
(2) $H(q) \le H(p) = \log \sqrt{2\pi e}$, and
(3) $H(q) = H(p)$ only if $q = p$;

in short, H is a "figure of merit".

Boltzmann (1860) introduced a variant of H into statistical mechanics where it plays the role of the thermodynamical entropy. There and more generally too, it has a subtle statistical meaning as an indicator of *disorder*, having two complementary aspects *vis-à-vis* the experimental observation of the value x of the variable $\mathbf{x}$ underlying the density q: *after* the trial, some amount of *information* is obtained; *before* the trial, the observer finds himself in some degree of uncertainty as to what will come out; and these should be the same. Think now of $H(p) = \log\sqrt{2\pi e}$ as the maximal amount of information/uncertainty for any variable with $\mathbb{E}(\mathbf{x}^2) = 1$ and think of $H(p) - H(q)$ as the degree to which the actual information/uncertainty of the experiment falls short of the ideal. In this language, CLT may be paraphrased as saying that for $\mathbb{E}(\mathbf{x}) = 0$, $\mathbb{E}(\mathbf{x}^2) = 1$, and a large number of independent trials, $n^{-1/2}(\mathbf{x}_1 + \cdots + \mathbf{x}_n)$ is becoming as disorderly as it possibly can. Naturally, this is all too vague, but I ask you to be patient until §9.1 in which information/uncertainty is fully explained. Better yet, take a look at §9.1 now if you have not already done so. As to CLT, Linnik [1949] showed that *either* $H_n = H[2^{-n/2}(\mathbf{x}_1 + \cdots + \mathbf{x}_{2^n})]$ is already close to $\log\sqrt{2\pi e}$ *or else* H_{n+1} is substantially larger, leading to what is surely the most complicated proof of CLT known to man, but also, perhaps, the most interesting. I don't reproduce it – it's too hard.

Exercise 5.5.1 Compute the entropies of the standard densities of §1.7.5:

(1) $\dfrac{e^{-x^2/2\sigma^2}}{\sigma\sqrt{2\pi}}$;

(2) $\dfrac{b}{\pi}\dfrac{1}{b^2 + x^2}$; and

(3) $\dfrac{xe^{-x^2/2t}}{\sqrt{2\pi t^3}}$.

Each of these is reproduced under convolution, with addition of the relevant parameter $\gamma = \sigma^2$ for (1), b for (2), and x for (3). Check that for n independent trials with $\gamma = 1$, $H = $ some constant $+ (\frac{1}{2}, 1, \text{ or } 2) \times \log n$ for (1), (2), (3) respectively. In short, for large n, the information/uncertainty doubles once and then once more as you go down the list.

5.5.2 Fisher's information

The English statistician R. A. Fisher (1930) introduced another figure of merit, closely related to H and I, *to wit,*

$$F(q) = \int \frac{(q' + xq)^2}{q} = \int \frac{(q')^2}{q} + 2 \int xq' + \int x^2 q = \int \frac{(q')^2}{q} - 1,$$

vanishing only if $q' + xq = 0$, which is to say $\log q = -\frac{x^2}{2} + \text{a constant}$, or what is the same, $q(x) = (2\pi)^{-1/2} e^{-x^2/2}$.

Actually, "information" is not quite the right word for F. It is rather, an "information production rate" as will now be explained. Update the underlying variable $\mathbf{x}$ as per $\mathbf{x} \to \mathbf{x}(t) = e^{-t}\mathbf{x} + \sqrt{1 - e^{-2t}}\mathbf{w}$, in which $\mathbf{w}$ is an independent Gaussian variable with $\mathbb{E}(\mathbf{w}) = 0$ and $\mathbb{E}(\mathbf{w}^2) = 1$. Then $\mathbb{E}(\mathbf{x}) = 0$ and $\mathbb{E}(\mathbf{x}^2) = 1$ still obtain; $\mathbf{x}(t)$ has the updated density

$$Q(t, x) = \int \frac{e^{-(x - e^{-t}y)^2/2\sigma^2}}{\sigma \sqrt{2\pi}} Q_0(y)\, dy \quad \text{with } \sigma^2 = 1 - e^{-2t} \text{ and } Q_0 = q;$$

and it's easy to check by hand that

$$\frac{\partial Q}{\partial t} = \frac{\partial^2 Q}{\partial x^2} + \frac{\partial(xQ)}{\partial x} \equiv KQ.$$

Now compute the "entropy production rate":

$$\begin{aligned}
\frac{d}{dt} H(Q) &= -\int KQ \times \log Q - \int KQ \\
&= \int (Q' + xQ) \frac{Q'}{Q} = \int \frac{(Q')^2}{Q} - 1 \\
&= F(Q).
\end{aligned}$$

I don't know any proof of CLT based solely upon F, but see §§11.5.3 and 11.7.3 where it turns up again.

5.5.3 Log Sobolev

Of these two figures of merit: $I(Q) = \log \sqrt{2\pi e} - H(Q)$ and $F(Q) = \int Q^{-1}(Q')^2 - 1$, it's pretty obvious that Fisher's is the more delicate, and there is the nice "logarithmic Sobolev inequality" $I(Q) \leq F(Q)/2$ to substantiate this. Technicalities aside, the proof is easy: with Q as above, $\mathrm{d}H/\mathrm{d}t = F$ as you know already, and

$$
\begin{aligned}
\frac{\mathrm{d}F(Q)}{\mathrm{d}t} &= 2\int \frac{Q'}{Q}(Q' + xQ)'' - \int \left(\frac{Q'}{Q}\right)^2 (Q' + xQ)' \\
&= -2\int (\log Q)''(Q'' + xQ' + Q) + 2\int \frac{Q'}{Q}(\log Q)''(Q' + xQ) \\
&= -2\int (\log Q)'' \left(\frac{Q''}{Q} - \left(\frac{Q'}{Q}\right)^2\right)Q - 2\int (\log Q)''Q \\
&= -2\int (\log Q)''^2 Q + 2\int \frac{(Q')^2}{Q} \\
&\leq -2\left(\int \frac{(Q')^2}{Q}\right)^2 + 2\int \frac{(Q')^2}{Q} \\
&\qquad\qquad \text{since } \int (\log Q)''^2 Q \int Q \geq \left(\int (\log Q)''Q\right)^2 \\
&\leq -2\int \frac{(Q')^2}{Q} + 2 \qquad \text{since } -2x^2 + 2x \leq -2x + 2, \\
&= -2F(Q),
\end{aligned}
$$

whence $F(Q) \leq \mathrm{e}^{-2t}F(Q_0)$ and $I(Q_0) = \int_0^\infty F(Q)\,\mathrm{d}t \leq F(Q_0)/2$, as advertised.

5.6* Gauss is prime

You know from §1.7 that

$$
\frac{\mathrm{e}^{-x^2/2\alpha^2}}{\alpha\sqrt{2\pi}}\,\frac{\mathrm{e}^{-x^2/2\beta^2}}{\beta\sqrt{2\pi}} = \frac{\mathrm{e}^{-x^2/2\gamma^2}}{\gamma\sqrt{2\pi}} \quad \text{with } \gamma = \sqrt{\alpha^2 + \beta^2},
$$

which is to say Gauss+independent Gauss = Gauss. P. Lévy (1930) conjectured and Cramér [1936] proved the remarkable fact that the converse is also true: If the sum of independent variables is Gaussian, then the individual variables are Gaussian, too, i.e. Gauss is "prime" if I may say it so. This seems natural in view of Gibbs's lemma: for fixed mean-square, Gauss has maximal information/uncertainty, and how could that be if the individual variables did not have maximal information/uncertainty, too. There *must* be a proof like that, but till now it has been necessary to use the Fourier transform – in the complex plane already. Very efficient to be sure, but providing *no statistical understanding at all!* Odd.

Proof [1]. The (independent) variables are $\mathbf{x}_1$ and $\mathbf{x}_2$ and the sum $\mathbf{x}_1 + \mathbf{x}_2$ is distributed by $(2\pi)^{-1/2}e^{-x^2/2}$, say. Then

$$e^{k^2/2} = \mathbb{E}(e^{k\mathbf{x}_1}) \times \mathbb{E}(e^{k\mathbf{x}_2}) \equiv f_1(k) \times f_2(k),$$

and this may be used, first for real and then for complex k, to prove that both f_1 and f_2 are root-free integral functions of $k \in \mathbb{C}$, of limited growth as in $|f(k)| \leq Ae^{B|k|^2}$. For example, with real k and $\mathbb{P}(|\mathbf{x}_2| \leq L) \geq 1/2$, you have

$$e^{k^2/2} = \left|\mathbb{E}(e^{k\mathbf{x}_1})\,\mathbb{E}(e^{k\mathbf{x}_2})\right| \geq |f_1(k)|e^{-|k|L} \times \frac{1}{2}.$$

Note also that the reciprocals, of either f_1 or f_2, have the same type of limited growth, as per $|1/f_1| = |f_2 e^{-k^2/2}| \leq Ae^{B|k|^2}$ with new constants A and B. (These keep changing during the proof.) A single-valued branch of $\log f = \log|f| + \sqrt{-1}\,$phase $f \equiv h + \sqrt{-1}\,\theta$ may now be chosen in the whole plane with $|h| \leq A + Br^2$, and a like estimate of θ would mean that $|\log f| \leq A + Br^2$, too. Then $\log f$ would be a polynomial in k of degree ≤ 2, and you would have $f(k) = e^{A + Bk + Ck^2}$ with

(1) $A = 0$ since $f(0) = 1$,

(2) $B = 0$ if $f'(0) = \mathbb{E}(\mathbf{x}) = 0$ which you may also assume, and

(3) $2C = f''(0) = \mathbb{E}(\mathbf{x}^2) > 0$,

so that $\mathbb{E}(e^{\sqrt{-1}\,k\mathbf{x}}) = e^{-Ck^2}$, as promised. Now for the estimate of θ. This function is related to $h = \log|f|$ by the Cauchy–Riemann equations, so $|\text{grad}\,h| = |\text{grad}\,\theta|$, and it would be enough to know that $|\text{grad}\,h| \leq A + Br$. To obtain this estimate, let h' be the derivative of h in some fixed direction e. Like h, this function is harmonic, so by the mean-value property of such functions,

$$h'(0) = \frac{1}{2\pi}\int_0^{2\pi} h'(re^{\sqrt{-1}\,\theta})\,d\theta \quad \text{for any radius } r$$

$$= \frac{1}{\pi R^2}\int_{|x|\leq R} h'(x)\,d\text{ area} \quad \begin{array}{l}\text{by averaging}\\ \text{from } r = 0 \text{ to } r = R.\end{array}$$

Inspection of the figure and a little thought leads easily to the estimate

$$|h'(0)| \leq \frac{1}{\pi R^2} \times 2 \max_{|k|=R}|h(k)| \times \text{the diameter } 2R$$

---

[1] If you need it, you will find the necessary complex function theory in Ahlfors [1979:208]. The proof is an instance of Hadamard's product theorem, so-called, but I prefer to do it by hand in a simple way.

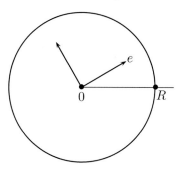

Figure 5.6.1

or, what is wanted here,

$$|\operatorname{grad} h(0)| \leq \frac{8}{\pi R} \times \max_{|k|=R} |h(k)|.$$

The origin is now displaced to a point of the circle $|k| = R$, and the same principle is applied to produce, from $|h| \leq A + BR^2$, an estimate of the same type for $\operatorname{grad} h$, *viz.*

$$\max_{|k|=R} |\operatorname{grad} h| \leq \frac{8}{\pi R} \max_{|k|\leq 2R} |h| \leq CR \quad \text{for large } R,$$

which is all you really need.

Exercise 5.6.1 The variable $\mathbf{x}$, with whole number values $n = 0, 1, 2, 3$ etc., is Poisson-distributed, with parameter $c > 0$, i.e. $\mathbb{P}(\mathbf{x} = n) = \frac{c^n}{n!} e^{-c} \equiv p_c(n)$. Obviously, the sum of two such independent variables is also Poisson, as per

$$p_a * p_b(n) = \sum_{i+j=n} p_a(i) p_b(j) = p_c(n) \quad \text{with } c = a + b;$$

and conversely, just as for Gauss: If $q_1 * q_2$ is Poisson with parameter c, then $q_1 = p_a$ and $q_2 = p_b$ for some $a + b = c$, i.e. Poisson is also prime. Details please.

5.7 The general iterated log

The result dates back to W. Döblin (1938). The conditions are the same as for CLT: with $\mathbb{E}(\mathbf{x}) = 0$ and $\mathbb{E}(\mathbf{x}^2) = 1$,

$$\mathbb{P}\left[\lim_{\substack{\sup \\ \inf}} \frac{\mathbf{x}_1 + \cdots + \mathbf{x}_n}{\sqrt{2n\log\log n}} = \pm 1\right] = 1,$$

just as for RW(1). The conventional proof runs along the lines of §2.3, but the estimates required are not easy in this generality, so I prefer to wait until §7.6 where a slick proof can be made with the help of a better machinery, *to wit*, the Brownian motion.

5.8 Sparre-Andersen's combinatorial method and the general arcsine law

It came as a surprise when Erdős–Kac [1947] showed that the arcsine law of §3.4 has little to do with RW(1): in fact, it holds for any random walk $\mathbf{s}_0 = 0$, $\mathbf{s}_n = \mathbf{x}_1 + \cdots + \mathbf{x}_n$, $n \geq 0$, with independent copies of a symmetrically distributed variable $\mathbf{x}$ and $\mathbb{P}(\mathbf{s}_n \geq 0) = \mathbb{P}(\mathbf{s}_n < 0) = 1/2$. More surprising still was the discovery of the Danish actuary E. Sparre-Andersen [1953, 1954] that it has little to do with probability at all. Really, it is a purely combinatorial business, as I will explain, following Feller [1966(2): Chapter 12] where a fuller account of this circle of ideas will be found.

5.8.1 Sparre-Andersen's combinatorial lemma

Fix real numbers $x_1, \ldots, x_n$ and form the usual partial sums $s_0 = 0$, $s_k = x_1 + \cdots + x_k$, $k \leq n$. The numbers x may be permuted, producing different sums, and a little probability space may be made by declaring each of the $n!$ permutations π of the n "letters" $x_1, \ldots, x_n$ to be equally likely. In this language, it is the remarkable discovery of Sparre-Andersen that the number $(\#/\#^*)$ of positive/non-negative sums among $s_1, \ldots, s_n$ is distributed the same as the index m/m^* of the first/last maximum among the sums $s_0 = 0, s_1, \ldots, s_n$.

Proof. For $n = 1$, $\# = m$ and $\#^* = m^*$ as in the table.

	#	m	$\#^*$	m^*
$x_1 < 0$	0	0	0	0
$x_1 > 0$	1	1	1	1
$x_1 = 0$	0	0	1	1

The rest of the proof is by induction on n.

Case 1. $x_1 + \cdots + x_n < 0$. Here, $m \le m^* < n$ and also $\# \le \#^* < n$, so it suffices to deal with $\mathbb{P}(m = \ell)$ for $\ell < n$. Now with a self-evident notation,

$$\mathbb{P}(m = \ell)$$

$$= \frac{1}{n!} \sum_\pi \mathbf{1}_{m=\ell}(\pi)$$

$$= \frac{1}{n!} \sum_\pi \mathbf{1}_{m=\ell}(\pi'\pi) \qquad \begin{array}{l} \text{for any permutation } \pi' \\ \text{of the first } n-1 \text{ letters} \end{array}$$

$$= \frac{1}{n!} \sum_\pi \frac{1}{(n-1)!} \sum_{\pi'} \mathbf{1}_{m=\ell}(\pi'\pi) \qquad \begin{array}{l} \text{by induction, thinking of } \pi \text{ as fixed} \\ \text{and } \pi' \text{ permuting the first } n-1 \\ \text{of the letters } x_{\pi 1}, x_{\pi 2}, \ldots, x_{\pi n} \end{array}$$

$$= \frac{1}{n!} \sum_\pi \mathbf{1}_{\#=\ell}(\pi) \qquad \text{by reversal of steps}$$

$$= \mathbb{P}(\# = \ell),$$

and similarly for m^* and $\#^*$.

Case 2. $x_1 + \cdots + x_n > 0$. Now $-x_1 - \cdots - x_n > 0$ and Case 1 tells you that the index of the first/last minimum is distributed like the number of negative/non-positive sums. But then, e.g., $\# =$ the number of positive sums, *viz.* n minus the number of non-positive sums, is distributed like n minus the index of the last minimum, and this, in turn, is distributed like the index of the first maximum, as may be seen by reading $x_1, \ldots, x_n$ backwards. The example for $n = 5$ in Figure 5.8.1 clarifies this.

Exercise 5.8.1   Check it out.

Case 3. $x_1 + \cdots + x_n = 0$. $\#$ and m are both $< n$, as in Case 1, and you may check that they are equidistributed in the same way as before. Here also $-x_1 - \cdots - x_n = 0$ and the reasoning of Case 2 confirms that the following are distributed alike:

  (1) the number of negative sums and the index of the first min,
  (2) the number of non-negative sums and $n -$ the index of the first min, and

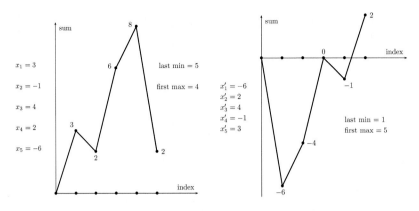

Figure 5.8.1

(3) n − the index of the first min and the index of the last max, by reversal.

The induction is finished.

Aside. The proof is slick but much is hidden – and how, I ask you, was the fact discovered? Sparre-Andersen's proof is more complicated but also more instructive; see Feller [1966(2)]. It is all quite remarkable.

5.8.2 Application to random walk

Now let $\mathbf{s}_0 = 0$ and $\mathbf{s}_n = \mathbf{x}_1 + \cdots + \mathbf{x}_n$ $(n \geq 1)$ with common distribution of the (independent) steps, and write $\mathbf{x} = (\mathbf{x}_1, \ldots, \mathbf{x}_n)$. Obviously, $\#(\mathbf{x})$ = the number of positive sums among $\mathbf{s}_1, \ldots, \mathbf{s}_n$ is (statistically speaking) insensitive to permutations of $\mathbf{x}$, and likewise the index $m(\mathbf{x})$ of the first maximum among $\mathbf{s}_0, \ldots, \mathbf{s}_n$, so Sparre-Andersen's combinatorial lemma implies

$$\mathbb{P}\left[\#(\mathbf{x}) = \ell\right] = \mathbb{P}\left[\#(\pi\mathbf{x}) = \ell\right] \qquad \text{for any permutation } \pi$$

$$= \mathbb{E}\left[\frac{1}{n!}\sum_\pi \mathbf{1}_\ell(\#(\pi\mathbf{x}))\right]$$

$$= \mathbb{E}\left[\frac{1}{n!}\sum_\pi \mathbf{1}_\ell(m(\pi\mathbf{x}))\right]$$

$$= \mathbb{P}\left[m(\mathbf{x}) = \ell\right]$$

$$= \mathbb{P}\left[\mathbf{s}_1, \ldots, \mathbf{s}_{\ell-1} < \mathbf{s}_\ell \text{ and } \mathbf{s}_{\ell+1}, \ldots, \mathbf{s}_n \leq \mathbf{s}_\ell\right].$$

This looks hopeful à propos of arcsine if only these probabilities could be evaluated.

Sparre-Andersen's identity does that for you. It reads:

$$\sum_{n=0}^{\infty} \gamma^n \, \mathbb{P}(\mathbf{s}_1, \ldots, \mathbf{s}_n \geq 0) = \exp\left[\sum_{n=1}^{\infty} \frac{\gamma^n}{n} \mathbb{P}(\mathbf{s}_n \geq 0)\right], \quad 0 \leq \gamma < 1,$$

the point being that the probabilities to the left are usually hard to compute while those to the right are, hopefully, more simple. The proof is postponed in favor of the application: if $\mathbb{P}(\mathbf{s}_n \geq 0) = 1/2$, then the right-hand exponent is $\frac{1}{2}\sum_{n=1}^{\infty} \frac{\gamma^n}{n} = -\log\sqrt{1-\gamma}$, so

$$\sum_{n=0}^{\infty} \gamma^n \, \mathbb{P}(\mathbf{s}_1, \ldots, \mathbf{s}_n \geq 0) = \frac{1}{\sqrt{1-\gamma}} = \sum_{0}^{\infty} \gamma^n \binom{2n}{n} 2^{-2n},$$

and with the additional assumptions $\mathbb{P}(\mathbf{s}_n \neq 0) = 1$ and symmetrical distribution of steps, you find

$$\mathbb{P}\big[\#(k \leq n : \mathbf{s}_k > 0) = \ell\big] = \mathbb{P}\big[\#(\mathbf{x}) = \ell\big]$$

$$= \mathbb{P}\big[\mathbf{s}_1, \ldots, \mathbf{s}_{\ell-1} > 0\big) \times \mathbb{P}\big[\mathbf{s}_1, \ldots, \mathbf{s}_{n-\ell} > 0\big) \quad \text{by symmetry}$$

$$= \binom{2\ell-2}{\ell-1} 2^{-2\ell+2} \binom{2n-2\ell}{n-\ell} 2^{-2n+2\ell} \quad \text{by Sparre-Andersen}$$

$$\simeq \frac{1}{\pi} \frac{1}{\sqrt{\ell(n-\ell)}} \quad \text{by Stirling}$$

if both ℓ and $n - \ell$ are large, confirming the ubiquity of the arcsine law: in detail, for $0 < a < b < 1$,

$$\mathbb{P}\left[a \leq \frac{1}{n}\#(k \leq n : \mathbf{s}_k > 0) < b\right] \simeq \sum_{a \leq \ell/n < b} \frac{1}{\pi} \frac{1}{\sqrt{\frac{\ell}{n}\left(1 - \frac{\ell}{n}\right)}} \times \frac{1}{n}$$

$$\simeq \frac{1}{\pi} \int_a^b \frac{\mathrm{d}x}{\sqrt{x(1-x)}}.$$

Here line 3 of the prior display looks wrong in comparison to §3.4, but it is not so: for RW(1), $\mathbb{P}(\mathbf{s}_n = 0)$ vanishes for odd, but not for even n.

5.8.3 Spitzer's identity

Sparre-Andersen's proof of his identity is combinatorial. It's difficult but see Feller [19662]:418] for a simpler way. For variety, I re-derive it from the deeper identity of Spitzer [1956], proved here by the pretty method of Wendel [1958]. The identity reads

$$\sum_{n=0}^{\infty} \mathbb{P}(\mathbf{s}_1, \ldots, \mathbf{s}_n \geq 0, \mathbf{s}_n \leq x) = \exp\left[\sum_{n=1}^{\infty} \frac{\gamma^n}{n} \mathbb{P}(\mathbf{s}_n \geq 0, \mathbf{s}_n \leq x)\right],$$

subject to interpretation: the left-hand side means just what it says, but the exponential at the right is taken in the convolution algebra of (sub-)distribution functions F with $F(-\infty) = 0$ and $F(\infty) \leq 1$, i.e. with $F^n = F * \cdots * F$ and $\exp(F) = \sum_{n=0}^{\infty} F^n/n! \equiv e^F$ for short, F^0 being the identity

$$I(x) = \begin{array}{ll} 0 & x < 0 \\ 1 & x \geq 0. \end{array}$$

Besides exp, introduce also the logarithm $-\log(I - F) \equiv \sum_{n=1}^{\infty} n^{-1} F^n$ in the case $F(\infty) < 1$ and the projection

$$F \to F^+(x)) = \begin{array}{ll} 0 & x < 0 \\ F(x) & x \geq 0. \end{array}$$

Then

$$e^{F^+} - I = e^{F^+ - F} - I + (I - e^{-F})e^{F^+},$$

by inspection, and projection produces

$$e^{F^+} - I = \left[(I - e^{-F})F^+\right]^+$$

since $F^+ - F$ lives on $x < 0$. Now substitute $-\log(I - F)$ in place of F. Then $I - e^{-F}$ is replaced by F itself, and you have

$$G \equiv e^{-(\log(I-F))^+} = I + \left[Fe^{-(\log(I-F))^+}\right]^+ = (I + FG)^+$$

$$= I + F^+ + (FF^+)^+ + (F(FF^+)^+)^+ + \text{etc.},$$

by iteration, making good sense in the convolution algebra if $F(\infty) < 1$. This identity is now applied to the step distribution F of the walk, with a factor $0 \leq \gamma < 1$ attached to guarantee it all makes sense, and out comes

$$\sum_{n=0}^{\infty} \gamma^n (F \cdots (FF^+)^+)^+ \text{ } n\text{-fold} = \exp\left[\sum_{n=1}^{\infty} \frac{\gamma^n}{n} (F^n)^+\right],$$

which is to be interpreted. The meaning of $(F^n)^+$ is pretty obvious: it is just the (sub-)distribution $\mathbb{P}(\mathbf{s}_n \leq x, \mathbf{s}_n \geq 0)$; also

$$(FF^+)^+ = \mathbb{P}(\mathbf{s}_2 \leq x, \mathbf{s}_1 \text{ and } \mathbf{s}_2 \geq 0),$$

$$(F(FF^+)^+)^+ = \mathbb{P}(\mathbf{s}_3 \leq x, \mathbf{s}_1, \mathbf{s}_2 \text{ and } \mathbf{s}_3 \geq 0), \quad \text{and so on.}$$

Spitzer's identity now stands before you, and Sparre-Andersen's identity follows by evaluation at $x = +\infty$, this being a homomorphism of the convolution algebra into the real numbers. The exponential is then taken in its naive numerical sense, *to wit*,

$$\sum_{n=0}^{\infty} \mathbb{P}(\mathbf{s}_1, \ldots, \mathbf{s}_n \geq 0) = \exp \left| \sum_{n=1}^{\infty} \frac{\gamma^n}{n} \mathbb{P}(\mathbf{s}_n \geq 0 \right|$$

as stated before.

5.9 CLT in dimensions 2 or more

This is an easy extension of §5.3 once the Gaussian distribution in several dimensions is understood.

5.9.1 Gaussian variables

The variable $\mathbf{x}$ with values $x \in \mathbb{R}^d$, $\mathbb{E}(\mathbf{x}) = 0$ and $\mathbb{E}(\mathbf{x}^2) < \infty$ is said to be Gaussian if it has a density (relative to the d-dimensional volume element) of the form $p(x) = Z^{-1} \exp\left(-\frac{1}{2} x \cdot K^{-1} x\right)$ in which K is a $d \times d$, symmetric, positive (-definite) matrix and Z is a normalizer to make $\int p \, d\,\mathrm{vol} = 1$. You can compute Z by means of the substitution $x = \sqrt{K} \, x'$ with the symmetric, positive root of K. The quadratic form $x \cdot K^{-1} x$ reduces to $(x')^2$, the Jacobian $|\partial x / \partial x'|$ is $\det \sqrt{K}$, and

$$Z = \int e^{-\frac{1}{2} x' \cdot K^{-1} x} \, d\,\mathrm{vol} = \int e^{-\frac{1}{2}(x')^2} \, d\,\mathrm{vol}' \times \det \sqrt{K} = (2\pi)^{d/2} \sqrt{\det K}.$$

The same device identifies K as the "correlation matrix" of $\mathbf{x}$: with the abbreviation $[x_i x_j : 1 \leq i; j \leq d] = x \otimes x$, you have $x \otimes x = \sqrt{K} \, x' \otimes x' \sqrt{K}$, so

$$\mathbb{E}(\mathbf{x} \otimes \mathbf{x}) = \sqrt{K} \, \mathbb{E}(\mathbf{x}' \otimes x') \sqrt{K} = \sqrt{K} \times \text{the identity} \times \sqrt{K} = K.$$

Exercise 5.9.1   Compute in the same style

$$\hat{p}(k) = \int e^{-\sqrt{-1} \, k \cdot x} p(x) \, d\,\mathrm{vol} = e^{-\frac{1}{2} k \cdot K k}.$$

Exercise 5.9.2   Let $\mathbf{x}_k$ ($k \leq n$) be independent Gaussian variables with common density $(2\pi)^{-1/2} e^{-x^2/2}$, conditioned so that $\mathbf{s}_n = \mathbf{x}_1 + \cdots + \mathbf{x}_n = 0$. The free (unconditioned) sums have correlation $K_{ij} = \mathbb{E}(\mathbf{s}_i \mathbf{s}_j) = \min(i, j)$, $1 \leq i, j \leq n$. The tied (conditioned) sums are also Gaussian. What is *their* correlation? *Answer*: For $i \leq j$, $K_{ij} = i(1 - j/n)$,

$1 \leq i \leq j \leq n$.

Hints: K_{ij} is not changed by permuting $\mathbf{x}_k$, $k \leq i$, and so reduces to $i\,\mathbb{E}(\mathbf{x}_1\mathbf{s}_j|\mathbf{s}_n = 0)$. The same idea will show that it is a constant $C \times i(n - j)$. Now find the conditional density for $\mathbf{x}_1$, and compute $\mathbb{E}(\mathbf{x}_1^2|\mathbf{s}_n = 0) = (n - 1)C$.

Exercise 5.9.3   With the free variables, find the correlations of the sums $\mathbf{s}'_k = \mathbf{s}_k - (k/n)\mathbf{s}_n$, $1 \leq k \leq n$. What does that tell you? These sums will come into play in §6.8 where the Kolmogorov–Smirnov statistics are discussed more carefully than in §5.2.

5.9.2 Gauss and independence

It is an immediate consequence of the form of the density

$$p(x) = \frac{1}{Z}e^{-\frac{1}{2}\,x \cdot K^{-1}x}$$

that if the variables $\mathbf{x}'_i$, $i \leq a$, and $\mathbf{x}''_j$, $j \leq b$, are jointly Gaussian but uncorrelated then they are independent. The correlations (K) are seen in Figure 5.9.1, and as K^{-1} has the same shape, so the quadratic form

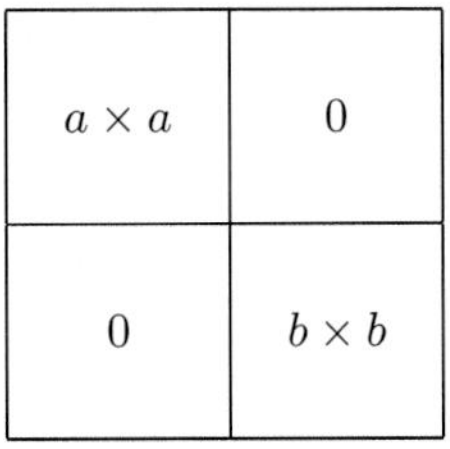

$a \times a$	0
0	$b \times b$

Figure 5.9.1

$x \cdot Kx$ splits in two, one part involving only the x's, the other only the x''s, so that the joint density $p(x)$ factors accordingly, into $p(x') \times p(x'')$, which is what independence is all about anyhow. This is remarkable: for example, if $\mathbb{E}(\mathbf{x}_1\mathbf{x}_2) = 0$ then also $\mathbb{E}(\mathbf{x}_1^4\mathbf{x}_2^3) = \mathbb{E}(\mathbf{x}_1^4)\,\mathbb{E}(\mathbf{x}_2^3)$ and the like. More remarkable still, it is only the Gaussian distribution that has this extraordinarily convenient property. The proof is easy: $\mathbb{E}(\mathbf{x}) = 0$ and $\mathbb{E}(\mathbf{x}^2) < \infty$ are assumed. Reduce to $K =$ the identity by the substitution $\mathbf{x} = \sqrt{K}\mathbf{x}'$, assuming also $\det K \neq 0$. $\mathbb{E}(\mathbf{x}'_1\mathbf{x}'_2) = 0$ is said to imply $\mathbf{x}'_1$ and $\mathbf{x}'_2$ are independent, and likewise for $\mathbf{x}'_1 \pm \mathbf{x}'_2$ since $\mathbb{E}(\mathbf{x}'_1 - \mathbf{x}'_2)(\mathbf{x}'_1 + \mathbf{x}'_2) = 0$ as well, so $\mathbf{x}'_1$ and $\mathbf{x}'_2$ are individually Gaussian by Exercise 5.3.3. That does it.

5.9.3 CLT itself

The correct statement is now self-evident: if $\mathbb{E}(\mathbf{x}) = 0$ and $\mathbb{E}(\mathbf{x}^2) < \infty$, and if also $K = \mathbb{E}(\mathbf{x} \otimes \mathbf{x})$ is positive, indicating that $\mathbf{x}$ is genuinely d-dimensional, then

$$\lim_{n\uparrow\infty} \mathbb{P}\left[\frac{\mathbf{s}_n}{\sqrt{n}} \in C\right] = \int_C \frac{e^{-\frac{1}{2}x \cdot K^{-1}x}}{(2\pi)^{d/2}\sqrt{\det K}} \, d\,\mathrm{vol}$$

for any reasonable figure $C \subset \mathbb{R}^d$. The proof is much the same as in §5.3 in either version. Think it over.

Example 5.9.4 RW(3) provides the simplest example. For paths starting at $\mathbf{x}(0) = 0$, $\mathbf{x}(n)$ is the sum of n independent copies of the variable $\mathbf{e}$ with values $(\pm 1\,0\,0)$, $(0 \pm 1\,0)$ and $(0\,0 \pm 1)$ with $\mathbb{E}(\mathbf{e} \otimes \mathbf{e}) = \frac{1}{3} \times$ the identity, so $\mathbf{x}(n)/\sqrt{n}$ is approximately Gaussian with density $(2\pi/3)^{-3/2} \exp(-3x^2/2)$, validating, in part, the discussion of the angular part of RW(3) in §4.2.

Example 5.9.5 $\mathbf{s}_n$ is now the sum of n independent copies of the variable $\mathbf{e}$ uniformly distributed on the surface of the unit sphere in $\mathbb{R}^3$. $\mathbb{E}(\mathbf{e}) = 0$ since $-\mathbf{e}$ is a copy of $\mathbf{e}$. Check that $\mathbb{E}(\mathbf{e} \otimes \mathbf{e}) = \frac{1}{3} \times$ the identity *without* computation. CLT now provides the solution to Lord Rayleigh's "problem of random flights" (1890) illustrated in Figure 5.9.2.

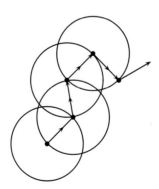

Figure 5.9.2

Compute also $\Phi(k) = \mathbb{E}(e^{\sqrt{-1}\,k \cdot \mathbf{e}})$ and check that $[\Phi(k/\sqrt{n})]^n \simeq e^{-k^2/6}$ for large n, as it should be.

Example 5.9.6 Make yourself a little universe inside a big ball of radius R in $\mathbb{R}^3$ by placing there N stars of unit mass, uniformly and independently distributed, with fixed density $D = N/\frac{4}{3}\pi R^3$. The force **f** acting on a test star placed at the origin is then the sum of $G \times \mathbf{e}/\mathbf{r}^2$, $\mathbf{x} = (\mathbf{r}, \mathbf{e})$ being the position of a typical star in spherical polar coordinates, and G the gravitational constant. $\mathbb{E}(\mathbf{e}/\mathbf{r}^2) = 0$, by symmetry, but $\mathbb{E}(\mathbf{e}^2/\mathbf{r}^4) = +\infty$, so one cannot expect any variant of CLT to work. But what *does* **f** look like for large N? Let's compute. In spherical polars, you have:

$$
\mathbb{E}\left[e^{\sqrt{-1}\,k\cdot\mathbf{f}}\right]
$$

$$
= \frac{1}{\frac{4}{3}\pi R^3} \int_0^R r^2\,dr \int_0^{2\pi} d\theta \int_0^\pi e^{\sqrt{-1}\,|k|\cos\varphi/r^2} \sin\varphi\,d\varphi
$$

$$
= \frac{3}{R^3} \int_0^R r^2\,dr \frac{\sin(|k|/r^2)}{|k|/r^2}
$$

$$
= 1 - \frac{3|k|^{3/2}}{2R^3} \int_{|k|/R^2}^\infty \left(1 - \frac{\sin x}{x}\right) \frac{dx}{x^{5/2}}, \quad \text{after putting } x = |k|/r^2.
$$

Here, $3/R^3 = 4\pi D/N$ and the density D is fixed, so for large N

$$
\mathbb{E}\left[e^{\sqrt{-1}\,k\cdot\mathbf{f}}\right] \simeq e^{-C|k|^{3/2}} \quad \text{with } C = 2\pi D \times \int_0^\infty \left(1 - \frac{\sin x}{x}\right) \frac{dx}{x^{5/2}}.
$$

★**Exercise 5.9.7** This is not all like the $e^{-k^2/2}$ per Gauss. Taking $G = 1$ for simplicity, the corresponding density, relative to the 3-dimensional value element $4\pi r^2\,dr$, may be written as

$$
p(x) = \frac{1}{2\pi^2} \int_0^\infty \frac{\sin kr}{kr} e^{-k^{3/2}} k^2\,dk \quad \text{with } r = |x|,
$$

and judicious partial integrations show that $p(x)$ is proportional to $|x|^{-7/2}$ far out. Then the radial density $4\pi|x|^2 \times p$ is then proportional to $|x|^{-5/2}$ and has neither a finite mean square nor even a finite mean. Pretty bad. Details please.

5.9.4 Maxwell's distribution

J. C. Maxwell (1859) encountered Gauss's law *à propos* of the distribution of velocities in a gas of like molecules of mass $m = 1$, say, in thermal (i.e. in statistical) equilibrium. The answer is very pretty, as is Maxwell's proof. If, as it is natural to suppose, the velocities in the three coordinate directions are independent, and if the distribution has a nice density, then this can only be

$$p(v) = \frac{e^{-v^2/2kT}}{(2\pi kT)^{3/2}},$$

with the interpretation, coming from the ideal gas law, of T as Kelvin's absolute temperature and k as Boltzmann's constant $(1.38+) \times 10^{-16} \times (\mathrm{cm})^2(\sec)^{-2}(\deg)^{-1}$ to get the numbers and the dimensions right.

Maxwell's line of thinking is based on the obvious remark that God does not care what coordinate frame is used: that is only an artefact of the description, not anything in Nature. Then $p(v) = p_1(v_1) \times p_2(v_2) \times p_3(v_3)$ (by independence) must be rotation-invariant; $p_1 = p_2 = p_3 \equiv p_0$ since, e.g. $v_1 v_2 v_3 \to v_2 v_1 v_3$ is a rotation; and $p_0(v_1) = p_0(-v_1)$ since $v_1 v_2 v_3 \to -v_1 v_2 v_3$ is a rotation, too, so with $p_0(x) = q(x^2)$, you find

$$q(v_1^2)q(v_2^2)q(v_3^2) = p(v) = q(v_1^2 + v_2^2 + v_3^2)q(0)q(0),$$

which is to say

$$\frac{q(v_1^2)}{q(0)} \times \frac{q(v_2^2)}{q(0)} \times \frac{q(v_1^3)}{q(0)} = \frac{q(v_1^2 + v_2^2 + v_3^2)}{q(0)},$$

of which the only permissible solution is $q(x^2) = A \exp(-Bx^2)$ with positive constants A and B, leading at once to $p(v) = Z^{-1} \exp(-v^2/2\,kT)$ with a positive constant kT and the correct normalizer $Z = (2\pi kT)^{3/2}$ so as to make $\int p\,d\,\mathrm{vol} = 1$.

Maxwell now resorts to the mechanical definition of pressure as the average momentum (per unit time, per unit area) imparted, by the molecules of the gas, to an imaginary plane, such as $x_3 = 0$, in position-space $\mathbb{R}^3$.

Let N be the number of molecules in the gas, these being more or less evenly spread out in some big but not unlimited vessel of volume V. Then the number of molecules in the slant cylinder seen in Figure 5.9.3, which hit the shaded patch of $x_3 = 0$ in a short time B is (more or less)

$$\left(\frac{N}{V} = \text{the molecular density}\right) \times (-v_3 AB = \text{the volume of the cylinder}),$$

A being the area of the patch. Each of these imparts to the plane an impulse $-v_3 \times$ the molecular mass $(= 1)$, once for coming, twice for departing after a perfect reflection, for a total of $-2v_3$, so the pressure is

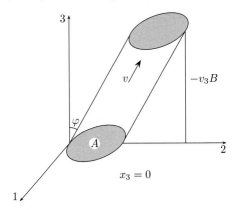

Figure 5.9.3

or ought to be

$$P = \int_{v_3<0} \frac{N}{V} \times (-v_3) \times (-2v_3) \frac{e^{-\frac{1}{2}v^2/2kT}}{(2\pi kT)^{3/2}} \, d\,\text{vol}$$

$$= \text{half the integral over } \mathbb{R}^3 \text{ of the same}$$

$$= \frac{1}{2}\frac{N}{V} \int_0^\infty r^2 \, dr \int_0^{2\pi} d\theta \int_0^\pi \sin\varphi \, d\varphi \; r^2 \cos^2\varphi \frac{e^{-r^2/2kT}}{(2\pi kT)^{3/2}}$$

$$\text{since } -v_3 = |v|\cos\varphi \text{ as per the figure}$$

$$= \frac{N}{V} kT \qquad \text{when you work it out.}$$

This is nothing but the ideal gas law $PV = NkT$, valid for ordinary gases when the density N/V is low and the temperature T is moderate so that the molecules do not collide too much. Maxwell's interpretation of k and T is dictated by that.

5.10 Measure in dimension $+\infty$

It is always amusing to meet Gauss's law in unexpected places. I ask you now: What is a sensible volume element in $\mathbb{R}^\infty$? Nothing like the ∞-dimensional Lebesgue measure $d^\infty x$ will do: for example, if $\mathbb{R}^\infty$ is provided with the Pythagorean distance $|x| = \sqrt{x_1^2 + x_2^2 + \text{etc.}}$, and if (unaccountably) $\text{vol}(|x| \leq 1) = 1$, then $\text{vol}(|x| \leq r)$ ought to be $= 0$ or $+\infty$ according as $r < 1$ or $r > 1$ – and worse – $|x| \leq 1$ is covered by countably many copies of $|x| \leq 1/2$, so $\text{vol}(|x| \leq 1)$ ought to vanish.

Too bad. Could it help to drop $|x| < \infty$ and look at the whole of $\mathbb{R}^\infty$? Obviously, no: it only makes trouble. Maybe it helps to make the space *smaller* and *nonlinear*. We'll see.

5.10.1 A better way

Now try this. Look at the $(n - 1)$-dimensional sphere of radius $\sqrt{n}$: $S^{n-1}(\sqrt{n}) = (x : x_1^2 + \cdots + x_n^2 = n)$, provided with the "round" volume element inherited from the ambient space $\mathbb{R}^n$, normalized to make the total volume 1, and let's see what happens as the dimension tends to $+\infty$. Guided by the picture, you find that Gauss comes in! In fact, for

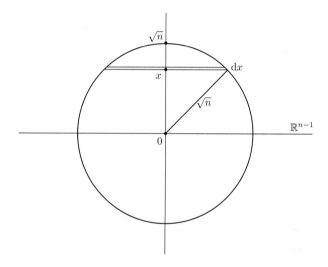

Figure 5.10.1

any finite $a < b$,

$$\mathbb{P}(a \le \mathbf{x}_1 < b) = \frac{\int_a^b (n - x^2)^{n/2-1} \, \mathrm{d}x}{\int_{-\sqrt{n}}^{\sqrt{n}} (n - x^2)^{n/2-1} \, \mathrm{d}x}$$

$$= \frac{\int_a^b (1 - \frac{x^2}{n})^{n/2-1} \, \mathrm{d}x}{\int_{-\sqrt{n}}^{\sqrt{n}} (1 - \frac{x^2}{n})^{n/2-1} \, \mathrm{d}x}$$

$$\simeq \int_a^b \frac{\mathrm{e}^{-x^2/2}}{\sqrt{2\pi}} \, \mathrm{d}x \quad \text{for large } n,$$

taking account of the estimates $0 \le 1 - \frac{x^2}{n} \le e^{-x^2/n}$ for $|x| \le \sqrt{n}$ and $(1 - \frac{x^2}{n})^{n/2-1} \ge e^{-x^2/2}$ for $|x| \le$ some fixed c. Indeed, more is found at little or no extra cost. The volume element is invariant under rotations of the sphere, so for fixed $m < n$, $k = (k_1, \dots, k_m, 0, \dots, 0)$, and an m-dimensional rotation taking k to $|k| \times (1, 0, \dots, 0)$, you have

$$\mathbb{E}(e^{\sqrt{-1}\,k \cdot \mathbf{x}}) = \mathbb{E}(e^{\sqrt{-1}\,|k|\mathbf{x}_1}) \simeq \int e^{\sqrt{-1}\,|k|x} \frac{e^{-x^2/2}}{\sqrt{2\pi}}\,dk = e^{-k^2/2},$$

which is to say that as $n \uparrow \infty$, the round probabilities $d\,\mathbb{P}_n$ on $S^{n-1}(\sqrt{n})$ tend to the joint distribution $d\,\mathbb{P} = \prod_{n=1}^{\infty} \frac{e^{-x_n^2/2}}{\sqrt{2\pi}}\,dx_n$ of independent Gaussian variables on some figure "$S^{\infty}(\sqrt{\infty})$" $\subset \mathbb{R}^{\infty}$. Then $d\,\mathbb{P}$ is the "round" volume element on this "∞-dimensional sphere of radius $\sqrt{\infty}$" (what ever that may be). This entertaining observation, usually attributed to Poincaré (1900) is due in fact to Mehler [1866]; compare McKean [1975] for part of what Mehler did and a bit more besides.

Aside.   The temptation to write $d\,\mathbb{P} = (2\pi)^{-\infty/2} \times e^{-x^2/2} \times d^{\infty}x$ is to be avoided: $(2\pi)^{-\infty/2}$ is silly, so is $x^2/2$ since $\mathbb{P}(\mathbf{x}^2 = +\infty) = 1$, and likewise $d^{\infty}x$ as you already know. In short, $(2\pi)^{-\infty/2}e^{-x^2/2}\,d^{\infty}x$ is not to be taken apart, though the whole is perfectly kosher and even a reliable guide to computation.

But what is this $S^{\infty}(\sqrt{\infty})$ where $\mathbb{P}$ lives? $\mathbb{P}\big[\lim_{n\uparrow\infty} \frac{1}{n}(\mathbf{x}_1^2 + \cdots + \mathbf{x}_n^2) = 1\big] = 1$ by LLN, so $S^{\infty}(\sqrt{\infty})$ lies within the linear subspace of $\mathbb{R}^{\infty}$ where

$$|\mathbf{x}| = \sup_{n \ge 1} \sqrt{n^{-1}(x_1^2 + \cdots + x_n^2)} < \infty.$$

That's pretty rough. What more?

5.10.2 Curvature

Well the radius $\sqrt{\infty}$ is big, so $S^{\infty}(\sqrt{\infty})$ is pretty flat – but how flat is that? To get a first idea, fix $y \in \mathbb{R}^{\infty}$ with $|y| = \sqrt{y_1^2 + y_2^2 + \text{etc.}} < \infty$ and look at the distribution $d\mathbb{Q} = \prod_{n=1}^{\infty}(2\pi)^{-1/2}e^{-(x_n-y_n)^2/2}\,dx_n$ of $\mathbf{x} + y$. This is or ought to be $d\mathbb{Q} = e^{\mathbf{x} \cdot y - y^2/2}\,d\,\mathbb{P}$, and that makes perfect sense: $\mathbf{x} \cdot y = \lim_{n\uparrow\infty}(\mathbf{x}_1y_1 + \mathbf{x}_2y_2 + \cdots + \mathbf{x}_ny_n)$ exists with probability 1 by Exercise 1.8.5, so $e^{\mathbf{x} \cdot y - y^2/2}$ must be the Radon–Nikodym derivative $d\mathbb{Q}/d\,\mathbb{P}$, the moral being that a "small" translation, with $|y| < +\infty$, cannot move $\mathbf{x}$ off $S^{\infty}(\sqrt{\infty})$. But what about a "big" translation, with $|y| = +\infty$? Then $\mathbf{x} \cdot y$ is divergent by Exercise 1.8.8, and $\mathbb{Q}$, which is a perfectly nice measure in itself, vanishes on $S^{\infty}(\sqrt{\infty})$, which is to say that $\mathbf{x}$ translated by y is moved off the sphere to some place else. In short, $S^\infty(\sqrt{\infty})$ is *not* wholly flat but has some curvature to it.

To see that dQ now vanishes on $S^\infty(\sqrt{\infty})$, introduce the fair game $\mathbf{z}_n = \exp[\mathbf{x}_1 y_1 + \cdots + \mathbf{x}_n y_n - (1/2)(y_1^2 + \cdots + y_n^2)]$. Obviously,

$$\int (\mathbf{z}_n)^{1/2} \, d\mathbb{P} = \int e^{\frac{1}{2}(\mathbf{x}_1 y_1 - \frac{1}{4} y_1^2 + \text{etc.})} \, d\mathbb{P} = e^{-(y_1^2 + \cdots + y_n^2)/8} \downarrow 0 \text{ as } n \uparrow \infty,$$

so $\mathbb{P}[\lim_{n \uparrow \infty} \mathbf{z}_n = 0] = 1$. Now write $d\mathbb{Q} = f \, d\mathbb{P}$ plus a part vanishing on $S^\infty(\sqrt{\infty})$. Then, with a self-evident notation,

$$\int (\mathbf{z}_n)^{1/2} \, d\mathbb{P} = \int (\mathbf{z}_n)^{1/2} \, d\mathbb{P}_n = \int (\mathbf{z}_n)^{-1/2} \, d\mathbb{Q}_n = \int (\mathbf{z}_n)^{-1/2} \, d\mathbb{Q}$$

$$\geq \int (\mathbf{z}_n)^{-1/2} f \, d\mathbb{P},$$

so

$$0 = \lim_{n \uparrow \infty} \int (\mathbf{z}_n)^{1/2} \, d\mathbb{P} \geq \int \lim_{n \uparrow \infty} (\mathbf{z}_n)^{-1/2} f \, d\mathbb{P} \quad \text{by Fatou's lemma,}$$

and $\mathbb{P}(f = 0) = 1$. That's the trick.

5.10.3 A more delicate description

$S^\infty(\sqrt{\infty})$ is part of $\mathbb{R}^\infty$ to be sure, but its tangent space is like the Pythagorean subspace where $\sup_{n \geq 1} \frac{1}{n}(x_1^2 + \cdots + x_n^2)$ is finite. More delicately,

$$\mathbb{P}\big[\mathbf{x}_1^2 + \cdots + \mathbf{x}_n^2 = n + \text{an absolute error} \leq \sqrt{5n \log n} \text{ as } n \uparrow \infty\big] = 1,$$

showing decisively that $S^\infty(\sqrt{\infty})$ is *not* a linear space.

Proof. Fix $\alpha > 0$. Then, with $0 < \beta < 1/2$, you find

$$\mathbb{P}\big[\mathbf{x}_1^2 + \cdots + \mathbf{x}_n^2 - n > \alpha\big] = \mathbb{P}\big[\beta(\mathbf{x}_1^2 + \cdots + \mathbf{x}_n^2) > \beta(n + \alpha)\big]$$

$$\leq \mathbb{E}\Big[e^{\beta(\mathbf{x}_1^2 + \cdots + \mathbf{x}_n^2)}\Big] \times e^{-\beta(n+\alpha)}$$

$$= (1 - 2\beta)^{-n/2} e^{-\beta(n+\alpha)}$$

$$= \Big(1 + \frac{\alpha}{n}\Big)^{n/2} e^{-\alpha/2} \quad \begin{array}{l} \text{by choice of the best} \\ \beta, \text{ namely } \frac{1}{2}\frac{\alpha}{\alpha+n} \end{array}$$

$$\simeq e^{-\alpha^2/4n} \quad \begin{array}{l} \text{by inspection of logarithms} \\ \text{if } \alpha = o(n) \end{array}$$

$$\leq n^{-5/4} \quad \text{by choice of } \alpha = \sqrt{5n \log n}.$$

That's half the battle.

Exercise 5.10.1   Finish the proof by a similar estimate of

$$\mathbb{P}[\mathbf{x}_1^2 + \cdots + \mathbf{x}_n^2 - n < -\alpha] = \mathbb{P}[-\beta(\mathbf{x}_1^2 + \cdots + \mathbf{x}_n^2) > \beta(\alpha - n)].$$

This queer $S^{\infty}(\sqrt{\infty})$ will be seen again, in §§6.11 and 11.7.7.

5.11 Prime numbers

Remarkably enough, LLN and CLT enter into the arithmetic of the prime numbers $p = 2, 3, 5$ etc. Recall, from Exercise 1.8.13, the conjecture of Gauss, that the number $\pi(n)$ of primes $p \leq n$ is about $n/\log n$, and let's begin with an elementary demonstration of that, adapted from the lovely book *What is Mathematics?* by Courant–Robbins [1951:482–486].

Fix a prime p and count the number of times p^k divides n. Obviously, it's about n/p^k since p^k divides only p^k, $2p^k$, $3p^k$ etc. Then the number of times p^k divides n but p^{k+1} does not is about $(n/p^k) \times \left(1 - \frac{1}{p}\right)$, and the total number of times p divides $n!$ is roughly

$$1 \times \left(\frac{n}{p} - \frac{n}{p^2}\right) + 2 \times \left(\frac{n}{p^2} - \frac{n}{p^3}\right) + 3 \times \left(\frac{n}{p^3} - \frac{n}{p^4}\right) + \text{etc.}$$

$$\simeq n \times \left(\frac{1}{1 - 1/p} - 1\right)$$

$$= \frac{n}{p - 1},$$

permitting the approximate expression of the logarithm of $n!$ as

$$\log n! \simeq \sum_{p \leq n} \frac{n}{p - 1} \log p.$$

Now you know already that $\log n! \simeq n \log n$, by Stirling's approximation, and if you now *assume* that

$$\pi(n) \simeq \int_2^n w(x)\,dx$$

with a nice, positive and as it should be, decreasing function w, then you could write

$$n \log n \simeq \sum_{p \leq n} \frac{n}{p - 1} \log p \simeq n \times \int_2^n \frac{\log x}{x - 1} w(x)\,dx$$

and *compute* $w(x) \simeq 1/\log x$ by "differentiation" with respect to n after cancellation of an n, right and left.

Of course, that is naive. The primes are quite erratically distributed among the whole numbers, so that the *existence* of such a weight w does not come for free. Indeed, the first real proof did not appear until

1890 or so, and it was not until some 60 years later that Selberg [1949] and Erdős [1949] discovered a purely arithmetical, elementary proof, avoiding the heavy complex-function machinery employed before and immediately objected to as being inappropriate in a purely arithmetical question[2] – and it must be said that the usual tools of probability did not help either. But they *do* help for *additive* number-theoretical functions such as $f(n) =$ the number of distinct prime divisors of the whole number n, *additive* meaning that $f(nm) = f(n) + f(m)$ if n and m have no common factor. This special function will serve to illustrate how LLN and CLT enter into arithmetic.

To begin with, probability enters if you think of the *variable* **n** as being uniformly distributed over the whole numbers $\mathbb{N} = 1, 2, 3$ etc. The probability of the event $Z \subset \mathbb{N}$ is declared to be the "density":

$$D(Z) = \lim_{N \uparrow \infty} \frac{1}{N} \#(n \le N : n \in Z),$$

provided this limit exists, and the class of such events gets a name: $\mathfrak{Z}$. *But slowly on.* Densities are *not* probabilities seeing as $D(\mathbb{N}) = 1$ is not the same as $D(1) + D(2) + D(3)$ etc. $= 0$ – *and worse.* It may be that A and B belong to $\mathfrak{Z}$ but $A \cap B$ does not, i.e. $\mathfrak{Z}$ is *not* a field.

Exercise 5.11.1 Make an example to that effect.
Hint: Let A be the even numbers $2, 4, 6$ etc. with density $1/2$ and let B be made of long, alternating stretches of (consecutive) even/odd numbers. Draw a picture of $A \cap B$ and you will see what to do.

These unfortunate facts mean that our customary tools, such as the two Borel–Cantelli lemmas, don't apply any more, so that most things have to be done *by hand*.

Now back to the special function $f(n)$. Write $e_p(n) = 1$ or 0 according as p divides n or not, so that $f(n) = \sum_{p \le n} e_p(n)$. Obviously,

$$\lim_{n \uparrow \infty} \frac{1}{n} \sum_{1}^{n} e_p(k) = \frac{1}{p} \qquad \text{for any prime } p$$

and

$$\lim_{n \uparrow \infty} \frac{1}{n} \sum_{1}^{n} e_p(k)e_q(k) = \frac{1}{p} \times \frac{1}{q} \qquad \text{for distinct primes } p \text{ and } q,$$

---

[2] Of course, *elementary* does not mean *simple* or *easy*, but see Nagell [1964:275–299] for a nice, accessible account.

suggesting that the *es* are independent. Then, hopeful or merely naive, you might think that LLN and CLT might enter into the description of

$$f(n) = \sum_{p \leq n} e_p(n) = \sum_{p \leq n} \frac{1}{p} + \sum_{p \leq n} \left[e_p(n) - \frac{1}{p} \right] = (1) + (2)$$

in that (1) might be the *bulk* effect as per LLN, and (2) might contribute (Gaussian) *fluctuations* as per CLT. This is indeed correct, but let's make a preliminary *demonstration*, pretending that the *es* are really honest, independent variables e^* with $\mathbb{P}(e_p^* = 1) = 1/p$.

At the bulk level, taking Gauss at his word, you find

$$m(n) \equiv \mathbb{E}[f^*(n)] = \sum_{p \leq n} \frac{1}{p} \simeq \int_2^n \frac{d\pi}{x} \simeq \int_2^n \frac{dx}{x \log x} = \log \log n,$$

and similarly, at the level of fluctuations,

$$\sigma^2(n) \equiv \mathbb{E} \left| f^*(n) - m(n) \right|^2 = \mathbb{E} \left[\sum_{p \leq n} \left(e_p(n) - \frac{1}{p} \right) \right]^2 = \sum_{p \leq n} \frac{1}{p} \left(1 - \frac{1}{p} \right)$$

$$\simeq \log \log n$$

as well. Then, at the *bulk* level,

(1) $\mathbb{P}\left[|f^*(n) - m(n)| \geq C(n)\sigma(n) \right] \leq \frac{1}{C^2(n)} \downarrow 0$ if $C(n) \uparrow \infty$ with n,

which is a refined sort of law of large numbers, first proved for the actual function $f(n)$ by Hardy–Ramanujan [1917]. Now, for the scaled variable $\mathbf{z}^*(n) = [f^*(n) - m(n)]/\sigma(n)$, you have

$$\mathbb{E} \left| e^{\sqrt{-1}k \mathbf{z}^*(n)} \right| = \prod_{p \leq n} \left[\frac{1}{p} e^{\sqrt{-1} \frac{k}{\sigma} (1 - \frac{1}{p})} + \left(1 - \frac{1}{p} \right) e^{-\sqrt{-1} \frac{k}{p\sigma}} \right]$$

$$\simeq \prod_{p \leq n} \left[1 - \frac{k^2}{2} \times \frac{1}{\sigma^2} \frac{1}{p} \left(1 - \frac{1}{p} \right) \right]$$

$$\simeq e^{-k^2/2}   \text{for large } n,$$

which is to say that at the level of *fluctuations*,

(2) $\lim_{n \uparrow \infty} \mathbb{P} \left[\dfrac{f^*(n) - \log \log n}{\sqrt{\log \log n}} \leq x \right] = \displaystyle\int_{-\infty}^x \frac{e^{-y^2/2} \, dy}{\sqrt{2\pi}},$

first proven by Erdős–Kac [1940] for the real, unstarred $f(n)$. This ends the demonstration and the proof is not far off.

(1) may be proved by hand – it is not hard. (2) employs the fact that for distinct primes p_1, p_2, etc. with product q, the empirical frequency

$$\frac{1}{n}\#\big(k \leq n : e_{p_1}(k)e_{p_2}(k) \text{ etc.} = 1\big)$$

differs from $\mathbb{P}\left[e_{p_1}^* e_{p_2}^* \text{ etc.} = 1\right]$ by at most $1/n$, independently of n and q. Then, with a little better appraisal of $m(n) = \sum_{p \leq n} \frac{1}{p}$, it is found that the true $\mathbf{z}(n)$s imitate the $\mathbf{z}(n)^*$s in that

$$\lim_{n \uparrow \infty} \frac{1}{n} \sum_{1}^{n} [\mathbf{z}(k)]^d \, \frac{(2d)!}{d!} \, n^{-2d} = \int_{-\infty}^{\infty} x^d \, \frac{e^{-x^2/2} \, \mathrm{d}x}{\sqrt{2\pi}}$$

for each $d = 1, 2, 3$ etc. separately. The rest is done by A.A. Markov's method of moments, introduced in §4.4 already, the moral being – as Hardy and Ramanujan put it – *that the typical whole number has about* $\log \log n$ *distinct prime divisors*, and this in a very precise sense. Remarkable! I stop here but see the amusing review of the whole subject by Kac [1959b].

Coda: Beyond CLT

Gauss's distribution occupies a special place. CLT shows its importance, and it is ubiquitous, *beyond* CLT when independence or even any intervention of chance is lacking in the small, but there is some mechanism producing an approximation of these in the large so that the big picture is as before, with a *bulk* effect on the scale n as per LLN, and Gaussian fluctuations on the scale $\sqrt{n}$ as per CLT, in e.g. *geometry*, in geodesic flow in negative curvature, illustrating Birkhoff's ergodic theorem in Chapter 8; in the *statistical mechanics* of J.W. Gibbs which occupies Chapters 10 and 11; and even in *arithmetic* as in §5.11 just above. Quite a variety, don't you think, and that's just a sample. Astonishing, really.

6

Brownian Motion

The Scottish botanist Robert Brown (1828) observed and was struck by the erratic motion of pollen in liquid suspension. You will have seen the same type of thing when a dust mote (made visible by a sun beam) is impelled this way and that by collisions with the (invisible) molecules of the air. Einstein (1905)[1] took up this "Brownian motion", connecting it with the heat equation $\partial w/\partial t = \frac{1}{2}\partial^2 w/\partial x^2$ and its elementary solution $e^{-x^2/2t}/\sqrt{2\pi t}$, interpreted as the probability density for, e.g. the vertical displacement of the particle at time $t > 0$. This not quite right: the mathematical "Brownian motion" underlying the heat equation is only e.g. the vertical velocity $\mathbf{v}(t) : t \geq 0$ and must be integrated once to obtain the displacement $\mathbf{x}(t) = \mathbf{x}(0) + \int_0^t \mathbf{v}\, dt'$. This was all explained from a physical point of view and reconciled with what Einstein said by Ornstein–Uhlenbeck [1930][2], but I won't go into that just now. What I want is Einstein's Brownian motion BM(1), and this only as an aid to proving limit theorems though it is endlessly fascinating in itself.

In §6.4, you will find a sort of rich man's CLT providing a link between RW(1) properly scaled, and the Brownian motion BM(1), newly constructed in §6.2. This is followed by a number of new proofs of things already seen, not that you need them, but the methods are versatile and very general: §6.6 is occupied with the arcsine law of §3.4; §6.7 includes a proof of the general law of the iterated logarithm, postponed from §5.8; §6.8 justifies the informal discussion of the Kolmogorov–Smirnov statistics in §5.2; §§6.9 and 6.10 contain an informal introduction to K. Itô's differential and integral calculus for Brownian paths which I will want in Chapter 11 and at the end, in Chapter 12.

---

[1] see Einstein [1956]
[2] see also Kac [1947]

6.1 Preview

I want you to have a preliminary idea of what the Brownian motion ought to be and what needs to be done to put it on a solid mathematical footing. Think of the dust mote, made visible by a sunbeam, described above.

If $\mathbf{x}(t)$ is, e.g. its height above the floor, and $\mathbf{v}(t) = \dot{\mathbf{x}}(t)$ its (vertical) velocity, then $m \times \dot{\mathbf{v}} = \text{mass} \times \text{acceleration} = $ the force $\mathbf{f}$ imparted to it by the "bath" *aka* the molecules of the air, and you may expect a few simple rules to hold:

(1) $\mathbf{v} : t \to \mathbb{R}$ should be continuous, and may be $\dot{\mathbf{v}}$ as well, though that is not so likely, $\mathbf{f}$ being quite wild;

(2) for fixed $T > 0$, the "future" $\mathbf{v}^+ : t \to \mathbf{v}(t + T) - \mathbf{v}(T)$ should be a copy of $\mathbf{v}$ itself, independent of the "past" $\mathbf{v}(t') : t' \le T$, the bath being (statistically) unchanged by the passage of time – without memory so to say;

(3) $\mathbb{E}(\mathbf{v}) = 0$, supposing $\mathbf{f}$ to be unbiased and $\mathbf{v}(0) = 0$, as tacitly assumed in (2);

(4) $\mathbb{E}(\mathbf{v}^2) < \infty$. Why not?

Let's see what comes of that.

To begin with, $\mathbb{E}(\mathbf{v}^2) \equiv \sigma^2(t)$ satisfies $\sigma^2(t + h) = \sigma^2(t) + \sigma^2(h)$ by (2) and (3), as you will check, and so must be a positive multiple of t.

(4′) $\mathbb{E}(\mathbf{v}^2) = t$ is now adopted as the simplest thing to do, to which I add a *technical* condition

(4″) $\mathbb{E}[\mathbf{v}^2, |\mathbf{v}| \ge \ell] = \mathrm{o}(t)$ for $t \downarrow 0$ and any fixed $\ell > 0$.

(1), (2), (3) and (4) lead directly to the heat equation and to the Gaussian distribution: in fact, with $\mathbf{v}(0) = 0$ as in (3), you find

(5) $$\mathbb{P}[a \le \mathbf{v}(t) < b] = \int_a^b \frac{e^{-v^2/2t}}{\sqrt{2\pi t}} \, dv \quad \text{for any } a < b.$$

This is already suggested by (2), (3), and (4′): $\mathbf{v}(t)$ is supposed to be the sum of n independent copies of $\mathbf{v}(t/n)$, so CLT should apply. It's not *quite* as easy as that, but you can imitate the second proof of CLT in §5.3. The line of thought goes back to Khinchine [1933:8–11].

To see what's going on, take $f \in C^3(\mathbb{R})$, vanishing far out, and put

$w(t, x) = \mathbb{E}[f(x + \mathbf{v}(t))]$. Then

$$w(t + h, x) - w(t, x)$$
$$= \mathbb{E}\big[f(x + \mathbf{v}(t) + \mathbf{y}) - f(x + \mathbf{v}(t))\big] \quad \text{with } \mathbf{y} = \mathbf{v}(t + h) - \mathbf{v}(t)$$
$$\simeq \mathbb{E}\Big[f'(x + \mathbf{v})\mathbf{y} + \frac{1}{2}f''(x + \mathbf{v})\mathbf{y}^2\Big]$$
$$= \frac{1}{2}\frac{\partial^2 w}{\partial x^2} \times h \qquad \text{by (3) and (4')},$$

with an (absolute) error in line 3 not more than the expectation of

$$\big|f''(x + \mathbf{v} + \mathbf{z}) - f''(x + \mathbf{v})\big| \times \mathbf{y}^2 \quad \text{for some } \mathbf{z} \text{ between 0 and } \mathbf{y}.$$

Now fix a small number $\ell > 0$ and divide the error into two parts according as $|\mathbf{y}| \leq \ell$ or not. The first part contributes at most $\max|f'''| \times \ell \times \mathbb{E}(\mathbf{y}^2) = h$, the second only $2\max|f''| \times \mathbb{E}\big[\mathbf{y}^2, |\mathbf{y}| > \ell\big]$ which is small compared to h by the technical condition (4''); in short, the *whole* error is small compared to h, which is to say

$$(6) \quad \frac{\partial w}{\partial t} = \frac{1}{2}\frac{\partial^2 w}{\partial x^2}.$$

But also, by (3) and (4'),

$$(7) \quad |w(h, x) - f(x)| \leq \max|f'| \times \mathbb{E}|\mathbf{v}(h)| = \mathrm{O}(\sqrt{h}) \text{ for } h \downarrow 0,$$

and

$$(8) \quad |w(t, x)| \leq \max|f| \times \mathbb{P}\big[\mathbf{v}(t) \geq x/2\big] = \mathrm{O}(x^{-2}) \text{ for } x \to \pm\infty \text{ since } f = 0 \text{ far out,}$$

and it is a fact that the only solution of (6), subject to (7) and (8) is

$$(9) \quad w(t, x) = \int_{-\infty}^{\infty} f(x + y)\frac{\mathrm{e}^{-y^2/2t}}{\sqrt{2\pi t}},$$

leading at once to (5). Take (7) and (8) as (easy) exercises. The rest is easy, too: I if (6) had a second solution of that type, you would have, by subtraction, a third solution, vanishing both at $t = 0$ and at $x = \pm\infty$, but *not vanishing identically*. This is not possible. Look at Figure 6.1.1. If this third solution w is, let's say, anywhere positive, you can choose the height $h > 0$ so that w has a positive maximum at some place (✗) inside or at the top. There $\partial^2 w/\partial x^2 \leq 0$ and $\partial w/\partial t = 0$ if ✗ is inside or else $\partial w/\partial t \geq 0$ if ✗ is at the top. That's not quite good enough, but $\mathrm{e}^{-t}w$ has also such a positive maximum, and *there* you find a contradiction:

$$0 \leq \frac{\partial}{\partial t}\mathrm{e}^{-t}w = -\mathrm{e}^{-t}w + \mathrm{e}^{-t}\frac{\partial w}{\partial t} = -\mathrm{e}^{-t}w + \mathrm{e}^{-t}\frac{1}{2}\frac{\partial^2 w}{\partial x^2} \leq -\mathrm{e}^{-t}w < 0.$$

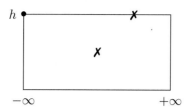

Figure 6.1.1

Exercise 6.1.1 Check by hand that (9) is the only solution of (5) subject to

(7′) $w(0+, \cdot) = f \in C(\mathbb{R})$ and

(8′) $|w| \le \max_{\mathbb{R}} |f|$.

Aside. Incidentally, do you know why $\partial w / \partial t = \frac{1}{2} \partial^2 w / \partial x^2$ is called the *heat equation*? Two simple facts are needed: first, that the *amount* (Q) of heat in a small interval is proportional to its length × the local temperature (T), and second, Newton's law of cooling, *to wit*, that at any place, the *flux* of heat (directed from hot to cold) is proportional to the gradient of temperature there, as in Figure 6.1.2. Then, up to constants of proportionality, $Q = \int_a^b T \, \mathrm{d}x$ changes with time as in

$$(b - a) \times \frac{\partial T}{\partial t} \simeq \int_a^b \frac{\partial T}{\partial t} \, \mathrm{d}x = \frac{\mathrm{d}Q}{\mathrm{d}t} = \text{flux} = \left. \frac{\partial T}{\partial x} \right|_a^b \simeq (b - a) \times \frac{\partial^2 T}{\partial x^2}.$$

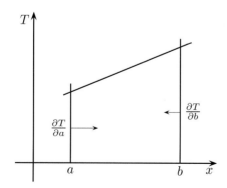

Figure 6.1.2

(5) combined with (2) now leads to the joint distribution of the several

observations $\mathbf{v}(t_1), \mathbf{v}(t_2), \ldots, \mathbf{v}(t_n)$: with $\mathbf{v}(0) = 0$ for simplicity, and $t_1 < t_2 < \cdots < t_n$,

$$\mathbb{P}\left[\bigcap_{i=1}^{n}\big(a_i \le \mathbf{v}(t_i) < b_i\big)\right] = \int_{\bigcap_{i=1}^{n}[a_i, b_i)} \frac{e^{-v_1^2/2t_1}}{\sqrt{2\pi t_1}} \frac{e^{-(v_2-v_1)^2/2(t_2-t_1)}}{\sqrt{2\pi(t_2-t_1)}} \text{ etc. } d^n v$$

seen already in Example 1.2.7. But now a serious question comes: Can these probabilities be extended from the class $\mathfrak{F}_0$ of "cylinder sets", *aka* "tame events",

$$A = \big[\mathbf{v} : \big(v(t_1), \ldots, v(t_n)\big) \in B\big]$$

with $n \ge 1$, $0 < t_1 < \cdots < t_n$, and $B \subset \mathbb{R}^n$ (which is not a field but where all is well) to the natural field $\mathfrak{F}$ for the space $C[0, \infty)$ of continuous paths required by (1)? And why is this troublesome? The matter may be decided by Carathéodory's lemma of §1.1.1: namely, the answer is *yes, it can be done* if $\mathbb{P}(A_n) \ge c > 0$ for decreasing cylinder sets $A_1 \supset A_2 \supset A_3$ etc. implies that the intersection $\bigcap_1^{\infty} A_n$ is not void, but the smaller the path space, the more difficult it becomes to find an actual path *there*, in that intersection. I will tell you the facts: looking only up to time $t = 1$, the extension can be made to the space

$$C^{\gamma}[0,1] = \Big[\mathbf{v} : |\mathbf{v}(t_2) - \mathbf{v}(t_1)| \le c(\mathbf{v})(t_2 - t_1)^{\gamma} \text{ for } 0 \le t_1 < t_2 \le 1\Big]$$

if $\gamma < 1/2$ but not if $\gamma \ge 1/2$; indeed, P. Lévy [1937] proved that the space of paths with $|v(t_2) - v(t_1)| \le c(\mathbf{v})\sqrt{2h|\log h|}$ for $h = t_2 - t_1$ and $0 \le t_1 < t_2 \le 1$ is just right, i.e. $\gamma = 1/2$ is just right with a little help from the logarithm. §6.2 provides the estimate for $\gamma = 1/3$. Carathéodory's lemma is avoided, but not really; the reasoning only looks different.

Notice that $\gamma = 1$ is not possible for elementary reasons already. If $|\mathbf{v}(t + h) - \mathbf{v}(t)| \le c(\mathbf{v})|h|$ for *some* $0 < t < 1$, $c(\mathbf{v}) < \infty$, and small h, then for large n and choice of i/n as close to t as possible,

$$\left|\mathbf{v}\Big(\frac{j+1}{n}\Big) - \mathbf{v}\Big(\frac{j}{n}\Big)\right| \le \left|\mathbf{v}\Big(\frac{j+1}{n}\Big) - \mathbf{v}(t)\right| + \left|\mathbf{v}\Big(\frac{j}{n}\Big) - \mathbf{v}(t)\right|$$

$$\le c(\mathbf{v})\Big(\frac{j+1}{n} - \frac{i}{n} + \frac{j}{n} - \frac{i}{n}\Big) \quad \text{if } j > i$$

$$\le c(\mathbf{v})\frac{7}{n} \quad \text{for } i < j \le i + 3,$$

so replacing $c(\mathbf{v})$ by a large whole number m, you find that $\mathbf{v}$ belongs to the set

$$\bigcup_{m=1}^{\infty} \bigcup_{\ell=1}^{\infty} \bigcap_{n \ge \ell} \bigcup_{i \le n} \bigcap_{i < j \le i+3} \left[\left|\mathbf{v}\Big(\frac{j+1}{n}\Big) - \mathbf{v}\Big(\frac{j}{n}\Big)\right| < \frac{7m}{n}\right].$$

But for fixed m and ℓ, and any $n \geq \ell$, the probability of the first intersection is overestimated by

$$\mathbb{P}\left[\bigcup_{i \leq n} \bigcap_{i < j \leq i+3} \left(\left| \mathbf{v}\left(\frac{j+1}{n} \right) - \mathbf{v}\left(\frac{j}{n} \right) \right| < \frac{7m}{n} \right) \right]$$

$$\leq n \left[\mathbb{P}\left(\left| \mathbf{v}\left(\frac{1}{n} \right) \right| < \frac{7m}{n} \right) \right]^3 \qquad \text{by Rule (2)}$$

$$= n \left[\int_{|v| < 7m/\sqrt{n}} \frac{\mathrm{e}^{-v^2/2}}{\sqrt{2\pi}} \, \mathrm{d}v \right]^3$$

$$< \text{a constant multiple of } n^{-1/2},$$

and $\mathbb{P}(C^1) = 0$ follows.

Exercise 6.1.2   Use the same method to show that $\mathbb{P}(C^\gamma) = 0$ for any $\gamma > 1/2$.

Afterthought   Did you know that a family of functions $v \in C[0,1]$ with $v(0) = 0$ is compact if (and only if) it is closed and $|v(t_2) - v(t_1)| \leq \omega(h)$ for one and the same function $\omega(h) > 0$ with $\omega(0+) = 0$, independently of v? That was the content of Exercise 1.7.1. Do it now if you skipped it.

6.2 Direct construction of BM(1)

I change notation from $\mathbf{v}$ to $\mathbf{x}$ – misleading in the present context but more conventional and more in keeping with RW(1).

6.2.1 P. Lévy's construction

I begin with P. Lévy's [1948:17–20] construction of BM(1). The numbers $0 \leq j/2^n \leq 1$ are organized by "level": 1 is placed at level 0 and $k/2^n : 0 < \text{odd } k < 2^n$ at level $n \geq 1$. The variables $\mathbf{x}(j/2^n)$ are supposed to be Gaussian with mean 0 and correlations

$$\mathbb{E}\left[\mathbf{x}(i/2^n)\mathbf{x}(j/2^n) \right] = \text{the smaller of } i/2^n \text{ and } j/2^n,$$

as you will please check by Rule (2) of §6.1. These should tell the whole story since everything is Gaussian. Now look at the Figure 6.2.1, from

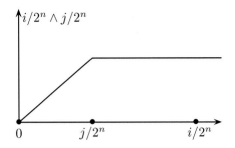

Figure 6.2.1

which you see that the variables

$$\mathbf{y}\left(\frac{i}{2^n}\right) = \mathbf{x}\left(\frac{i}{2^n}\right) - \frac{1}{2}\left[\mathbf{x}\left(\frac{i-1}{2^n}\right) + \mathbf{x}\left(\frac{i+1}{2^n}\right)\right] \quad \text{and} \quad \mathbf{x}\left(\frac{j}{2^n}\right)$$

are uncorrelated, and so also independent if $j < i$, and likewise if $j > i$.
It follows that the variables $\mathbf{y}(k/2^n) : 0 < \text{odd } k < 2^n$ at level n are

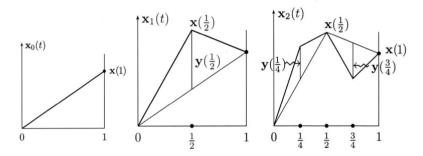

Figure 6.2.2

independent among themselves, the distance between odd i and j being
≥ 2, and also independent of what went before, at levels $0 \leq l < n$,
i.e. independent of $\mathbf{x}(j/2^n)$ for even $j \leq 2^n$, since such a $j/2^n$ must lie
either to the left of $(k-1)/2^n$ or to the right of $(k+1)/2^n$, k being
odd. Figure 6.2.2 shows what is happening at levels 0, 1, and 2, the
observations $\mathbf{x}(j/2^n) : 0 \leq j \leq 2^n$ being interpolated by a broken line
$\mathbf{x}_n(t) : 0 \leq t \leq 1$. Now at level n, $\mathbf{x}_n$ is obtained from $\mathbf{x}_{n-1}$ by adding
the little non-overlapping tents seen in Figure 6.2.3, and this is repeated
indefinitely to produce the sum, indicated schematically by

$$\mathbf{x}_\infty(t) = t\mathbf{x}(1) + \sum_{n=1}^{\infty} \sum_{\text{odd } k < 2^n} \mathbf{y}(k/2^n) \times \bigtriangleup_{k/2^n},$$

which should be the Brownian motion up to time $t = 1$.

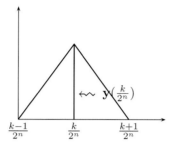

Figure 6.2.3

OK. You see the plan. To put $\mathbf{x}_\infty$ on a solid footing, note that

$$\mathbb{E}\left| \mathbf{y}\left(\frac{k}{2^n}\right)\right|^2$$

$$= \mathbb{E}\left| \frac{1}{2}\left[\mathbf{x}\left(\frac{k}{2^n}\right) - \mathbf{x}\left(\frac{k-1}{2^n}\right)\right] - \frac{1}{2}\left[\mathbf{x}\left(\frac{k+1}{2^n}\right) - \mathbf{x}\left(\frac{k}{2^n}\right)\right]\right|^2$$

$$= 2^{-n-1}$$

and replace $\mathbf{x}(1)$ by $\mathbf{z}(0)$ and $\mathbf{y}(k/2^n)$ by $2^{-(n+1)/2}\mathbf{z}(k/2^n)$ with independent unit Gaussian variables $\mathbf{z}$, constructed by Wiener's trick (§1.4). $\mathbf{x}_\infty$ has now an unimpeachable status, built as it is over the unit interval with its honest probabilities. But does it make sense? Well, yes: the inner sum over the little non-overlapping tents at level n is not more than $\pm 2^{-(n+1)/2} \times \max|\mathbf{z}(k/2^n)|$, and

$$\mathbb{P}\left[\max\left|\mathbf{z}\left(\frac{k}{2^n}\right)\right| > \sqrt{2n}\right] < 2^n \int_{\sqrt{2n}}^\infty \frac{e^{-x^2/2}}{\sqrt{2\pi}}\,\mathrm{d}x < (2/e)^n \text{ by Mills's ratio,}$$

so the sum over the levels $n \geq 1$ converges like $\sum_1^\infty 2^{-(n+1)/2}\sqrt{2n}$ to a continuous path $\mathbf{x}_\infty(t) : 0 \leq t \leq 1$ which is evidently Gaussian, with correlations $\mathbb{E}[\mathbf{x}_\infty(i/2^n)\mathbf{x}_\infty(j/2^n)] = $ the smaller of $i/2^n$ and $j/2^n$ because the observations $\mathbf{x}_{n'}(k/2^n)$ did not change since level $n' = n$. The rest is done in a line or two:

$$\mathbb{E}\left|\mathbf{x}_\infty(j/2^n) - \mathbf{x}_\infty(i/2^n)\right|^2 = 2^{-n}(j - i) \text{ if } i < j,$$

so $\mathbb{E}\left|\mathbf{x}_\infty(t_2) - \mathbf{x}_\infty(t_1)\right|^2 \leq t_2 - t_1$ if $t_1 < t_2$, by Fatou and the continuity of paths, and $\mathbb{E}\left|\mathbf{x}_\infty(t_1)\mathbf{x}_\infty(t_2)\right| = $ the smaller of t_1 and t_2 generally, i.e. the correlations are as they should be and so also the distributions. The full Brownian motion $\mathbf{x}(t) : t \geq 0$ is now produced by placing indepen-dent copies $\mathbf{x}'_\infty$, $\mathbf{x}''_\infty$, etc. of $\mathbf{x}_\infty$ and to end as in

$$\begin{aligned} \mathbf{x}(t) &= \mathbf{x}_\infty(t) & \text{for } 0 \leq t \leq 1, \\ &= \mathbf{x}(1) + \mathbf{x}'_\infty(t-1) & \text{for } 1 \leq t \leq 2, \\ &= \mathbf{x}(2) + \mathbf{x}''_\infty(t-2) & \text{for } 2 \leq t \leq 3, \\ &\quad \text{and so on,} \end{aligned}$$

and checking that the correlations are right.

Exercise 6.2.1 For fixed $c > 0$, show $c\,\mathbf{x}(t/c^2) : t \geq 0$ is a copy of $\mathbf{x}$. This is "Brownian scaling".

Exercise 6.2.2 Show $t\mathbf{x}(1/t) : t \geq 0$ is also a copy of $\mathbf{x}$. *Warning*: Be careful about $t = 0+$.

6.2.2 Wiener's construction

Wiener's construction [1930:220–222] was a little different, being a direct proof that all is well in the space $C^\gamma[0,1]$ for $\gamma = 1/4$. I do it for $\gamma = 1/3$.

The first step is to extend the probabilities to the totality of variables $\mathbf{x}(j/2^n)$ with $0 \leq j/2^n \leq 1$ and $n \geq 0$. That's not the interesting part. The rest hinges upon the estimate

$$\mathbb{P}\left[\left|\mathbf{x}\left(\frac{j}{2^n}\right) - \mathbf{x}\left(\frac{i}{2^n}\right)\right| > \left(\frac{j-i}{2^n}\right)^{1/3} \text{ for some } 0 \leq i < j \leq 2^n \right.$$
$$\left. \text{with } \frac{j-i}{2^n} \leq \frac{1}{64n^3}\right]$$

$$\leq 2^{2n}(\text{over-counting the number of pairs } i,j)\times$$

$$\times \mathbb{P}\left[\mathbf{x}\left(\frac{j-i}{2^n}\right) > \left(\frac{j-i}{2^n}\right)^{1/3}\right]$$

$$= \mathbb{P}\left[\mathbf{x}(1) > \left(\frac{j-i}{2^n}\right)^{1/6}\right] \qquad \text{by Brownian scaling}$$

$$\leq \mathbb{P}\left[|\mathbf{x}(1)| > \sqrt{4n}\right] \qquad \text{since } 2^{-n}(j-i) \leq \frac{1}{64n^3}$$

$$< \mathrm{e}^{-2n} \qquad \text{by Mills's ratio}$$

$$= \left(\frac{2}{\mathrm{e}}\right)^{2n},$$

whence

$$\left|\mathbf{x}\left(\frac{j}{2^n}\right) - \mathbf{x}\left(\frac{i}{2^n}\right)\right| \leq \left(\frac{j-i}{2^n}\right)^{1/3} \quad \text{for } 0 \leq i < j \leq 2^n \text{ if } \frac{j-i}{2^n} \leq \frac{1}{64n^3}$$

provided n is sufficiently large, i.e. if n is larger than some path-dependent $n_0(\mathbf{x})$. This number n_0 may always be increased so as to make $2^{-n_0+1} \leq 1/(64n^3)$. Now take $0 \leq t_1 < t_2 \leq 1$, subject to $t_2 - t_1 \leq 2^{-n_0+1}$. Then $2^{-n} < t_2 - t_1 \leq 2^{-n+1}$ for some $n \geq n_0$, and you can find $i \leq j$ such that $t_1 \leq i/2^n \leq j/2^n \leq t_2$ with $i/2^n$, respectively $j/2^n$, as close as possible to t_1, respectively t_2. Write $t_2 = j/2^n + e_1/2^{n+1} + e_2/2^{n+1} + $ etc. in which the es are either 1 or 0 and form the sum

$$\mathbf{x}(t_2) = \mathbf{x}\left(\frac{j}{2^n}\right) + \mathbf{x}\left(\frac{j}{2^n} + \frac{e_1}{2^{n+1}}\right) - \mathbf{x}\left(\frac{j}{2^n}\right)$$
$$+ \mathbf{x}\left(\frac{j}{2^n} + \frac{e_1}{2^{n+1}} + \frac{e_2}{2^{n+2}}\right) - \mathbf{x}\left(\frac{j}{2^n} + \frac{e_1}{2^{n+1}}\right)$$
$$+ \text{ etc.}$$

This converges well since

$$\sum_{k=1}^{+\infty} \left| \mathbf{x}\left(\frac{j}{2^n} + \frac{e_1}{2^{n+1}} + \cdots + \frac{e_k}{2^{n+k}}\right) - \mathbf{x}\left(\frac{j}{2^n} + \frac{e_1}{2^{n+1}} + \cdots + \frac{e_{k-1}}{2^{n+k-1}}\right) \right|$$
$$\leq \sum_{k=1}^{+\infty} \left(\frac{e_k}{2^{n+k}}\right)^{1/3} \leq 2^{-n/3} \times K \equiv \sum_{k=1}^{\infty} 2^{-k/3};$$

it *defines* $\mathbf{x}(t_2)$ if t_2 is not already of the form $k/2^n$; and by the same reasoning applied to t_1 and $i/2^n$,

$$\left| \mathbf{x}(t_2) - \mathbf{x}(t_1) \right| \leq \left| \mathbf{x}\left(\frac{j}{2^n}\right) - \mathbf{x}\left(\frac{i}{2^n}\right) \right| + 2^{-\frac{n}{3}} \times 2K \leq (1 + 2K)(t_2 - t_1)^{\frac{1}{3}}$$

since $n \geq n_0$ and $2^{-n} < t_2 - t_1$. The rest is obvious: $\mathbf{x}(t) : 0 \leq t \leq 1$ is (1) Gaussian, (2) of class $C^{1/3}[0,1]$, (3) $\mathbb{E}[\mathbf{x}(t_1)\mathbf{x}(t_2)] = $ the smaller of t_1 and t_2 may be proved as before, and all is well.

Exercise 6.2.3 $\gamma = 1/3$ can be improved: any $\gamma < 1/2$ will do, but the mode of estimation fails (as it must) for $\gamma = 1/2$. Check it out.

6.3 Markov property and passage times

The discussion imitates §§3.2–3.3 – a little repetitious but that's not bad. Write $\mathbb{P}_0$ for the Brownian probabilities of §6.2 to indicate the starting point $\mathbf{x}(0) = 0$, and $\mathbb{P}_x$ for the allied probabilities for paths starting at $\mathbf{x}(x) = x$: namely, $\mathbb{P}_x(B) = \mathbb{P}_0(x + \mathbf{x} \in B)$ for $B \in \mathfrak{F}$, as for RW(1).

6.3.1 The simple Markov property

The simple Markov property knits these several probabilities together. For fixed $t > 0$, let $\mathfrak{F}_t$ be the field descriptive of the path up to time t, i.e. the smallest field containing all the elementary events $(a \leq \mathbf{x}(t') < b)$ for $a \leq b$ and $t' \leq t$, $\mathfrak{F}_\infty = \mathfrak{F}$ being the smallest field containing all of these. Now look at Figure 6.3.1. The simple Markov property for BM(1) states that if $\mathbf{x}^+ : t \to \mathbf{x}(t + T)$ is the continuation of the path after the fixed time T, then

$$\mathbb{P}_a(\mathbf{x}^+ \in B | \mathfrak{F}_T) = \mathbb{P}_b(B) \quad \text{with } b = \mathbf{x}(T),$$

which is to say that the Brownian traveler has *no memory*: having

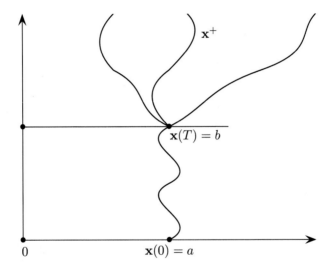

Figure 6.3.1

come to $\mathbf{x}(T) = b$, it forgets how it got there, commencing a new Brownian motion *starting from scratch at that place*. The proof is immediate: $\mathbf{x}^+(t) = \mathbf{x}(t + T) - \mathbf{x}(T) + \mathbf{x}(T)$ and $t \to \mathbf{x}(t + T) - \mathbf{x}(T)$ is a Brownian motion starting at 0, *independent of the past* $\mathbf{x}(t') : t' \leq T$ described by the field $\mathfrak{F}_T$. That's all there is to it.

6.3.2 The strict Markov property

It is desired to extend this simple Markov property to "stopping times" T. These are functions of the path $T : \mathbf{x} \to [0, \infty)$ with this characteristic feature: that for every fixed time $t > 0$, the event $(T < t)$ is measurable over $\mathfrak{F}_t$. For example, the passage time $T = \min(t : \mathbf{x}(t) = 1)$ is OK, but the last root $T = \max(t \leq 1 : \mathbf{x}(t) = 0)$ is not. The associated field $\mathfrak{F}_T$ is now declared to be the class of events $A \in \mathfrak{F}$ such that $A \cap (T < t) \in \mathfrak{F}_t$ for every $t > 0$. Then the statement is as before: in the event that $\mathbb{P}(t < \infty) = 1$,

$$\mathbb{P}_a(\mathbf{x}^+ \in B|\mathfrak{F}_T) = \mathbb{P}_b(B) \quad \text{with } b = \mathbf{x}(T).$$

Proof. This is a little more elaborate than for RW(1). Take $A \in \mathfrak{F}_T$, $B \in \mathfrak{F}$, and define $T_n = k/2^n$ when $(k-1)/2^n \leq T < k/2^n$. This is a stopping time, decreasing to T as n tends to infinity, so with a self-evident notation,

$$\mathbb{P}_a\Big[A \cap \big(\mathbf{x}(\,\cdot\,+T_n) \in B\big)\Big]$$

$$= \sum_{k=1}^{\infty} \mathbb{P}_a\Big[A \cap \Big(T_n = \frac{k}{2^n}\Big) \cap \big(\mathbf{x}_{k/2^n} \in B\big)\Big]$$

$$= \sum_{k=1}^{\infty} \mathbb{E}_a\Big[\mathbb{P}_a\Big(A \cap \Big(T_n = \frac{k}{2^n}\Big) \cap \big(\mathbf{x}_{k/2^n} \in B\big)\Big|\mathfrak{F}_{k/2^n}\Big)\Big]$$

$$= \sum_{k=1}^{\infty} \mathbb{E}_a\Big[A \cap \Big(T_n = \frac{k}{2^n}\Big), \mathbb{P}_{\mathbf{x}(k/2^n)}(B)\Big] \quad \begin{array}{l} \text{by the simple Markov property} \\ \text{since } A \cap (T_n = k/2^n) \in \mathfrak{F}_{k/2^n} \end{array}$$

$$= \mathbb{E}_a[A, \mathbb{P}_b(B)] \quad \text{where now } b = \mathbf{x}(T_n).$$

It remains to make $n \uparrow \infty$ to obtain the correct statement with T in place of T_n. Notice that both the start and the finish of the display are countably additive in respect to B, with the implication that it is enough to think about tame events (= cylinder sets) $C = \big[\mathbf{x} : \big(\mathbf{x}(t_1), \ldots, \mathbf{x}(t_n)\big) \in D\big]$ with $0 < t_1 < \cdots < t_n$ and $D \subset \mathbb{R}^n$. But

$$\mathbb{P}_x(C) = \int_D \frac{e^{-(x_1-x)^2/2t_1}}{\sqrt{2\pi t_1}} \frac{e^{-(x_2-x_1)^2/2(t_2-t_1)}}{\sqrt{2\pi(t_2-t_1)}} \text{ etc. } d^n x$$

is manifestly a continuous function of $x \in \mathbb{R}$, so $T_n \downarrow T$ and the continuity of paths do the rest. Now translate the present statement into the language of conditional probabilities.

6.3.3 Passage times

As for RW(1) in §3.3, these make nice examples of stopping times, in continual use below. Take, for example, $(T = \min(t : \mathbf{x}(t) = 1)$ for paths starting at $\mathbf{x}(0) = 0$. The requirement $(T < t) \in \mathfrak{F}_t$ is obvious, but is it true that that $\mathbb{P}_0(T < \infty) = 1$. I'd like that. The proof is easy and illustrates a useful principle:

$$
\begin{aligned}
\mathbb{P}_0(T = \infty) &= \mathbb{P}_0\big[\mathbf{x}(t) < 1 \text{ for } t \geq 0\big] \\
&= \mathbb{P}_0\big[c\,\mathbf{x}(t/c^2) < 1 \text{ for } t \geq 0\big] && \text{by Brownian scaling} \\
&= \mathbb{P}_0\Big[\mathbf{x}(t) < \frac{1}{c} \text{ for } t \geq 0\Big] \\
&= \mathbb{P}_0\big[\mathbf{x}(t) \leq 0 \text{ for } t \geq 0\big] && \text{by making } c \uparrow \infty \\
&\leq \mathbb{P}_0\Big[\bigcap_{h>0}(\mathbf{x}(t) \leq 0 \text{ for } t \leq h)\Big].
\end{aligned}
$$

Now this last event belongs to the "germ field" $\mathfrak{F}_{0+} = \bigcap_{h>0}\mathfrak{F}_h$; as such, it is independent of $\mathbf{x}(\,\cdot\, + h) - \mathbf{x}(h)$ for any $h > 0$ and so also of the whole path $\mathbf{x}$ itself. Then it must be independent of itself and have probability 0 or 1 – in short, $\mathfrak{F}_{0+}$ is trivial. The proof that $\mathbb{P}_0(T < \infty) = 1$ is finished by noting that $\mathbb{P}_0[\mathbf{x}(h) \leq 0] = 1/2$.

Now let's find the distribution of T or, more generally, the distribution of $T_x = \min(t : \mathbf{x}(t) = x)$ for $x > 0$, taking $\mathbf{x}(0) = 0$ as before. This can be done in various ways. Here's the quick way, imitating the reflection principle of D. André of §3.4. Look at Figure 6.3.2. Obviously, $T_x \leq t$ if

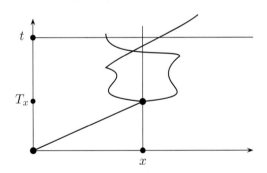

Figure 6.3.2

$\mathbf{x}(t) \geq x$; also T_x is a stopping time, and there is self-evident symmetry here: $\mathbf{x}(T_x) = x$ being known, $\mathbf{x}(\,\cdot\, + T_x) - x$ is a Brownian motion starting at 0, independent of the past $\mathbf{x}(t') : t' \leq T_x$ and so equally likely to be positive or negative, with the result that

$$\mathbb{P}_0\big(T_x \leq t\big) = \mathbb{P}_0\big[T_x \leq t, \mathbf{x}(t) \geq x\big] + \mathbb{P}_0\big[T_x \leq t, \mathbf{x}(t) < x\big],$$
$$= 2\,\mathbb{P}_0\big[T_x \leq t, \mathbf{x}(t) \geq x\big] = 2\,\mathbb{P}_0\big[\mathbf{x}(t) \geq x\big]$$
$$= 2\int_x^\infty \frac{\mathrm{e}^{-y^2/2t}}{\sqrt{2\pi t}}\,\mathrm{d}y,$$

providing at once the distribution of the passage time and the distribution of the maximum $\mathbf{m} = \max\big[\mathbf{x}(t') : t' \leq t\big]$ since $T_x \leq t$ is the same as saying $\mathbf{m} \geq x$.

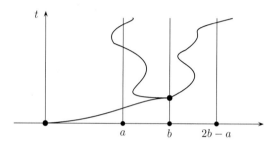

Figure 6.3.3

Exercise 6.3.1 $\mathbb{P}_0(T_x \leq t)$ is to be expressed more naturally as $\int_0^t (2\pi s)^{-3/2} \times \mathrm{e}^{-x^2/2s}\,\mathrm{d}s$, in which you see the law of distant passage for RW(1) from §3.2, as you might have guessed.

Exercise 6.3.2 Compute the joint density for $a = \mathbf{x}(t)$ and $b = \max_{t' \leq t} \mathbf{x}(t')$ by looking at Figure 6.3.3. *Answer:* $\frac{2(2b-a)}{\sqrt{2\pi t^3}}\,\mathrm{e}^{-(2b-a)^2/2t}$.

Exercise 6.3.3 Here's a slicker way for the passage times: $\mathbf{z}(t) = \exp\big[\alpha\mathbf{x}(t) - \frac{1}{2}\alpha^2 t\big]$ is a fair game which is bounded up to time T_x if α is positive. Stop it then and see what comes out, with reference to §1.7.5. *Note*: Everything that was said in §§1.3.4 and 3.2.3 about fair and favorable games carries over intact. The proofs do not offer any difficulties so I do not pause for them.

6.3.4 Two-sided passage or the Gambler's Ruin

Take $-\infty < a < b < +\infty$ and $\mathbf{x}(0) = x$ between, and look at the exit time $T_a \wedge T_b \equiv T$. The chief facts are these:

(1) $\mathbb{P}_x(T_a < T_b) = \dfrac{b - x}{b - a}.$

(2) $\mathbb{E}_x(T) = d^2 - (x - c)^2$, c being the midpoint $(a + b)/2$ and d the half-length $(b - a)/2$.

(3) $\mathbb{E}_x\left(e^{-\alpha T}\right) = \dfrac{\cosh \sqrt{2\alpha}(x - c)}{\cosh \sqrt{2\alpha}\, d}$.

Exercise 6.3.4 These things can also be done by stopping fair games. For (1), the game is $\mathbf{x}(t)$ itself; for (2), use $\mathbf{x}^2(t) - t$, but note that this is *not* bounded up to the stopping time, so you will have to deal with that somehow; for (3), use $e^{-\alpha^2 t/2} \times \cosh[\alpha \mathbf{x}(t)]$ a little modified.

Exercise 6.3.5 Take $a = -1$, $b = +1$, and $x = 0$ in (3) so that $\mathbb{E}_0\left(e^{-\alpha T}\right) = 1/\cosh \sqrt{2\alpha}$. Prove that $\mathbb{P}_0(T > t)$ is over-estimated by $e^{-\gamma t}/\cos \sqrt{2\gamma}$ for any positive number $\gamma < \pi^2/8$, showing how fast BM(1) exits from $-1 \le x \le +1$ in comparison to the very long tail of $\mathbb{P}_0(T_1 > t) \simeq \sqrt{2/\pi t}$.

6.4 The invariance principle

This is, so to say, the rich man's CLT, expressing the fact that RW(1), properly scaled, approximates BM(1); like CLT itself, it is a weak law only. Donsker [1951] proved it first. The present proof is due to Knight [1962]. Its simplicity is due to the effective use of the Brownian motion – and I will make a second proof, of a different character, as a reprise.

Write $\mathbf{s}(n) : n \ge 0$ for the random walk, starting at $\mathbf{s}(0) = 0$; speed it up (jump time $1/n$); scale it back (jump size $1/\sqrt{n}$); and let $\mathbf{x}_n(t) : 0 \le t \le 1$ be the broken line seen in Figure 6.4.1 joining the

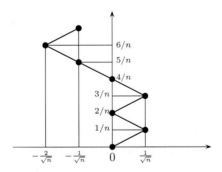

Figure 6.4.1

points $t = k/n$ and $x = \mathbf{s}(k)/\sqrt{n}$ for $0 \le k \le n$. Take now a function

$F\colon C[0,1] \to \mathbb{R}$ which is continuous – relative to the natural distance $d(\mathbf{x}, \mathbf{y}) = \max_{t\leq 1}|\mathbf{x}(t) - \mathbf{y}(t)|$ – at a set of Brownian paths of full probability 1. Then with a self-evident notation, the invariance principle (IP) says that

$$\lim_{n\uparrow\infty} \mathrm{RW}\big[F(\mathbf{x}_n) \leq c\big] = \mathrm{BM}\big[F(\mathbf{x}) \leq c\big]$$

at every value of c where the right side does not jump.

6.4.1 Proof

Part 1. Figure 6.4.2 depicts the Brownian path starting at $\mathbf{x}(0) = 0$. The exit time T_1 is

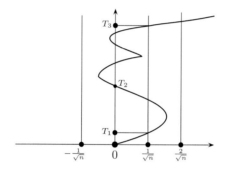

Figure 6.4.2

$$\min\Big(t : |\mathbf{x}(t)| = \frac{1}{\sqrt{n}}\Big),$$

T_2, the next exit time, is

$$\min\Big(t \geq T_1 : |\mathbf{x}(t) - \mathbf{x}(T_1)| = \frac{1}{\sqrt{n}}\Big),$$

T_3 is

$$\min\Big(t \geq T_2 : |\mathbf{x}(t) - \mathbf{x}(T_2)| = \frac{1}{\sqrt{n}}\Big),$$

and so on.

Obviously, the exit places $\mathbf{x}(0), \mathbf{x}(T_1), \mathbf{x}(T_2)$, etc. may be viewed as a copy of RW(1) on the scale $1/\sqrt{n}$: each step is $\pm 1/\sqrt{n}$ with equal probabilities, and these are independent by the strict Markov property. In short, the broken line $\mathbf{x}_n(t) : 0 \leq t \leq 1$ joining the points $t = k/n$ and $x = \mathbf{x}(T_k)$ for $0 \leq k \leq n$ is a copy of RW(1), except that the time scales –

k/n for $\mathbf{x}_n$ and T_k for $\mathbf{x}$ – are not the same. The proof is finished by reconciling these, the fact being that for large n, $\max_{k\leq n}\left|T_k - \frac{k}{n}\right| \leq n^{-1/6}$ so that $\max_{t\leq 1}|\mathbf{x}_n(t) - \mathbf{x}(t)| \leq c(\mathbf{x})\sqrt[3]{n^{-1/6}} = c(\mathbf{x})n^{-1/18}$, by Wiener's estimate in §6.2. That's all you need.

Part 2.   The reconciliation is made with the help of Doob's inequality from §1.8: $\mathbf{z}(k) = T_k - k/n$, $0 \leq k \leq n$, is a fair game and x^4 is convex, so $\mathbf{z}^4(k)$, $0 \leq k \leq n$, is a favorable game, and Doob tells you that

$$\mathbb{P}\left[\max_{k\leq n}|\mathbf{z}(k)| > n^{-1/6}\right] = \mathbb{P}\left[\max_{k\leq n}\mathbf{z}^4(k) > n^{-2/3}\right]$$
$$\leq n^{2/3}\,\mathbb{E}(\mathbf{z}^4(n))$$
$$= n^{2/3} \times \left[n\,\mathbb{E}(\mathbf{z}^4(1)) + n(n-1)\big(\mathbb{E}(\mathbf{z}^2(1))\big)^2\right]$$
$$\leq n^{2/3} \times n^2\,\mathbb{E}(\mathbf{z}^4(n)).$$

Now $T_1 = \min(t : |\mathbf{x}(t)| = 1/\sqrt{n})$ looks like $\frac{1}{n} \times T_0 \equiv \min(t : |\mathbf{x}(t)| = 1)$ by the Brownian scaling $\mathbf{x}(t) \to c\,\mathbf{x}(t/c^2)$ with $c = 1/\sqrt{n}$; and $\mathbb{E}(T_0^4)$ is finite as per $\mathbb{E}_0(e^{-\alpha T}) = 1/\cosh\sqrt{2\alpha}$ from §6.3, so

$$\mathbb{P}\left[\max_{k\leq n}|\mathbf{z}(k)| > n^{-\frac{1}{6}}\right] \leq n^{\frac{2}{3}} \times n^2 \times \text{a constant multiple of } n^{-4} = \mathrm{O}(n^{-\frac{4}{3}})$$

does the job.

Note   This proof makes IP look like an *individual*, rather than a *weak* law, but it is not so: for each n, the broken line joining the points $t = k/n$ and $x = \mathbf{x}(T_k)$ is based upon a *different* copy of RW(1).

Exercise 6.4.1   Take $n = 1$ in Figure 6.4.2. Then $\mathbf{x}(T_k) : k \geq 0$ is a copy of RW(1) plain, to which the old law of the iterated logarithm of §2.3 applies. Use that to prove

$$\mathbb{P}_0\left[\limsup_{t\uparrow\infty} \frac{\mathbf{x}(t)}{\sqrt{2t\log\log t}} = +1\right] = 1$$

for BM(1). See it? It's two lines. See also §6.7 below for a proof in this style of the general law of the iterated logarithm stated in §5.7.

Attention.   In view of IP, it is clear that everything in Chapter 3 must have its counterpart for BM(1): for example, the fair game $\exp\left[\alpha\mathbf{x}(t) - \alpha^2 t/2\right]$ employed in Exercise 6.3.3 is the Brownian cousin of $(\cosh\alpha)^{-n}\exp\left[\alpha\mathbf{x}(n)\right]$ similarly employed for RW(1) in §3.2.2. Keep this in mind as the Brownian story unfolds.

6.4.2 Reprise

Here is another take on IP for later use, to be compared also to Wiener's estimate in §6.2. Begin like this, $\mathbf{s}(n) : n \geq 0$ being the standard walk, $\mathbf{x}(t) : t \geq 0$ the corresponding broken line, and $\mathbb{P}_n$ the corresponding probabilities:

$$\mathbb{P}_n\left[\mathbf{x}\left(\frac{j}{n}\right) - \mathbf{x}\left(\frac{i}{n}\right) > n^{-\alpha} \text{ for some } 0 \leq i < j \leq n \text{ with } j - i \leq n^{\beta}\right]$$

$$\leq n\,\mathbb{P}_n\left[\max_{k \leq n^{\beta}} \mathbf{s}(k) > n^{1/2-\alpha}\right]$$

$$\leq 2n\,\mathbb{P}_n\left[\mathbf{s}(\ell) > n^{1/2-\alpha}\right] \quad \begin{array}{l}\text{with } \ell = \text{the last } k \leq n^{\beta}, \text{ this} \\ \text{by the reflection principle of D. André}\end{array}$$

$$\leq 2n \exp\left(-n^{1-2\alpha}/3\ell\right) \quad \text{as in §1.8}$$

$$\leq 2n \exp(-n^{\gamma}/3) \quad \text{with } \gamma = 1 - 2\alpha - \beta.$$

This is the general term of a convergent sum if $\gamma > 0$, e.g. if $\alpha = 1/3$ and $\beta = 1/4$, and since the estimate applies equally to $\mathbf{x}$ and to $-\mathbf{x}$, you have

$$\left|\mathbf{x}\left(\frac{j}{n}\right) - \mathbf{x}\left(\frac{i}{n}\right)\right| \leq n^{-1/3} \text{ for } 0 \leq i < j \leq n \text{ and } j - i \leq n^{1/4}$$

for large n with probability 1. Let $n_0(\mathbf{x}) < +\infty$ be the index n at which this starts to work; take $0 \leq t_1 < t_2 \leq 1$ with $t_2 - t_1 \leq n_0^{-3/4}$ and also $n > n_0$ so that $n^{-3/4} < t_2 - t_1 \leq (n+1)^{-3/4}$, and choose $i/n \geq t_1$ and $j/n \leq t_2$ as close as possible to t_1, respectively t_2. Then

$$|\mathbf{x}(t_2) - \mathbf{x}(t_1)|$$

$$\leq \left|\mathbf{x}\left(\frac{i}{n}\right) - \mathbf{x}\left(\frac{i-1}{n}\right)\right| + \left|\mathbf{x}\left(\frac{j}{n}\right) - \mathbf{x}\left(\frac{i}{n}\right)\right| + \left|\mathbf{x}\left(\frac{j+1}{n}\right) - \mathbf{x}\left(\frac{j}{n}\right)\right|$$

$$\leq 3n^{-1/3} \leq 3(t_2 - t_1)^{4/9}, \quad \mathbf{x} \text{ being a broken line.}$$

This means that when h is small, the probabilities $\mathbb{P}_n$ induced in the path space $C[0,1] \cap (\mathbf{x}(0) = 0)$ by the broken line are nearly concentrated on the compact

$$K = \left(\mathbf{x} : |\mathbf{x}(t_2) - \mathbf{x}(t_2)| \leq 3(t_2 - t_1)^{4/9} \text{ for } t_2 - t_1 \leq h\right)$$

seeing as $\mathbb{P}_n(K) \geq \mathbb{P}(n_0 \leq h^{-4/3})$, and it is a general principle, illustrated in its simplest form in §1.7, that in such a case, it is possible to select $n = n_1 < n_2 <$ etc. $\uparrow +\infty$ so as to make $\mathbb{P}_n$ tend to an honest countably additive measure $\mathbb{P}_\infty$ on $\mathbb{X} = C[0,1]$ in the sense that $\mathbb{E}_n(f)$ tends to $\mathbb{E}_\infty(f)$ for every function $f \in C(\mathbb{X} \to \mathbb{R})$. The identification of

$\mathbb{P}_\infty$ is immediate: by CLT, it is Gaussian, with the Brownian correlations, so what can it be but the probabilities for the Brownian motion $\mathbf{x}(t) : t \leq 1$ starting at $\mathbf{x}(0) = 0$. This outcome has nothing to do with the selection of any particular $n_1 < n_2 <$ etc. so that can dispensed with, and you deduce IP in the alternative form:

$$\lim_{n \uparrow \infty} \mathrm{RW}(\mathbf{x}_n \in A) = \mathrm{BM}(\mathbf{x} \in A) \quad \text{provided } \mathrm{BM}(\partial A) = 0,$$

where ∂A denotes the boundary of A in the path space $C[0, 1]$.

Exercise 6.4.2 Think it over, with help from e.g. Lax [2002] if help is needed.

6.5 Volume RW(1) (reprise)

The invariance principle is used to reprove the formula

$$\lim_{n \uparrow \infty} \mathrm{RW}\left[\max_{k \leq n} \mathbf{x}(k) - \min_{k \leq n} \mathbf{x}(k) \leq \sqrt{nx}\right] = \int_0^x \sum_{n \in \mathbb{Z}} (-1)^{n-1} 4n^2 \frac{e^{-n^2 y^2 / 2}}{\sqrt{2\pi}} \, dy$$

from §3.5 for the number of distinct sites occupied by $\mathrm{RW}(1)$ up to time n. The proof is really the same, but seen under a different aspect. IP converts the left side into

$$\mathrm{BM}\left[\max_{t \leq 1} \mathbf{x}(t) - \min_{t \leq 1} \mathbf{x}(t) \leq x\right],$$

but as this probability has still to be found you may ask what has been gained? The answer is that the whole machinery of the heat equation $\partial w / \partial t = \frac{1}{2} \partial^2 w / \partial x^2$ can now be brought to bear.

The computation is made by a device of Lord Kelvin seen before in §3.5, but first two simpler examples of the idea.

Example 6.5.1 The reflecting Brownian motion $\mathbf{x}^+(t) = |\mathbf{x}(t)|$ is Markovian, both simple and strict, with transition density

$$p^+(t, x, y) = \frac{1}{\sqrt{2\pi t}}\left[e^{-(x-y)^2/2t} + e^{-(x+y)^2/2t}\right] \quad \text{for } x, y \geq 0$$

since for $\mathbf{x}^+$ to go from $x \geq 0$ to $y \geq 0$ is the same as for the free Brownian motion $\mathbf{x}$ to go from x, either to y or to $-y$.

Example 6.5.2 The example is now the absorbing Brownian motion $\mathbf{x}^-$: starting at $\mathbf{x}(0) = x > 0$, it follows the free Brownian motion $\mathbf{x}$ up to the passage time $T = \min(t : \mathbf{x}(t) = 0)$ when it is "killed" – Death comes and it is not seen again.

Obviously, $\mathbf{x}^-$ is also Markovian, and you may read off *its* transition density

$$p^-(t,x,y) = \frac{1}{\sqrt{2\pi t}}\left[e^{-(x-y)^2/2t} - e^{-(x+y)^2/2t}\right] \quad \text{for } x,y \geq 0$$

with the help of the reflection principle.

These are instances of Kelvin's Method of Images: The free Brownian transition density $p(t,x,y) = (2\pi t)^{-1/2}e^{-(x-y)^2/2t}$ describes the diffusion of a unit lump of heat placed at $y = x$ at time $t = 0$. For the reflecting motion, Kelvin places a positive image at $y = -x$; for the absorbing motion, it is the same, only now the image is negative.

Coming back to $\max\mathbf{x} - \min\mathbf{x}$, it is desired to express the transition density[3]

$$p_{ab}(t,x,y) = \mathbb{P}_x\left[\min_{t'\leq t}\mathbf{x}(t') > -a, \max_{t'\leq t}\mathbf{x}(t') < b, \mathbf{x}(t) = y\right]$$
$$= \mathbb{P}_x\left[T_{-a} \wedge T_b > t, \mathbf{x}(t) = y\right]$$

with killing at both $-a < 0$ and $b > 0$, by placement of signed images of $p(t,x,y)$. Figure 6.5.1 shows a grid ($\circ$), with spacing $a+b$, and images ($\bullet$) of alternating sign starting with $+1$ at $y = x$, each being the reflection of its last neighbor across the intervening grid point.

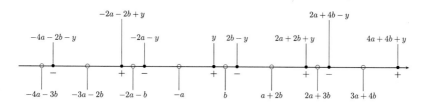

Figure 6.5.1

Obviously, you must first place negative images at the reflection $2b-y$ of y across b and at its reflection $-2a - y$ across $-a$, but that is not

---

[3] Here and below, I use this type of sloppy but self-evident notation for densities, as in $\mathbb{P}_0[\mathbf{x}(t) = x] = (2\pi t)^{-1/2}e^{-x^2/2t}$.

enough: each image must still be reflected (with the opposite sign) across every subsequent point of the grid, as in

$$p_{ab}(t,x,y) = \sum_{n \in \mathbb{Z}} \frac{1}{\sqrt{2\pi t}} \left[e^{-(x-y-2na-2nb)^2/2t} - e^{-(x+y-2na-(2n+2)b)^2/2t} \right].$$

This is not yet proved – it's only a suggestion. It could be done by hand, probabilistically, in the manner of §6.3.1, but let's use the heat equation instead. I write p for p_{ab}, q for the sum, and make a rough proof in three steps for fixed $-a < y < b$.

Step 1.   The sum q converges very fast and solves the heat equation

(1) $\partial w/\partial t = \frac{1}{2}\partial^2 w/\partial x^2$ for $t > 0$ and $-a < x < b$;
(2) q also vanishes at $x = -a$ and at $x = b$, by inspection;
(3) and q reduces to $(2\pi t)^{-1/2}e^{-(x-y)^2/2t}+o(1)$ for $t \downarrow 0$ and $-a < x < b$.

These are the points to be considered.

Step 2.   The same is true for p. Let's check (1). Write $T = T_{-a} \wedge T_b$ for short. Then for $t > 0$, $-a < x < b$ and small $h > 0$,

$$p(t+h,x,y)$$
$$= \mathbb{E}_x\left[T > h, \mathbb{P}_{\mathbf{x}(h)}\big(T > t, \mathbf{x}(t) = y\big) \right] \text{by the (simple) Markov property}$$
$$= \mathbb{E}_x\left[T > h, p\big(t, \mathbf{x}(h), y\big) \right]$$
$$= \mathbb{E}_x\left[T > h, p + \Delta\mathbf{x}\frac{\partial p}{\partial x} + (\Delta\mathbf{x})^2 \frac{1}{2}\frac{\partial^2 p}{\partial x^2} + \mathrm{O}(\Delta\mathbf{x})^3 \right]$$

where $\Delta\mathbf{x} = \mathbf{x}(h) - x$, assuming what is fact, that $p = p(t,x,y)$ is smooth in $-a \leq x \leq b$. Now let c be the shortest distance from x to $-a$ or b. Then

$$\mathbb{P}_x(T \leq h) \leq 2\,\mathbb{P}_0(T_c \leq h) = 2 \int_0^h \frac{c\,e^{-c^2/2t}}{\sqrt{2\pi t^3}}\,dt$$

is very small compared to $h \downarrow 0$, i.e. for small times, the Brownian motion scarcely knows it will be killed at the exit time T, so the proviso $T > h$ can be neglected, with the outcome that

$$p(t+h,x,y) = p(t,x,y) + h \times \frac{1}{2}\frac{\partial^2}{\partial x^2}p(t,x,y) + o(h);$$

in short, (1) holds. (2) is also correct since

$$p(2t,x,y) \leq \mathbb{P}_x(T > t) \times \max_{-a \leq x' \leq b} p(t,x',y),$$

by line three of the big display with $h = t$, and $\mathbb{P}_x(T > t) \leq \mathbb{P}_d(T_0 > t) \downarrow$

0 as $d =$ the distance from x to $-a$ or b tends to zero. Likewise (3) since the proviso $T > h$ can be neglected for small h and $-a < x < b$, as before.

Step 3. This step recapitulates the reasoning of §6.1. By Steps 1 and 2, $w = p - q$ solves (1) and vanishes on the sides and at the bottom of the box $(0 \leq t \leq 1) \times (-a \leq x \leq b)$. It is also continuous there, so if it were ever positive, the modified function $v = \mathrm{e}^{-t}w$ would have a positive maximum inside the box or on its (open) top. But at that place, $0 \leq \partial v/\partial t = -v + \frac{1}{2}\partial^2 v/\partial x^2 \leq -v < 0$ is contradictory, and since the same applies to $-w$, you have $p \equiv q$.

Now integrate

$$p_{ab}(1,0,y) = \sum_{n\in\mathbb{Z}} \frac{1}{\sqrt{2\pi}}\left[\mathrm{e}^{-(y+2na+2nb)^2/2} - \mathrm{e}^{-(y+2na+(2n-2)b)^2/2}\right]$$

from $y = -a$ to b to obtain

$$\mathbb{P}_0\left[\min_{t\leq 1}\mathbf{x}(t) > -a \ \& \ \max_{t\leq 1}\mathbf{x}(t) < b, \mathbf{x}(t) = y\right]$$
$$= \sum_{n\in\mathbb{Z}}\left(\int_{(2n-1)a+2nb}^{2na+(2n+1)b} - \int_{(2n-1)a+(2n-2)b}^{2na+(2n-1)b}\right)\frac{\mathrm{e}^{-y^2/2}}{\sqrt{2\pi}}$$
$$\equiv f(a,b).$$

The joint density for min and max is then $\partial^2 f/\partial a\partial b$, whence

$$\mathbb{P}_0(\max-\min \leq x) = \int_{a+b\leq x}\frac{\partial^2 f}{\partial a\partial b}\,\mathrm{d}a\,\mathrm{d}b = \int_0^x \frac{\partial f}{\partial a}(a,x-a)\,\mathrm{d}a,$$

and this may be reduced (with tears) to the answer:

$$\mathbb{P}_0(\max-\min \leq x) = \int_0^x \sum_{n\in\mathbb{Z}}(-1)^{n-1}4n^2 \frac{\mathrm{e}^{-n^2y^2/2}}{\sqrt{2\pi}}\,\mathrm{d}y.$$

Exercise 6.5.3 Try it if you want.

Exercise 6.5.4 Think over Steps 1 and 2, filling in what details you like.

6.6 Arcsine (reprise)

The function $F(\mathbf{x}) = \mathrm{meas}(t \leq 1 : \mathbf{x}(t) > 0)$ is continuous at almost every Brownian path since $\mathfrak{Z} = (t \leq 1 : \mathbf{x}(t) = 0)$ has measure 0.

Exercise 6.6.1 Think it over.

The invariance principle may then be used to replace the evaluation

$$\lim_{n \uparrow \infty} \mathrm{RW}\left[\frac{1}{n}\,\#(k \le n : \mathbf{s}(k) > 0) \le c\right] = \frac{2}{\pi}\arcsin\sqrt{x}$$

of §3.4 by $\mathrm{BM}[F(\mathbf{x}) \le c]$. But how is that to be computed?

6.6.1 Feynman–Kac (FK)

Much as in §6.5, a new machinery allied to the heat equation comes in to help, but first an introductory story. I follow Kac [1949] but the idea is due to Feynman in a quantum mechanical context, so his name comes first. Think about the free Brownian motion $\mathbf{x}$, subject to hazard (death) described by a (piecewise continuous) function $k(x) \ge 0$, the rule being that if the Brownian traveller lives to reach the place $\mathbf{x}(t)$, then the probability that he (or she) is killed in the next little lapse of time is $k \circ \mathbf{x}(t) \times \mathrm{d}t$. Then the probability of surviving up to the fixed time $t > 0$ would be the product of $1 - k(\mathbf{x})\,\mathrm{d}t'$ for $t' \le t$, and what could that be but $\exp(-\int_0^t k(\mathbf{x})\,\mathrm{d}t')$, which is to say, that conditional on the path $\mathbf{x}$, the "killing time" T is distributed as in

$$\mathbb{P}(T > t|\mathbf{x}) = \mathrm{e}^{-\mathfrak{K}(t)} \quad \text{with } \mathfrak{K}(t) = \int_0^t k(\mathbf{x})\,\mathrm{d}t'.$$

The transition density – to come safely, from $x = \mathbf{x}(0)$ to $y = \mathbf{x}(t)$ – is then

$$p(t,x,y) = \mathbb{E}_x\left[T > t, \mathbf{x}(t) = y\right] = \mathbb{E}_x\left[\mathrm{e}^{-\mathfrak{K}(t)}, \mathbf{x}(t) = y\right],$$

and it is easy to check that, for any function f, of class $C^2(\mathbb{R})$ or better,

$$w(t,x) = \mathbb{E}_x\left[T > t, f \circ \mathbf{x}(t)\right] = \mathbb{E}_x\left[\mathrm{e}^{-\mathfrak{K}(t)} \times f \circ \mathbf{x}(t)\right]$$

solves $\partial w/\partial t = \frac{1}{2}\partial^2 w/\partial x^2 - kw \equiv \mathfrak{G}w$ with $w(0+, \cdot\,) = f$, subject to the proviso that if k jumps at $x = 0$ say, then $w(0-) = w(0+)$ and $\frac{1}{2}[w'(0+) - w'(0-)] = [k(0+) - k(0-)]w(0)$ to keep $\mathfrak{G}w$ in $C(\mathbb{R})$.

The alternative statement of Kac [1951] that for fixed $\alpha > 0$, $\hat{w} = \int_0^\infty \mathrm{e}^{-\alpha t}w\,\mathrm{d}t$ solves $(\alpha - \mathfrak{G})\hat{w} = f$ subject to the same proviso, is more convenient at this juncture. The recipe requires two positive solutions of $\mathfrak{G}h = \alpha h$, the one (h_-) vanishing at $-\infty$, the other (h_+) vanishing at $+\infty$, adjusted to make their (constant) wronskian $h'_- h_+ - h_- h'_+$ identically 2. Then

$$\hat{w}(x) = h_+(x)\int_{-\infty}^x h_- f + h_-(x)\int_x^\infty h_+ f,$$

as you will please check. Figure 6.6.1 shows the preliminary solution $h_c(x)$ for negative $c \leq x \leq 0$ with $h_c(c) = 0$ and $0 < h'_c(c)$ adjusted to make $h_c(0) = 1$. There you will see that $h_a < h_b$ for $b < a$ and that $h_c(x) < e^{\sqrt{2\alpha}x}$ for any $c < 0$. Once this is understood, just take $h_-(x) = \lim_{c\downarrow-\infty} h_c(x)$ for $x \leq 0$ and extend it to $x > 0$, subject to the proviso. The rest will be plain.

The recipe is now applied to the proof of arcsine. The function k is the indicator of the right half-line, multiplied by a fixed number $\beta \geq 0$, and f is identically 1. By inspection, $h_-(x) = e^{\sqrt{2\alpha}x}$ for $x \leq 0$,

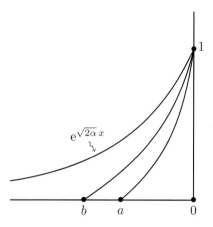

Figure 6.6.1

$h_+(x) = e^{-\sqrt{2(\alpha+\beta)}x}$ for $x \geq 0$, and $h'_- h_+ - h_- h'_+ = \sqrt{2\alpha} + \sqrt{2(\alpha+\beta)}$ at $x = 0$ (which must be adjusted) with the result that

$$
\begin{aligned}
\hat{w}(0) &= \int_0^\infty e^{-\alpha t}\, dt\, \mathbb{E}_c \left[e^{-\beta\, \text{meas}(t' \leq t : x(t) \leq 0)} \right] \\
&= \frac{h_+(0) \int_{-\infty}^0 h_- + h_-(0) \int_0^\infty h_+}{\frac{1}{2}\left[\sqrt{2\alpha} + \sqrt{2(\alpha+\beta)} \right]} \\
&= \frac{1}{\sqrt{\alpha(\alpha+\beta)}}.
\end{aligned}
$$

That's all you need. Just use $\int_0^\infty e^{-\gamma t}(\pi t)^{-1/2}\, dt = \gamma^{-1/2}$ to invert in respect to α to obtain

$$
\mathbb{E}_0 \left[e^{-\beta\, \text{meas}(s \leq t : x(s) \geq 0)} \right] = \frac{1}{t} \int_0^t e^{-\beta s} \frac{ds}{\sqrt{(t-s)s}},
$$

put $t = 1$, and invert *that* in respect to β. Pretty neat, don't you think?

6.6.2 Proof by Brownian paths

Feynman–Kac is often indispensable, but is not really needed here. Sparre-Andersen's combinatorial method of §5.9, promoted via IP from RW(1) to BM(1), states that $\Re(1) = \text{meas}(t \leq 1 : \mathbf{x}(t) > 0)$ is distributed the same as the first time $\mathfrak{z}$ at which the maximum (**b**) of $\mathbf{x}(t) : t \leq 1$ is reached.

Let's think about that for a minute. The fact is that $\mathbf{x}(t) = \mathbf{b}$ at just one time $0 < \mathfrak{z} < 1$. It cannot happen at $t = 0$, nor at $t = 1$ since $\mathbf{x}(t) \leq \mathbf{x}(1)$ for $t \leq 1$ is the same as to say that the reversed Brownian motion $\mathbf{x}^{\downarrow}(t) = \mathbf{x}(1) - \mathbf{x}(1 - t)$ is non-negative – and if the maximum should be seen, as in Figure 6.6.2, *first* at time $\mathfrak{z}$, and *again* at a later time $\mathfrak{z}' < 1$, there would be an intermediate time k/n at which $\mathbf{a} = \mathbf{x}(k/n) < \mathbf{b}$, whereupon $\mathbf{x}$ would make the passage from $\mathbf{a}$ to $\mathbf{b}$ *before time 1 and immediately overshoot*, contradicting $\mathbf{b} = \max_{t \leq 1} \mathbf{x}(t)$.

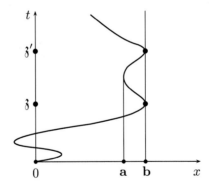

Figure 6.6.2

Now that's settled, it's clear that IP may be used to justify

$$\mathbb{P}\left[\text{meas}(t' \leq 1 : \mathbf{x}(t') > 0) \leq t\right] = \mathbb{P}(\mathfrak{z} < t).$$

The rest is a brief computation: $\mathfrak{z} < t$ is the same as saying that $\mathbf{b} = \max_{t' \leq t} \mathbf{x}(t')$ is larger than

$$\max_{t \leq t' \leq 1} \mathbf{x}(t') = \mathbf{x}(t) + \max_{t \leq t' \leq 1}\left[\mathbf{x}(t' + t) - \mathbf{x}(t)\right] = \mathbf{a} + \max_{t' \leq 1 - t} \mathbf{x}^{+}(t')$$

with a new Brownian motion $\mathbf{x}^{+}$, independent of $\mathbf{a} = \mathbf{x}(t)$ and $\mathbf{b}$, and you know both the joint density

$$\frac{2(2b-a)}{\sqrt{2\pi t^3}}\,e^{-(2b-a)^2/2t} \quad \text{for } \mathbf{a} \text{ and } \mathbf{b} \text{ from Exercise 6.3.2}$$

and also the density

$$\frac{2e^{-c^2/2t}}{\sqrt{2\pi t}} \quad \text{for } \mathbf{c} = \max_{t'\le t}\mathbf{x}(t'),$$

so

$$\mathbb{P}(\mathbf{z} < t) = \int_{\substack{a<b\\b,c>0\\b>a+c}} \frac{2(2b-a)}{\sqrt{2\pi t^3}}\,e^{-(2b-a)^2/2t}\,2\,\frac{e^{-c^2/2(1-t)}}{\sqrt{2\pi(1-t)}}$$

$$= \frac{2}{\pi}\int_0^\infty e^{-c^2/2}\,\mathrm{d}c\int_c^\infty \frac{e^{-b^2/2}}{\sqrt{(1-t)/t}}\,\mathrm{d}b$$

when you work it out, and the density for $\mathbf{z}$ is found to be

$$\frac{2}{\pi}\int_0^\infty e^{-c^2/2}e^{-c^2(1-t)/2t}c\,\mathrm{d}c \times \frac{\mathrm{d}}{\mathrm{d}t}\sqrt{\frac{1-t}{t}} = \frac{1}{\pi}\,\frac{1}{\sqrt{t(1-t)}},$$

as it should be.

Aside. P. Lévy [1948:234] discovered the remarkable fact that for Brownian paths starting at $\mathbf{x}(0) = 0$, $\mathbf{x}^+(t) = \max_{t'\le t}\mathbf{x}(t') - \mathbf{x}(t)$ is a copy of the reflecting Brownian motion $|\mathbf{x}|$. Then $\mathfrak{z}$ is seen to be the last root $t \le 1$ of $\mathbf{x}^+(t) = 0$, or what is distributed the same, the last root of $\mathbf{x}(t) = 0$, from which you see that $\mathfrak{z} < t$ is just as likely as that $\mathbf{x}(t')$ should have no root for $1 \ge t' \ge t$. In short,

$$\mathbb{P}_0(\mathfrak{z} < t) = 2\int_0^\infty \frac{e^{-x^2/2t}}{\sqrt{2\pi t}}\,\mathrm{d}x\,\mathbb{P}_x(T_0 > 1-t)$$

$$= 2\int_0^\infty \frac{e^{-x^2/2t}}{\sqrt{2\pi t}}\,\mathrm{d}x\int_{1-t}^\infty \frac{x\,e^{-x^2/2s}}{\sqrt{2\pi s^3}}\,\mathrm{d}s$$

$$= \frac{2}{\pi}\int_{\sqrt{(1-t)/t}}^\infty \frac{\mathrm{d}s}{1+s^2} \qquad \text{after integrating out the } x$$

$$= \frac{2}{\pi}\left(\frac{\pi}{2} - \arctan\sqrt{\frac{1-t}{t}}\right)$$

$$= \frac{2}{\pi}\arcsin\sqrt{t},$$

yet one more time!

6.6.3★ Still another way by Brownian paths

Not that you need another proof, but it's nice.

Step 1.   Look at Figure 6.6.3. The $\mathfrak{z} = \max(t' \leq t : \mathbf{x}(t') = 0)$ is the last

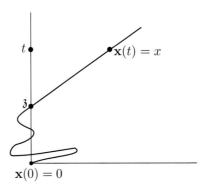

Figure 6.6.3

root before time t. It is *not* a stopping time, but still the past $\mathbf{x}(t') : t' \leq \mathfrak{z}$ and the future $\mathbf{x}(t') : \mathfrak{z} \leq t' \leq t$ are independent, conditional upon $\mathfrak{z}$, i.e. $\mathfrak{z}$ is a "splitting time", so-called. This is most easily understood by running the path backwards: start at x, proceed up to what is now the passage time $t - \mathfrak{z} = \min(t' \leq t : \mathbf{x}(t - t') = 0)$, start over from scratch, and require (by conditioning) that the path end at 0. This backwards attitude leads at once to the joint density for $\mathfrak{z}$ and $\mathbf{x}(t)$: with my sloppy but self-evident notation,

$$\mathbb{P}_0\left[\mathfrak{z} = s, \mathbf{x}(t) = \pm x\right] = \frac{xe^{-x^2/2(t-s)}}{\sqrt{2\pi(t-s)^3}} \times \frac{1}{\sqrt{2\pi s}}$$

for $0 \leq s \leq t$ and $x > 0$; in particular, integrating out the x, you see that $\mathfrak{z}$ is arcsine-distributed:

$$\mathbb{P}_0(\mathfrak{z} = s) = \frac{1}{\pi} \frac{1}{\sqrt{s(t-s)}}.$$

Step 2.   Now let $\mathfrak{K}(t) = \mathrm{meas}(t' \leq t : \mathbf{x}(t') \geq 0)$. The motion $\mathbf{x}^*(t) = \mathbf{x} \circ \mathfrak{K}^{-1}(t)$ is seen in Figure 6.6.4: the substitution $t \to \mathfrak{K}^{-1}(t)$ is cutting out the left-hand excursions of $\mathbf{x}$ and pushing the (shaded) right-hand excursions down. The reflecting Brownian motion $\mathbf{x}^+ : t \to |\mathbf{x}(t)|$ of Example 6.5.1 has a similar portrait: in fact, to make a copy of $\mathbf{x}$ from $\mathbf{x}^+$ you have only to assign independent, equally likely signatures

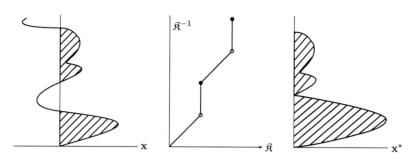

Figure 6.6.4

± 1 to the absolute excursions. This will, I hope, make it clear that $\mathbf{x}^* \colon t \to \mathbf{x} \circ \mathfrak{K}^{-1}(t)$ is statistically identical to $\mathbf{x}^+$.

Step 3.   Now fix $\beta > 0$ and look at the fair game $\mathbf{z}(t) = \mathrm{e}^{\beta \mathbf{x}(t) - \beta^2 t/2}$. The quantity $T = \mathfrak{K}^{-1}(t) = \min(t' : \mathfrak{K}(t') = t)$ is a stopping time, as you will check, and

$$\max_{t' \le \mathfrak{K}^{-1}(t)} \mathbf{x}(t') = \max_{t' \le t} \mathbf{x}^*(t') \quad \text{by inspection of Figure 6.6.4,}$$

from which you see, first, that $\displaystyle\max_{t' \le \mathfrak{K}^{-1}(t)} \mathbf{z}(t')$ is summable; second, that $\mathbb{E}_0[\mathbf{z} \circ \mathfrak{K}^{-1}(t)] = 1$; and third, that

$$\frac{1}{\alpha} = \int_0^\infty \mathrm{e}^{-\alpha t}\, \mathbb{E}_0\left[\mathrm{e}^{\beta \mathbf{x}(\mathfrak{K}^{-1}) - \beta^2 \mathfrak{K}^{-1}/2}\right]\, \mathrm{d}t$$

$$= \mathbb{E}_0\left[\int_0^\infty \mathrm{e}^{-\alpha t} \mathrm{e}^{\beta \mathbf{x}(\mathfrak{K}^{-1}) - \beta^2 \mathfrak{K}^{-1}/2}\, \mathrm{d}t\right]$$

$$= \mathbb{E}_0\left[\int_0^\infty \mathrm{e}^{-\alpha \mathfrak{K}} \mathrm{e}^{\beta \mathbf{x}(t) - \beta^2 t/2}\, \mathbf{1}_{0\infty}(\mathbf{x})\, \mathrm{d}t\right]$$

by the substitution $t \to \mathfrak{K}(t)$ under the expectation – and careful inspection of Figure 6.3.3.

Step 4.   Now $\mathfrak{K}(t) = t - \mathfrak{z} + \mathfrak{K}(\mathfrak{z})$ for $\mathbf{x}(t) > 0$ and $\mathfrak{z}$ is a splitting time, so the display immediately above may be continued as follows:

$$\int_0^\infty \mathrm{d}t\, \mathbb{E}_0\left[\mathrm{e}^{-\alpha(t-\mathfrak{z})} \mathrm{e}^{-\alpha \mathfrak{K}(\mathfrak{z})} \mathrm{e}^{\beta \mathbf{x} - \beta^2 t/2}\, \mathbf{1}_{0\infty}(\mathbf{x})\right]$$

$$= \int_0^\infty \mathrm{e}^{-\gamma t}\, \mathrm{d}t\, \mathbb{E}_0\left[\mathbb{E}_0\left(\mathrm{e}^{-\alpha \mathfrak{K}(\mathfrak{z})} \big| \mathfrak{z}\right) \mathrm{e}^{\alpha \mathfrak{z}} \mathrm{e}^{\beta \mathbf{x}}\, \mathbf{1}_{0\infty}(\mathbf{x})\right] \quad \text{with } \gamma = \alpha + \beta^2/2.$$

Think of the inner expectation from the viewpoint of the backward path:

$\mathfrak{z}$ is its passage time to $x = 0$ and it is conditioned to stop at $\mathbf{x}(0) = 0$, so by scaling,

$$\mathbb{E}_0\big(e^{-\alpha\mathfrak{K}(\mathfrak{z})}\big|\mathfrak{z}\big) = \mathbb{E}_0\big(e^{-\alpha\mathfrak{K}(1)}\big|\mathbf{x}(1) = 0\big) \quad \text{computed for fixed } \mathfrak{z},$$

permitting the display to be continued still more as

$$\int_0^\infty e^{-\gamma t}\,dt \int_0^\infty dx \int_0^t ds \frac{e^{-x^2/2(t-s)}}{\sqrt{2\pi(t-s)^3}} \frac{1}{\sqrt{2\pi s}}\, \mathbb{E}_0\big(e^{-\alpha s\mathfrak{K}(1)}\big|\mathbf{x}(1) = 0\big)e^{\alpha s}e^{\beta x}$$

by Step 1.

Exercise 6.6.2 Reduce all this to

$$\mathbb{E}_0\left(\frac{1}{\sqrt{\mathfrak{K}(1) + c}}\bigg|\mathbf{x}(1) = 0\right) = 2\big(\sqrt{c+1} - \sqrt{c}\big) = \int_0^1 \frac{dx}{\sqrt{x+c}}$$

with $c = \beta^2/2\alpha$.

Exercise 6.6.3 Deduce $\mathbb{P}_0\big(\mathfrak{K}(1) \leq x \,\big|\, \mathbf{x}(1) = 0\big) = x$ or, what is the same, $\mathbb{P}_0\big(\mathfrak{K}(\mathfrak{z}) \leq x|\mathfrak{z}\big) = x/\mathfrak{z}$ for $0 \leq x \leq \mathfrak{z}$.

Now for the punch-line.

$$\mathfrak{K}(1) = \begin{array}{l} \mathfrak{K}(\mathfrak{z}) \;\; \text{if } \mathbf{x}(1) < 0 \\[4pt] 1 - \text{meas}(t \leq 1 : \mathbf{x}(t) < 0) = 1 - \text{a copy of } \mathfrak{K}(\mathfrak{z}) \;\; \text{if } \mathbf{x}(1) > 0; \end{array}$$

also $\mathbb{P}_0\big(\mathbf{x}(1) < 0\,|\,\mathfrak{z}\big) = \mathbb{P}_0\big(\mathbf{x}(1) > 0\,|\,\mathfrak{z}\big) = 1/2$, so with the usual sloppy notation for densities,

$$\mathbb{P}_0\big(\mathfrak{K}(1) = x\big)$$
$$= \mathbb{P}_0\big(\mathfrak{K}(\mathfrak{z}) = x, \mathbf{x}(1) < 0\big) + \mathbb{P}_0\big(1 - \mathfrak{K}(\mathfrak{z}), \mathbf{x}(1) > 0\big)$$
$$= \frac{1}{2}\,\mathbb{E}_0\big[\mathbb{P}_0\big(\mathfrak{K}(\mathfrak{z}) = x\,\big|\,\mathfrak{z}\big] + \frac{1}{2}\,\mathbb{E}_0\big[\mathbb{P}_0\big(\mathfrak{K}(\mathfrak{z}) = 1 - x\,\big|\,\mathfrak{z}\big)\big]$$
$$= \frac{1}{2}\,\mathbb{E}_0\left(\frac{1}{\mathfrak{z}}, \mathfrak{z} \geq x\right) + \frac{1}{2}\,\mathbb{E}_0\left(\frac{1}{\mathfrak{z}}, \mathfrak{z} \geq 1 - x\right)$$
$$= \frac{1}{2}\int_x^1 \frac{1}{z}\frac{1}{\pi}\frac{dz}{\sqrt{z(1-z)}} + s\frac{1}{2}\int_{1-x}^1 \frac{1}{z}\frac{1}{\pi}\frac{dz}{\sqrt{z(1-z)}} \quad \text{by Step 1}$$
$$= \frac{1}{\pi}\sqrt{\frac{1-x}{x}} + \frac{1}{\pi}\sqrt{\frac{x}{1-x}}$$
$$= \frac{1}{\pi}\frac{1}{\sqrt{x(1-x)}}\;!$$

6.7 Skorokhod embedding

F. B. Knight's way of imitating RW by sampling BM, employed in §6.4, is an instance of A. V. Skorokhod's "embedding" to be explained here in general and illustrated by still another proof of CLT (not that you need this either) and by a simple proof of the general law of the iterated logarithm, postponed from §5.8.

Let the variable $\mathbf{z}$ have mean 0, mean-square 1, and distribution function $\mathbb{P}(\mathbf{z} \leq c) = F(c)$; let $\mathbf{x}(t) : t \geq 0$ be the Brownian motion starting at $\mathbf{x}(0) = 0$; and let $T = \min(t : \mathbf{x}(t) = \mathbf{a}$ or $\mathbf{b})$ with random $\mathbf{a} \leq 0 < \mathbf{b}$ distributed, independently of $\mathbf{x}$, by the rule

$$\mathbb{P}(\mathbf{a} \in da \text{ and } \mathbf{b} \in db) = \frac{1}{Z}\,dF(a)\,dF(b)(b - a)$$

with normalizer $Z = -\int_{-\infty}^{0} a\,dF(a) = \int_{0}^{\infty} b\,dF(b)$ reflecting the fact that $\mathbb{E}(\mathbf{z}) = 0$. Then it is easy to see that $\mathbf{x}(T)$ is a copy of $\mathbf{z}$. That's Skorokhod's idea. The proof is simplicity itself. For $T_{ab} = \min(t : \mathbf{x}(t) = a$ or $b)$ with fixed $a < 0 < b$, you know from §6.3 that

(1) $\mathbb{P}_0(T_a < T_b) = \dfrac{b}{b - a}$ and $\mathbb{P}_0(T_b < T_a) = -\dfrac{a}{b - a}$,

(2) $\mathbb{E}_0(T_{ab}) = \left(\dfrac{b - a}{2}\right)^2 - \left(\dfrac{b + a}{2}\right)^2 = -ab,$

and

(3) $\mathbb{E}_0(e^{-\alpha T_{ab}}) = \dfrac{\cosh \sqrt{2\alpha}\left(\dfrac{a + b}{2}\right)}{\cosh \sqrt{2\alpha}\left(\dfrac{b - a}{2}\right)}.$

so by (1),

$$\mathbb{P}_0[\mathbf{x}(t) \leq c] = \frac{1}{Z}\int_{-\infty}^{0+} dF(a)\int_{0+}^{\infty} dF(b)(b - a)\,\mathbb{P}_0[\mathbf{x}(T_{ab}) \leq c]$$

$$= \frac{1}{Z}\int_{-\infty}^{0+} dF(a)\int_{0+}^{\infty} dF(b)\Big[\mathbf{1}_{a \leq c} \times b + \mathbf{1}_{b \leq c} \times (-a)\Big] \text{ by (1)}$$

$$= \int_{-\infty}^{0+} \mathbf{1}_{a \leq c}\,dF(a) + \int_{0+}^{\infty} \mathbf{1}_{b \leq c}\,dF(b)$$

$$= F(c),$$

as advertised, and similarly, by (2),

$$
\begin{aligned}
\mathbb{E}_0(T) &= \frac{1}{Z} \int_{-\infty}^{0+} dF(a) \int_{0+}^{\infty} dF(b)(b-a) \times (-ab) \\
&= \int_{0+}^{\infty} b^2 \, dF(b) + \int_{-\infty}^{0+} a^2 \, dF(a) \\
&= \mathbb{E}(\mathbf{z}^2) \\
&= \mathbb{E}[\mathbf{x}^2(T)].
\end{aligned}
$$

(3) is reserved for later use.

6.7.1 CLT

Now $T = \min(t : \mathbf{x}(t) = \mathbf{a} \text{ or } \mathbf{b})$ is a stopping time in a self-evident extended sense, so the sum $\mathbf{s}(n) = \mathbf{z}_1 + \cdots + \mathbf{z}_n$ of independent copies of $\mathbf{z}$ is statistically identical to $\mathbf{x}(T_1 + \cdots + T_n)$ with $T_1 = T$, $T_2 = T$ recomputed for the shifted path $\mathbf{x}(t + T_1) - \mathbf{x}(T_1)$, $T_3 = T$ recomputed for $\mathbf{x}(t + T_1 + T_2) - \mathbf{x}(T_1 + T_2)$, and so on, these new times, T_2, T_3, etc. being independent copies of $T = T_1$. At this point, the unwary will say:

$$
\begin{aligned}
\frac{\mathbf{s}_n}{\sqrt{n}} &= \frac{1}{\sqrt{n}} \mathbf{x}(T_1 + \cdots + T_n) \\
&= \mathbf{x}\left(\frac{T_1 + \cdots + T_n}{n} \right) \qquad && \text{by Brownian scaling} \\
&\simeq \mathbf{x}(1) && \text{by LLN since } \mathbb{E}_0(T) = 1,
\end{aligned}
$$

but that's fake since line 3 requires a new Brownian motion for each choice of n; indeed, you *know* that CLT is only a *weak* law, so something is wrong. Let's fix that as follows: For large n, $T_1 + \cdots + T_n$ lies between $n(1 - h)$ and $n(1 + h)$ for any $h > 0$ fixed in advance, so

$$
\begin{aligned}
&\mathbb{P}\left[\frac{\mathbf{s}(n)}{\sqrt{n}} \le c \right] \\
&= \mathbb{P}\left[\frac{\mathbf{x}(T_1 + \cdots + T_n)}{\sqrt{n}} \le c, |T_1 + \cdots + T_n - n| \le nh \right] \\
&\quad + \mathbb{P}\left[\frac{\mathbf{x}(T_1 + \cdots + T_n)}{\sqrt{n}} \le c, |T_1 + \cdots + T_n - n| > nh \right] \\
&\le \mathbb{P}\left[\min_{|t-n| \le nh} \frac{\mathbf{x}(t)}{\sqrt{n}} \le c \right] \\
&= \mathbb{P}\left[\min_{|t-1| \le h} \mathbf{x}(t) \le c \right] \qquad \text{by Brownian scaling } \mathbf{x}(t) \to \sqrt{n}\mathbf{x}(t/n),
\end{aligned}
$$

with an error in last two lines not more than

$$\mathbb{P}\left[\left|\frac{T_1 + \cdots + T_n}{n} - 1\right| > h\right] = o(1) \quad \text{for } n \uparrow \infty.$$

In short,

$$\limsup_{n\uparrow\infty} \mathbb{P}\left[\frac{\mathbf{s}(n)}{\sqrt{n}} \le c\right] \le \mathbb{P}\left[\min_{|t-1|\le h} \mathbf{x}(t) \le c\right] \downarrow \mathbb{P}\left[\mathbf{x}(1) \le c\right] \text{ as } h \downarrow 0.$$

Now it's OK after a similar estimate from below with *max* in place of *min*.

6.7.2 The iterated log

This takes just a little more work. The statement is

$$\mathbb{P}\left[\limsup_{n\uparrow\infty} \frac{\mathbf{s}(n)}{\sqrt{2n\log\log n}} = 1\right] = 1,$$

and the cheap half ($\limsup \le 1$) is easy: Exercise 6.4.1 states that

$$\mathbf{x}(t) < \sqrt{(2+)t\log\log t} \quad \text{for } t \uparrow \infty,$$

and you know that $T_1 + \cdots + T_n$ is comparable to n by LLN, so

$$\mathbf{s}(n) = \mathbf{x}(T_1 + \cdots + T_n) < \sqrt{(2+)n\log\log n} \quad \text{for } n \uparrow \infty \text{ with probability 1.}$$

For the rest ($\limsup \ge 1$) look at the variables

$$\mathbf{o}_n = \text{the oscillation osc} = \max - \min \text{ of } \mathbf{x}(t) \text{ for}$$
$$T_1 + \cdots + T_n \le T_1 + \cdots + T_{n+1}.$$

These are independent copies of $\mathbf{o} = \text{osc}\left[\mathbf{x}(t) : t \le T_{\mathbf{a}} \wedge T_{\mathbf{b}}\right]$ of size not more than $\sqrt{n}$ if n is large, this in view of

$$\lim_{a\uparrow\infty} \frac{1}{n}\left(\mathbf{o}_1^2 + \mathbf{o}_2^2 + \cdots + \mathbf{o}_n^2\right) = \mathbb{E}(\mathbf{o}^2)$$

provided the expectation is finite. Grant me that for now and recall the prediction of Exercise 6.4.1 that $\mathbf{x}(t) > \sqrt{(2-)t\log\log t}$ infinitely often as $t \uparrow \infty$. then by choice of such ts and of $n \uparrow \infty$ with $T_1 + \cdots + T_n \le t < T_1 + \cdots + T_{n+1}$, you find

$$\mathbf{o}(n) = \mathbf{x}(T_1 + \cdots + T_n) \ge \mathbf{x}(t) - \mathbf{o}_n > \sqrt{(2-)t\log\log t} - \sqrt{n}$$
$$> \sqrt{(2-)n\log\log n}$$

with a new $(2-)$ a little smaller than the old.

It remains to show that $\mathbb{E}(\mathbf{o}^2)$ is finite. Fix $\mathbf{a} = a \leq 0 < \mathbf{b} = b$. Then

$$\mathbf{o} = \operatorname*{osc}_{t \leq T_{ab}} \mathbf{x}(t) = \max_{t \leq T_a} \mathbf{x}(t) - a \quad \text{or} \quad b - \min_{t < T_b} \mathbf{x}(t)$$

according as $T_a < T_b$ or $T_b < T_a$. But now

$$\mathbb{P}\left[\max_{t \leq T_a} \mathbf{x}(t) < c, T_a < T_b\right] = \mathbb{P}(T_a < T_c) = \frac{c}{c - a} \quad \text{for } c \text{ between } 0 \text{ and } b$$

so

$$\mathbb{E}(\mathbf{o}^2, T_a < T_b) = \int_0^b (c - a)^2 \, \mathrm{d}\frac{c}{(c - a)} = -ab,$$

and likewise $\mathbb{E}(\mathbf{0}^2, T_b < T_a) = -ab$. It remains only to drop the conditioning of $\mathbf{a}$ and $\mathbf{b}$ and to compute $\mathbb{E}(\mathbf{o}^2) = \mathbb{E}(-2\mathbf{ab}) = 2\,\mathbb{E}(\mathbf{z}^2) = 2$, which I leave to you.

6.8 Kolmogorov–Smirnov (reprise)

The tied Brownian motion is BM(1) starting at $\mathbf{x}(0) = 0$, conditioned to end back at $\mathbf{x}(1) = 0$. Here, it is used together with a variant of IP to *prove* rather than to *demonstrate*, and also to interpret, the Kolmogorov–Smirnov formulas of §5.2

$$(1) \quad \lim_{n \uparrow \infty} \mathbb{P}(\sqrt{n}\mathbf{D}_N^+ \geq x) = \mathrm{e}^{-2x^2}$$

and

$$(2) \quad \lim_{n \uparrow \infty} \mathbb{P}(\sqrt{n}\mathbf{D}_N \leq x) = \sum_{n \in \mathbb{Z}} (-1)^n \mathrm{e}^{-2n^2 x^2}$$

for the statistics

$$\mathbf{D}_N^+ = \max_{x \in \mathbb{R}} \left[\frac{1}{n}\#(k \leq n : \mathbf{x}_k \leq x) - F(x)\right]$$

and

$$\mathbf{D}_N = \max_{x \in \mathbb{R}} \left|\frac{1}{n}\#(k \leq n : \mathbf{x}_k \leq x) - F(x)\right|,$$

describing the discrepancy between the empirical frequencies for independent trials of a variable $\mathbf{x}$ and its a priori distribution function F.

6.8.1 Tied Brownian motion

I write $\mathbb{P}_{00}$ and $\mathbb{E}_{00}$ for the tied Brownian probabilities and expectations. Recall that P. Lévy's interpolation scheme, described in §6.2, starts by specifying $\mathbf{x}(1)$. The rest is independent of that, so the tied Brownian motion can be expressed as the free motion $\mathbf{x}$ with "level 0" removed, i.e. as $\mathbf{x}(t) - t\mathbf{x}(1) : 0 \leq t \leq 1$. That's Gaussian with mean 0, and you read off the correlations

$$\mathbb{E}_{00}\big[\mathbf{x}(t_1)\mathbf{x}(t_2)\big] = t_1(1 - t_2) \quad \text{for } 0 \leq t_1 \leq t_2 \leq 1.$$

Here is also an invariance principle for the tied random walk of $2n$ steps, starting at $\mathbf{s}(0) = 0$ and conditioned to end at $\mathbf{s}(2n) = 0$, proved at little expense by a simple trick: if A is any event descriptive of the first n steps of the walk and if n is big, then

$$\mathbb{P}_{00}(A) = \sum_x \frac{\mathbb{P}_0[A, \mathbf{s}(n) = x]\,\mathbb{P}_x[\mathbf{s}(n) = 0]}{\mathbb{P}_0[\mathbf{s}(2n) = 0]} \lesssim \sqrt{2}\,\mathbb{P}_0(A);$$

for example, if n is even, it is easy to see that $\mathbb{P}_x[\mathbf{s}(n) = 0]$ peaks at $x = 0$, so that

$$\frac{\mathbb{P}_x[\mathbf{s}(n) = 0]}{\mathbb{P}_0[\mathbf{s}(2n) = 0]} \leq \frac{\binom{n}{n/2}2^{-n}}{\binom{2n}{n}2^{-2n}} \simeq \sqrt{2} \quad \text{by Stirling's formula,}$$

the implication being that the broken line $\mathbf{x}$ joining the points $t = k/2n$ and $x = \mathbf{s}(k)/\sqrt{2n}$ for the tied, scaled walk inherits from the free walk a compactness of paths as in §6.3 (reprise), by application of this principle to the two like segments $\mathbf{s}(k) : k \leq n$ and $\mathbf{s}(2n - k) : k \leq n$ of the present tied walk. The rest goes as before: CLT, plus a little extra work for the conditioning, says that $\mathbf{x}$ is nearly Gaussian, so you have only to check the correlation: with $0 < t_1 < t_2 < 1$, $i/2n \simeq t_1$, and $j/2n \simeq t_2$, you find

$$\mathbb{E}\left[\mathbf{x}\Big(\frac{i}{2n}\Big)\mathbf{x}\Big(\frac{j}{2n}\Big)\right] \text{ for the scaled walk}$$

$$= \frac{\mathbb{P}_0\big[(2n)^{-1/2}s(i)(2n)^{-1/2}\mathbf{s}(j), \mathbf{s}(2n) = 0\big]}{\mathbb{P}_0\big[\mathbf{s}(2n) = 0\big] \simeq (2\pi n)^{-1/2}}$$

$$\simeq \int x_1 x_2 \frac{e^{-x_1^2/2t_1}}{\sqrt{2\pi t_1}}\,\frac{e^{-(x_2-x_1)^2/2(t_2-t_1)}}{\sqrt{2\pi(t_2 - t_1)}}\,\frac{e^{-x_2^2/2(1-t_2)}}{\sqrt{2\pi(1 - t_2)}}$$

$$\times \sqrt{2\pi} \text{ from the conditioning}$$

$$= \mathbb{E}_{00}\big[\mathbf{x}(t_1)\mathbf{x}(t_2)\big] \text{ for the tied Gaussian motion.}$$

In short, the tied, scaled walk imitates the tied Brownian motion for large n.

6.8.2 Tied Poisson walk

In §5.2, (1) and (2) were reduced to the computation of

$(1')$ $\lim\limits_{n\uparrow\infty} \mathbb{P}\left[\dfrac{1}{\sqrt{2n}}\max\limits_{k\le 2n}\mathbf{s}(k)\ge x \,\Big|\, \mathbf{s}(2n)=0\right]$

and

$(2')$ $\lim\limits_{n\uparrow\infty} \mathbb{P}\left[\dfrac{1}{\sqrt{2n}}\max\limits_{k\le 2n}|\mathbf{s}(k)|\le x \,\Big|\, \mathbf{s}(2n)=0\right]$

for the Poisson walk $\mathbf{s}(0)=0$, $\mathbf{s}(k)=\#_1+\cdots+\#_k-k$ ($1\le k\le 2n$) with independent copies of the variable $\#$ distributed by $\mathbb{P}(\#=j)=1/ej!$, and then *evaluated* by switching from this walk to the standard walk RW(1), hoping for the best. This may now be justified, or rather circumvented, by an invariance principle for the tied Poisson walk, leading to the tied Brownian motion much as for RW(1) above. The ingredients are the same: compactness of paths, Gaussian character in the limit as $n\uparrow\infty$, and tied Brownian correlations.

I begin with the free Poisson walk and the corresponding broken line $\mathbf{x}(t): 0\le t\le 1$ with $\mathbf{x}(k/n)=\mathbf{s}(k)/\sqrt{n}$ for $0\le k\le n$. The walk is a fair game, so Doob's inequality applies to the 6th powers as follows: up to unnecessary constants,

$$\mathbb{P}\left[\left|\mathbf{x}\left(\dfrac{j}{n}\right)-\mathbf{x}\left(\dfrac{i}{n}\right)\right|>n^{-\alpha}\text{ for some }0\le i<j\le n\text{ with }j-i<n^\beta\right]$$
$$\le (n\text{ for choice of }i)\times\mathbb{P}\left[\max\limits_{k\le n^\beta}|\mathbf{s}(k)|>n^{1/2-\alpha}\right]$$
$$\le n\times n^{-3+6\alpha}\times\mathbb{E}[\mathbf{s}^6(\ell)]\quad \ell\text{ being the biggest whole number }\le n^\beta$$
$$\le n^{-2+6\alpha+3\beta}$$

since the main part of the expectation comes from $3\binom{\ell}{3}\le\ell^3$ choices of 3 unequal pairs. This guarantees compactness of paths as in §6.3 (reprise) by choice of $\alpha=1/12$ and $\beta=1/8$, in which case $2-6\alpha+3\beta=-9/8$, and this compactness may now be transferred from the free walk to the tied walk by the type of estimate seen before:

$$\dfrac{\mathbb{P}_x[\mathbf{s}(n)=0]}{\mathbb{P}_0[\mathbf{s}(2n)=0]}\lesssim\sqrt{2}\quad\text{for large }n\text{, independently of }x.$$

I leave that to you. The rest of the proof is the same as in §6.3 (reprise).

6.8.3 Evaluations

This invariance principle permits the expression of (1) and (2) in terms of the tied Brownian motion as

(1') $\mathbb{P}_{00}\left[\max_{0 \le t \le 1} \mathbf{x}(t) \ge x\right]$

and

(2') $\mathbb{P}_{00}\left[\max_{0 \le t \le 1} |\mathbf{x}(t)| \le x\right]$

which have now to be evaluated.

(1') is easy: with the free Brownian motion $\mathbf{x}$ and the usual sloppy notation for densities

$$\mathbb{P}_0\left[\max_{s \le t} \mathbf{x}(s) < b, \mathbf{x}(t) = a\right]$$

$$= \mathbb{P}_0\left[T_b > t, \mathbf{x}(t) = a\right]$$

$$= \mathbb{P}_b\left[T_0 > t, \mathbf{x}(t) = b - a\right] \quad \begin{array}{l} \text{by reflection of Figure 6.8.1(a)} \\ \text{across } b/2 \text{ to obtain Figure 6.8.1(b)} \end{array}$$

$$= \frac{e^{-a^2/2t}}{\sqrt{2\pi t}} - \frac{e^{-(2b-a)^2/2t}}{\sqrt{2\pi t}} \quad \text{by Exercise 6.5.2.}$$

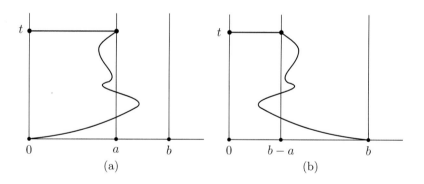

Figure 6.8.1

Now put $t = 1$, $a = 0$, and $b = x$ and multiply by $\sqrt{2\pi}$ for the conditioning $\mathbf{x}(1) = 0$ to produce

$$\mathbb{P}_{00}\left[\max_{0 \le t \le 1} \mathbf{x}(t) < x\right] = 1 - e^{-2x^2}.$$

($2'$) is computed by Kelvin's method of images by taking $t = 1$, $x = 0$, $y = 0$, and $a = b = x$ in the identity of §6.5:

$$\mathbb{P}_x\left[\min_{s\le t}\mathbf{x}(s) > -a, \max_{s\le t}\mathbf{x}(s) < b, x(t) = y\right]$$

$$= \sum_{n\in\mathbb{Z}}\frac{1}{\sqrt{2\pi t}}\left[e^{-(x-y-2na-2nb)^2/2t} - e^{-(x+y-2na-(2n-2)b)^2/2t}\right]$$

to produce

$$\mathbb{P}_{00}\left[\max_{t\le 1}|\mathbf{x}(t)| < x\right] = \mathbb{P}_0\left[\max_{t\le 1}|\mathbf{x}(t)| < x, \mathbf{x}(1) = 0\right] \times \sqrt{2\pi}$$

$$= \sum_{n\in\mathbb{Z}}\left[e^{-(4nx)^2/2} - e^{-((4n+2)x)^2/2}\right]$$

$$= \sum_{n\in\mathbb{Z}}(-1)^n e^{-2n^2x^2}$$

as it should be.

Exercise 6.8.1   The tied Brownian motion is written $\mathbf{x}(t) - t\mathbf{x}(1)$ with a free Brownian motion $\mathbf{x}$. Then the event $\max_{t\le 1}[\mathbf{x}(t) - t\mathbf{x}(1)] \ge x$ has the same probability as $\max_{t\le 1} t\mathbf{x}(1/t) - t\mathbf{x}(1) \ge x$, by the inversion of Exercise 6.2.2, which is to say that the new free Brownian motion $\mathbf{x}^+(t) = \mathbf{x}(t+1) - \mathbf{x}(1)$ rises above $(t+1)x$ for some $t \ge 0$. Now use the fair game $e^{\alpha\mathbf{x}^+(t)-\alpha^2t/2}$ with $\alpha > 2x$ to verify $\mathbb{E}_0(e^{-\beta T}) = \exp\left(-x(x+\sqrt{2\beta + x^2})\right)$ for $\beta > 0$ and $T = \min\left(t : \mathbf{x}^+(t) = (t+1)x\right)$, and to conclude that $\mathbb{P}_{00}\left[\max_{t\le 1}\mathbf{x}(t) \ge x\right] = \mathbb{P}_0(T < \infty) = e^{-2x^2}$, as before.

6.9 Itô's lemma

BM(1) has a lovely differential/integral calculus, very helpful for computation and easy to use. I will want it in connection with statistical mechanics out of equilibrium (Chapter 11) and again Chapter 12, §12.4. Mostly, it is the invention of K. Itô (1940), though P. Lévy (1930) had some idea of it, and W. Döblin (1940) really understood it, too, as found in private papers, recently opened (2000) long after his unfortunate death (1940). You may wish to look at McKean [1969] for a more advanced but hopefully still accessible account. I write throughout $\mathbf{b}(t) : t \ge 0$ for the Brownian motion starting at $\mathbf{b}(0) = 0$.

6.9.1 Brownian integrals and differentials

BM(1) is *nowhere differentiable*: in fact, it is not even true that $|\mathbf{b}(t_2) - \mathbf{b}(t_1)| \le c(\mathbf{b})(t_2 - t_1)^{1/2}$ for $0 \le t_1 < t_2 \le 1$, as seen already in §6.2.

Evidently, the familiar calculus you know from school cannot be applied, and you will think it more than queer when I tell you that the Brownian differential $\mathrm{d}\mathbf{b}(t)$ squared is $\mathrm{d}t$, as in the table!

	db	dt
db	dt	0
dt	0	0

6.9.2 An example

This will show that I'm not talking nonsense. The integral $\int_0^1 \mathbf{b}(t)\,\mathrm{d}\mathbf{b}(t)$ is declared to be

$$\lim_{n\uparrow\infty} \sum_{1\le k\le 2^n} \mathbf{b}\left(\frac{k-1}{2^n}\right)\left[\mathbf{b}\left(\frac{k}{2^n}\right) - \mathbf{b}\left(\frac{k-1}{2^n}\right)\right],$$

in which you observe that the "differential" $\mathbf{b}(k/2^n) - \mathbf{b}((k-1)/2^n)$ sticks out into the future and so is independent of the "integrand" $\mathbf{b}((k-1)/2^n)$, but more of that below. Now Itô's rule says that

$$\mathrm{d}(\mathbf{b}^2) = (\mathbf{b} + \mathrm{d}\mathbf{b})^2 - \mathbf{b}^2 = 2\mathbf{b}\,\mathrm{d}\mathbf{b} + (\mathrm{d}\mathbf{b})^2 = 2\mathbf{b}\,\mathrm{d}\mathbf{b} + \mathrm{d}t,$$

as per the table, leading, not to the naive $\int_0^1 \mathbf{b}\,\mathrm{d}\mathbf{b} = \frac{1}{2}\mathbf{b}^2(1)$, but to $\frac{1}{2}\mathbf{b}^2(1) - \frac{1}{2}$. Here's why. The sum is multiplied by 2 and re-arranged as in

$$\sum_{1\le k\le 2^n}\left[\mathbf{b}^2\left(\frac{k}{2^n}\right) - \mathbf{b}^2\left(\frac{k-1}{2^n}\right)\right] - \sum_{1\le k\le 2^n}\left[\mathbf{b}\left(\frac{k}{2^n}\right) - \mathbf{b}\left(\frac{k-1}{2^n}\right)\right]^2$$

$$= b^2(1) - \left[b^2(0) = 0\right] + \sum_{1\le k\le 2^n}|\Delta\mathbf{b}|^2 \quad \text{with an obvious notation,}$$

so you need the sum $\sum_{1\le k\le 2^n}|\Delta\mathbf{b}|^2$ to tend to 1 as $n \uparrow \infty$. Now $\sum_{1\le k\le 2^n}\left[|\Delta\mathbf{b}|^2 - 2^{-n}\right]$ is the sum of 2^n independent copies of $\mathbf{b}^2(2^{-n}) - 2^{-n}$. This, in turn, is a copy of $2^{-n}[\mathbf{b}^2(1) - 1]$ by Brownian scaling, so

208   *Brownian Motion*

ordinary Chebyshev says that

$$\mathbb{P}\left[\left|\sum_1^{2^n}\left[|\Delta\mathbf{b}|^2-2^{-n}\right]\right|\geq n\,2^{-n/2}\right]\leq\frac{2^n}{n^2}\,\mathbb{E}\left|\sum_1^{2^n}\left[|\Delta\mathbf{b}|^2-2^{-n}\right]\right|^2$$

$$=\frac{2^n}{n^2}\times 2^n\times 2^{-2n}\,\mathbb{E}\left|\mathbf{b}^2(1)=1\right|^2$$

$$=\frac{2}{n^2}\,,$$

and Borel–Cantelli does the rest:

$$\mathbb{P}\left[\left|\sum_1^{2^n}\left[|\Delta\mathbf{b}|^2-2^{-n}\right]\right|<n\,2^{-n/2}\text{ for }n\uparrow\infty\right]=1$$

Exercise 6.9.1   Try this improvement:

$$\mathbb{P}\left[\sum_{k/2^n\leq t}|\Delta\mathbf{b}|^2=t\quad\begin{array}{l}\text{with an absolute error }<n\,2^{-n/2},\\\text{independently of }0\leq t\leq 1.\end{array}\right]=1$$

Hint: Look at $\mathbf{z}(t)=\sum_{k/2^n\leq t}|\Delta\mathbf{b}|^2-2^{-n}+|\mathbf{b}(t)-\mathbf{b}(l/2^n)|^2-(t-l/2^n)$ in which l is the last $k\leq 2^nt$, this being a fair game.

6.9.3 Itô's lemma

Now try this: with $f\in C^2(\mathbb{R})$, you should have

$$\mathrm{d}f(\mathbf{b})=f(\mathbf{b}+\mathrm{d}\mathbf{b})-f(\mathbf{b})=f'(\mathbf{b})\,\mathrm{d}\mathbf{b}+\frac{1}{2}f''(\mathbf{b})\times\left[(\mathrm{d}\mathbf{b})^2=\mathrm{d}t\right],$$

by Itô's way of thinking, and so also

$$f\circ\mathbf{b}(t)=f\circ\mathbf{b}(0)+\int_0^t f'(\mathbf{b})\,\mathrm{d}\mathbf{b}+\frac{1}{2}\int_0^t f''(\mathbf{b})\,\mathrm{d}t',$$

in which the Brownian integral is taken in the same style as for $f(\mathbf{b})=\mathbf{b}$ above, with the differential $\mathrm{d}\mathbf{b}$ sticking out into the future. That's what I call Itô's lemma.

Proof for $t=1$.   Take $1\leq k\leq 2^n$ and write $\Delta\mathbf{b}=\mathbf{b}\big(\frac{k}{2^n}\big)-\mathbf{b}\big(\frac{(k-1)}{2^n}\big)$ as before. Then

$$f\left[\mathbf{b}\left(\frac{k}{2^n}\right)\right]-f\left[\mathbf{b}\left(\frac{k-1}{2^n}\right)\right]\qquad\qquad=(1)$$

$$=f'\left[\mathbf{b}\left(\frac{k-1}{2^n}\right)\right]\times\Delta\mathbf{b}+\frac{1}{2}f''\left[\mathbf{b}\left(\frac{k-1}{2^n}\right)\right]\times(\Delta\mathbf{b})^2\qquad=(2)+(3),$$

with an error not more than

$$(\Delta \mathbf{b})^2 \times \frac{1}{2} \max \left| f''(\mathbf{x}) - f'' \left[\mathbf{b} \left(\frac{k-1}{2^n} \right) \right] \right| \quad \begin{array}{l} \text{for some } \mathbf{x} \text{ between} \\ \mathbf{b} \left(\frac{k-1}{2^n} \right) \text{and } \mathbf{b} \left(\frac{k}{2^n} \right) \end{array} = (4).$$

As to the sum $f[\mathbf{b}(1)] - f[\mathbf{b}(0) = 0]$ of the terms in (1), (4) contributes not more than $\mathrm{o}(1) \times \sum (\Delta \mathbf{b})^2$ by continuity of paths, and so may be neglected, and (3) contributes

$$\sum_{k \le 2^n} \frac{1}{2} f'' \left[\mathbf{b} \left(\frac{k-1}{2^n} \right) \right] \times \frac{1}{2^n} \simeq \frac{1}{2} \int_0^1 f''(\mathbf{b}) \, \mathrm{d}t',$$

with an additional error

$$\mathbf{e} = \sum_{k \le 2^n} \frac{1}{2} f'' \left[\mathbf{b} \left(\frac{k-1}{2^n} \right) \right] \times \left[(\Delta \mathbf{b})^2 - \frac{1}{2^n} \right],$$

which is also small in view of

$$\mathbb{P}(\mathbf{e} > n \, 2^{-n/2}) \le \frac{2^n}{n^2} \, \mathbb{E}(\mathbf{e}^2)$$

$$= \frac{2^n}{n^2} \frac{1}{4} \sum_{k \le 2^n} \mathbb{E} \left| f'' \left[\mathbf{b} \left(\frac{k-1}{2^n} \right) \right] \right|^2 \times \left[(\Delta \mathbf{b})^2 - \frac{1}{2^n} \right]^2$$

$$\le \frac{1}{n^2} \times \max_{\mathbb{R}} |f''|^2.$$

This leaves only (2) which contributes

$$\lim_{n \uparrow \infty} \sum_{k \le 2^n} f' \left[\mathbf{b} \left(\frac{k-1}{2^n} \right) \right] \left[\mathbf{b} \left(\frac{k}{2^n} \right) - \mathbf{b} \left(\frac{k-1}{2^n} \right) \right] + \frac{1}{2} \int_0^1 f''(\mathbf{b}) \, \mathrm{d}t,$$

the existence of the limit – now *declared* to be $\int_0^1 f'(\mathbf{b}) \, \mathrm{d}\mathbf{b}$ – being guaranteed by the rest.

Note 6.9.2　Observe that it is essential in dealing with (3) and (4) that the differential $\mathbf{b}(k/2^n) - \mathbf{b}((k-1)/2^n)$ should stick out into the future and so be independent of, e.g. $f'' [\mathbf{b}((k-1)/2^n)]$.

Note 6.9.3　Obviously, the extra differential $\partial f / \partial t \times \mathrm{d}t$ must be included if f depends not only on the path, but (smoothly) on the time as well, as in $\mathrm{d}(\mathbf{b}^2 - t) = 2\mathbf{b} \, \mathrm{d}\mathbf{b} + \frac{1}{2} \times 2(\mathrm{d}\mathbf{b})^2 - \mathrm{d}t = 2\mathbf{b} \, \mathrm{d}\mathbf{b}$. Now think about that special case this way: $\int_0^t \mathrm{d}\mathbf{b}(t_1) \int_0^{t_1} \mathrm{d}\mathbf{b}(t_2) = \int_0^t \mathbf{b} \, \mathrm{d}\mathbf{b}$ should have been $\mathbf{b}^2(t)/2$, but is not – it's "$\mathbf{b}^2/2$" $= \mathbf{b}^2/2 - t/2$ instead. What then is "$\mathbf{b}^3/3!$"? It ought to be

$$\int_0^t \mathrm{db}(t_1) \int_0^{t_1} \mathrm{db}(t_2) \int_0^{t_2} \mathrm{db}(t_3) = \int_0^t \left(\frac{\mathbf{b}^2}{2} - \frac{t'}{2}\right) \mathrm{db}(t')$$

$$= \frac{\mathbf{b}^3(t)}{3!} + \text{corrective powers of } t \text{ and } \mathbf{b}(t):$$

in fact,

$$\mathrm{d}(\mathbf{b}^3/6 - t\mathbf{b}(t)/2) = \frac{1}{2}\mathbf{b}^2\,\mathrm{db} + \frac{1}{2}\mathbf{b}(\mathrm{db})^2 - \frac{1}{2}\,\mathrm{dt}\,\mathbf{b} - \frac{1}{2}\,\mathrm{db} = \frac{1}{2}(\mathbf{b}^2 - t)\,\mathrm{db}:$$

so that's what it is – "$(\mathbf{b}^3/3!)$" $= \mathbf{b}^3/6 - t\mathbf{b}/2$. More generally, the differential of $\mathbf{z}(t) = \exp\left(\alpha\mathbf{b}(t) - \frac{1}{2}\alpha^2 t\right)$ is $\mathbf{z} \times \left(\alpha\,\mathrm{db} + \frac{1}{2}\alpha^2(\mathrm{db})^2 - \frac{1}{2}\alpha^2\,\mathrm{dt}\right) = \alpha\mathbf{z}\,\mathrm{db}$ so $\mathbf{z}$ is, so to say, "$\exp(\alpha\mathbf{b})$", and you have

$$\mathbf{z}(t) = \mathbf{z}(0) = 1 + \int_0^t \alpha\mathbf{z}\,\mathrm{db} = 1 + \sum_{n=1}^\infty \alpha^n \frac{\text{"}\mathbf{b}^n(t)\text{"}}{n!},$$

with

$$\frac{\text{"}\mathbf{b}^n(t)\text{"}}{n!} = \int_0^t \mathrm{db}(t_1) \int_0^{t_1} \mathrm{db}(t_2) \cdots \int_0^{t_{n-1}} \mathrm{db}(t_n)$$

by a formal iteration. That's all perfectly correct *pathwise* with fast convergence of the sum, but let's pass on. I write it out only to give you some feeling for how this new Brownian calculus looks.

Note 6.9.4   The integral is easily generalized to *non-anticipating* integrands $e(t, \mathbf{b})$ depending, for fixed t, only on the past $\mathbf{b}(t') : t' \le t$. The existence of

$$\int_0^t e\,\mathrm{db} = \lim_{n\uparrow\infty} \sum_{k/2^n \le t} e\left(\frac{k-1}{2^n}\right)\left[\mathbf{b}\left(\frac{k}{2^n}\right) - \mathbf{b}\left(\frac{k-1}{2^n}\right)\right]$$

simultaneously for all $t \le 1$ is easy to come by if, e.g. e is continuous in t when the Brownian path is fixed, and here you notice the essential *technical* point which I repeat for emphasis: that the so-called differential $\mathbf{b}(k/2^n) - \mathbf{b}((k-1)/2^n)$ *sticks out into the future and so is independent of* $e((k-1)/2^n)$.

Exercise 6.9.5   Convince yourself that $\mathbb{E}\left(\int_0^1 e\,\mathrm{db}\right) = 0$ and $\mathbb{E}\left(\int_0^1 e\,\mathrm{db}\right)^2 = \mathbb{E}\left(\int_0^1 e^2\,\mathrm{dt}\right)$ provided this last expectation is finite. Just argue it formally, from the display just above.

Note 6.9.6   The special case, when e is constant, is already informative. Then $\int_0^t e\,\mathrm{db} = e\mathbf{b}(t) = \mathbf{b}^*(e^2 t)$ with a new Brownian motion $\mathbf{b}^*$, by Brownian scaling, suggesting more generally that $\int_0^t e\,d\mathbf{b} = \mathbf{b}^*(C)$ with a new Brownian $\mathbf{b}^*$, run with the clock $C(t) = \int_0^t e^2\,dt'$.

Exercise 6.9.7 Check this by hand in the case $e(t, \mathbf{b})$ is mostly flat, changing with time only by occasional jumps at $t = k/2^n$, say.

Exercise 6.9.8 The relation between $\mathbf{b}$ and $\mathbf{b}^*$ is too complicated to help with computation, but the idea is useful. Already here, it shows, e.g. that $\int_0^1 e\,d\mathbf{b}$ cannot be defined for almost all Brownian paths unless $\mathbb{P}\left[\int_0^1 e^2\,dt < \infty\right] = 1$. See why?

Note 6.9.9 Now let's start the other way around, from the differential equation $d\mathbf{x} = d\mathbf{b} - k\mathbf{x}\,dt$ with constant $k > 0$, to be solved for the unknown $\mathbf{x}$ in terms of $\mathbf{b}$. Evidently, $\mathbf{x}$ is just the free Brownian motion, pulled back towards the origin by a restoring drift proportional to displacement, as if it were attached by a simple spring as in Figure 6.9.1.

Figure 6.9.1

Obviously,

$$d e^{kt}\mathbf{x} = e^{kt}(k\,dt\,\mathbf{x} + d\mathbf{x}) = e^{kt}\,d\mathbf{b},$$

or, what is the same,

$$\mathbf{x}(t) = e^{-kt}\mathbf{x}(0) + e^{-kt}\int_0^t e^{kt'}\,d\mathbf{b}(t'),$$

and it is equally obvious that $\mathbf{x}$ is Markovian: in fact, for fixed $T > 0$, $\mathbf{b}^+(t) \equiv \mathbf{b}(t+T) - \mathbf{b}(T)$, and $\mathbf{x}^+(t) \equiv \mathbf{x}(t+T)$, you have

$$\mathbf{x}^+(t) = e^{-kt}\mathbf{x}^+(0) + e^{-kt}\int_0^t e^{kt'}\,d\mathbf{b}^+(t'),$$

by inspection, which is to say that at time T, $\mathbf{x}$ starts from scratch at the place $\mathbf{x}(T)$ with a new Brownian motion $\mathbf{b}^+$, independent of the past $\mathbf{x}(t') : t' \leq T$. That's the process (OU) of Ornstein–Uhlenbeck (1930). It will reappear in §§6.11 and 11.3.

Note 6.9.10 BM(1) is associated with the "infinitesimal operator" $\mathfrak{G} = \frac{1}{2}\partial^2/\partial x^2$ in the sense that $e(t, x, y) = (2\pi t)^{-1/2}\exp\left[-(x-y)^2/2t\right]$ is at once the elementary solution of the heat equation $\partial w/\partial t = \mathfrak{G}w$ and the transition density of BM(1), and you may ask: What is the analogue of that for OU? Itô's lemma provides a quick way in. For f of class $C^2(\mathbb{R})$,

$$w(t,x) = \mathbb{E}[f \circ \mathbf{x}(t)|\mathbf{x}(0) = x] = \mathbb{E}_0\left[f\left[e^{-kt}x + e^{-kt}\int_0^t e^{kt'}\, d\mathbf{b}(t')\right]\right]$$

is also of class C^2, and $\partial w/\partial t = \mathfrak{G}w$ with $\mathfrak{G} = \frac{1}{2}\partial^2/\partial x^2 - kx\partial/\partial x = $ the Brownian operator $\frac{1}{2}\partial^2/\partial x^2$, modified by the drift.

Exercise 6.9.11 Check $df(\mathbf{x}) = f'(\mathbf{x})\,d\mathbf{b} + \mathfrak{G}f(\mathbf{x})\,dt$ and integrate back, to conclude that $\partial w/\partial t = \mathfrak{G}u$ at $t = 0$. Then use the (simple) Markovian character of $\mathbf{x}$ to promote this up to $\partial w/\partial t = \mathfrak{G}w$ for all $t \geq 0$.

Exercise 6.9.12 Write $\mathbf{x}(t) = e^{-kt}\mathbf{x}(0) + e^{-kt}\mathbf{b}^*\left[(e^{2kt} - 1)/2k\right]$ with a new Brownian motion $\mathbf{b}^*$, run with the clock $(e^{2kt} - 1)/2k$; read off the transition density $e(t,x,y) = (2\pi\sigma^2)^{-1/2}\exp\left[-(y - e^{-kt}x)^2/2\sigma^2\right]$ with $\sigma^2 = (1 - e^{-2kt})/2k$, and check (by hand) that e really is the elementary solution of $\partial w/\partial t = \mathfrak{G}w$. Note that e, taken with $k = 1/2$ and the time doubled, is the kernel employed in §5.5 for the proof of the logarithmic Sobolev inequality.

Important Note. Write $w(0+, \cdot) = f$. Obviously, $\dot{w} = \partial w/\partial t$ is the solution of $\partial\dot{w}/\partial t = \mathfrak{G}\dot{w}$ with $\dot{w}(0+, \cdot) = \mathfrak{G}f$, so with a subscript to indicate how $\mathfrak{G}$ acts,

$$\int(\mathfrak{G}_x e)f\,dy = \mathfrak{G}\int ef\,dy = \dot{w} = \int e\mathfrak{G}f\,dy = \int(\mathfrak{G}_y^\dagger e)f\,dy,$$

where $\mathfrak{G}^\dagger$ is the dual or transposed of $\mathfrak{G}$, viz. $\mathfrak{G}^\dagger = \frac{1}{2}\partial^2/\partial y^2 + k(\partial/\partial y)y$, produced by partial integration, and since $\dot{w}$ may also be expressed as $\int(\partial e/\partial t)f\,dy$, you see that e obeys both the *backward* equation $\partial e/\partial t = \mathfrak{G}_x e$, so called because $x = \mathbf{x}(0)$ is where you came from, and also the *forward* equation $\partial e/\partial t = \mathfrak{G}_y^\dagger e$, so called because $y = \mathbf{x}(t)$ is where you go to. Then $\mathfrak{G}$ is the *backward* and $\mathfrak{G}^\dagger$ the *forward* infinitesimal operator. For $\mathfrak{G} = \frac{1}{2}\partial^2/\partial x^2$, it makes no difference, but in the present case and more importantly in Chapter 11, it *does*, and it's best to keep things clear. Just keep in mind the very simplest case, of translation of speed 1, where $y = \mathbf{x}(t) = \mathbf{x}(0) + t = x + t$, so that $\partial/\partial t$ acts backwards like $\partial/\partial x$ and forwards like $(\partial/\partial x)^\dagger = -\partial/\partial x$.

Exercise 6.9.13 Now look at OU after some large epoch T, writing $\mathbf{x}(t + T) = \mathbf{x}_T(t)$. Compute the correlation $\mathbb{E}[\mathbf{x}_T(t_1)\mathbf{x}_T(t_2)]$. Conclude that as $T \uparrow \infty$, $\mathbf{x}_T$ tends to the shift-invariant Gaussian process

$\mathbf{x}_\infty$, with correlation $\mathbb{E}[\mathbf{x}_\infty(t_1)\mathbf{x}_\infty(t_2)] = (2k)^{-1}e^{-k|t_2-t_1|}$, modeled by $e^{-kt}\mathbf{b}(e^{2kt}/2k)$. This is the stationary OU process, proceeding so to say in equilibrium, having forgotten where it came from. It will be wanted in §6.11.3.

6.9.4 Robert Brown and Einstein

This will be a good place to explain, what was promised at the start of this chapter, how Ornstein-Uhlenbeck [1930] reconciled Einstein's (1905) picture of the Brownian motion as *displacement* with the fact that it is the *velocity* of Robert Brown's pollen. Write $\mathbf{x}$ for displacement and $\mathbf{v}$ for velocity to keep things clear. Newton says:

$$(\text{mass} = 1 \text{ say}) \times \left(\text{acceleration} = \frac{d^2\mathbf{x}}{dt^2}\right) = \text{force} = -k \times \left(\mathbf{v} = \frac{d\mathbf{x}}{dt}\right) + \frac{d\mathbf{b}}{dt},$$

in which k is a positive constant, $k\mathbf{v}$ is the drag (air restance) and $d\mathbf{b}/dt$ caricatures the force due to the molecules of the air in their incessant thermal agitation, colliding with Robert Brown's pollen. Don't worry about the non-existent $d\mathbf{b}/dt$. It can take care of itself: put a factor k in front of $d\mathbf{b}/dt$ so that the two species of force are on the same footing, and let k be large. Integrate twice for $\mathbf{x}(0) = 0$ and $\mathbf{v}(0) = 0$ to obtain $\mathbf{x}(t) = \mathbf{b}(t) + e^{-kt}\int_0^t e^{ks}\,d\mathbf{b}(s)$, exhibiting the discrepancy between displacement and the true Brownian motion as a copy of OU, of which this little identity is the historical origin. Now if t is large then kt is all the more so, and

$$\left|e^{-kt}\int_0^t e^{ks}\,d\mathbf{b}(s)\right| = e^{-kt}\left|\mathbf{b}^*\left(\frac{e^{2kt}-1}{2k}\right)\right|$$

$$< e^{-kt}\sqrt{(2+)\frac{e^{2kt}-1}{2k}\text{loglog}\frac{e^{2kt}-1}{2k}} \quad \begin{array}{l}\text{by the iterated}\\ \text{logarithm}\end{array}$$

$$\simeq \sqrt{\frac{(2+)}{2k}\log(t)}.$$

That's pretty good tracking and makes the point I wanted. For example, if $k = 10^{12}$, the (absolute) error is $< 10^{-3}$ for $t < \exp(10^6)$!

This will be enough to go on with. Just keep it continually in mind that all differentials must be computed up to terms in dt, $d\mathbf{b}$, and $(d\mathbf{b})^2 = dt$. More varied and more interesting applications of Itô's lemma will be seen in the next section.

6.10 Brownian motion in dimensions ≥ 2

The d-dimensional Brownian motion $\mathrm{BM}(d)$ is just the joint motion $\mathbf{b} = (\mathbf{b}_1, \mathbf{b}_2,$ etc.) of d independent copies of $\mathrm{BM}(1)$, standing in the same relation to $\mathrm{RW}(d)$ as $\mathrm{BM}(1)$ does to $\mathrm{RW}(1)$, *to wit*, it is what comes out when $\mathrm{RW}(d)$ is speeded up and scaled back as in $n^{-1/2}\mathbf{x}([nt])$ and n is taken to $+\infty$, so it's pretty evident that BM will recapitulate the whole story of RW; indeed, it's somewhat simpler, lacking as it does the combinatorial complications of $\mathbb{Z}^d$. It is included here to provide further and better illustrations of Itô's lemma and, incidentally, for future use, notably in §12.5. I hope that these will show you how efficient Itô's lemma is and how easy to use. Dimensions 2 and 3 will suffice. Nothing much changes after $d = 3$. Obviously, a Markovian character, both simple and strict, is inherited from $\mathrm{BM}(1)$. The infinitesimal operator is now half the Laplacian

$$\Delta = \frac{\partial^2}{\partial x_1^2} + \frac{\partial^2}{\partial x_2^2} \quad \left(+ \frac{\partial^2}{\partial x_3^2} \text{ in dimension } 3 \right);$$

and the transition density is $(2\pi t)^{-d/2} \exp\left[-\frac{1}{2t}|x-y|^2\right]$. Note in passing that Δ may be expressed in polar coordinates as

$$\frac{\partial^2}{\partial r^2} + \frac{1}{r}\frac{\partial}{\partial r} + \frac{1}{r^2}\frac{\partial^2}{\partial \theta^2} \qquad \text{in dimension 2}$$

and as[4]

$$\frac{\partial^2}{\partial r^2} + \frac{2}{r}\frac{\partial}{\partial r} + \frac{1}{r^2}\left(\frac{1}{\sin\varphi}\frac{\partial}{\partial \varphi}\sin\varphi\frac{\partial}{\partial \varphi} + \frac{1}{\sin^2\varphi}\frac{\partial^2}{\partial \theta^2} \right) \quad \text{in dimension 3.}$$

Exercise 6.10.1 If you have not made this computation before, please do it now.

6.10.1 Itô's lemma

Obviously, $(\mathrm{db}_1)^2 = (\mathrm{db}_2)^2$ etc. $= \mathrm{d}t$. But what about $\mathrm{db}_1 \, \mathrm{db}_2$ and the like? That's easy: $\mathbf{b}_0 = \frac{1}{\sqrt{2}}(\mathbf{b}_1 + \mathbf{b}_2)$ is a copy of $\mathrm{BM}(1)$, so you must have

$$\mathrm{d}t = (\mathrm{db}_0)^2 = \frac{1}{2}(\mathrm{db}_1 + \mathrm{db}_2)^2 = \mathrm{d}t + \mathrm{db}_1\,\mathrm{db}_2;$$

in short, the multiplication table for $\mathrm{BM}(1)$ needs only to be supplemented by the new rule $\mathrm{db}_1\,\mathrm{db}_2 = 0$ in dimension 2, and similarly, by

---

[4] $r = |x|$, $\varphi =$ colatitude, and $\theta =$ longitude as usual.

$\mathrm{db}_2\,\mathrm{db}_3 = \mathrm{db}_3\,\mathrm{db}_1$, in dimension 3. The rest carries over in the obvious way: for $f \in C^2$, you have

$$\mathrm{d}f(\mathbf{b}) = \sum \frac{\partial f}{\partial x_k}\,\mathrm{db}_k + \frac{1}{2}\sum_{i,j} \frac{\partial^2 f}{\partial x_i \partial x_j}\,\mathrm{db}_i\,\mathrm{db}_j$$

$$= \operatorname{grad} f(\mathbf{b}) \cdot \mathrm{db} + \frac{1}{2}\Delta f(\mathbf{b})\,\mathrm{d}t,$$

and if f depends (smoothly) upon t as well, then $\partial f/\partial t \times \mathrm{d}t$ must be added as before.

Note that integrals such as $\int_0^1 e\,\mathrm{db}$ come in several different flavors. For example, in dimension 3, e might take its values in $\mathbb{R}^3$ and $e\,\mathrm{db}$ might be the inner product $e \cdot \mathrm{db}$ or the cross-product $e \times \mathrm{db}$; or again, e might take its values in the rotation group $\mathrm{SO}(3)$ of 3×3 orthogonal matrices with determinant $+1$ and $e\,\mathrm{db}$ would take its values in $\mathbb{R}^3$ once more.

Exercise 6.10.2 In particular, Note 6.9.4 carries over, showing, e.g. that $\int_0^1 e\,\mathrm{db}$ is a copy of $\mathrm{BM}(3)$ when e takes values in $\mathrm{SO}(3)$. Think it over from scratch, on the level of sample paths, why this should be so.

6.10.2 BM(2): some details

$\mathrm{BM}(1)$ started at the origin is (statistically speaking) invariant under the reflection $\mathbf{b} \to -\mathbf{b}$. That is why the reflecting one-dimensional Brownian motion $\mathbf{r} = |\mathbf{b}|$ is Markovian, and similarly for $\mathrm{BM}(2)$: statistically speaking, it is invariant under action of the rotation group $\mathrm{O}(2)$ of 2×2 orthogonal matrices. Then just as in dimension 1, its radial part $\mathbf{r} = |\mathbf{b}|$ is Markovian, too. This is the process $\mathrm{BES}(2)$, so-called after the German astronomer Bessel (1830). It will reappear in §12.4, and the analogous $\mathrm{BES}(3)$ as well, in §12.6.

Exercise 6.10.3 Think it over – the Markovian character of $\mathrm{BES}(2)$.

Itô's lemma is now applied to $\mathrm{BES}(2)$: $\operatorname{grad} r = x/r \equiv e$ and $\Delta r = 2/r$, so with $\mathbf{b} = r\mathbf{e}$ in polar coordinates, you have

$$\mathbf{r}(t) = \mathbf{r}(0) + \int_0^t \mathbf{e}\cdot\mathrm{db} + \int_0^t (2\mathbf{r})^{-1}\,\mathrm{d}t',$$

in which the Brownian integral is a copy of $\mathrm{BM}(1)$ by Exercise 6.10.2 and the second imparts to $\mathbf{r}$ an extra drift of rate $1/2\mathbf{r}$, in conformity with the expression $\partial^2/\partial r^2 + r^{-1}\partial/\partial r$ for the radial part of Δ. That's the Bessel operator in dimension 2, *aka* $2 \times$ the infinitesimal operator of $\mathrm{BES}(2)$.

Exercise 6.10.4   You may worry that $\int_0^t r^{-1}\,dt'$ might be infinite, but it is not so. Just compute its expectation for, e.g. $t = 1$, in the worst case $\mathbf{r}(0) = 0$.

But what about the angular part $\mathbf{e} = \mathbf{b}/r$ of BM(2), moving on the circle $|e| = 1$, which would seem to have the infinitesimal operator $r^{-2} \times \frac{1}{2}\partial^2/\partial\theta^2$? Evidently, $\frac{1}{2}\partial^2/\partial\theta^2$ is the infinitesimal operator of the *circular* Brownian motion $\mathbf{o}(t) : t \geq 0$, by which I mean BM(1) considered *modulo* 2π, and the presence of $1/r^2$ in front indicates that $\mathbf{e}$ is nothing but $\mathbf{o}$ run with the clock $\int_0^t \mathbf{r}^{-2}\,dt'$. That's kosher if $\mathbf{b}(0) \neq 0$: such paths cannot hit the origin, as will be seen shortly. For paths starting *at* the origin, the clock *is* divergent, but never mind. Let's go on.

The next item is the analogue of BM(2) of the Gambler's Ruin for BM(1). Fix $0 < a < b < \infty$, take $a < \mathbf{r}(0) = r < b$ and compute the differential of $\log r$ by Itô's lemma: $\operatorname{grad}\log r = x/r$ and $\Delta\log r = 0$, so

$$\log\mathbf{r}(t) = \log\mathbf{r}(0) + \int_0^t \frac{\mathbf{b}\cdot d\mathbf{b}}{\mathbf{r}} = \log\mathbf{r}(0) + \int_0^t \mathbf{e}\cdot d\mathbf{b} \text{ with } \mathbf{e} = \mathbf{b}/r.$$

This is or ought to be a fair game when stopped at the exit time $T = T_a \wedge T_b = \min(t : \mathbf{r}(t) = a \text{ or } b)$, leading at once to

$$\mathbb{P}(T_a < T_b) = \mathbb{E}\left[\frac{\log b - \log\mathbf{r}(T)}{\log b - \log a}\right] = \frac{\log b - \log r}{\log b - \log a}.$$

That's the rule for the Gambler's Ruin. In particular, $\mathbb{P}(T_a < \infty) = 1$ is obtained by making $b \uparrow \infty$, which means, e.g. that BM(2), started at the origin, passes from $r = 1$ to $r = 2$ and back infinitely often, since BM(1), and so also BM(2), makes large excursions. That's a *loop*. Now take any small disc and a large concentric circle enclosing the origin, as in Figure 6.10.1. BM(2) hits the big circle, starts over from scratch by the strict Markov property, and so must hit the little circle infinitely often, by the previous reasoning, from which you see that BM(2) hits *every* disc infinitely often with probability 1, just as RW(2) visits every place in $\mathbb{Z}^2$ infinitely often. But note that it cannot hit the *origin* if it starts elsewhere since $\mathbb{P}(T_0 < \infty) = 0$ as you see from the Gambler's Ruin by making $a \downarrow 0$ and $b \uparrow \infty$, in that order; in particular, BM(2), starting at the origin, leaves at once and never comes back. This is already obvious, from the fact that $\log\mathbf{r}(t) = \log\mathbf{r}(0) + \int_0^t \mathbf{e}\cdot d\mathbf{b}$: the Brownian integral is a copy of BM(1) and cannot run off to $-\infty$ as it would if $\mathbf{r}$ were to hit the origin.

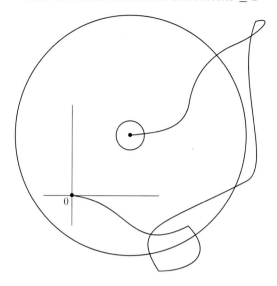

Figure 6.10.1

6.10.3 BM(3) and how it goes to ∞

BM(3) is similar, except that it runs off to ∞ like RW(3). In spherical polar coordinates, it is written $\mathbf{b} = r\mathbf{e}$, in which $r = |\mathbf{b}|$ is a copy of BES(3), with infinitesimal operator $(1/2) \times$ the 3-dimensional Bessel operator $\partial^2/\partial r^2 + 2r^{-1}\partial/\partial r$. Here, it is $1/r$ that plays the role of $\log r$, and the Gambler's Ruin, for paths starting at $a < \mathbf{r}(0) = r < b$, is now expressed as

$$\mathbb{P}(T_a < T_b) = \frac{(1/r) - (1/b)}{(1/a) - (1/b)},$$

in particular, BM(3) starting at the origin, is bound to hit the sphere of radius 2, but returns to the sphere of radius 1 with probability $1/2$, only. Obviously, it cannot do this infinitely often, which is to say that, ultimately, BM(3) leaves the little sphere and never comes back, and, since there is nothing special about the radii 1 and 2, it is plain that BM(3) runs off to ∞ with probability 1. But in what way does it do this? That is the subject of the next few articles.

Speed

Obviously, $|\mathbf{b}(t)| \leq \sqrt{(6+) \times t \log\log t} < t^{(1/2)+}$ by three applications of the law of the iterated logarithm for BM(1), and it is not hard to get the lower estimate $|\mathbf{b}(t)| > t^{(1/2)-}$ as well. Note first that

$$\mathbb{P}_0\left[\min_{t \geq n}|\mathbf{b}(t)| \leq n^{1/3}\right]$$

$$= \mathbb{P}_0\big[|\mathbf{b}(n)| \leq n^{1/3}\big] + \mathbb{E}_0\left[|\mathbf{b}(n)| > n^{1/3},\ \mathbb{P}_{\mathbf{b}(n)}\big(T_{n^{1/3}} < \infty\big)\right]$$

$$= \mathbb{P}_0\big[|\mathbf{b}(n)| \leq n^{1/3}\big] + \mathbb{E}_0\left[|\mathbf{b}(n)| > n^{1/3},\ \frac{n^{1/3}}{|\mathbf{b}(n)|}\right] \quad \begin{array}{l} \text{by the} \\ \text{Gambler's Ruin} \end{array}$$

$$= \mathbb{P}_0\big[|\mathbf{b}(1)| \leq n^{-1/6}\big] + \mathbb{E}_0\left[|\mathbf{b}(1)| > n^{-1/6},\ \frac{n^{-1/6}}{|\mathbf{b}(1)|}\right] \quad \begin{array}{l} \text{by Brownian} \\ \text{scaling} \end{array}$$

$$\leq O(n^{-1/2}) + n^{-1/6} \times 4\pi \int_0^\infty \frac{e^{-r^2/2}}{(2\pi)^{3/2}}\, r\, dr = O(n^{-1/6}),$$

with the result that

$$\mathbb{P}_0\left[\min_{t \geq n^7}|\mathbf{b}(t)| \leq n^{7/3}\right] = O(n^{-7/3})$$

and so also

$$\mathbb{P}_0\left[\mathbf{b}(t) \geq t^{7/3} \text{ as } t \uparrow \infty\right] = 1$$

or very nearly so: in fact, $\min_{t \geq n^7}|\mathbf{b}(t)| > n^{7/3}$ as $n \uparrow \infty$, by cheap Borel–Cantelli, so far large t and $n^7 \leq t < (n+1)^7$, you find

$$|\mathbf{b}(t)| > n^{7/3} > t^{1/3} \times \text{a nuisance factor} \left(\frac{n}{n+1}\right)^{7/3} \simeq 1.$$

Better still, a little more attention to the proof shows that there is nothing special about the exponent $\frac{1}{3}$: any exponent smaller than $\frac{1}{2}$ will do.

Exercise 6.10.5   Details please.

Exercise 6.10.6   Redo the lower estimate $|\mathbf{x}|(n) \geq n^{(1/2)-}$ for RW(3) of §4.4.2 in this style, using the hitting probabilities of §4.7 to replace the Gambler's Ruin. That's much less laborious than the old proof.

Inspection of Δ, written out in spherical polar coordinates, shows that the angular part $\mathbf{e} = \mathbf{b}/r$ of BM(3), moving on the unit sphere, looks like the spherical Brownian motion $\mathbf{o} : t \geq 0$, with infinitesimal operator

$$(1/2) \times \text{the "spherical Laplacian"} \frac{1}{\sin\varphi}\frac{\partial}{\partial\varphi}\sin\varphi\frac{\partial}{\partial\varphi} + \frac{1}{\sin^2\varphi}\frac{\partial^2}{\partial\theta^2}\,,$$

run, much as for BM(2), with the clock $K(t) = \int_0^t \mathbf{r}^{-2}\, dt'$. I will come back to this spherical Brownian motion presently, but first let's see how fast the clock is running. Obviously, it slows down since $\mathbf{r}(t)$ is (very roughly) $\sqrt{t}$, and you might think that it must be comparable to $\log t$. That's not so clear, but Itô's lemma makes it so: In fact,

$$\mathrm{d}\log\mathbf{r} = \mathbf{r}^{-1} \times \text{ the differential of a copy } \mathbf{b}_0 \text{ of } \mathrm{BM}(1) + \mathrm{d}K$$

when you work it out, and integrating back, you find

$$(0) = \log\mathbf{r}^2(t) = \log\mathbf{r}^2(0) + 2\int_0^t \mathbf{r}^{-1}\,\mathrm{d}\mathbf{b}_0 + K(t) = (1) + (2) + (3)$$

in which $(0) \simeq \log t$, (1) can be ignored, and $(2) = 2 \times$ yet another copy of $\mathrm{BM}(1)$ run with the same clock K – this by Note 6.9.4 – and, as such, not much worse than $\sqrt{K}$. That does it.

Now back to the angular part $\mathbf{e}$ of $\mathrm{BM}(3)$ with its representation as the spherical Brownian motion $\mathbf{o}$, run with the clock $K(t) = \int_0^t \mathbf{r}^{-2}\,\mathrm{d}t'$, supposing that $\mathbf{o}$ has an "individual" law of large numbers, by which I mean that for any reasonable spherical function f,

$$\mathbb{P}\left[\lim_{T\uparrow\infty}\frac{1}{T}\int_0^T f(\mathbf{o})\,\mathrm{d}t = \int_{|e|=1} f(e)\frac{\mathrm{d}\,\mathrm{area}}{4\pi}\right] = 1.$$

Then with $f =$ the indicator of the spherical cap C, you would have

$$\frac{(\text{spherical}) \text{ area } C}{4\pi} = \lim_{T\uparrow\infty}\frac{1}{K(T)}\int_0^{K(T)} f(\mathbf{o})\,\mathrm{d}t$$

$$= \lim_{T\uparrow\infty}\frac{1}{\log T}\int_0^T f(\mathbf{o}(K))\,\mathrm{d}K$$

$$= \lim_{T\uparrow\infty}\frac{1}{\log T}\int_1^T f(\mathbf{e})\frac{\mathrm{d}t}{t},$$

with a quick, hopefully imperceptible sleight of hand in the third line, which I really do not know how to justify, but serves to make my point by analogy, that this is just the type of logarithmic equidistribution of angles put forward in §4.4.2 for $\mathrm{RW}(3)$.

I have still to justify the individual law of large numbers stated above, as I now do for the simpler Brownian motion $\mathbf{o}$ on a circle of perimeter 1, *aka* $\mathbb{R}/\mathbb{Z}$. Once this is understood, it will be clear what to do in the spherical case. The transition density $p(t, x, y)$ is now

$$\sum_{\mathbb{Z}}(2\pi t)^{-1/2}e^{-(x-y+n)/2t} = \sum_{\mathbb{Z}}e^{2\pi\sqrt{-1}n(x-y)}e^{-2\pi^2 n^2 t}$$

by Jacobi's identity of §1.6.2, and the thing to notice is that $p = 1 + O(e^{-2\pi^2 t})$ for $t\uparrow\infty$, independently of x and y – not so clear from the left side, but the right side makes it plain. This is more than one really needs, but let's use it anyhow.

Let f be any reasonable circular function, i.e. a function of period 1 on $\mathbb{R}$, with mean value $\int_0^1 f(x)\,\mathrm{d}x = 0$, and replace $\mathrm{e}^{-2\pi^2 t}$ by e^{-t} for brevity. Then

$$\mathbb{E}_0\left[\int_0^T f(\mathbf{o})\,\mathrm{d}t\right]^2$$

$$= 2\int_0^T \mathrm{d}t \int_0^t \mathrm{d}t' \iint p(t',0,x)p(t-t',x,y)f(x)f(y)\,\mathrm{d}x\,\mathrm{d}y$$

$$= 2\int_0^T \mathrm{d}t \int_0^t \mathrm{d}t' \iint [1+\mathrm{O}(\mathrm{e}^{-t'})]f(x)\,\mathrm{d}x \times [1+\mathrm{O}(\mathrm{e}^{-(t-t')})]f(y)\,\mathrm{d}y$$

$$\leq \text{an unimportant multiple of } \int_0^T \mathrm{e}^{-t}\,\mathrm{d}t \int_0^t \mathrm{d}t' \leq \int_0^\infty t\mathrm{e}^{-t}\,\mathrm{d}t = 1,$$

independently of T. Consequently,

$$\mathbb{P}_0\left[\left|\int_o^{n^2} f(\mathbf{o})\,\mathrm{d}t\right| > x^{(1/2)+}\right] \leq n^{-(1+)}$$

and so also

$$\mathbb{P}_0\left[\lim_{T\uparrow\infty} \frac{1}{T} \int_0^T f(\mathbf{o})\,\mathrm{d}t = 0\right] = 1$$

since $\int_{n^2}^T f(\mathbf{o}) = \mathrm{O}(n)$ for $n^2 \leq T < (n+1)^2$. Easy enough.

As to the spherical case, there is a Jacobi-type formula for the associated transition density, taken relative to the volume element $\mathrm{d}o = (4\pi)^{-1}\,\mathrm{d}\,\mathrm{area}$, *to wit*,

$$p(t,x,y) = \sum_{n=0}^\infty \sum_{|l|\leq n} e_n^l(x)e_n^l(y)\mathrm{e}^{-n(n+1)t/2}$$

where, for each $n \geq 0$, the es are real eigenfunctions of the spherical Laplacian, appearing in little subspaces of dimension $2n+1$, with common eigenvalue $-n(n+1)$, starting with $e \equiv 1$ for $n = 0$. This is not the place to say more about these "spherical harmonics", but see §6.4.2 for their concrete description and Dym–McKean [1972:§4.19] if you want more. See also §6.11.2 where they appear in a different guise due to J.C. Maxwell Here, I want only to make it plausible that $p = 1+\mathrm{O}(\mathrm{e}^{-t})$ for $t \uparrow \infty$. As before, that's more than is really needed. It is enough to know that if $\int f\,\mathrm{d}o = 0$ then $\int f\Delta f\,\mathrm{d}o \leq -\gamma \int f^2\,\mathrm{d}o$ with a fixed positive number γ. That's not hard to see and the rest of the proof is the same.

The same type of individual laws of large numbers will be found in

Chapter 7, on Markov chains, and in Chapter 8, on Birkhoff's ergodic theorem. The proofs are entirely different.

Coda

Well, I could go on and on, but I want to get back to limit theorems, and I've told you everything you need to know about Brownian motion for *that*, and a good deal more besides.

6.11 $S^\infty(\sqrt{\infty})$ revisited

I remind you of that queer object from §5.10 with its round volume element

$$\mathrm{d}G(x) = \prod_1^\infty \frac{\mathrm{e}^{-x_n^2/2}}{\sqrt{2\pi}}\, \mathrm{d}x_n, \text{ informally written } \frac{\mathrm{e}^{-x^2/2}}{(2\pi)^{\infty/2}}\, \mathrm{d}^\infty x.$$

There you see the joint distribution of an infinite number of independent Gaussian variables $\mathbf{x}_n : n \geq 1$ which may be used in P. Lévy's style to construct the one-dimensional Brownian motion $\mathbf{b}(t) : 0 \leq t \leq 1$ and conversely: the point $\mathbf{x} = (\mathbf{x}_1, \mathbf{x}_2, \text{etc.}) \in S^\infty(\sqrt{\infty})$ may be read off from the Brownian path as in Figure 6.2.2. In short, you may also think of $S^\infty(\sqrt{\infty})$ as the Brownian path space, and it is a pretty story how much can be done there by analogy with $S^{n-1}(\sqrt{n})$. A large part of this was foreseen by Mehler [1866] and then forgotten; further references will be given below.

6.11.1 div, grad, and all that

$S^{n-1}(\sqrt{n})$ inherits from the ambient space $\mathbb{R}^n$ a nice, rotation-invariant "spherical" Laplacian Δ, spelled out in §6.10 for $n = 2$ and 3, and it is natural to ask what happens to it as the dimension n goes to ∞ and what that means in the present circumstances. A formal answer is obtained via the identity

$$\frac{\partial^2}{\partial x_1^2} + \frac{\partial^2}{\partial x_2^2} + \text{etc} = \frac{\partial^2}{\partial r^2} + \frac{n-1}{r}\frac{\partial}{\partial r} + \frac{1}{r}\Delta$$

by writing out the radial part in the ambient coordinates of $\mathbb{R}^n$, keeping in mind that $x_1^2 + \cdots + x_n^2 = n$ is fixed. This produces

$$\frac{1}{n}\Delta = \sum \frac{\partial^2}{\partial x_i^2} - \frac{1}{n}\sum x_i x_j \frac{\partial^2}{\partial x_i \partial x_j} - \frac{n-1}{n}\sum x_k \frac{\partial}{\partial x_k},$$

in which in number of active derivatives $\partial^2/\partial x_i \partial x_j$ is controlled, independently of $n \uparrow \infty$ when the operator is applied to *tame functions only*, suggesting that the infinite-dimensional spherical Laplacian is, or ought to be

$$\Delta = \sum_1^\infty \left(\frac{\partial^2}{\partial x_n^2} - x_n \frac{\partial}{\partial x_n} \right) = \text{div} \cdot \text{grad}$$

with

$$\text{grad} = \left(\frac{\partial}{\partial x_1}, \frac{\partial}{\partial x_2}, \text{etc.} \right)$$

and

$$\text{div} = \left(\frac{\partial}{\partial x_1} - x_1, \frac{\partial}{\partial x_2} - x_2, \text{etc.} \right).$$

This is all quite natural in retrospect: you want Δ to be negative and symmetric in $L^2 \left[S^\infty(\sqrt{\infty}), dG \right]$, as in the weighted space $L^2 \left[\mathbb{R}, (2\pi)^{-1/2} e^{-x^2/2} \, dx \right]$ where $-\partial/\partial x + x$ is dual to $\partial/\partial x$ so that

$$\Delta = -\sum_1^\infty \left(\frac{\partial}{\partial x_n} \right)^\dagger \frac{\partial}{\partial x_n}.$$

A test of this suggestion is to ask if there is not a Gauss/Green/divergence-type theorem

$$\int_D \text{div} \, v \, d\,\text{vol} = \int_{\partial D} v \cdot n \, d\,\text{area}$$

for reasonable vector fields $v = (v_1, v_2, \text{etc.})$ in nice regions $D \subset S^\infty(\sqrt{\infty})$, of co-dimension 1, with smooth boundary ∂D and outward-pointing unit normal n, as indeed there is. The simplest choice of D would be the half-space $x_1 \geq 0$, in which case

$$\int_D \text{div} \, v \, d\,\text{vol} = \int_{x_1 \geq 0} \sum_1^\infty \left(\frac{\partial}{\partial x_n} - x_n \right) v_n \, dG$$

$$= \int_0^\infty \left(\frac{\partial}{\partial x_1} - x_1 \right) \bar{v}_1 \frac{e^{-x_1^2/2}}{\sqrt{2\pi}} \, dx_1 \text{ with } \bar{v}_1 = \mathbb{E}(v_1 | \mathbf{x}_1 = x_1)$$

$$+ \sum_2^\infty \int_{-\infty}^{+\infty} \left(\frac{\partial}{\partial x_n} - x_n \right) \bar{v}_n \frac{e^{-x_n^2/2}}{\sqrt{2\pi}} \, dx_n$$

$$\text{with } \bar{v}_n = \mathbb{E}(v_n | \mathbf{x}_n = x_n)$$

$$= -\frac{1}{\sqrt{2\pi}} \bar{v}_1(0) \qquad \qquad \text{as you will please check}$$

$$= \frac{1}{\sqrt{2\pi}} \int_{\partial D} v \cdot n \, d\,\text{area},$$

the integral over ∂D being the expectation, conditional on $\mathbf{x}_1 = 0$ if the nuisance factor $\sqrt{2\pi}$ be ignored.

6.11.2 Hermite and polynomial chaos

The versatile operator $\mathfrak{G} = \partial^2/\partial x^2 - x\partial/\partial x$ appeared already in §6.9 in connection with Ornstein–Uhlenbeck; and in its dual form $\partial^2/\partial x^2 + (\partial/\partial x)x$ in §5.3. Its natural setting is $L^2[\mathbb{R}, (2\pi)^{-1/2}e^{-x^2/2}\,dx]$. There it is symmetric, as noted before, with mutually perpendicular eigenfunctions $H_n : n \geq 1$ spanning the whole space. These are the Hermite polynomials $H_n(x) = (-1)^n e^{x^2/2}(\partial/\partial x)^n e^{-x^2/2} : n \geq 0$, with eigenvalues $-n$, as you may read off from the action of the operators of

$$creation \quad -\frac{\partial}{\partial x} + x : H_n \to H_{n+1} \text{ and } annihilation \quad \frac{\partial}{\partial x} : H_n \to nH_{n-1}.$$

Exercise 6.11.1   Check the rules for creation and annihilation with the help (if needed) of the elementary identity $\sum_0^\infty \frac{y^n}{n!} H_n(x) = e^{xy - y^2/2}$. Check also that $(2\pi)^{-1/2}\int H_n H_m e^{-x^2/2}\,dx = n!$ or 0 according as $n = m$ or not.

Hermite's polynomials implement a stylish decomposition of

$$\mathfrak{Z} = L^2[S^\infty(\sqrt{\infty}), dG]$$

into mutually perpendicular eigenspaces of the infinite-dimensional Laplacian Δ, *viz.*

$$\mathfrak{Z} = \mathfrak{Z}^0 \oplus \mathfrak{Z}^1 \oplus \mathfrak{Z}^2 \oplus \text{etc.},$$

$\mathfrak{Z}^n$ being the polynomial chaos of weight n, so-called[5], spanned by products $H_{p_1}(x_1)H_{p_2}(x_2)$etc. of weight or total degree $p_1 + p_2 + \text{etc.} = n$. These are eigenfunctions of Δ with common eigenvalue $-n$, in which language $\Delta = (0)\oplus(-1)\oplus(-2)\oplus(-3)$ etc., and it is easy to trace them back to $S^{n-1}(\sqrt{n})$. The eigenfunctions of the spherical Laplacian in dimension 3 appeared already in §6.10 in connection with the angular motion of BM(3) as it runs off to ∞. They appear in mutually perpendicular eigenspaces of "weight n" as it is said, of dimension $2n+1$, with common eigenvalue $-n(n+1)$, and may be written out in angular coordinates, but that is not the best format when n is large. Better to copy Maxwell [1892] who noticed that in dimension 3, the eigenspace of weight n is spanned

---

[5] The name comes from Wiener [1923, 1930] who reinvented this scheme; see also Cameron–Martin [1947] and McKean [1973]. Mehler [1866] knew all this but was forgotten, by myself along with everyone else.

by the functions $\partial^n r^{-1}/\partial x_1^{n_1}\partial x_2^{n_2}\partial x_3^{n_3}$ with $n_1+n_2+n_3 = n$, restricted to S^2. This may be imitated in high dimension: just replace r^{-1} by r to the power (2 minus the dimension) and you will see products of Hermite polynomials coming out as Mehler [1866] already knew.

Exercise 6.11.2   Try it if you like; McKean [1973] spells it out.

Exercise 6.11.3   Comparison with RW(1) is really too simple, but a look in passing can do no harm. $S^\infty(\sqrt{\infty})$ is replaced by the unit interval equipped with its Lebesgue measure, and $2\mathbf{x} - 1$ is written $\sum_1^\infty 2^{-n}\mathbf{x}_n$, where the variables $\mathbf{x}_n$ are independent, with common distribution $\mathbb{P}(\mathbf{x}_n = \pm 1) = 1/2$, in imitation of RW(1). These do not *span* $\mathfrak{Z} = L^2[0,1]$, but their products *do* if the supplementary variable $\mathbf{x}_0 \equiv 1$ is added: in fact, $\mathfrak{Z}$ is decomposed into the sum of mutually perpendicular subspaces

$\mathfrak{Z}^0 =$ the constants,

$\mathfrak{Z}^1 =$ the span of $\mathbf{x}_n : n > 0$,

$\mathfrak{Z}^2 =$ the span of $\mathbf{x}_{n_1}\mathbf{x}_{n_2} : n_2 > n_1 > 0$,

and so on, and if you now imagine that each $\mathbf{x}_n$ performs it own, private random walk in the space ± 1, with 1-step probabilities $1/2$ to stay and $1/2$ to move, then you will see that $\mathfrak{Z}^n$ is the eigenspace, for eigenvalue $-n$, of the corresponding operator

$$\frac{1}{2}\Delta : w(x) \to \sum_1^\infty \left[\frac{1}{2}w(-x_n) + \frac{1}{2}w(x_n) - w(x_n)\right] ;$$

useless but amusing.

6.11.3 The Brownian format

I mentioned before that the point $\mathbf{x} \in S^\infty(\sqrt{\infty})$ may be identified with the Brownian path, $\mathbf{b}(t) : 0 \le t \le 1$ starting at $\mathbf{b}(0) = 0$ via P. Lévy's construction from §6.2. To clarify this Brownian–Hermite connection, I repeat the set-up at level $n \ge 1$ in the present Figure 6.11.1: $k < 2^n$ is odd, and the discrepancies

$$\mathbf{b}\left(\frac{k}{2^n}\right) - \frac{1}{2}\left[\mathbf{b}\left(\frac{k-1}{2^n}\right) + \mathbf{b}\left(\frac{k+1}{2^n}\right)\right]$$

are Lévy's variables $\mathbf{z}(k/2^n)$ multiplied by $2^{-(n+1)/2}$, to which must be adjoined the supplementary variable $\mathbf{z}(0) = \mathbf{b}(1)$ at level 0. These

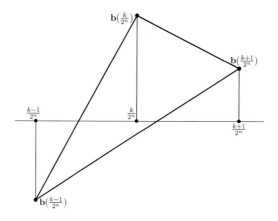

Figure 6.11.1

may be expressed by means of the functions $e_{k/2^n}$ seen in Figure 6.11.2, supplemented by $e_0 = 1$ at level 0. Indeed,

$$\mathbf{z}(0) = \int_0^1 e_0 \, \mathrm{d}\mathbf{b}$$

and

$$\mathbf{z}(k/2^n) = 2^{(n-1)/2} \int_0^1 e_{k/2^n} \, \mathrm{d}\mathbf{b}.$$

The $\mathbf{z}$s are now matched to the coordinates $\mathbf{x}_n : n \geq 1$ of $S^\infty(\sqrt{\infty})$, and it doesn't really matter how that is done: the functions $e_0 = 1$ and $e_{k/2^n} : k$ odd $< 2^n$, $n \geq 1$, together, make unit perpendicular frame in $L^2[0,1]$, and[6]

$$\mathbf{b}(t) = \int_0^t e_0 \, \mathrm{d}t \times \int_0^1 e_0 \, \mathrm{d}\mathbf{b} + \sum_{n=1}^\infty \sum_{\text{odd } k < 2^n} \int_0^t e_{k/2^n} \, \mathrm{d}t' \times \int_0^1 e_{k/2^n} \, \mathrm{d}\mathbf{b},$$

but any other such frame will do. Roughly speaking, the sum is just the expression of $\mathbf{b}(t) : 0 \leq t \leq 1$ in the frame in hand: and changing the frame only changes the coordinates on $S^\infty(\sqrt{\infty})$, and this by a rotation in response to the rotation of frames in $L^2[0,1]$, preserving the volume element $\mathrm{d}G = (2\pi)^{-\infty/2} e^{-x^2/2} \, \mathrm{d}^\infty x$.

A nice example of this Brownian–Hermite connection is provided by

---

[6] Ciesielski [1961] was the first to notice this nice way of writing Lévy's sum.

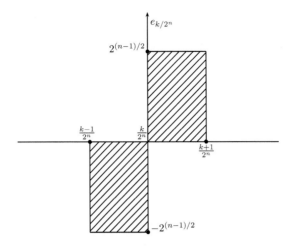

Figure 6.11.2

the "powers"

$$\text{“}b^n(1)\text{”} = n! \int_0^1 \mathrm{db}(t_1) \int_0^{t_1} \mathrm{db}(t_2) \cdots \int_0^{t_{n-1}} \mathrm{db}(t_n)$$

already seen in §6.9 together with the identity

$$1 + \sum_1^\infty \frac{y^n}{n!} \text{“}\mathbf{b}^n(1)\text{”} = \text{“}e^{y\mathbf{b}(1)}\text{”}$$

$$= e^{y\mathbf{b}(1) - y^2/2}$$

$$= \sum_0^\infty \frac{y^n}{n!} H_n \circ \mathbf{b}(1) \quad \text{by Exercise 6.11.1,}$$

with the pleasing recognition that Itô's "powers" are the same as Hermite's polynomials.

6.11.4 Back to Δ

Now any well-thought-up person, seeing a so-called manifold such as $S^\infty(\sqrt{\infty})$ with a (half)-Laplacian such as $\Delta/2$ on it, must ask: What's the Brownian motion? The point $\mathbf{x} \in S^\infty(\sqrt{\infty})$ already encodes the Brownian path $\mathbf{b}(t) : 0 \le t \le 1$, so I'm talking *Brownian motion of Brownian motion* – "big Brownian motion" if I may call it so – and it's obvious what it must be in the Hermite format: each of the little Gaussian variables in $\mathbf{x} = (\mathbf{x}_1, \mathbf{x}_2$, etc.) must perform its own, private, stationary Ornstein–Uhlenbeck process with (backward) infinitesimal operator $\frac{1}{2}(\partial^2/\partial x^2 - x\partial/\partial x)$ – and this independently of all the others – preserving its Gaussian character and so also the whole volume element dG, just as the spherical Brownian motion of §6.10 does on the sphere $S^{n-1}(\sqrt{n})$.

But what is that in the *big* Brownian format?

The simplest answer employs the "Brownian sheet", so-called. This is produced by a Gaussian set-function on the figure $[0, 1] \times [0, \infty)$ of two time parameters, "Roman" time $0 \le t \le 1$ and "Greek" time $0 \le \tau < \infty$, attaching to each rectangle $A \times B$, a Gaussian variable $\mathfrak{g}(A \times B)$ with mean 0 and mean-square $|A| \times |B|$, the variables assigned to non-overlapping rectangles being uncorrelated and so also independent[7]. The Brownian sheet is now declared to be

$$\mathbf{b}_\tau(t) = e^{-\tau/2}\mathfrak{g}\big([0, t] \times [0, e^\tau]\big) \text{ for } 0 \le t \le 1 \text{ and } 0 \le \tau < \infty,$$

and it is a very interesting object indeed. To begin with,

$$\mathbb{E}\big[\mathbf{b}_\tau(t_1)\mathbf{b}_\tau(t_2)\big] = t_1 \wedge t_2$$

by inspection of Figure 6.11.3, showing that $\mathbf{b}_\tau(t) : 0 \le t \le 1$ is a copy

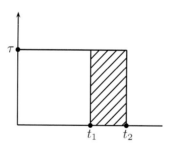

Figure 6.11.3

of $\mathbf{b}(t) : 0 \le t \le 1$ for each fixed $\tau \ge 0$, which is to say that $\mathbf{b}_\tau : \tau \ge 0$ parametrizes a volume-preserving flow on $S^\infty(\sqrt{\infty})$, issuing from the general point $\mathbf{b}_0(t) = \mathfrak{g}([0, t] \times [0, 1])$. Besides, for fixed σ and variable

---

[7] Don't worry how to make such a thing. It's not hard, but that's not my purpose here.

$\tau \geq 0$, you have, what is really nice,

$$\mathbf{b}_{\tau+\sigma}(t)$$
$$= e^{-\tau/2}e^{-\sigma/2}\mathfrak{g}\left[(0,t) \times (0, e^{\sigma})\right]$$
$$+ e^{-\tau/2}e^{-\sigma/2}\mathfrak{g}\left[(0,t) \times (e^{\sigma}, e^{\sigma+\tau})\right]$$
$$= e^{-\tau/2}\mathbf{b}_{\sigma}(t)$$
$$+ e^{-\tau/2} \times \text{a copy of } \mathfrak{g}\left[(0,t) \times (0, e^{\tau})\right] \quad \text{independent of } \mathbf{b}_{\sigma'} : \sigma' \geq \sigma$$
$$= \text{a copy of } \mathbf{b}_{\tau} : \tau \geq 0 \quad \text{starting from scratch at } \mathbf{b}_{\sigma}.$$

In short, the big Brownian motion on $S^{\infty}(\sqrt{\infty})$ is Markovian and also stationary, preserving as it does, the volume element dG.

OK, but how does this big Brownian motion look in the Hermite format? It is just as you would hope, its infinitesimal operator being exactly $(1/2) \times$ our old round Laplacian Δ. To see this take any unit, perpendicular frame $e_n : n \geq 1$ for $L^2[0,1]$, and let's ask how the corresponding coordinates $\mathbf{x}_n = \int_0^1 e_n \, d\mathbf{b}$ move. Obviously, correlations will tell, so it is enough to compute

$$\mathbb{E}\left[\mathbf{x}_i(\tau_1)\mathbf{x}_j(\tau_2)\right] = e^{-\tau_1/2}e^{-\tau_2/2}\,\mathbb{E}\left[\int_0^1 e_i(t)\,d\mathfrak{g}\left[(0,t) \times (0, e^{\tau_1})\right] \times \right.$$
$$\left. \times \int_0^1 e_j(t)\,d\mathfrak{g}\left[(0,t) \times (0, e^{\tau_2})\right]\right]$$
$$= e^{-|\tau_2 - \tau_1|/2} \times 1 \text{ or } 0 \quad \text{according as } i = j \text{ or not,}$$

to understand that each coordinate $\mathbf{x}_n$ performs its own private, stationary OU process, independently of the rest, in conformity with the big Brownian motion proposed at the start of this article.

6.11.5 Drift and Jacobian

As a final act of the play, that ubiquitous actor $\exp\left[m\mathbf{b}(t) - m^2t/2\right]$ appears in a new costume. It is the Brownian cousin of the fair game $(\cosh\alpha)^{-n}\exp[\alpha\mathbf{x}(n)]$ used in §4.9.4 to impart a drift to RW(1), and may be used in the same way to impart a drift to BM(1). The computation is easy: with my usual sloppy notation,

$$\mathbb{E}_x\left[e^{m[\mathbf{b}(t)-\mathbf{b}(0)]-m^2t/2},\mathbf{b}(t)=y\right]$$

$$=e^{m(y-x)-m^2t/2}\frac{e^{-(y-x)^2/2t}}{\sqrt{2\pi t}}$$

$$=\frac{e^{-(y-mt-x)^2/2t}}{\sqrt{2\pi t}}$$

$$=\mathbb{P}_x\left[\mathbf{x}(t)=y\right], \qquad \text{where } \mathbf{x}(t)=\mathbf{b}(t)+mt,$$

from which you see that

$$\mathbb{P}_0(\mathbf{x}\in A)=\mathbb{E}_0\left[e^{m\mathbf{b}(1)-m^2/2},\mathbf{b}\in A\right] \qquad \begin{array}{l}\text{for any event } A\\ \text{descriptive of the}\\ \text{path up to time 1,}\end{array}$$

or what is the same,

$$\frac{\mathrm{d}\,\mathbb{P}^{\mathbf{x}}}{\mathrm{d}\,\mathbb{P}^{\mathbf{b}}}=e^{m\mathbf{b}(1)-m^2/2},$$

exhibiting $\exp[m\mathbf{b}(1)-m^2/2]$ as the Jacobian of the change of Brownian coordinates from $\mathbf{b}(t):0\le t\le 1$ to $\mathbf{x}(t)=\mathbf{b}(t)+mt:0\le t\le 1$.

Exercise 6.11.4 There are some simple details to fill in here. I leave them to you.

Exercise 6.11.5 Take $t=1$ in the big display and differentiate lines 1 and 5 at $m=0+$ to obtain $\mathbb{E}_0[f'\circ\mathbf{b}(1)]=\mathbb{E}_0[\mathbf{b}(1)f\circ\mathbf{b}(1)]$. Obviously, that could have been done by hand, but I wanted you to see it this way. It's *integration by parts* in path space, in which the differentiation is moved over onto the Jacobian and so made harmless.

Exercise 6.11.6 The same reasoning shows that if $y(t):0\le t\le 1$ is any nice fixed function vanishing at $t=0$, then the Jacobian of the change of coordinates $\mathbf{b}\to\mathbf{x}=\mathbf{b}+y$ is $\exp\left[\int_0^1\dot{y}\,\mathrm{d}\mathbf{b}-\frac{1}{2}\int_0^1(\dot{y})^2\,\mathrm{d}t\right]$, as could have been foretold from the Jacobian

$$\frac{\mathrm{d}G(\mathbf{x}-y)}{\mathrm{d}G(\mathbf{x})}=\prod_1^\infty\frac{e^{-(\mathbf{x}_n-y_n)^2/2}}{e^{-\mathbf{x}_n^2/2}}=e^{\mathbf{x}\cdot y-y^2/2}$$

for the change of coordinates $\mathbf{x}\to\mathbf{x}+y$ in $S^\infty(\sqrt{\infty})$, seen in §5.10.2. This goes back to Cameron–Martin [1944].

Question. One more thing in case you didn't notice. Why is the Brownian path space round?

Answer:

$$\mathbb{P}_0 \left[\lim_{n \uparrow \infty} \sum_{k/2^n \leq 1} \left| \mathbf{b}\left(\frac{k}{2^n}\right) - \mathbf{b}\left(\frac{k-1}{2^n}\right) \right|^2 = 1 \right] = 1 \quad \text{as per §6.9.}$$

7
Markov Chains

These may be regarded as the prototype of the most general Markovian motion that, like RW(1), forgets the past, starting afresh at the place it happens to be. The figure shows a simple case, with arrows and numbers

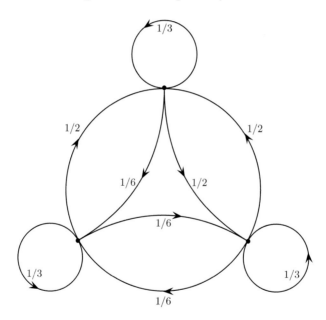

Figure 7.0.1

to indicate the probabilities of staying put or moving to a new place in one step. The starting place is then forgotten and the motion continues by the same rule. For the most part, I keep to what is really needed: for information and coding in Chapter 9, and for the statistical mechanics

in Chapters 10 and 11. The subject goes back to A.A. Markov (1910).
Feller [1966(2)] is the place to look for a more complete account.

7.1 Set-up and the Markov property

Think of the space $\mathbb{X}$ of paths $\mathbf{x}\colon n \geq 0 \to \mathbf{x}(n)$ taking values in a finite collection of "places" of any kind, labeled $x = 1, 2, 3, \ldots, d$ *without
any implication as to proximity*; introduce the fields $\mathfrak{F}_n$ descriptive of
the path up to "time" n, and also the smallest field $\mathfrak{F}$ containing all of
these; assign to each pair $1 \leq x, y \leq d$ a number $1 > p_{xy} > 0$, pronounced "the probability to come from x to y in one step", subject to
$\sum_y p_{xy} = 1$ for any x since you have to go *some place*; and to each x,
each $n \geq 0$, and each string of places $x_0, x_1, \ldots, x_n$ assign probabilities

$$\mathbb{P}_x\big[\mathbf{x}(1) = x_1, \mathbf{x}(2) = x_2, \ldots, \mathbf{x}(n-1) = x_{n-1}, \mathbf{x}(n) = x_n\big]$$
$$= p_{x_0 x_1} p_{x_1 x_2} \cdots p_{x_{n-1} x_n} \text{ with } x_0 = x.$$

$\mathbb{P}_x$, pronounced "probability starting at x", is defined thereby in a con-
sistent way on the fields $\mathfrak{F}_0 \subset \mathfrak{F}_1 \subset \mathfrak{F}_2$ etc. and extends to a countably
additive measure on the big field $\mathfrak{F}$. An easy variant of Carathéodory's
lemma of §1.1.1 supplies the proof. Try it if you like.

The triple $(\mathbb{X}, \mathfrak{F}, \mathbb{P}_\bullet)$ is a Markov chain. It is simpler than $\mathrm{RW}(1)$ in
that the number (d) of places is finite, but more complicated, too, in its
lack of any symmetries such as $p_{xy} = p_{yx}$ or of any protocol restricting
steps to nearest neighbors. The condition $d < \infty$ may and will be re-
laxed, and likewise the requirement that p_{xy} be strictly positive, but I
keep them for now to make things as simple as they can be. Really, all
you need is that each place (x) can communicate with any other (y), i.e.
that $\mathbb{P}_x[\mathbf{x}(n) = y] > 0$ for some $n \geq 1$.

The simple Markov property, knitting together the several probabil-
ities $\mathbb{P}_\bullet$, is built in: if $N \geq 1$ fixed, if $A \in \mathfrak{F}_N$ is any event descriptive of $\mathbf{x}(1), \ldots, \mathbf{x}(N)$ only, if $B \in \mathfrak{F}$, and if $\mathbf{x}^+$ is the shifted path
$\mathbf{x}^+\colon n \to \mathbf{x}(n+N)$, then

$$\mathbb{P}_x\big[A, \mathbf{x}(N) = y, \mathbf{x}^+ \in B\big] = \mathbb{P}_x\big[A, \mathbf{x}(N) = y\big]\, \mathbb{P}_y(B)$$

or, what is the same,

$$\mathbb{P}_x(\mathbf{x}^+ \in B \mid \mathfrak{F}_N) = \mathbb{P}_y(B) \quad \text{with } y = \mathbf{x}(N).$$

Exercise 7.1.1   Just write it out and you'll see, first for "tame" $B \in \mathfrak{F}_n$
and then in general.

Exercise 7.1.2 Check also the strict Markov property. It's automatic. N is now a stopping time, meaning, as it should, that $(N = n) \in \mathfrak{F}_n$ for $n \geq 0$, and $\mathfrak{F}_N$ is the class of events $B \in \mathfrak{F}$ with $B \cap (N = n) \in \mathfrak{F}_n$ for any $n \geq 0$.

Exercise 7.1.3 Notice that $\mathfrak{F}_N$ is just the field of the stopping time N and the stopped path $\mathbf{x}^\circ \colon n \to \mathbf{x}(n \wedge N)$. Proof please.

7.2 The invariant distribution

It is important to understand the long-time (statistical) equilibration of the chain. The fact is that as $n \uparrow \infty$, the n-step probability $\mathbb{P}_x[\mathbf{x}(n) = y]$ tends rapidly to a positive number π_y, independent of x, i.e. the starting place $\mathbf{x}(0) = x$ is forgotten, and a statistical balance or equilibrium is obtained, as in

$$\sum_x \pi_x p_{xy} = \lim_{n \uparrow \infty} \sum_x \mathbb{P}_\bullet[\mathbf{x}(n-1) = x] p_{xy} = \lim_{n \uparrow \infty} \mathbb{P}_\bullet[\mathbf{x}(n) = y] = \pi_y,$$

whence the name "invariant distribution" for π.

Exercise 7.2.1 More generally, the chain is (statistically) stationary relative to the equilibrium probabilities $P_\pi = \sum \pi_x P_x$, i.e. for $N \geq 1$ and $\mathbf{x}^+ \colon n \to \mathbf{x}(n+N)$ as before, you have $P_\pi(\mathbf{x}^+ \in C) = P_\pi(C)$. $(\mathbb{X}, \mathfrak{F}, P_\pi)$ is the "equilibrium chain." The latter may be extended to 2-sided paths $\mathbf{x} \colon n \in \mathbb{Z} \to \mathbf{x}(n)$ in a simple way. Spell it out.

Exercise 7.2.2 Now write $\mathbf{x}^\uparrow$ for this 2-sided chain and $\mathbf{x}^\downarrow$ for the backward chain $\mathbf{x}^\downarrow(n) = \mathbf{x}^\uparrow(-n)$. The backward chain $\mathbf{x}^\downarrow$ is also Markovian, with $p_{xy}^\downarrow = \pi_y p_{yx}^\uparrow / \pi_x$, as you will check, and it may happen that the chain is "reversible", meaning that $\mathbf{x}^\downarrow$ is a copy of $\mathbf{x}^\uparrow$. This is always so if $d = 2$, but usually fails if $d = 3$.

I will prove the existence of π in three ways, each with its advantages, both here and in more general circumstances.

7.2.1 Geometrical proof

Take, for simplicity, $d = 3$, write p for the 3×3 matrix $(p_{ij} : 1 \leq i, j \leq 3)$, and think of the totality of probability distributions $\gamma = (\gamma_1, \gamma_2, \gamma_3)$ as the triangle $\triangle$ seen in Figure 7.2.1. The invariant distribution satisfies $\pi p = \pi$, i.e. it is a fixed point of the map $\gamma \to \gamma p$ (also written $p^\dagger \gamma$) of the triangle $\triangle$ into itself. Apply the map once: since all the little p_{ij} are positive, each corner moves inside, producing the smaller shaded triangle

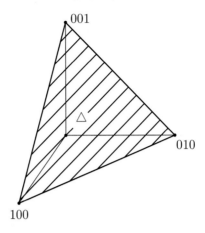

Figure 7.2.1

$p^\dagger \triangle$ seen in Figure 7.2.2. This may be enclosed in a somewhat larger

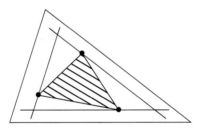

Figure 7.2.2

triangle by drawing lines through *its* corners parallel to the sides of $\triangle$. Viewed in the plane of $\triangle$, *this* triangle is a translate of $\triangle$ shrunk by a factor $\vartheta < 1$, i.e. $p^\dagger \triangle \subset \vartheta \triangle + c$. Now repeat: $(p^\dagger)^2 \triangle \subset \vartheta^2 \triangle +$ a new c and so on, and since translation does not affect diameters, you have $\operatorname{diam}(p^\dagger)^n \triangle \leq \vartheta^n \times (\operatorname{diam} \triangle = \sqrt{2})$, which is to say that the nested triangles $\triangle \supset p^\dagger \triangle \supset (p^\dagger)^2 \triangle$ etc. shrink to a *single point*, fixed by the map. That's the invariant distribution.

Exercise 7.2.3 A simple numerical example is instructive. Here, $d = 2$ and $p = \begin{pmatrix} 1/2 & 1/2 \\ 2/3 & 1/3 \end{pmatrix}$ as per Figure 7.2.3. Compute $p^4 = 6^{-4} \begin{pmatrix} 741 & 555 \\ 740 & 556 \end{pmatrix}$ and $\pi = (4/7, 3/7)$ and compare $741/6^4$ to $4/7$ numerically. You see how quickly it goes!

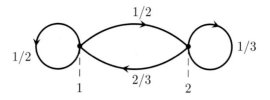

Figure 7.2.3

7.2.2 Analytical proof

This is the same, really, but with a more explicit estimate supplied. Take now any number of places and any two distributions α and β, let γ be their (signed) difference $\alpha - \beta$ with positive/negative part γ^+/γ^-, and let J be the set of indices j for which $(\gamma^+ p)_j > (\gamma^- p)_j$. If J is void, then $(\gamma^+ p)_j \leq (\gamma^- p)_j$ without exception, and with the notation $|\gamma| = \sum_1^d |\gamma_j|$, you have

$$|\gamma p| = \sum_{j=1}^d [(\gamma^- p)_j - (\gamma^+ p)_j] = \sum_1^d (\gamma_i^- - \gamma_i^+) = -\sum_1^d \gamma_i = 0,$$

and similarly if the complement J' is void. Otherwise neither J nor J' is void, and

$$\cdot \ |\gamma p| = \sum_{j=1}^d |(\gamma^+ p)_j - (\gamma^- p)_j|$$

$$\leq \sum_J (\gamma^+ p)_j + \sum_{J'} (\gamma^- p)_j$$

$$\leq \sum \gamma_i^+ \times \max_i \sum_J p_{ij} + \sum \gamma_i^- \times \max_i \sum_{J'} p_{ij}$$

$$\leq \vartheta |\gamma|$$

with the fixed number

$$\vartheta = \max_{J' \neq \varnothing} \max_i \sum_J p_{ij} < 1.$$

The rest will be plain.

7.2.3 Probabilistic proof

This is the most interesting argument, providing as it does a pleasing statistical interpretation of π. The idea is due to Döblin (1937), for which compare §3.3. Fix a starting place $\mathbf{x}(0)$, denoted by 0 for neutrality, and let R be the return- or loop-time $\min(n \geq 1 : \mathbf{x}(n) = 0)$. Then

$$\mathbb{P}_0(R > n) = \mathbb{P}_0\big[\mathbf{x}(1), \mathbf{x}(2), \ldots, \mathbf{x}(n) \neq 0\big]$$

$$= \sum_x \mathbb{P}_0\big[\mathbf{x}(1), \ldots, \mathbf{x}(n-1) \neq 0, \mathbf{x}(n-1) = x\big](1 - p_{x0})$$

$$\leq \mathbb{P}_0(R > n - 1)\vartheta \qquad \text{with } \vartheta = \max(1 - p_{x0}) < 1$$

$$\leq \vartheta^n \qquad\qquad \text{by repetition,}$$

showing that $\mathbb{P}_0(R < \infty) = 1$, $\mathbb{E}_0(R) < \infty$, and more. The loop $\mathbf{x}(n)$: $0 < n \leq R$ is then of finite duration, and as R is a stopping time and the terminus $\mathbf{x}(R) = 0$ is known, so the motion starts afresh, a new *independent* loop is made, and so on, over and over. Now introduce the expected number of visits to x per loop:

$$e_x = \mathbb{E}_0\Big[\#(0 < n \leq R : \mathbf{x}(n) = x)\Big] \leq \mathbb{E}_0(R) < \infty,$$

and let's check that $ep = e$:

$$e_x = \sum_{n=1}^{\infty} \mathbb{P}_0[R \geq n, \mathbf{x}(n) = x],$$

so

$$(ep)_y = \sum_{n=1}^{\infty} \sum_x \mathbb{P}_0\big[\mathbf{x}(1), \ldots, \mathbf{x}(n-1) \neq 0, \mathbf{x}(n) = x, \mathbf{x}(n+1) = y\big]$$

$$= \sum_{n=1}^{\infty} \sum_x \mathbb{P}_0\big[R \geq n, \mathbf{x}(n) = x\big] p_{xy}$$

$$= \sum_{n=1}^{\infty} \mathbb{P}_0\big[R \geq n, \mathbf{x}(n+1) = y\big] \qquad \text{since } (R \geq n) \in \mathfrak{F}_n$$

$$= \sum_{n=1}^{\infty} \mathbb{P}_0\big[R = n, \mathbf{x}(n+1) = y\big] + \sum_{n=1}^{\infty} \mathbb{P}_0\big[R > n, \mathbf{x}(n+1) = y\big]$$

$$= \mathbb{P}_0\big[\mathbf{x}(R+1) = y\big] + \sum_{n=2}^{\infty} \mathbb{P}_0\big[R \geq n, \mathbf{x}(n) = y\big]$$

$$= \sum_{n=1}^{\infty} \mathbb{P}_0\big[R \geq n, \mathbf{x}(n) = y\big]$$

$$= e_y$$

since, in line 5, $\mathbb{P}_0[\mathbf{x}(R+1) = y] = \mathbb{P}_0[\mathbf{x}(1) = y]$ is just the term $(n = 1)$ missing from the sum. But this means that e is proportional to π, the latter being the only normalized solution of $pe = e$. In short,

$$\pi_x = \frac{1}{Z} \, \mathbb{E}_0 \big| \#(0 < n \le R : \mathbf{x}(n) = x) \big| \quad \text{with normalizer } Z = \mathbb{E}_0(R).$$

Simple as that and, in fact, more simple still. No computation is needed. Only think: the next step after the first loop is the first step of the independent, statistically identical second loop.

Markov chains with an infinite number of states may or may not have invariant probabilities. The next two exercises are illustrative.

Exercise 7.2.4 Check that RW(1) has as (positive) invariant measure only $\pi \equiv 1$ (or some multiple of it). Now take the chain of Figure 7.2.4, al-

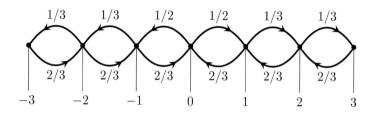

Figure 7.2.4

ready seen in §3.4.2, with a "drift" pulling the traveller back towards the origin. Here, you *do* have invariant probabilities. Obviously, $\pi_x = \pi_{-x}$ which cuts the work in half. Write out $ep = e$ and solve. You will find only one solution with $\sum e_x < \infty$. Normalize it. *Answer:* $\pi_0 = 1/4$, $\pi_x = (3/8)2^{-x}$ $(x > 0)$.

Exercise 7.2.5 Make a new random walk, this time on $\mathbb{R}$, according to the rule $\mathbf{x}(n) = m\mathbf{x}(n-1) + \mathbf{y}(n)$ for $n > 0$, with fixed $0 < m < 1$ and independent, unit Gaussian "innovations" $\mathbf{y}$. Here, too, the introduction of $m\mathbf{x}(n-1)$ acts as a restoring drift. The invariant distribution is, or ought to be $\lim_{n \uparrow \infty} \mathbb{P}_x[\mathbf{x}(n) \le y]$. Compute that and check it for invariance.

7.3 LLN for chains

The statement is that for any function $f : x \to \mathbb{R}$,

$$\mathbb{P}_\bullet \left[\lim_{N \uparrow \infty} \frac{1}{N} \sum_0^{N-1} f \circ \mathbf{x}(n) = \pi f \right] = 1,$$

in which πf is the invariant mean $\sum_x \pi_x f(x)$. The law of large numbers for visiting times, *to wit*

$$\mathbb{P}_\bullet \left[\sum_{N\uparrow\infty} \frac{1}{N} \#(n \le N : \mathbf{x}(n) = x) = \pi_x \text{ for every } x \right] = 1$$

is equivalent.

Proof Döblin's method of loops is the thing to use. Fix $\mathbf{x}(0) = 0$ and let R_1, R_2 etc. be the durations of the successive loops. These are independent copies of $R_1 \equiv R$; the loops themselves are independent copies of the first loop; and likewise for the sums $F_1 = \sum_{0 < n \le R_1} f\mathbf{ox}(n)$, $F_2 = \sum_{R_1 < n \le R_1 + R_2} f\mathbf{ox}(n)$, etc. so by the general law of large numbers of §5.1,

$$\lim_{n\uparrow\infty} \frac{1}{n} \sum_1^n F_k = \mathbb{E}_0(F_1)$$

$$= \sum_x f(x) \sum_1^\infty \mathbb{P}_0 \left[R \ge n, \mathbf{x}(n) = x \right]$$

$$= \sum_x f(x) e_x$$

$$= \pi f \times \text{ the normalizer } Z = \mathbb{E}_0(R).$$

The proof is finished by a second application of LLN. Let the (sure) number N be large and pick the random number $\mathbf{n}$ so that $R_1 + \cdots + R_\mathbf{n} \le N < R_1 + \cdots + R_{\mathbf{n}+1}$. Here, $\mathbf{n}$ increases to $+\infty$ with N,

$$\lim_{N\uparrow\infty} \frac{N}{\mathbf{n}} = \lim_{n\uparrow\infty} \frac{R_1 + \cdots + R_\mathbf{n}}{\mathbf{n}} = \mathbb{E}_0(R) = Z \quad \text{by LLN,}$$

$R_{\mathbf{n}+1}$ being small compared to $\mathbf{n}$ (see why?), and since $F_1 + \cdots + F_\mathbf{n} = f\mathbf{ox}(0) + \cdots + f\mathbf{ox}(N)$ with an error comparable to $R_{\mathbf{n}+1}$, you have

$$\frac{1}{N} \sum_0^{N-1} f\mathbf{ox}(n) \simeq \frac{\mathbf{n}}{N} \times \frac{F_1 + \cdots + F_\mathbf{n}}{\mathbf{n}} \simeq \frac{1}{Z} \times \pi f \times Z = \pi f,$$

as promised. Döblin makes it so easy!

7.3.1 LLN improved

A simple modification of the chain provides a big improvement. Look at the doubled chain of pairs $\mathbf{x}(0)\mathbf{x}(1)$, $\mathbf{x}(1)\mathbf{x}(2)$, $\mathbf{x}(2)\mathbf{x}(3)$ etc. relative to the equilibrium probabilities $\mathbb{P} = \mathbb{P}_\pi$ and expectations $\mathbb{E} = \mathbb{E}_\pi$. This is Markovian with 1-step probabilities

$$\mathbb{P}\big[\mathbf{x}(1) = c, \mathbf{x}(2) = d \,\big|\, \mathbf{x}(0) = a, \mathbf{x}(1) = b\big] = \begin{array}{ll} p_{bc}p_{cd} & b = c \\ 0 & b \neq c. \end{array}$$

As these are not all positive, you may ask if that makes trouble. It doesn't really, but to keep within the rules of §7.1, look separately at the *even* chain $\mathbf{x}(0)\mathbf{x}(1), \mathbf{x}(2)\mathbf{x}(3)$, etc. and the *odd* chain $\mathbf{x}(1)\mathbf{x}(2), \mathbf{x}(3)\mathbf{x}(4)$, etc. These are themselves Markovian: in fact, they are copies of each other, with common positive 1-step probabilities

$$\mathbb{P}\big[\mathbf{x}(2) = c, \mathbf{x}(3) = d \,\big|\, \mathbf{x}(0) = a, \mathbf{x}(1) = b\big]$$
$$= p_{bc}p_{cd} = \mathbb{P}\big[\mathbf{x}(3) = c, \mathbf{x}(4) = d \,\big|\, \mathbf{x}(1) = a, \mathbf{x}(2) = b\big]$$

and common invariant distribution $\pi_{ab} = \pi_a p_{ab}$, so if f is any function of the pair ab, then

$$\frac{1}{N} \sum_0^{N-1} f \circ \mathbf{x}(n)\mathbf{x}(n+1) = \frac{1}{N} \sum_{\text{even } n} + \frac{1}{N} \sum_{\text{odd } n}$$
$$\simeq \left(\frac{1}{2} + \frac{1}{2}\right) \times \sum \pi_a p_{ab} f(ab)$$
$$= \mathbb{E}(f).$$

The same idea works upon redoubling the chain and splitting the sum in four according as $n = 0, 1, 2,$ or $3 \mod 4$, and so on, with the more general result:

$$\lim_{N \uparrow \infty} \frac{1}{N} \sum_0^{N-1} f(T^n \mathbf{x}) = \mathbb{E}(f),$$

in which T is the shift $\mathbf{x}(0)\mathbf{x}(1)\mathbf{x}(2)$ etc $\to \mathbf{x}(1)\mathbf{x}(2)\mathbf{x}(3)$ etc, of the whole chain and $f\colon \mathbf{x} \to \mathbb{R}$ is any "tame" function depending upon a limited number of observations $\mathbf{x}(0)\mathbf{x}(1)\dots\mathbf{x}(\ell)$, say – and even this can be improved: It is correct for any summable $f\colon \mathbf{x} \to \mathbb{R}$, tame or not, as will be seen in the next chapter. Note for future reference that the shift is measure-preserving in the sense that $\mathbb{P}(T^{-1}A) = \mathbb{P}(Tx \in A) = \mathbb{P}(A)$ for any $A \in \mathfrak{F}$.

7.3.2 Mixing

Take tame A depending on $\mathbf{x}(0)\mathbf{x}(1)\ldots\mathbf{x}(\ell)$ as above and also a general event B, tame or not. Then for large n,

$$
\begin{aligned}
\mathbb{P}\big(A \cap T^{-n}B\big) &= \sum_{xy} \mathbb{P}\big[A, \mathbf{x}(\ell) = x\big]\, \mathbb{P}_x\big[\mathbf{x}(n-\ell) = y\big]\, \mathbb{P}_y(T^{-n+\ell}B) \\
&\simeq \sum_x \mathbb{P}\big[A, \mathbf{x}(\ell) = x\big] \sum_y \pi_y\, \mathbb{P}_y(T^{-n+\ell}B) \\
&= \mathbb{P}(A)\,\mathbb{P}(T^{-n+\ell}B) \\
&= \mathbb{P}(A)\,\mathbb{P}(B),
\end{aligned}
$$

with an absolute error not more than ϑ^n for some $0 < \vartheta < 1$, and this carries over to the general event $A \in \mathfrak{F}$ (without the error estimate): just approximate A by a tame event C, making $\mathbb{P}(A \cap C') + \mathbb{P}(A' \cap C)$ small[1]; note that

$$
\begin{aligned}
&\mathbb{P}(A \cap B) - \mathbb{P}(C \cap B) \\
&= \mathbb{P}(A \cap B \cap C) + \mathbb{P}(A \cap B \cap C') - \mathbb{P}(A \cap B \cap C) - \mathbb{P}(A' \cap B \cap C)
\end{aligned}
$$

is bounded *above* by $\mathbb{P}(A \cap C')$ and *below* by $-\mathbb{P}(A' \cap C)$; and replace B by $T^{-n}B$. In short,

$$
\lim_{n\uparrow\infty} \mathbb{P}(A \cap T^{-n}B) = \mathbb{P}(A)\,\mathbb{P}(B) \quad \text{for any events } A \text{ and } B,
$$

expressing an approximate independence of $\mathbf{x}$ and $T^n\mathbf{x}$ for large n. It is called "mixing" and will play a role in Chapter 8.

7.3.3 McMillan's theorem

This pretty illustration of LLN for the doubled chain describes the statistics of long strings of observations $\mathbf{x}(0)\mathbf{x}(1)\ldots\mathbf{x}(n)$. A variant will be seen in §8.5.5; see also §9.5 for what McMillan [1952] really did. Here is only the simplest case, but it's already very beautiful. The language information/entropy is used, for which see §9.1.

Begin with the information conveyed by observation of $\mathbf{x}(0)\mathbf{x}(1)\ldots\mathbf{x}(n)$ for some fixed starting place $\mathbf{x}(0)$. This is the quantity (note $\log_2$ is con-

---

[1] This comes, via Carathéodory's lemma, from the fact that $\mathfrak{F}$ is the smallest field containing all tame events. Think it over.

ventional in this particular business)

$$H_n = -\sum \mathbb{P}_\bullet\big[\mathbf{x}(0) = x_0, \mathbf{x}(1) = x_1, \ldots, \mathbf{x}(n) = x_n\big] \times$$

$$\times \log_2 \mathbb{P}_\bullet\big[\mathbf{x}(0) = x_0, \mathbf{x}(1) = x_1 \text{ etc.}\big]$$

$$= -\mathbb{E}_\bullet\big[\log_2 p_{\mathbf{x}(0)\mathbf{x}(1)} + \log_2 p_{\mathbf{x}(1)\mathbf{x}(2)} + \text{etc.}\big]$$

$$= -\sum_{k=1}^{n}\sum_{xy} \mathbb{P}_\bullet\big[\mathbf{x}(k-1) = x\big]\, p_{xy}\log_2 p_{xy}$$

$$\simeq n \times -\sum_{xy} \pi_x p_{xy}\log_2 p_{xy} \quad \text{by §7.3.1,}$$

from which you read off the "information production rate":

$$h = \lim_{n\uparrow\infty} \frac{1}{n} H_n = -\mathbb{E}_\pi\big[\log_2 p_{\mathbf{x}(0)\mathbf{x}(1)}\big],$$

and looking back at line 3 of the big display with LLN for the doubled chain in mind, you recognize that

$$\mathbb{P}_\bullet\left[\lim_{n\uparrow\infty} -\frac{1}{n}\log_2 p_{\mathbf{x}(0)\mathbf{x}(1)}p_{\mathbf{x}(1)\mathbf{x}(2)} \cdots p_{\mathbf{x}(n-1)\mathbf{x}(n)} = h\right] = 1.$$

This is the "individual" McMillan theorem. What it says is that long strings of observations $\mathbf{x}(0)\mathbf{x}(1)\mathbf{x}(2)\ldots\mathbf{x}(n)$ fall into two classes: the first is a little junk heap of negligible probability; the second contains $N \simeq 2^{nh}$ strings of more or less equal probability

$$p_{\mathbf{x}(0)\mathbf{x}(1)}p_{\mathbf{x}(1)\mathbf{x}(2)} \cdots p_{\mathbf{x}(n-1)\mathbf{x}(n)} \simeq 2^{-nh} \simeq 1/N.$$

Very simple, very beautiful. This type of equidistribution of probability plays an indispensable role in §9.5 in connection with Shannon's ideas about the transmission of information over a noisy channel. It is odd that such a simple but remarkable fact is not a commonplace, but you won't find it in the standard books, not even in Feller [1966].

7.4 CLT for chains

This is not quite so easy as LLN. The statement is as you would expect:

$$\lim_{N\uparrow\infty} \mathbb{P}_\bullet\left[\frac{\sum_1^N f\circ\mathbf{x}(n) - N\pi f}{\sqrt{N}} \le c\right] = \int_{-\infty}^{c} \frac{e^{-x^2/2\sigma^2}}{\sigma\sqrt{2\pi}}\, dx$$

with a positive number σ to be identified below. The mechanism can be traced to the mixing of §7.3.2: any approximate independence of the past and the remote future can only be favorable to CLT.

Proof for $\pi f = 0$. Döblin's loops are used once more as in §7.3.2, with a fixed starting place $\mathbf{x}(0) = 0$, a large whole number N, and $R_1 + \cdots + R_\mathbf{n} \leq N < R_1 + \cdots + R_{\mathbf{n}+1}$. Fix a small positive number α and write $\mathbb{E}_0(R) = Z$ as before. Then with overwhelming probability, $(1 - \alpha)Z\mathbf{n} < N < (1 + \alpha)Z\mathbf{n}$, or what is the same,

$$N_- < \frac{N}{(1 + \alpha)Z} < \mathbf{n} < \frac{N}{(1 - \alpha)Z} < N_+,$$

N_+/N_- being the smallest/largest whole numbers possible. Take $M =$ the whole number nearest to $\frac{1}{2}(N_+ + N_-) \simeq N/Z$, α being small, and let's compare $\sum_1^N f \circ \mathbf{x}(n)$ to $F_1 + \cdots + F_M$. There are errors of two types, indicated in the display:

$$\sum_1^N f \circ \mathbf{x}(n) = F_1 + \cdots + F_\mathbf{n} + \sum_{R_1 + \cdots + R_\mathbf{n} + 1}^N f \circ \mathbf{x}(n)$$

$$= F_1 + \cdots + F_M$$

$$+ F_{M+1} + \cdots + F_\mathbf{n} \qquad \text{if } \mathbf{n} > M$$

$$- F_{\mathbf{n}+1} + \cdots + F_M \qquad \text{if } \mathbf{n} < M \qquad \text{type 1}$$

$$+ \sum_{R_1 + \cdots + R_\mathbf{n} + 1}^N f \circ \mathbf{x}(n) \qquad\qquad \text{type 2}.$$

Here $\mathbb{E}_0(F) = 0$, so Kolmogorov's inequality may be used to control errors of type 1: for example,

$$\mathbb{P}_0 \left[|F_{M+1} + \cdots + F_\mathbf{n}| > \beta\sqrt{N}, \mathbf{n} > M \right]$$

$$\leq \mathbb{P}_0 \left[\max_{M < n \leq N_+} |F_{M+1} + \cdots + F_n| > \beta\sqrt{N} \right]$$

$$\leq \frac{(N_+ - M) \, \mathbb{E}_0(F^2)}{\beta^2 N}$$

is comparable to α/β^2 and is made small, independently of N, by choice of $\beta = \sqrt[3]{\alpha}$, say. The error of type 2 is not more than $\max|f| \times R_{\mathbf{n}+1}$ and so may be controlled by

$$\mathbb{P}_0(R_{\mathbf{n}+1} > \beta\sqrt{N})$$

$$\leq \mathbb{P}_0 \left(\max_{N_- < n \leq N_+} R_n > \beta\sqrt{N} \right)$$

$$\leq (N_+ - N_-) \, \mathbb{P}_0(R > \beta\sqrt{N})$$

$$\leq \frac{(N_+ - N_-) \, \mathbb{E}_0(R^2)}{\beta^2 N},$$

which is also small, independently of N, with the same β as before. The upshot is, that for all practical purposes,

$$F_1 + \cdots + F_M - 2\beta\sqrt{N} \leq \sum_1^N f \circ \mathbf{x}(n) \leq F_1 + \cdots + F_M + 2\beta\sqrt{N}$$

with a small number β, and since $N/M \simeq Z$, CLT follows by application of the ordinary CLT to $F_1 + \cdots + F_M$. If that was a bit too swift, think it over carefully.

7.4.1★ Kubo's formula

Two versions of σ^2 follow from this argument, *to wit,*

$$(1) \quad \lim_{N\uparrow\infty} \frac{1}{N} \mathbb{E}_0 (F_1 + \cdots + F_M)^2 = \lim_{N\uparrow\infty} \frac{M}{N} \mathbb{E}_0(F^2) = \frac{\mathbb{E}_0(F^2)}{\mathbb{E}_0(R)}$$

and

$$(2) \quad \lim_{N\uparrow\infty} \frac{1}{N} \mathbb{E}_0 |f\circ\mathbf{x}(1) + \cdots + f\circ\mathbf{x}(N)|^2 = \pi f^2 + 2\sum_1^\infty \mathbb{E}_\pi\big[f\circ\mathbf{x}(0)f\circ\mathbf{x}(n)\big],$$

the latter being a simple case of Kubo's more general formula (1956). The computation of (2) relies upon the estimate $\mathbb{E}_0[f \circ \mathbf{x}(n)] \simeq \mathbb{E}_\pi[f \circ \mathbf{x}(0)] = 0$ with an absolute error not worse than ϑ^n for some $0 < \vartheta < 1$, as per §7.3.2: In detail,

$$(2) = \frac{1}{N} \mathbb{E}_0 |f\circ\mathbf{x}(1) + \cdots + f\circ\mathbf{x}(N)|^2$$

$$\simeq \pi f^2 + \frac{2}{N} \sum_{1\leq i<j\leq N} \mathbb{E}_0\big[f\circ\mathbf{x}(i)f\circ\mathbf{x}(j)\big]$$

$$= \pi f^2 + \frac{2}{N} \sum_{1\leq i<j\leq N} \sum_x f(x)\,\mathbb{P}_0[\mathbf{x}(i) = x]\,\mathbb{E}_x[f\circ\mathbf{x}(j-i)],$$

in which the error from replacing $\mathbb{P}_0[\mathbf{x}(i) = x]$ by π_x is not worse than a constant multiple of

$$\frac{1}{N} \sum_{1\leq i<j\leq N} \vartheta^i\vartheta^{j-i} < \frac{1}{N}\frac{\vartheta^2}{(1-\vartheta)^2},$$

so that the computation can be continued in the modified form

$$(2) \simeq \pi f^2 + \frac{2}{N} \sum_{1\leq i<j\leq N} \sum_x f(x)\pi_x\,\mathbb{E}_x\big[f\circ\mathbf{x}(j-i)\big]$$

$$\simeq \pi f^2 + \frac{2}{N} \sum_{1\leq i<j\leq N} \mathbb{E}_\pi\big[f\circ\mathbf{x}(0)f\circ\mathbf{x}(j-i)\big],$$

producing the right-hand side of (2) as $N \uparrow \infty$.

Exercise 7.4.1   It is a pleasant exercise to convert (1) directly into (2) without such intermediate steps. Try it.

Exercise 7.4.2   (1) is hard to compute in all but the very simplest cases. (2) is better though not really simpler either. Work both of them out for two places ± 1, symmetrical one-step probabilities, $1/3$ to stay, $2/3$ to go, and $f(\pm) = \pm 1$. *Answer*: $\sigma^2 = (1) = (2) = 2$.

7.4.2 CLT improved

The improvement of LLN to cover tame functions may be modified to make a like improvement of CLT. The chief point is that the doubled chain inherits from the single chain an infinite number of independent, identically distributed loops of finite mean duration. Once this is known the rest is much as before.

Proof.   Fix the starting place of the doubled chain at $\mathbf{x}(0)\mathbf{x}(1) = 00$ say. The loop time is now $R = \min(n \geq 2 : \mathbf{x}(n-1)\mathbf{x}(n) = 00)$ which may be described in terms of the successive loop times R_1, R_2 etc. of the single chain: for $n \geq 2$, $\mathbf{x}(n-1) = 0$ is followed by $\mathbf{x}(n) = 0$ *for the first time* if and only if $n - 1 = R_1 + \cdots + R_j$ for some $j \geq 1$ and $R_{j+1} = 1$ in the event that $R_{i+1} \neq 1$ for $1 \leq i < j$. Notice that $\mathbb{P}(R < \infty) = 1$ since $R = \infty$ requires that the first step of the nth loop is not 0 for any $n \geq 2$, violating the second Borel–Cantelli lemma, so now you may write

$$R = 1 + \sum_{j=1}^{\infty}\sum_{k=1}^{j} R_k \times \mathbf{e}_2 \cdots \mathbf{e}_j (1 - \mathbf{e}_{j+1}) \qquad \begin{array}{l}\text{with } \mathbf{e}_n = \text{the indicator} \\ \text{of } \mathbf{x}(1 + R_1 + \cdots + R_n) \neq 0\end{array}$$

$$\leq 1$$

$$+ \sum_{j=1}^{\infty} R_1 \mathbf{e}_2 \cdots \mathbf{e}_j \qquad\qquad\qquad \text{from } k = 1$$

$$+ \sum_{j=2}^{\infty} (j-1) \text{ copies of } R_2 \mathbf{e}_3 \cdots \mathbf{e}_j \quad \text{from } k \geq 2.$$

$\mathbb{E}_{00}(R) < \infty$ follows by inspection from the independence of loops since $\mathbb{E}(\mathbf{e}) < 1$. The same idea applies to the tripled chain, etc. and if you like to LLN as well. Easy enough.

7.5 Real time

The chain is a little modified for use in connection with statistical mechanics out of equilibrium in Chapter 11. The path $\mathbf{x} \colon t \to \mathbf{x}(t)$ takes values $x = 1, 2, \ldots, d$ as before and it is required that $\mathbf{x}(t+) = \mathbf{x}(t)$ for $t \geq 0$. Fix $\mathbf{x}(0) = 0$. Then $\mathbf{x}(0+) = 0$ and so also $\mathbf{x}(t) = 0$ for a little while. How long does it wait? Well, if you want to keep the simple Markov property, then the "holding time" $T = \min(t : \mathbf{x}(t) \neq 0)$ must satisfy

$$\mathbb{P}_0(T > t_1 + t_2) = \mathbb{P}_0\big[\mathbf{x}(t') = 0 \ (t' \leq t_1) \ \& \ \mathbf{x}(t_1 + t'') = 0 \ (t'' \leq t_2)\big]$$
$$= \mathbb{P}_0\big[\mathbf{x}(t') = 0 \ (t' \leq t_1)\big] \times \mathbb{P}_0\big[\mathbf{x}(t'') = 0 \ (t'' \leq t_2)\big]$$
$$= \mathbb{P}_0(T > t_1) \times \mathbb{P}_0(T > t_2).$$

The only way to do that is to have $\mathbb{P}_0(T > t) = \mathrm{e}^{-\lambda t}$ for some non-negative number λ, and you may as well take $\lambda > 0$ since $\mathbb{P}_0[\mathbf{x}(t) \equiv \mathbf{x}(0), t \geq 0] = 1$ is not interesting. The "rates" λ may be different for different starting points $\mathbf{x}(0)$, but that's all, and the motion goes like this. Start at $\mathbf{x}(0) = x_0$; wait for such a holding time T_0 with the rate λ_0 assigned to that place; and at time T_0, jump to the place $\mathbf{x}(T_0)$ with the old, pre-assigned probabilities $\mathbb{P}_0[\mathbf{x}(T_0) = y | \mathbf{x}(0) = x] = p_{xy}$, as in §7.1. Then repeat, waiting at $\mathbf{x}(T_0)$ for an independent holding time T_1, with the rate λ_1 assigned to *that* place, jump to $\mathbf{x}(T_0 + T_1)$ by the same rule, and so on. Little really changes: the Markov property holds, both simple and strict; a modified invariant distribution is found; LLN may be proved by Döblin's loops, and likewise CLT. I spell it out.

7.5.1 The Markov property

Introduce the natural fields $\mathfrak{F}_t$ measuring $\mathbf{x}(t') : t' \leq t$ for each $t \geq 0$. For fixed $T \geq 0$, you want $\mathbb{P}_\bullet\big[\mathbf{x}(t + T) = y \,\big|\, \mathfrak{F}_T\big] = P_x[\mathbf{x}(t) = y]$ with $x = \mathbf{x}(T)$. It's easy to check. $\mathfrak{F}_T$ measures

(1) the number $\mathbf{n}$ of jumps before time T,

(2) the jump times $T_0, T_0 + T_1, \ldots, T_0 + \cdots + T_{\mathbf{n}-1}$,

and

(3) the successive places $\mathbf{x}(T_0) = \mathbf{x}_1$, $\mathbf{x}(T_0 + T_1) = \mathbf{x}_2, \ldots, \mathbf{x}(T_0 + \cdots + T_{\mathbf{n}-1}) = \mathbf{x}_\mathbf{n}$.

Keep in mind that the fixed number T itself cannot be a jump time (see why?) and that the "skeleton" $\mathbf{x}(T_0 + \cdots + T_{\mathbf{n}-1}) = \mathbf{x}_\mathbf{n}$ is just a copy of the old chain of §7.1 with the old one-step probabilities p_{xy}. Now the shifted path $\mathbf{x}^+(t) = \mathbf{x}(t+T)$, $t \geq 0$, looks like the original path starting at $\mathbf{x}(T)$ except, perhaps, in one aspect: you need the next holding time $T^+ = T_0 + \cdots + T_\mathbf{n} - T$ to be distributed with the correct rate assigned to the place $\mathbf{x}(T)$. That's the only thing that could go wrong. But look:

246 *Markov Chains*

the conditioning fixes $\mathbf{n}$, $T_0, \ldots, T_{\mathbf{n}-1}$, and $\mathbf{x}_1, \ldots, \mathbf{x}_\mathbf{n}$, and you have $T_0 + \cdots + T_{\mathbf{n}-1} \leq T < T_0 + \cdots + T_\mathbf{n}$; $\mathbf{x}(T) = \mathbf{x}_\mathbf{n}$; and $\mathbb{P}_\bullet(T_\mathbf{n} > t \,|\, \mathfrak{F}_T) = e^{-\lambda t}$, so

$$\mathbb{P}_\bullet(T^+ > t \,|\, \mathfrak{F}_T)$$

$$= \mathbb{P}\big[T_0 + \cdots + T_\mathbf{n} - T > t \,\big|\, T_0 + \cdots + T_\mathbf{n} > T\big] \quad \begin{array}{l}\text{with fixed} \\ T_0 + \cdots + T_{\mathbf{n}-1} \equiv S \\ \text{smaller than } T \text{ and} \\ \text{independent } T_\mathbf{n}\end{array}$$

$$= \frac{\mathbb{P}(T_\mathbf{n} > t + T - S)}{\mathbb{P}(T_\mathbf{n} > T - S)}$$

$$= e^{-\lambda t},$$

as it is supposed to be.

Exercise 7.5.1 Note 6.9.2 speaks of the forward and backward infinitesimal operators $\mathfrak{G}^\dagger$ and $\mathfrak{G}$ for the Brownian motion with drift. What are these for the chain?

The strict Markov property is pretty easy, too. T is now a stopping time in the sense that $(T \leq t) \in \mathfrak{F}_t$ for every $t \geq 0$, $\mathfrak{F}_T$ is the field measuring T and the stopped path $\mathbf{x}^-: t \in [0, \infty) \to \mathbf{x}(t \wedge T)$; $\mathbf{x}^+: t \to \mathbf{x}(t+T)$ as usual; and you want

$$\mathbb{P}_\bullet\big[\mathbf{x}^+ \in B \,\big|\, \mathfrak{F}_T\big] = \mathbb{P}_x(B) \quad \text{with } x = \mathbf{x}(T).$$

Exercise 7.5.2 Think it over in imitation of §6.3.3.

7.5.2 Loops and the invariant distribution

Fix $\mathbf{x}(0) = 0$. $R = \min(t \geq T_0 : \mathbf{x}(t) = 0)$ is the loop time, $\mathbf{x}(t) : 0 \leq t < R$ is the loop, and with a self-evident notation,

$$\mathbb{E}_0(R) = \sum_{n=0}^\infty \mathbb{E}_0\big(T_0 + \cdots + T_{n-1}, \mathbf{x}_1, \ldots, \mathbf{x}_{n-1} \neq 0, \mathbf{x}_n = 0\big)$$

$$= \sum_{n=0}^\infty \mathbb{E}_0\bigg[\int_0^\infty \lambda_0 t e^{-\lambda_0 t}\, dt + \cdots +$$

$$+ \int_0^\infty \lambda_{n-1} t e^{-\lambda_{n-1} t}\, dt, \, \mathbf{x}_1, \ldots, \mathbf{x}_{n-1} \neq 0, \mathbf{x}_n = 0\bigg]$$

$$\leq \frac{1}{\min \lambda} \sum_0^\infty n\, \mathbb{E}_0(\mathbf{x}_1, \ldots, \mathbf{x}_{n-1} \neq 0, \mathbf{x}_n = 0)$$

$$= \frac{1}{\min \lambda} \times \text{the mean loop time for the skeleton.}$$

The invariant distribution π is then $e_x = \mathbb{E}_0\big[\text{measure}\big(t < R : \mathbf{x}(t) = x\big)\big]$ = the expected occupation time at x, per loop, normalized by $Z = \mathbb{E}_0(R)$. The verification is not quite so simple as in §7.2: you have

$$e_x = \int_0^\infty \mathbb{P}_0(R > t, \mathbf{x}(t) = x)\,dt,$$

and what is to be proved is that for fixed $T > 0$,

$$\mathbb{P}_e\big[\mathbf{x}(T) = y\big] \equiv \sum_x e_x\,\mathbb{P}_x\big[\mathbf{x}(T) = y\big] = \int_0^\infty \mathbb{P}_0\big[(R > t, \mathbf{x}(t+T) = y\big]\,dt$$

$$= e_y.$$

Not so easy to handle as it stands, but try this: with $\alpha > 0$,

$$\int_0^\infty e^{-\alpha t}\,\mathbb{P}_0\big[R > t, \mathbf{x}(t+T) = y\big]\,dt$$

$$= \mathbb{E}_0\left[\int_0^R e^{-\alpha t}\,\mathbf{1}_{\mathbf{x}(t+T)=y}\,dt\right]$$

$$= \mathbb{E}_0\left(\int_0^\infty\right) - \mathbb{E}_0\left(\int_R^\infty\right)$$

$$= \mathbb{E}_0\left(\int_0^\infty\right) - \mathbb{E}_0\left[e^{-\alpha R}\int_0^\infty e^{-\alpha t}\,\mathbf{1}_{\mathbf{x}(t+T+R)=y}\,dt\right]$$

$$= \mathbb{E}_0\left(\int_0^\infty\right) - \mathbb{E}_0(e^{-\alpha R})\,\mathbb{E}_0\left[\int_0^\infty e^{-\alpha t}\,\mathbf{1}_{\mathbf{x}(t+T)=y}\,dt\right] \quad \begin{array}{l}R \text{ being a stopping} \\ \text{time and } \mathbf{x}(R) = 0\end{array}$$

$$= \big[1 - \mathbb{E}_0(e^{-\alpha R})\big] \times \int_0^\infty e^{-\alpha t}\,\mathbb{P}_0\big[\mathbf{x}(t+T) = y\big]\,dt$$

$$\simeq \alpha\,\mathbb{E}_0(R) \times e^{\alpha T}\int_T^\infty e^{-\alpha t}\,\mathbb{P}_0\big[\mathbf{x}(t) = y\big]\,dt \qquad \text{for } \alpha \downarrow 0$$

$$\simeq \alpha\,\mathbb{E}_0(R) \times \int_0^\infty e^{-\alpha t}\,\mathbb{P}_0\big[\mathbf{x}(t) = y\big]\,dt,$$

showing that $\mathbb{P}_e\big[\mathbf{x}(T) = y\big]$ does not depend on $T \geq 0$. The rest is obvious.

LLN is now easy as can be: with successive loop times R_n, $n \geq 1$, large T, $R_1 + \cdots + R_\mathbf{n} \leq T < R_1 + \cdots + R_{\mathbf{n}+1}$, and $F_k = \int f\circ\mathbf{x}(t)\,dt$ taken over the kth loop, you have $nZ \simeq T$ and

$$\frac{1}{T}\int_0^T f\circ\mathbf{x}(t)\,dt \simeq \frac{F_1 + \cdots + F_n}{nZ} \simeq \frac{\mathbb{E}_e(F)}{Z} = \pi f$$

just as in §7.3.

Exercise 7.5.3   Think it over.

CLT is easy, too:

$$\lim_{T\uparrow\infty} \mathbb{P}_\bullet \left[\frac{\int_0^T f\circ\mathbf{x}(t)\,dt}{\sqrt{T}} \le c \right] = \int_{-\infty}^c \frac{e^{-x^2/2\sigma^2}}{\sigma\sqrt{2\pi}}\,dx \quad \text{if } \pi f = 0$$

with suitable $\sigma > 0$.

Exercise 7.5.4   Think this over as well.

7.6 The standard Poisson process

This falls outside the present format, but I report on it here as it is the simplest chain in real time with $d = \infty$ and will be wanted in Chapter 11, on statistical mechanics out of equilibrium.

The path $\mathbf{n}\colon t \ge 0 \to \mathbf{n}(t)$ is now a "counter", recording the number of "arrivals" up to time $t \ge 0$ of, e.g. customers at the ticket window, calls at the trader's desk, electrons at the detector – what you will – and the (more or less realistic) rules are these:

(1) $\mathbf{n}(t) = \mathbf{n}(t+) \in \mathbb{N} = 0, 1, 2$ etc.,

(2) $\mathbf{n}(0) = 0$ and $\mathbf{n}(t) \uparrow \infty$,

(3) for fixed $T > 0$, the "future" $\mathbf{n}^+\colon t \to \mathbf{n}(t+T) - \mathbf{n}(t)$ is a copy of $\mathbf{n}$ itself, independent of the "past" $\mathbf{n}(t') : t' \le T$,

to which is added the technical rule

(4) $\mathbb{P}[\mathbf{n}(t) \ge 2] = o(t)$ for $t \downarrow 0$.

The first item of business is to show that $\mathbf{n}$ must be Poisson-distributed as in

$$\mathbb{P}[\mathbf{n}(T) = n] = \frac{(\lambda T)^n}{n!}\,e^{-\lambda T} \text{ for some number } 0 < \lambda < \infty.$$

Proof.   By rule (3),

$$\mathbb{P}[\mathbf{n}(t_1 + t_2) = 0] = \mathbb{P}[\mathbf{n}(t_1) = 0, \mathbf{n}(t_1 + t_2) - \mathbf{n}(t_1) = 0]$$
$$= \mathbb{P}[\mathbf{n}(t_1) = 0] \times \mathbb{P}[\mathbf{n}(t_2) = 0],$$

so $\mathbb{P}[\mathbf{n}(t) = 0] = e^{-\lambda t}$, with $0 < \lambda < \infty$ in view of rule (2), and $\mathbb{P}[\mathbf{n}(t) = 1] \simeq \lambda t$ for $t \downarrow 0$ in view of rule (4). Now for fixed $T \ge 0$ and large N, the event $\mathbf{n}(t) = n$ may be realized by requiring $\mathbf{n}(k\Delta) - \mathbf{n}((k-1)\Delta) = 1$ for $\Delta = T/N$ and n choices of k between 1 and N. A second application of rule (3) permits the interpretation of this event as the occurrence of n successes in N independent Bernoulli trials with success probability $\mathbb{P}[\mathbf{n}(\Delta) = 1] \simeq \lambda\Delta$, leading at once to

$$\mathbb{P}[\mathbf{n}(t) = n] \gtrsim \binom{N}{n}(\lambda\Delta)^n(1 - \lambda\Delta)^{N-n} = \frac{(\lambda T)^n}{n!}e^{-\lambda T}$$

in the limit when N is taken to $+\infty$. This looks wide of the mark, being only an underestimate, but it is not so: both the left- and right-hand sides add up to 1, so equality prevails for each n separately.

The picture now comes into focus. The path is seen in Figure 7.6.1, the successive times T_1, T_2 etc. being (hopefully) independent copies of

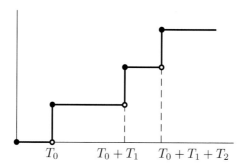

$$T_0 \qquad\qquad T_0 + T_1 \qquad T_0 + T_1 + T_2$$

Figure 7.6.1

$T_0 = \min(t : \mathbf{n}(t) \neq 0)$ with common distribution

$$\mathbb{P}(T_0 > t) = \mathbb{P}[\mathbf{n}(t) = 0] = e^{-\lambda t};$$

in short, $\mathbf{n}$ is, or ought to be, the (real-time) Markov chain on $\mathbb{N} = 0, 1, 2, 3$ etc. with common rates λ and $p_{xy} = 1$ or 0 according as $y = x+1$ or not. Evidently, rules (1), (2), and (4) obtain, and it's not hard to check rule (3) directly.

Proof. Fix $T \geq 0$ and look at the event $\mathbf{n}^+(t) = \mathbf{n}(t + T) - \mathbf{n}(T) = n$, conditional on the field $\mathfrak{F}_T$ of $\mathbf{n}(t') : t' \leq T$. The conditioning fixes

(1) the number (m) of jumps before time T,
(2) the last jump time $S = T_0 + \cdots + T_{m-1}$ before time T,

and

(3) $x(T) = x(S) = m$.

and *nothing else to do with* $\mathbf{n}^+(t) = n$ *besides the stipulation* $S + T_m > T$, so that

$$\mathbb{P}\big[\mathbf{n}^+(t) = n \,\big|\, \mathfrak{F}_T\big]$$
$$= \mathbb{P}\big[S + T_0' + \cdots + T_{n-1}' \le t + T \le S + T_0' + \cdots + T_n' \,\big|\, S + T_0' > T\big]$$

with fixed S and independent copies $T_0' (\equiv T_m), T_1', \ldots, T_n'$ of T_0. Here, the conditioning does not affect $T_1', \ldots, T_n'$, and

$$\mathbb{P}\big[S + T_0' - T > t \,\big|\, S + T_0' > T\big] = \mathrm{e}^{-\lambda t}$$

as well, which is to say that $S + T_0' - T, T_1', \ldots, T_n'$ may be regarded as independent copies of T_0, with the result that

$$\mathbb{P}\big[\mathbf{n}^+(t) = n \,\big|\, \mathfrak{F}_T\big] = \mathbb{P}\big[T_0 + \cdots + T_{n-1} \le t < T_0 + \cdots + T_n\big] = \frac{(\lambda t)^n}{n!}\, \mathrm{e}^{-\lambda t}$$

when you work it out, showing at one stroke that $\mathbf{n}^+$ is independent of $\mathfrak{F}_T$ and correctly distributed, too.

Exercise 7.6.1   Details please.

Exercise 7.6.2   Prove the law of large numbers

$$\mathbb{P}\left[\lim_{T \uparrow \infty} T^{-1}\mathbf{n}(T) = \lambda\right] = 1 :$$

another interpretation of the rate.

Exercise 7.6.3   Go back to Exercise 1.5.3. What does it say and how should it now be proved without the detailed use of Stirling's formula?

7.7 ⋆ Large deviations

The discussion of large deviations, commenced in §2.4 with Cramér's estimates for atypical behavior of independent Bernoulli trials, is now extended to the Markov chains of §7.1. Here you will see, in miniature, the wide scope of which Cramér's original idea is capable.

7.7.1 Setup and simplest examples of the main result

Here and below, all probabilities and expectations are reckoned for some fixed choice of $\mathbf{x}(0)$, it does not matter which. $\overline{Q}$ is the simplex of all probability distributions $q = (q_x : x = 1, \ldots, d)$, Q is its interior, and ∂Q its boundary where one or more of the individual probabilities vanish.

The principal role is played by the empirical distribution, at the "second level" of §2.4.2:

$$\mathbf{f}_x(n) = \frac{1}{n}\#(1 \le k \le n : \mathbf{x}(k) = x), \quad x = 1, 2, \ldots, d.$$

Of course,

$$\mathbb{P}\Big[\lim_{n\uparrow\infty} \mathbf{f}(n) = \pi\Big] = 1 \quad \text{by LLN},$$

but what about *large* deviations from such *typical* behavior? The answer may be expressed with the help of the function $I \colon \overline{Q} \to [0, \infty)$ defined by

$$I(q) = \sup_{f \in Q} J(f)$$

with

$$J(f) = \sum_x q_x \log \frac{f_x}{(pf)_x} \quad \text{and } p = [p_{xy} : 1 \le x, y \le d];$$

to wit: if $A \subset Q$ is an (open) neighborhood of π at positive distance from ∂Q and if B is its (compact) complement in $\overline{Q}$, then

$$\lim_{n\uparrow\infty} \frac{1}{n} \log \mathbb{P}[\mathbf{f}(n) \notin A] = -\min_B I(q),$$

which is to say that the probability of *atypical* behavior is exponentially small. Roughly speaking,

$$\mathbb{P}[\mathbf{f}(n) \notin A] \simeq \exp\Big[-n \times \min_B I(q)\Big].$$

The fundamental character of the "rate function" $I(q)$ is the remarkable discovery of Donsker–Varadhan [1975/76]. The present account follows them; see also Kac's amusing sketch [1980] and Varadhan's excellent survey [1984] of the whole subject of large deviations.

Exercise 7.7.1 The chain is simplest when it is in perfect equilibrium, i.e. when $p_{xy} = \pi_y$. Then the steps are independent, with common distribution π, and there is a self-evident reduction to the situation of §2.3. Check that in such a case $I(q) = \sum_x q_x \log(q_x/p_x)$ as in that section, where now $p_x = \pi_x$.

Exercise 7.7.2 Take just two states ± 1 and $\pi = (1/2, 1/2)$. Then

$$\mathbb{P}\Big[\Big|\mathbf{f}_+(n) - \frac{1}{2}\Big| + \Big|\mathbf{f}_-(n) - \frac{1}{2}\Big| \ge \varepsilon\Big] = \mathbb{P}_0\big[|\mathbf{x}(n)| \ge n\varepsilon\big] \quad \text{for RW(1)},$$

which ought to be something like $e^{-n\varepsilon^2/2}$, by CLT. Compare this rate $(\varepsilon^2/2)$ to the exact rate

$$\min\left[I(q) : \left|q_+ - \frac{1}{2}\right| + \left|q_- - \frac{1}{2}\right| \geq \varepsilon\right],$$

both for small ε and for ε near 1. *Partial answer*: For small ε, the exact rate is a tiny bit better; for large ε, it is *much* better ($\log 2 \simeq 0.69+ > 1/2$).

Exercise 7.7.3   Take $p = \begin{pmatrix} 1/3 & 2/3 \\ 2/3 & 1/3 \end{pmatrix}$ with $\pi = (1/2, 1/2)$ as before. I is now more complicated but the pattern is similar. Write $q_1 = q$ and $q_2 = 1 - q$, with a temporary abuse of notation, and compute

$$I(q) = \log 3 + q \log q + (1 - q) \log(1 - q) - q \log(2 - 3q + 2R)$$
$$- (1 - q) \log\big(2 - 3(1 - q) + 2R\big)$$

in which $R = \sqrt{1 - 3q + 3q^2}$. Check that $I(q) \simeq 4(q - 1/2)^2$ near $q = 1/2$, but reduces to $\log 3 > 1$ at $q = 1$. Deduce

$$\mathbb{P}\left[\big|\mathbf{f}(n) - \pi\big| \leq \sqrt{\log n/2n} \text{ for } n \uparrow \infty\right] = 1.$$

7.7.2 Preliminaries about I

Here are the facts you'll need. Cover up the proofs and take them as exercises if you want. For brevity of notation, think of non-negative vectors as functions of the index $1 \leq x \leq d$ so that, e.g. $(ab)_x = a_x b_x$, $(\log c)_x = \log c_x$, and so on. $J(f)$ is then written $q \cdot \log(f/pf)$ in which the spot ($\cdot$) signifies the inner product. $\mathbf{1}$ is the vector $(1, 1, \ldots, 1)$.

Fact 1 $J(f)$ is non-negative since $J(\mathbf{1}) = 0$.

Fact 2 $I(\pi) = 0$. Indeed, $\log pf \geq p \log f$ by Jensen's inequality, so $J(f) \leq \pi \cdot \log f - \pi \cdot p \log f = 0$ since $\pi p = \pi$. Item 1 does the rest.

Fact 3 $I(q)$ is positive otherwise. Evidently, $I(q) = 0$ only if $J(f) \leq 0$ for every $f \in Q$. Replace f by $\mathbf{1} + \varepsilon f$ with a small number ε of either sign. Then J vanishes at $\varepsilon = 0$, and so also grad $J \cdot f = 0$ there, which is to say $q \cdot f = q \cdot pf$ or what is the same, $p^\dagger q = q$.

Fact 4 $I(q)$ is convex. The proof is too simple to write.

Fact 5 $I(q)$ is continuous on $\overline{Q}$.

Proof. If q_n tends to q_∞ in $\overline{Q}$, then

$$\liminf_{n\uparrow\infty} I(q_n) \geq q_\infty \cdot \log(f/pf) \text{ for any } f \in Q$$

and this approximates $I(q_\infty)$ by choice of f.

That's the easy half. Now estimate from above: For any $\varepsilon > 0$,

$$I(q_n) - \varepsilon < J(f) = q_n \cdot \log \frac{f}{pf} \qquad \begin{array}{l} \text{by choice of } f \in Q \\ \text{depending upon } n \end{array}$$

$$< q_n \cdot \left[\log \frac{f + \varepsilon \mathbf{1}}{pf + \varepsilon \mathbf{1}} + \log \frac{pf + \varepsilon \mathbf{1}}{pf}\right]$$

so that

$$\limsup_{n\uparrow\infty} I(q_n) \leq \varepsilon + q_\infty \cdot \log \frac{f + \varepsilon \mathbf{1}}{pf + \varepsilon \mathbf{1}} + \frac{\varepsilon}{\min p_{xy}} \text{ for some new } f \in \overline{Q}.$$

$$\simeq q_\infty \cdot \log \frac{f}{pf} \text{ for } \varepsilon \downarrow 0$$

$$\leq I(q).$$

The rest will be plain.

Fact 6 For fixed q, $J(f)$ achieves its maximum at some place $f \in Q$ since $J(f) = -\infty$ on ∂Q.

Fact 7 = Fact 6 continued. Let f be any maximizer of $J(f)$. Then q is the invariant distribution π' of the "primed" chain with $p'_{xy} = p_{xy}f_y/(pf)_x$. This fact plays a decisive role below.

Proof. $\partial J/\partial f = 0$ at such a place, the constraint $f \cdot \mathbf{1} = 1$ being irrelevant, as you will please check, and that is nothing but $qp' = q$ when it's written out:

$$\frac{\partial J}{\partial f_x} = \frac{q_x}{f_x} - \sum_y q_y \frac{p_{yx}}{(pf)_y} = \frac{1}{f_x} \times \left[q_x - (qp')_x\right] = 0.$$

Fact 8 = Fact 7 continued. $J(f)$ achieves its maximum at just one place in Q.

Proof. Compute the Hessian of J at one of its critical points in Q: *viz.*

$$\frac{\partial^2 J}{\partial f_x \partial f_y} = -\frac{q_x}{f_x^2} \text{ on diagonal} + \sum_z \frac{q_z}{(pf)_z^2} p_{zx} p_{zy}.$$

This is, of course, symmetric in x and y and also *negative* definite. In detail, if e is any vector tangent to Q at f, then $e \cdot \mathbf{1} = 0$ and the quadratic form $e \cdot \partial^2 J/\partial f^2 \, e$ may be written

$$-\sum_y q_y \frac{e_y^2}{f_y^2} + \sum_x q_x \frac{(pe)_x^2}{(pf)_x^2} = -(1) + (2).$$

Now

$$(1) = \sum_y \frac{e_y^2}{f_y^2} \sum_x q_x \frac{p_{xy} f_y}{(pf)_x} \qquad \text{by Fact 7}$$

$$= \sum_x \frac{q_x}{(pf)_x^2} \left[\sum_y p_{xy} \frac{e_y^2}{f_y} \times \sum_y p_{xy} f_y \right]$$

$$\geq \sum_x q_x \frac{(pe)_x^2}{(pf)_x^2} = (2) \qquad \begin{array}{l}\text{by Schwarz's inequality} \\ \text{applied to the inner sums,}\end{array}$$

and here the equality cannot hold since any proportionality between $e/\sqrt{f}$ and $\sqrt{f}$ must violate the fact that e changes sign. In short, every critical point of J is a *local* maximum. Now let $d = 3$ (dimension 2) for simplicity, and suppose $J(f)$ takes on its *global* maximum at two different places, indicated in the figure by the black spots. Thinking of $J(f)$ as altitude above the plane of Q, it is clear that between two such

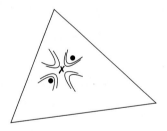

Figure 7.7.1

mountain peaks, there must be a mountain pass, and that at the height of land or saddle, marked by the cross (✗), there must be two ridges rising on either hand (in the direction $\pm e_+$) and two valleys falling away (in the direction $\pm e_-$). There, $J(f)$ is flat (or better to say, critical), but the quadratic form is ≥ 0 in the direction e_+ and ≤ 0 in the direction e_-, i.e. it is *in*definite. That's the proof. It's a little sketchy, but the idea I hope is clear. The general rule in dimensions ≥ 3 is the "mountain pass lemma", for which see, e.g. Milnor [1965].

Exercise 7.7.4 Give a direct proof that for $q = \pi$, $J(f) = I(q)$ only if $f = \mathbf{1}/d$.

7.7.3 Proof of the main result

Easy part: $\displaystyle \liminf_{n\uparrow\infty} \frac{1}{n} \log \mathbb{P}\big[\mathbf{f}(n) \notin A\big] \geq -\min_B I(q)$.

Proof Recall from Fact 7 that $q \in Q$ is the invariant distribution of the primed chain with $p'_{xy} = p_{xy} f_y/(pf)_x$, f being *the* maximizer of $J(f)$. Keep this in mind in the form $p_{xy}/p'_{xy} = (pf)_x/f_y$. Now fix a small, closed simplicial cell C centered at q and observe that

$$\mathbb{P}\big[\mathbf{f}(n) \in C\big]$$

$$= \sum_{f \in C} p'_{x_0 x_1} p'_{x_1 x_2} \cdots p'_{x_{n-1} x_n} \times \frac{p_{x_0 x_1} p_{x_1 x_2} \cdots p_{x_{n-1} x_n}}{p'_{x_0 x_1} p'_{x_1 x_2} \cdots p'_{x_{n-1} x_n}}$$

with $x_0 =$ the common starting point fixed at the beginning and $f =$ the empirical distribution of $x_1, x_2, \ldots, x_n$.

$$= \sum_{f \in C} p'_{x_0 x_1} p'_{x_1 x_2} \cdots p'_{x_{n-1} x_n} \times \frac{(pf)_{x_0} (pf)_{x_1} \cdots (pf)_{x_{n-1}}}{f_{x_1} f_{x_2} \cdots f_{x_n}}$$

$$= \mathbb{E}' \left[(pf)_{\mathbf{x}(0)} \exp\left[-\sum_1^n \log\left(\frac{f}{pf}\right)_{\mathbf{x}(k)} \right] \frac{1}{(pf)_{\mathbf{x}(n)}}, \mathbf{f}_n \in C \right]$$

with the expectation for the primed chain.

By Fact 7 and LLN for the primed chain, the sum under the exponential approximates $-n \times [J(f) = I(q)]$ as $n \uparrow \infty$, so

$$\mathbb{P}\big[\mathbf{f}(n) \in C\big] \gtrsim \frac{\min f_x}{\max f_x} \times \exp\big[-nI(q)\big],$$

which is to say

$$\liminf_{n\uparrow\infty} \frac{1}{n} \log \mathbb{P}\big[\mathbf{f}_n \in C\big] \geq -I(q).$$

The open neighborhood A of π is now enlarged to an (open) polygonal figure and its (closed, reduced) complement B is divided into a finite number of small, closed, simplicial cells C of the above type. Then

$$\liminf_{n\uparrow\infty} \frac{1}{n} \log \mathbb{P}\big[\mathbf{f}_n \in B\big] \geq \max_C \liminf_{n\uparrow\infty} \frac{1}{n} \log \mathbb{P}\big[\mathbf{f}_n \in C\big]$$

$$= -\min_C I(q)$$

$$\simeq -\min_B I(q)$$

if the subdivision is really fine: this for the reduced B. For the original B, line 1 is only larger, and Fact 5 takes care of the rest.

Not so easy part: $\displaystyle \limsup_{n\uparrow\infty} \frac{1}{n} \log \mathbb{P}\big[\mathbf{f}(n) \notin A\big] \leq -\min_B I(q)$.

Proof. For fixed $f \in Q$,

$$1 = \mathbb{E}\left[\frac{f_{\mathbf{x}(1)}}{(pf)_{\mathbf{x}(0)}}\right] = \mathbb{E}\left[\frac{f_{\mathbf{x}(1)} f_{\mathbf{x}(2)}}{(pf)_{\mathbf{x}(0)} (pf)_{\mathbf{x}(1)}}\right] \quad \text{and so on}$$

$$= \mathbb{E}\left[\frac{1}{(pf)_{\mathbf{x}(0)}} \exp\left[\sum_1^n \log \frac{f_{\mathbf{x}(k)}}{(pf)_{\mathbf{x}(k)}}\right] f_{\mathbf{x}(n+1)}\right]$$

$$\geq \frac{\min f_x}{\max f_x} \mathbb{E}\left[\exp\left[n \sum_y \mathbf{f}_y(n) \log \frac{f_y}{(pf)_y}\right], \mathbf{f}(n) \in B\right]$$

$$\geq \frac{\min f_x}{\max f_x} \exp\left[n \min_{q \in B} \sum_y q_y \log \frac{f_y}{(pf)_y}\right] \times \mathbb{P}\left[\mathbf{f}(n) \in B\right],$$

so

$$\limsup_{n\uparrow\infty} \frac{1}{n} \log \mathbb{P}\left[\mathbf{f}(n) \in B\right] \leq -\sup_{f \in Q} \min_{q \in B} J(f),$$

which is what was wanted except that the *sup* and the *min* are *in the wrong order*. To reverse them, modify A and B as before, only now *reducing A* and *enlarging B*, and dividing B into a number of small, closed simplicial cells C in which q does not change much. Then

$$\limsup_{n\uparrow\infty} \frac{1}{n} \log \mathbb{P}\left[\mathbf{f}(n) \in B\right]$$

$$\leq \max_C \limsup_{n\uparrow\infty} \frac{1}{n} \log \mathbb{P}\left[\mathbf{f}(n) \in C\right] \quad \begin{array}{l} \text{by Jensen's inequality with} \\ \mathbb{P}[\mathbf{f} \in B] \text{ replaced by } \sum \mathbb{P}[\mathbf{f} \in C] \\ \text{divided by the number of cells} \end{array}$$

$$\leq \max_C \left[-\sup_{f \in Q} \min_{q \in C} J(f)\right] \quad \begin{array}{l} \text{by the previous} \\ \text{reasoning.} \end{array}$$

But for fixed f, $J(f)$ does not change much for $q \in C$ if the subdivision is fine enough. This means that the *min* may be ignored at little cost and the *sup* taken to produce a value of I, differing only a little from $\min_{q \in C} I(q)$. In short,

$$\limsup_{n \uparrow \infty} \frac{1}{n} \log \mathbb{P}\left[\mathbf{f}(n) \in B\right] \lesssim \max_{C}\left[-\min_{q \in C} I(q)\right] \leq -\min_{B} I(q),$$

first for the enlarged B, and then for the original B itself.

Roughly speaking, what's been said is that $\mathbb{P}[\mathbf{f}(n) \in C] \simeq \exp[-nI(q)]$ for some q in any small closed cell C not containing π. For example, if $F \in C(\overline{Q})$ and if $F - I$ does not take its maximum in some open neighborhood A of π, then with a subdivision of $B = A'$ into small cells C and $\mathbf{f}(n) = \mathbf{f}$ for brevity, you should have

$$\mathbb{E}\left[e^{nF(\mathbf{f})}\right] \simeq \sum_{C} e^{nF(q)} \times e^{-nI(q)} \qquad \begin{array}{l}\text{with a suitable choice} \\ \text{of } q \text{ in each cell}\end{array}$$

$$\simeq \exp\left[n \max_{\overline{Q}}(F - I)\right] \qquad \begin{array}{l}\text{if the cells are} \\ \text{really small,}\end{array}$$

and that is correct when stated in a more cautious way, *to wit*,

$$\lim_{n \uparrow \infty} \frac{1}{n} \log \mathbb{E}\left[e^{nF(f)}\right] = \max_{\overline{Q}}(F - I).$$

Exercise 7.7.5 Try to make this "demonstration" into a real proof. The special case, $F = V \cdot q$ with $V \in \mathbb{R}^d$, will be helpful in the next article.

7.7.4 Legendre duality

There is some pretty geometry here, involving the Legendre duality of convex functions introduced in §2.4, but first a simple fact providing, incidentally, still another proof of the existence of the equilibrium distribution of the Markov chain.

Lemma 7.7.6 Any matrix $K = [K_{xy} : 1 \leq x, y \leq d]$ with all positive entries has just one positive eigenvector f. Naturally, the corresponding eigenvalue Λ is positive, too. It is the top of the spectrum of K.

Proof after Wielandt [1950]. Take $0 < m \leq K_{xy} \leq M$, $f \in Q$, and look at the form

$$\min \frac{Kf}{f} = \min_{x} \sum_{y} \frac{K_{xy} f_y}{f_x}$$

This is bounded, above by dM and below by dm, and so takes its minimum value Λ at some $f \in Q$. There $Kf \geq \Lambda f$, and *either $Kf = \Lambda f$, or else $Kf = \Lambda f + e$* where $e \geq 0$ does not vanish. But in the second case, $K^2 f = \Lambda Kf + Ke > \Lambda Kf$, which is contradictory. In short, $Kf = \Lambda f$ and so also $\Lambda > 0$ & $f \in Q$. Note the general principle here: if $Kf \geq \Lambda f$, then either $f = 0$ or $Kf = \Lambda f$. Now let $\Lambda' > 0$ and $f' \in Q$ be another such pair. Then $af \leq f' \leq bf$, by choice of $0 < a < b$, and $\Lambda' = \Lambda$ in view of

$$a\Lambda^n f = K^n af \leq (\Lambda')^n f' \leq K^n bf = b\Lambda^n f.$$

Next, split $f' - f$ into its positive and negative parts, as in $f^+ - f^-$, note that $Kf^+ \geq (Kf)^+ = \Lambda f^+$, and conclude as follows: *either $f^+ = 0$, $f' - f \leq 0$ and $f' = f$, or else $Kf^+ = \Lambda f^+$, $f^+ > 0$, $f^- = 0$, and $f' = f$* once more. Neat, no? I leave the easy proof that $\Lambda = \max \operatorname{spec} K$ to you. Note that the lemma supplies yet another proof of the existence of π – not that you need it.

Now to business. Take $V \in \mathbb{R}^d$ and solve $e^V pf = \Lambda f$ for $f \in Q$ and $\Lambda = \Lambda(V)$ as per the lemma. Then V may be written as $\log(f/pf) + \log \Lambda$ in only one way, and $V \cdot q - \log \Lambda(V) = J(f)$, with the immediate result that

(1) $\displaystyle\max_{V \in \mathbb{R}^d} \left[V \cdot q - \log \Lambda(V) \right] = \max_{f \in Q} J(f) = I(q)$ for fixed q;

in short, $I(q)$ is the Legendre dual of $\log \Lambda(V)$, and you would hope to have

(2) $\displaystyle\max_{q \in Q} \left[V \cdot q - I(q) \right] = \log \Lambda(V)$ for fixed V

as well, just as in Exercise 2.4.7. Here, (1) is maximized at the critical value of $J(f)$, but what about (2)? The answer is: only at $q = $ the invariant distribution of the primed chain. How neatly it all fits together!

Exercise 7.7.7   (2) implies that $\log \Lambda(V)$ is convex, $I(q)$ being such (Fact 4). Give a direct proof based solely upon $e^V pf = \Lambda f$.

As to (2), $V \cdot q - I(q) = J(f) - I(q) + \log \Lambda(V) \leq \log \Lambda(V)$ is obvious, as is the existence of the maximum taken over $\overline{Q}$, the which may be identified as $\log \Lambda(V)$ by use of Exercise 7.7.5 with $F(q) = V \cdot q$: in detail,

$$\max_{q \in \overline{Q}} \left[V \cdot q - I(q) \right]$$

$$= \lim_{n \uparrow \infty} \frac{1}{n} \log \mathbb{E} \left[\exp nV \cdot \mathbf{f}(n) \right]$$

$$= \lim_{n \uparrow \infty} \frac{1}{n} \log \mathbb{E} \left[\exp \sum_{1}^{n} V \circ \mathbf{x}(k) \right]$$

$$= \lim_{n \uparrow \infty} \frac{1}{n} \log \mathbb{E} \left[\sum_{12\cdots n} p_{01} e^{V_1} p_{12} e^{V_2} \cdots p_{n-1\,n} e^{V_n} \right]$$ with a self-evident abbreviation of notation

$$= \lim_{n \uparrow \infty} \frac{1}{n} \log \left[\sum_{y} (pe^{V})^n_{xy} \right]$$

$$\geq \lim_{n \uparrow \infty} \frac{1}{n} \log \left[\sum_{y} (pe^{V})^n_{xy} \frac{f_y}{\max f_x} \right]$$

$$\geq \lim_{n \uparrow \infty} \frac{1}{n} \log \left[\Lambda^n \frac{\min f}{\max f} \right]$$

$$= \log \Lambda(V).$$

It remains to rule out the possibility that $V \cdot q - I(q) = \log \Lambda(V)$ at some place $q \in \partial Q$. That's easy: for such q and $V = \log(f/pf) + \log \Lambda(V)$,

$$V \cdot q - I(q) - \log \Lambda(V) \leq \sum q \cdot \log \frac{f}{pf} - \sum q \cdot \log \frac{f'}{pf'},$$

for any $f' \in Q$, and is made strictly negative, by choice of $f'_x = f_x$ where $q_x > 0$ and $f'_x = 0$ elsewhere, since pf' is then strictly less than pf. The identification $q = \pi'$ has still to be made. That's easy, too: $I(q) = I'(q) + \sum q \log(f/pf)$ with the I-function for the primed chain, as an easy computation shows, so $0 \geq V \cdot q - I(q) - \log \Lambda(V) = I'(q)$, and that is known to vanish only if $q = \pi'$, by Exercise 7.7.4.

8

The Ergodic Theorem

G.D. Birkhoff's ergodic theorem [1931] has its roots in classical Hamiltonian mechanics and in the statistical mechanics of Boltzmann, Kelvin, Maxwell, and Gibbs. The next two sections will fill you in.

8.1 Hamiltonian mechanics

Hamilton's format subsumes Newton's *force = mass × acceleration* in all its classical varieties. Not much is needed here. Kepler's problem, of the motion of a planet about the sun is the simplest, really interesting example and will serve to set the stage: Figure 8.1.1 shows the planet (∘) moving in its orbit about the sun (•). The latter being so heavy, it

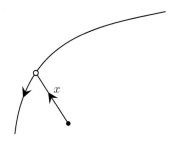

Figure 8.1.1

is effectively fixed, and any other celestial bodies have only a negligible influence. Kepler (1600) found, by endless patient study of Tycho's observations of the then known planets, three remarkable facts:

(1) that the orbit is an ellipse with one focus at the sun;

(2) that the rate of area production swept out by the vector x is a constant of motion; and

(3) that the period of revolution is a (universal) multiple of the $3/2$ power of the major axis of the ellipse;

and it was an astonishing thing when Newton (1690) showed that all this could be deduced from his $f = ma$ law. Briefly put by A. Pope, the English poet: "Nature and Nature's laws lay hid in night: God said: *Let Newton be!* and all was light." Newton's computation can be found in H. Pollard's elegant little book, *Celestial Mechanics* [1976]. I do not reproduce it, wanting only to describe the set-up, first in Newton's format, and then in Hamilton's.

Kepler's problem, of the motion of a planet about the sun, is the simplest example. Look at the picture. The sun (of mass M) is fixed at the origin, say, and the planet (of mass m) moves in a plane about it. The force on the planet (of attraction by the sun) is $GmM/|x|^2$, acting in the direction $e = -x/|x|$, G being the gravitational constant, so Newton's equation of motion is[1]

$$-\frac{GmMx}{|x|^3} = m\ddot{x},$$

which you may integrate once after taking the inner product with $\dot{x}$ to obtain the constant of motion:

$$H = \frac{1}{2}m(\dot{x})^2 - \frac{GmM}{|x|}.$$

Now write $x = Q$ and $m\dot{x} = P$. Then

$$H = \frac{P^2}{2m} - \frac{GmM}{|Q|},$$

and Newton's equation of motion takes the equivalent form:

$$\dot{Q} = \frac{P}{m} = \frac{\partial H}{\partial P} \quad \text{and} \quad \dot{P} = -\frac{GmMx}{|x|^3} = -\frac{\partial H}{\partial Q}.$$

These are Hamilton's equations of motion: Q is position, P is momentum, and $H = H(P,Q)$ is the total energy of the system = the *kinetic* energy (of motion) $P^2/2m$ plus the *potential* energy (of position) $-GmM/|Q|$.

More generally, one has some, possibly very large number ($\#$) of bodies, celestial or not, situated in $\mathbb{R}^3$ say, with positions Q and momenta P –

---

[1] $\cdot$ signifies differentiation by time.

six degrees of freedom apiece – with total energy H. These move accord-
ing to Hamilton's equations of motion $\dot{Q} = \partial H/\partial P$ and $\dot{P} = -\partial H/\partial Q$,
providing a flow in the big phase space $\mathbb{R}^{6\#}$. The energy H is a con-
stant of motion as before since $\dot{H} = (\partial H/\partial Q)\dot{Q} + (\partial H/\partial P)\dot{P} = 0$, inner
products understood. The phase space is then "foliated" by surfaces of
constant energy $(H = h)$ of co-dimension 1 – in Kepler's problem, there
are the individual planetary orbits $(h < 0$ for proper planets, $h > 0$ for
comets) – and the general flow will be confined to submanifolds of still
lower dimension in the presence of further constants of motion (of which
there are seldom very many). That should help a little, but really it's all
too complicated to think about. Look. A liter of air contains some 10^{26}
like molecules. To a very rough approximation, these may be thought of
as point masses, with 6 degrees of freedom apiece – 3 for position (Q),
3 for momentum (P), and more if there are *internal* degrees of freedom.
Ignoring the latter, there are still some 6×10^{26} degrees of freedom, and
Hamilton's recipe requires us to solve 6×10^{26} coupled, highly nonlinear
ordinary differential equations. Hopeless. This discouraging observation
prompted Boltzmann (1870), Kelvin (1880), Maxwell (1860), and most
notably Gibbs (1900) to introduce statistical methods, with a wonderful
success, to be seen in Chapters 10 and 11.

8.2 Gibbs, Birkhoff, and the statistical method

For now, only the briefest indications of how statistical methods come in
in a natural way, but see Chapter 10 for more, and also Thompson [1972]
for the best introduction to this business for beginners.

8.2.1 Gibbs's canonical ensemble

Gibbs [1902] begins by asking: What is the best, most natural statisti-
cal distribution $\mathrm{d}\Gamma = W(QP)\,\mathrm{d}\,\mathrm{vol}$ on the phase space[2] when the mean
energy $\int H\,\mathrm{d}\Gamma = h$ is known *and nothing more*. Put $G = Z^{-1}\mathrm{e}^{-H/kT}$
with fixed constant k, adjustable $T > 0$, and normalizer Z to make[3]
$\int G\,\mathrm{d}\,\mathrm{vol} = 1$. Here T is Kelvin's absolute temperature (up to the factor

---

[2] Q and P now signify the totality of the individual Qs and Ps, $\mathrm{d}\,\mathrm{vol} = \mathrm{d}Q\,\mathrm{d}P$
    being the volume element in phase space.
[3] Naturally, $\int \mathrm{e}^{-H/kT}$ must be finite. This may require the individual Qs to be
    confined to some limited region in $\mathbb{R}^3$, but that's not important now.

$k = $ Boltzmann's constant) and is (hopefully) fixed by $\int H\,\mathrm{G}\,\mathrm{d\,vol} = h$. Then, by Gibbs' lemma,[4] you have

$$0 \le \int \left(\frac{W}{\mathrm{G}} \log \frac{W}{\mathrm{G}} - \frac{W}{\mathrm{G}} + 1 \right) \mathrm{G} = \int W \log W - \int W \log \mathrm{G},$$

and since $\int W \log \mathrm{G} = -\log Z - h/T$ is the same with G in place of W, so $-\int W \log W \le -\int \mathrm{G} \log \mathrm{G}$, with the strict inequality unless $W = \mathrm{G}$. Now $-\int W \log W$ is entropy, *aka* information or uncertainty, as indicated in §5.5 and fully explained in §9.1, and maximal entropy signifies maximal randomness, so it appears that Gibbs's "canonical ensemble" with $\mathrm{d}\Gamma = Z^{-1} e^{-H/kT}\,\mathrm{d\,vol}$ is special and should play a special role. Note that G is preserved by the flow and that the flow is incompressible, preserving the volume element $\mathrm{d\,vol}$ since the vector field $QP \to \dot{Q}\dot{P}$ is divergence-free, as per $\operatorname{div} \dot{Q}\dot{P} =$ the trace of $\partial^2 H/\partial Q \partial P - \partial^2 H/\partial P \partial Q = 0$. Then the whole distribution $\mathrm{d}\Gamma$ is also preserved, which is nice.

8.2.2 Time averages

Next, let $F(QP)$ be some mechanical quantity depending upon the moving phase QP. Birkhoff [1931] proved the *existence* of the time average:

$$\overline{F} = \lim_{T \uparrow \infty} \frac{1}{T} \int_0^T F(QP)\,\mathrm{d}t,$$

what no practical man would dispute, but mathematicians must be humored.[5] OK. But what is $\overline{F}$? Well, it would be nice if it were the same as Gibbs's canonical phase average $\mathbb{E}(F) = \int F\,\mathrm{d}\Gamma$. This would be, so to say, a law of large numbers, justifying Gibbs' choice of $Z^{-1}e^{-H/kT}$ as statistically right for the job. That cannot be correct as it stands: $\overline{H} = H$ is not $\mathbb{E}(H)$ in the large, unless QP does not move at all, so, at the very best, you must cut down to a surface of constant energy. There, the inherited "microcanonical ensemble" with probabilities $\mathrm{d}\Gamma$ proportional to $|\operatorname{grad} H|^{-1} \times \mathrm{d\,area}$ is still flow-invariant, and after successive descents of this type in the presence of additional constants of motion, you can hope. But in practice, $\overline{F} = \mathbb{E}(F)$ is desperately hard to prove in realistic cases; see, for example, Sinai [1960] where it's done (with tears) for hard balls bouncing about in a box. Too bad.

---

[4] $x \log x - x + 1 \ge 0$ for $x \ge 0$, with equality only at $x = 1$.
[5] Birkhoff's proof uses a direct mechanical format, so it's all quite understandable though not easy. The present proof is postponed to §8.4. It is quick, slick, and you understand *nothing* but it's correctness. Still, better to get it over with.

8.2.3 H. Weyl's example

H. Weyl's example [1916] is the very simplest illustration of $\overline{F} = \mathbb{E}(F)$. The phase space is the product of the torus $\mathbb{R}^2/\mathbb{Z}^2$ of points $Q \in \mathbb{R}^2$ considered *modulo* $\mathbb{Z}^2$ and the directions $P = \omega$ from the circle $|\omega| = 1$, the Hamiltonian being $H = P^2/2$. Then $\dot{Q} = \partial H/\partial P = \omega$ and $\dot{P} = -\partial H/\partial Q = 0$, describing straight-line motion $Q(t) = Q(0) + \omega t$ at constant speed, wrapped around the torus. The momentum $P = \omega$ is then a constant of motion, and the fact is that

$$\lim_{T\uparrow\infty} \frac{1}{T} \int_0^T F(Q)\,dt = \int_{\mathbb{R}^2/\mathbb{Z}^2} F(x)\,d^2x$$

not for all, but for typical ω.

Case 1. $n \cdot \omega = 0$ for some $n \in \mathbb{Z}^2 \setminus 0$. Then

$$Q(t) = Q(0) \pm (n_2, -n_1)/\sqrt{n_1^2 + n_2^2} \times t \quad \text{taken} \quad \mod \mathbb{Z}^2$$

is of period $\sqrt{n_1^2 + n_2^2}$, and that's no good.

Case 2. $n \cdot \omega \neq 0$ for any $n \in \mathbb{Z}^2 \setminus 0$. This is better: if $F \in C^3(\mathbb{R}^2/\mathbb{Z}^2)$, then

$$F(x) = \sum_{\mathbb{Z}} \hat{F}(n) e^{2\pi\sqrt{-1}\, n \cdot x} \quad \text{with} \quad \hat{F}(n) = \int e^{-2\pi\sqrt{-1}\, n \cdot x} F(x)\,d^2x.$$

These $\hat{F}(n)$s decay like $|n|^{-3}$ so that $\sum_{n\neq 0}|\hat{F}(n)|$ is comparable to $\int_1^\infty r^{-3}r\,dr < \infty$, and if also $\int_{\mathbb{R}^2/\mathbb{Z}^2} F = \hat{F}(0) = 0$, then with $Q(0) = 0$ (it plays no role), you have

$$\frac{1}{T} \int_0^T F(Q)\,dt = \sum_{\mathbb{Z}^2\setminus 0} \hat{F}(n) \times \frac{1}{T} \int_0^T e^{2\pi\sqrt{-1}\, n \cdot \omega t}\,dt$$

$$= \sum_{\mathbb{Z}^2\setminus 0} \hat{F}(n) \times \frac{e^{2\pi\sqrt{-1}\, n \cdot \omega T} - 1}{2\pi\sqrt{-1}\, n \cdot \omega T}$$

$$\simeq 0 = \int_{\mathbb{R}^2/\mathbb{Z}^2} F(x)\,d^2x \quad \text{for large } T$$

since the individual summands are small (line 2) and dominated by $\hat{F}(n)$ (line 1).

8.3 A more general set-up

Take a space $\mathbb{X}$, a field $\mathfrak{F}$, probabilities $\mathbb{P}$, and a "shift" $T\colon \mathbb{X} \to \mathbb{X}$ which is measure-preserving in the sense that $\mathbb{P}(T^{-1}A) = \mathbb{P}(A)$ for any $A \in \mathfrak{F}$. The shift need not be invertible, so T^{-1} is a map of events only: for example, the shift $T\colon \mathbf{e}_1\mathbf{e}_2\mathbf{e}_3$ etc. $\to \mathbf{e}_2\mathbf{e}_3\mathbf{e}_4$ etc. of the Rademacher functions is measure-preserving in the present sense but only $2 : 1$. In this context, the appropriate variant of Birkhoff's theorem states that if $F\colon \mathbb{X} \to \mathbb{R}$ is summable, then the time average

$$\lim_{N\uparrow\infty} \frac{1}{N} \sum_{0}^{N-1} F(T^n\mathbf{x}) = \overline{F}(\mathbf{x})$$

exists with probability 1. Obviously, $\overline{F}(T\mathbf{x}) = \overline{F}(\mathbf{x})$, and since $\mathbf{1}_B(T\mathbf{x}) = \mathbf{1}_B(\mathbf{x})$ is the same as to say $T^{-1}B = B$, so $\overline{F}$ is measurable over the field $\overline{\mathfrak{F}}$ of such "shift-invariant" events B – in fact, $\overline{F}$ is just the conditional expectation $\mathbb{E}(F|\overline{\mathfrak{F}})$, amplifying Birkhoff's statement of mere existence. The agreement of the time average $\overline{F}$ with the phase average $\mathbb{E}(F)$ now comes down to the triviality of the field $\overline{\mathfrak{F}}$, meaning that $\mathbb{P}(B) = 0$ or 1 for every $B \in \overline{\mathfrak{F}}$. That's where the real work goes, of which you will see samples in two interesting cases later on – for continued fractions (§8.5) and for geodesic flow on surfaces of constant negative curvature (§8.6). The proof of the existence of $\overline{F} = \mathbb{E}(F|\overline{\mathfrak{F}})$ is postponed to §8.4 in favor of some illustrative remarks and exercises.

Usage   The shift is said to be "ergodic" when $\overline{F} = \mathbb{E}(F|\overline{\mathfrak{F}})$ – not a happy choice, coming as it does from the old Greek for "work", and what has work to do with it, I ask? What's wanted is for the empirical frequency

$$\lim_{N\uparrow\infty} \frac{1}{N} \#\big(0 \le n < N : T^n\mathbf{x} \in A\big)$$

to agree with the a priori frequency $\mathbb{P}(A)$, which is something else, so I prefer to say "metrically transitive" and will use that qualifier mostly.

8.3.1 Metric transitivity and mixing

Metric transitivity implies

$$\lim_{N\uparrow\infty} \frac{1}{N} \sum_{0}^{N-1} \mathbb{P}(A \cap T^{-n}B) = \lim_{N\uparrow\infty} \mathbb{E}\left[\mathbf{1}_A \times \frac{1}{N} \sum_{0}^{N-1} \mathbf{1}_B(T^n\mathbf{x})\right]$$

$$= \mathbb{P}(A)\,\mathbb{P}(B) \tag{1}$$

indicating some sort of independence of the present and the remote future. The narrower condition

$$\lim_{n\uparrow\infty} \mathbb{P}(A \cap T^{-n}B) = \mathbb{P}(A)\,\mathbb{P}(B) \tag{2}$$

is called "mixing".

Exercise 8.3.1   Check that, conversely, (1) implies metric transitivity.

The simplest variant of Weyl's example (§8.2.3) is the fact that for $\mathbb{X} = \mathbb{R}$ mod $1 = $ the circle $0 \le x < 1$ and $d\mathbb{P} = dx$, the shift $T\colon x \to x + \omega$ is metrically transitive if ω is irrational. The proof will be obvious.

Exercise 8.3.2   Prove the triviality of $\overline{\mathfrak{F}}$ directly in that case, without the help of Weyl.
Hint: $\mathbb{P}[B \cap (B + n\omega)] = \mathbb{P}(B)$ for $n < 0$ if B is invariant; also the numbers $n\omega$, taken mod 1, are dense in the circle (see why?). Then $\mathbb{P}[B \cap (B + x)] = \mathbb{P}(B)$ for any $0 \le x < 1$ (and once more, why?). Take it from there.

Exercise 8.3.3 = Exercise 8.3.2 continued. Is the shift $x \to x + \omega$ mixing in the sense of (2)? The answer should be *no*, don't you think?

Exercise 8.3.4   The shift $T\colon \mathbf{x}(0)\mathbf{x}(1)\mathbf{x}(2)$ etc. $\to \mathbf{x}(1)\mathbf{x}(2)\mathbf{x}(3)$ etc. for a simple Markov chain is measure-preserving under the equilibrium probabilities $\mathbb{P}_\pi$, and if it were metrically transitive as well, you'd get another proof of LLN. Do it that way – and more, by checking that T mixes really fast.

Speaking of LLN, the shift $T\colon \mathbf{x}_0\mathbf{x}_1\mathbf{x}_2\mathbf{x}_3$ etc. $\to \mathbf{x}_1\mathbf{x}_2\mathbf{x}_3\mathbf{x}_4$ etc. of independent copies of a summable variable $\mathbf{x}_0$ is measure-preserving, so Birkhoff guarantees the existence of $\lim_{N\uparrow\infty}\frac{1}{N}(\mathbf{x}_0 + \cdots + \mathbf{x}_{N-1}) = \overline{\mathbf{x}}_0$. This must be constant, by Kolmogorov's 01 law, i.e. $\overline{\mathbf{x}}_0 = \mathbb{E}(\mathbf{x}_0)$ – and there's mixing, too, no? Now, you've got three proofs of LLN: Kolmogorov's (§5.1.2) and Doob's (§5.1.3), and this, by Birkhoff.

The present version of Birkhoff's theorem covers the mechanical time averages of §8.2: $\mathbb{X}$ is the big phase space of all the Qs and Ps, $d\mathbb{P} = Z^{-1}e^{-H/kT}\,d\,\mathrm{vol}$ if you want (it's invariant so the flow is measure-preserving), and if F is some summable function of the moving phase QP, then

$$\lim_{T\uparrow\infty}\frac{1}{T}\int_0^T F(QP)\,\mathrm{d}t = \mathbb{E}(F|\overline{\mathfrak{F}})$$

where $\overline{\mathfrak{F}}$ is now the field of flow-invariant events.[6]

Exercise 8.3.5   Details please.

[6] Too many Ps and too many Ts. Forgive me, but it should be clear.

8.3.2 Poincaré recurrence

Now back to the general picture with a measure-preserving shift T. If $\mathbb{P}(A) > 0$, then A, $T^{-1}A$, $T^{-2}A$ etc. cannot be all disjoint since they have the same probability and $\mathbb{P}(\mathbb{X}) = 1$. But if $T^{-i}A \cap T^{-j}A$ is not void for some $i < j$, then neither is $A \cap T^n A$ for $n = j - i > 0$, which is to say that there are points $\mathbf{x} \in A$ that come back to A in n steps. This is hopeful but very far from metric transitivity which requires $T^n \mathbf{x}$ not only to come back infinitely often, but with the correct empirical frequency. That's asking a lot. But suppose it's true and that the shift is invertible. Then the (forward) return time $R^+ = \min(n > 0 : T^n \mathbf{x} \in A)$ is finite with probability 1, and likewise the backward return time $R^- = \min(n \geq 0 : T^{-n}\mathbf{x} \in A)$, so with $n \geq 1$, $B_n = A \cap (R^+ = n)$, and $0 \leq k < n$, the events

$$T^{-k}B_n = (R^+ = n - k) \cap (R^- = k)$$

are disjoint and fill up almost the whole of $\mathbb{X}$, as a look at Figure 8.3.1 shows, leading at once to the pretty formula of Kac [1947b]:

$$\int_A R^+ \, d\mathbb{P} = \sum_{n=1}^{\infty} n \, \mathbb{P}(B_n) = \sum_{n=1}^{\infty} \sum_{k=0}^{n-1} \mathbb{P}(T^k B_n)$$
$$= \mathbb{P}(\mathbb{X}) = 1,$$

or, what is the same,

$$\mathbb{E}(R^+|A) = \frac{1}{\mathbb{P}(A)},$$

if you like to say it that way.

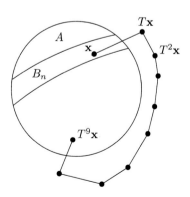

Figure 8.3.1

Exercise 8.3.6   Do it also this way: $R^+ \geq 1$ is automatic, and for $n \geq 1$, the event $(R^+ > n)$ is the complement of $\bigcup_0^{n-1}(T^{k-n}A)$, as you will please check. Conclude that

$$
\int_A R^+ \, dP = \sum_0^\infty \mathbb{P}[(R^+ > n) \cap A]
$$
$$
= \sum_0^\infty \mathbb{P}\left[T^n A \cap \text{complement} \bigcup_0^{n-1}(T^k A) \right]
$$
$$
= \mathbb{P}\left[\bigcup_0^\infty (T^n A) \right]
$$
$$
= 1.
$$

I owe this elegant argument to Ms. Rebecca Solas-Boni.

Exercise 8.3.7   Let $\mathbb{N}$ be the natural numbers $1, 2, 3$ etc. and let the "density" of $A \subset \mathbb{N}$ be

$$
\mathbb{D}(A) = \lim_{N \uparrow \infty} \frac{1}{N} \#(n \leq N : n \in A), \text{ as in §5.11.}
$$

Introduce also the corresponding expectation

$$
\mathbb{E}(F) = \lim_{N \uparrow \infty} \frac{1}{N} \sum_0^{N-1} F(n)
$$

and apply this language to the function $F(T^{\mathbf{n}}\mathbf{x})$. If this function were measurable with respect to $\mathbf{n}$ and $\mathbf{x}$ and also summable with respect to $\mathbb{D} \times \mathbb{P}$, then, by Fubini, it would be summable in respect to $\mathbf{n}$ for almost all $\mathbf{x}$, and you would have for free the existence of Birkhoff's average

$$
\lim_{N \uparrow \infty} \frac{1}{N} \sum_0^{N-1} F(T^n \mathbf{x}) = \overline{F}(\mathbf{x})
$$

with probability 1. That's fake, which is too bad. Tell me why.

8.4 Riesz's lemma and Garsia's trick

The proof of Birkhoff's theorem as stated in §8.3, may be reduced to the "maximal ergodic lemma" of F. Riesz [1945]. Write $S_n(F) =$

$\sum_0^{n-1} F(T^k \mathbf{x})$, changing capital "en" to the less cumbersome small "en". Then for summable F,

$$\int_{\substack{\sup S_n > 0 \\ n \geq 1}} F \, d\mathbb{P} \text{ is non-negative.}$$

That's the lemma. The proof follows Garsia [1965], *à propos* of which Garsia submitted a paper to a journal E. Hopf was editing. Hopf didn't want it all but said he *would* like to publish the footnote. Here it is.

Proof. I write a superior $+$ to indicate the shift, as in $F^+(\mathbf{x}) = F(T\mathbf{x})$. Define $M_n = \max[0, S_1, \ldots, S_n]$ noting that it's always ≥ 0. If $M_n > 0$, then it agrees with $\max[S_1, \ldots, S_n]$, and

$$
\begin{aligned}
S_1 + M_n^+ &= S_1 + \max[0, S_1^+, \ldots, S_n^+] \\
&= S_1 + \max[0, S_2 - S_1, \ldots, S_{n+1} - S_n] \\
&= \max[S_1, S_2, \ldots, S_{n+1}] \\
&= \max[0, S_1, \ldots, S_{n+1}] \qquad \text{keeping } M_n > 0 \text{ in mind} \\
&= M_{n+1},
\end{aligned}
$$

i.e. $S_1 = M_{n+1} - M_n^+$, from which you see that

$$
\begin{aligned}
\mathbb{E}(S_1, M_n > 0) &= \mathbb{E}(M_{n+1}, M_n > 0) - \mathbb{E}(M_n^+, M_n > 0) \\
&\geq \mathbb{E}(M_{n+1}, M_n > 0) - \mathbb{E}(M_n^+) \\
&= \mathbb{E}(M_{n+1}, M_n > 0) - \mathbb{E}(M_n) \qquad \text{yes?} \\
&= \mathbb{E}(M_{n+1} - M_n, M_n > 0) \qquad \text{right?} \\
&\geq 0 \qquad \text{since } M_{n+1} \geq M_n,
\end{aligned}
$$

or, what is really wanted,

$$0 \leq \mathbb{E}\left[S_1, \lim_{n \uparrow \infty} M_n = \sup_{n \geq 1} S_n > 0\right] = \int_{\substack{\sup S_n > 0 \\ n \geq 1}} F \, d\mathbb{P}.$$

That's it for Riesz.

To which you may say: So what? Well, here's what:

$$\mathbb{P}\left[\lim_{n \uparrow \infty} S_n/n \text{ exists}\right] = 1.$$

Proof. Otherwise, the event

$$Z: \liminf_{n \uparrow \infty} \frac{1}{n} S_n < a < b < \limsup_{n \uparrow \infty} \frac{1}{n} S_n$$

would have positive probability for some $-\infty < a < b < \infty$ and since $Z = (x : Tx \in Z) = T^{-1}Z$ is shift-invariant, you could restrict $d\mathbb{P}$ to Z, renormalize it, and so reduce to the case $\mathbb{P}(Z) = 1$. Now look at the modified sums $S'_n = S_n - nb$ for F replaced by $F - b$. Obviously, sup S'_n is positive with probability 1, so $\int(F - b)\,d\mathbb{P} \geq 0$, by Riesz's lemma – and similarly for the sums $S''_n = na - S_n$ for $a - F$: sup S''_n is also positive, so $\int(a - F)\,d\mathbb{P} \geq 0$. But then $a \geq \mathbb{E}(F) \geq b$, contradicting $a < b$. It follows that $\mathbb{P}(Z) = 0$ and with it comes the existence of $\lim_{n\uparrow\infty} S_n/n = \overline{F}$.

It remains only to identify $\overline{F}$ as $\mathbb{E}(F|\overline{\mathfrak{F}})$.

Proof. If $A \in \overline{\mathfrak{F}}$ and B is any event at all, then

$$\int_A \mathbf{1}_B(Tx)\,d\mathbb{P} = \int_X \mathbf{1}_A \mathbf{1}_{T^{-1}B}\,d\mathbb{P} = \mathbb{P}\big(T^{-1}(A \cap B)\big) = \mathbb{P}(A \cap B)$$

$$= \int_A \mathbf{1}_B\,d\mathbb{P}.$$

This extends at once from $F = \mathbf{1}_B$ to the general, summable, non-negative function F, so by Fatou,

$$\int_A \overline{F} \leq \lim_{n\uparrow\infty} \frac{1}{n}\sum_0^{n-1}\int_A F(T^k\mathbf{x}) = \int_A F = \int_A \mathbb{E}(F|\overline{\mathfrak{F}}),$$

i.e. $\overline{F} \leq \mathbb{E}(F|\overline{\mathfrak{F}})$, both functions being measurable over $\overline{\mathfrak{F}}$.

Exercise 8.4.1 Finish the proof.
Hint: Take $F \geq 0$ and look at $F \wedge N$ for large N.

8.5 Continued fractions

These provide one of the nicest applications of Birkhoff's theorem, metric transitivity included, with surprising geometrical consequences for geodesic flow in constant negative curvature to be seen in §8.6.

8.5.1 The set-up

The number $0 < x \leq 1$ may be written as a continued fraction

$$x = \cfrac{1}{n_1 + \cfrac{1}{n_2 + \cfrac{1}{n_3 \text{ etc.}}}}$$

in whole numbers n_1, n_2 etc. ≥ 1, determined by the rule:

1 $n_1 =$ the integral part of $\dfrac{1}{x}$,

2 $n_2 =$ the integral part of $\left(\dfrac{1}{x} - n_1\right)^{-1}$,

3 $n_3 =$ the integral part of $\left(\left(\dfrac{1}{x} - n_1\right)^{-1} - n_2\right)^{-1}$,

and so on. Figure 8.5.1 shows more clearly what is happening: $(0, 1]$ is

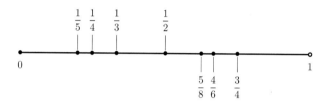

Figure 8.5.1

divided into intervals

$$\left(\frac{1}{n+1}, \frac{1}{n}\right]$$

and $n_1 = n_1(x)$ says to which of these x belongs, as per

$$\frac{1}{n_1 + 1} < x \leq \frac{1}{n_1};$$

if $x = \frac{1}{n_1}$, *stop*; if not divide

$$\left(\frac{1}{n_1 + 1}, \frac{1}{n_1}\right]$$

into intervals

$$\left(\frac{1}{n_1 + \frac{1}{n}}, \frac{1}{n_1 + \frac{1}{n+1}}\right],$$

and record $n_2 = n_2(x)$ to say to which of *these* x belongs, as per

$$\frac{1}{n_1 + \frac{1}{n_2}} < x \leq \frac{1}{n_1 + \frac{1}{n_2 + 1}};$$

if $x = \dfrac{1}{n_1 + \frac{1}{n_2}}$, *stop* – and so on. The process *does* stop if x is rational.
These being excluded, the successive fractions

$$\cfrac{1}{n_1 + \cfrac{1}{n_2 + \cfrac{1}{n_3 + \text{etc.}}}} \equiv \frac{1|}{|n_1} + \frac{1|}{|n_2} + \frac{1|}{|n_3} + \cdots + \frac{1|}{|n_N}$$

approximate x to within 2^{-N} or better, and taking N to $+\infty$, you may write

$$x = \frac{1|}{|n_1} + \frac{1|}{|n_2} + \frac{1|}{|n_3} + \text{etc.}$$

Exercise 8.5.1 $\frac{1|}{|1} + \frac{1|}{|1} + \frac{1|}{|1} + \text{etc.} = \frac{1}{2}(\sqrt{5} - 1)$; similarly $\frac{1|}{|2} + \frac{1|}{|2} + \frac{1|}{|2} + \text{etc.} = \sqrt{2} - 1$.

8.5.2 Birkhoff applied

Now introduce the shift $T: x \to 1/x$ less its integral part, acting by the rule $n_2 = n_1(Tx)$, $n_3 = n_2(Tx)$, etc. Gauss (1800) noticed that this map preserves the probabilities $d\,\mathbb{P} = \frac{1}{\log 2}(1 + x)^{-1}\,dx$: in fact, for any $0 < a < b < 1$,

$$T^{-1}[a, b) = \bigcup_1^\infty \Big(\frac{1}{b+n}, \frac{1}{a+n}\Big],$$

so

$$\begin{aligned}
\int_{T^{-1}[a,b)} \frac{dx}{1+x} &= \sum_1^\infty \log \frac{1 + \frac{1}{a+n}}{1 + \frac{1}{b+n}} \\
&= \sum_1^\infty \Big[\log \frac{a+n+1}{b+n+1} - \log \frac{a+n}{b+n}\Big] \\
&= \log \frac{b+1}{a+1} \\
&= \int_a^b \frac{dx}{1+x}.
\end{aligned}$$

Much later Knopp (1928) proved that T is metrically transitive, and (unaccountably much later still) Döblin (1940) noticed that this fact supplies cheap proofs of three lovely limit laws that Khinchine [1923/24] and Lévy [1937] had obtained by brute force, without the help of Birkhoff's theorem: *viz.* with probability one,

(1) $\lim_{N\uparrow\infty} \dfrac{1}{N} \#(k \le N : \mathbf{n}_k = n) = \dfrac{1}{\log 2} \log \dfrac{(n+1)^2}{n(n+2)}$,

(2) $\lim_{N\uparrow\infty} \dfrac{\mathbf{n}_1 + \cdots + \mathbf{n}_N}{N} = +\infty$,

and

(3) $\lim_{N\uparrow\infty} \sqrt[N]{\mathbf{n}_1 \cdots \mathbf{n}_N} = e^{\gamma}$ with $\gamma = \displaystyle\sum_{1}^{\infty} \dfrac{\log n}{\log 2} \log \dfrac{(n+1)^2}{n(n+2)}$.

The number γ is "Khinchine's constant". Its exact value is not known, but I have the numerical value $\gamma = .9878+$ for 10^5 summands, kindness of A. Yilmaz.

Exercise 8.5.2   Check (1), (2), (3).

8.5.3 Proof of metric transitivity

I follow Billingsley [1965: 144–45].

Step 1.   Fix $N \ge 1$ and also $n_1, \ldots, n_N \ge 1$, and define

$$n(x) = \frac{1|}{|n_1} + \frac{1|}{|n_2} + \frac{1|}{|n_3} + \cdots + \frac{1|}{|n_N + x}.$$

This is a decreasing/increasing function of x according as N is odd/even, of the form $(Ax+B)/(Cx+D)$ with whole numbers A, B, C, D subject to

$$0 \le A \le B; \ 1 \le C \le D; \ AD - BC = \pm 1, \ \text{according to the parity of } N.$$

Indeed, if $N = 1$, then $A = 0$, $B = 1$, $C = 1$, and $D = n_1$, while if $N \ge 2$, if $n(x)$ is of the stated form at the previous level, and if $n_N = n$, then

$$n(x) = \frac{A\dfrac{1}{n+x} + B}{C\dfrac{1}{n+x} + D} = \frac{Bx + A + nB}{Dx + C + nD} = \frac{A'x + B'}{C'x + D'}$$

with

$$0 \le A' = B \le A + nB = B',$$
$$1 \le C' = D \le C + nD = D',$$

and

$$A'D' - B'C' = B(C + nD) - (A + nB)D = -AD + BC.$$

Step 2.   Now let I be the interval between $n(0)$ and $n(1)$. For varying $n_1, n_2, \ldots, n_N$, these form a (disjoint) subdivision of $(0, 1]$ which is refined by increasing N. Take also another interval $J = (a, b]$. Then with the notation $|K|$ for the length of the interval K, you have the underestimate $|I \cap T^{-N} J| \geq \frac{1}{2}|I| \times |J|$ independently of N.

Proof.   To begin with, $x \in I$ means that $x = \mathbf{n}(y)$ for some $0 < y \leq 1$. Then $x \in T^{-N} J$ is the same as saying $y = T^N x \in J$, so $I \cap T^{-N} J = n(J)$ is of total length

$$
\begin{aligned}
|n(b) - n(a)| &= \left| \frac{Ab + B}{Cb + D} - \frac{Aa + B}{Ca + D} \right| \\
&= \frac{|AD - BC|(b - a)}{(Ca + D)(Cb + D)} \\
&\geq \frac{|J|}{(C + D)^2} \, .
\end{aligned}
$$

But also

$$
|I| = |n(1) - n(0)| = \left| \frac{A + B}{C + D} - \frac{B}{D} \right| = \frac{1}{D(C + D)} \leq \frac{2}{(C + D)^2}
$$

since $D \geq C$, with the desired result:

$$
|I \cap T^{-N} J| \geq \frac{1}{2}|I| \times |J|.
$$

Step 3. T^{-1} preserves the disjointness of intervals, so $|I \cap T^{-N} B| \geq \frac{1}{2}|I| \times |B|$ for any event B, where $|C|$ is now the Lebesgue measure of C, and if B were invariant, you would have $|I \cap B| \geq \frac{1}{2}|I| \times |B|$. Here N is still at your disposal, permitting the progressive refinement of the Is and the extension of the estimate to any event A: $|A \cap B| \geq \frac{1}{2}|A| \times |B|$. Then with $A = B'$, you see that $0 = |B' \cap B| \geq \frac{1}{2}|B'| \times |B|$, which is to say, $\mathbb{P}(B) = 0$ or 1. That's metric transitivity.

8.5.4 Mixing

A variant of the above shows that the shift is mixing as well, *viz.*

$$
\lim_{N \uparrow \infty} \mathbb{P}(A \cap T^{-N} B) = \mathbb{P}(A)\, \mathbb{P}(B)
$$

for any two events A and B.

Proof after Billingsley [1965:121].   Introduce the fields $\mathfrak{F}_N$ for the variables $n_{N+1}(\mathbf{x})$, $n_{N+2}(\mathbf{x})$, etc. These are decreasing and $\mathfrak{F}_\infty = \bigcap_1^\infty \mathfrak{F}_N$ is trivial. To see why, note first that $x \in T^{-N} B$ is the same as to say $T^N x \in B$, from which you will understand that $\mathfrak{F}_N = T^{-N} \mathfrak{F}_0$. In other words, $A \in \mathfrak{F}_\infty$ is of the form $T^{-N}B$ for any fixed N. Now with I and $n(x)$ as above,

$$\mathbb{P}(I \cap T^{-N}B) = \mathbb{P}[\mathbf{n}(B)]$$

$$= \frac{1}{\log 2} \int_B \frac{1}{1 + n(x)} \, dn(x) \qquad \text{see why?}$$

$$= \frac{1}{\log 2} \int_B \frac{1}{1 + n(x)} \frac{1}{(Cx + D)^2} \, dx$$

$$\geq \frac{1}{\log 2} \int_B \frac{dx}{1 + x} \times \frac{1}{2} \frac{1}{D(C + D)} \qquad \begin{array}{l} \text{since } n(x) \leq 1 \\ \text{and } 1 \leq C \leq D \end{array}$$

$$= \frac{1}{2} \mathbb{P}(B) \times \frac{1}{D(C + D)}$$

$$\geq \frac{1}{2} \log 2 \, \mathbb{P}(I) \, \mathbb{P}(B) \qquad \begin{array}{l} \text{since} \\ \mathbb{P}(I) \;\leq\; \frac{1}{\log 2} |n(1) - n(0)| \\ \phantom{\mathbb{P}(I)} = \frac{1}{\log 2} \frac{1}{D(C+D)} , \end{array}$$

so reverting to $A = T^{-N}B$, it appears that $\mathbb{P}(I \cap A) \geq \frac{1}{2} \log 2 \, \mathbb{P}(I) \, \mathbb{P}(A)$ with N and the intervals I still at your disposal. That takes care of the triviality of $\mathfrak{F}_\infty$ just as in Step 3 above. The rest is easy: for *any* $B \in \mathfrak{F}$

$$\left| \mathbb{P}(A \cap T^{-N}B) - \mathbb{P}(A)\,\mathbb{P}(B) \right|$$

$$= \left| \int_{T^{-N}B} [\mathbb{P}(A|\mathfrak{F}_N) - \mathbb{P}(A)] \, d\mathbb{P} \right| \qquad \text{since } T^{-N}B \in \mathfrak{F}_N$$

$$\leq \int_0^1 \left| \mathbb{P}(A|\mathfrak{F}_N) - \mathbb{P}(A) \right| d\mathbb{P},$$

and this tends to zero as $N \uparrow \infty$ since $\mathbb{P}(A|\mathfrak{F}_N)$ tends to $\mathbb{P}(A|\mathfrak{F}_\infty) = \mathbb{P}(A)$ by P. Lévy's first rule $(4')$ of §1.3.

Lévy [1937] and Döblin (1940) proved more by more sophisticated means, *viz.*

$$\left| \mathbb{P}(A \cap T^{-N}B) - \mathbb{P}(A)\,\mathbb{P}(B) \right| \leq e^{-kN}$$

with a universal constant k. Incidentally, such rapid mixing says you do not have to wait too long before the future $T^N\mathbf{x}$, $T^{N+1}\mathbf{x}$, etc. ($N \uparrow \infty$) is very nearly independent of the past $\mathbf{x}, T\mathbf{x}, \ldots, T^n\mathbf{x}$ (n fixed). This should favor a CLT-type correction to $\overline{F} = \mathbb{E}(F)$ (*aka* LLN) as, in fact, Lévy and Döblin found.

8.5.5 ★ Information rate (McMillan's theorem)

The probabilities

$$p(n_1, n_2, \ldots, n_N) = \mathbb{P}[n_1(\mathbf{x}) = n_1, \ldots, n_N(\mathbf{x}) = n_N]$$

are computed in a convenient approximate form for large N. Identifying $\frac{1|}{|n_1} + \frac{1|}{|n_2} + \cdots$ with the typical irrational number $x \in (0, 1]$, you have

$$p(n_1, n_2, \ldots, n_N) \simeq \frac{1}{\log 2} \int_0^1 \frac{dn(x)}{1 + n(x)} \simeq \frac{1}{\log 2} \frac{|n(1) - n(0)|}{1 + x} \text{ for large } N,$$

so $|n(1) - n(0)|$ is what counts on a logarithmic scale. Let's compute that for $N = 1, 2, 3$ in a new way with the more explicit notation

$$n(x) = \frac{1|}{|n_1} + \frac{1|}{|n_2} + \cdots + \frac{1|}{|n_N + x} \equiv [n_1, n_2, \ldots, n_N](x) :$$

for $N = 1$: $|n(1) - n(0)| = \dfrac{1}{n_1} - \dfrac{1}{n_1 + 1} = \dfrac{1}{n_1} \dfrac{1}{n_1 + 1} = [n_1](0)[n_1](1)$;

for $N = 2$: $|n(1) - n(0)| = \dfrac{1}{n_1 + \frac{1}{n_2+1}} - \dfrac{1}{n_1 + \frac{1}{n_2}}$

$$= \frac{\frac{1}{n_2} - \frac{1}{n_2+1}}{(n_1 + \frac{1}{n_2})(n_1 + \frac{1}{n_2+1})}$$

$$= [n_1, n_2](0)[n_1, n_2](1)[n_2](0)[n_2](1);$$

for $N = 3$: $|n(1) - n(0)| = \dfrac{1}{n_1 + \dfrac{1}{n_2 + \frac{1}{n_3}}} - \dfrac{1}{n_1 + \dfrac{1}{n_2 + \frac{1}{n_3+1}}}$

$$= \frac{\dfrac{1}{n_2 + \frac{1}{n_3+1}} - \dfrac{1}{n_2 + \frac{1}{n_3}}}{\left(n_1 + \dfrac{1}{n_2 + \frac{1}{n_3}}\right)\left(n_1 + \dfrac{1}{n_2 + \frac{1}{n_3+1}}\right)}$$

$$= [n_1, n_2, n_3](0)[n_1, n_2, n_3](1)$$
$$\times [n_2, n_3](0)[n_2, n_3](1)[n_3](0)[n_3](1).$$

The pattern is clear and easily checked by induction, with the general result:

$$|n(1) - n(0)| = \prod_{k=1}^N [n_k, \ldots, n_N](0)[n_k, \ldots, n_N](1).$$

Here, both $[n_k, \ldots, n_N](0)$ and $[n_k, \ldots, n_N](1)$ are within 2^{-N+k-1} of $T^{k-1}x$, so the general term of the sum for $\log |n(1) - n(0)|$ is about

$2 \times \log(T^{k-1}x)$, with the suggestion that

$$\lim_{N\uparrow\infty} \frac{1}{N} \log p(\mathbf{n}_1, \mathbf{n}_2, \ldots, \mathbf{n}_N) = \lim_{N\uparrow\infty} \frac{2}{N} \sum_0^{N-1} \log(T^k\mathbf{x}) = 2\,\mathbb{E}(\log\mathbf{x})$$

$$= -\frac{\pi^2}{3\log 2}$$

with probability 1, by Birkhoff.

To pin this down, two simple estimates are needed, from above and from below. The former is easy: $\mathbf{x}_k^+$ = the larger of $\big[\mathbf{n}_k, \ldots, \mathbf{n}_N\big](0)$ and $\big[\mathbf{n}_k, \ldots, \mathbf{n}_N\big](1)$ is ≤ 1 and not more than $T^{k-1}\mathbf{x} + 2^{-x+k-1}$, so

$$\sum_1^N \log\mathbf{x}_k^+ \leq \sum_0^{N-\sqrt{N}} \log(T^k\mathbf{x} + 2^{-\sqrt{N}}) \simeq N \times \mathbb{E}(\log\mathbf{x}).$$

The lower estimate needs a little more care:

$$\mathbf{x}_k^- = \text{ the smaller of } \big[\mathbf{n}_k, \ldots, \mathbf{n}_N\big](0) \text{ and } \big[\mathbf{n}_k, \ldots, \mathbf{n}_N\big](1)$$

is not less than $T^{k-1}\mathbf{x} - 2^{-N+k-1}$, but that could be very small or even negative. To circumvent this awkwardness, fix a small number $c > 0$, take N so large that $c > 2^{-\sqrt{N}}$, and neglect $\log\mathbf{x}_k^-$ unless $T^{k-1}x \geq c$ and $k - 1 \leq N - \sqrt{N}$. The remaining part of $\sum \log\mathbf{x}_k^-$ cannot then be larger than

$$\sum_{0 \leq k \leq N - \sqrt{N}} \log(T^k\mathbf{x} - 2^{-\sqrt{N}}) \times \text{ the indicator of the event } (T^k x \geq c)$$

$$\simeq N \times \mathbb{E}\big[\log\mathbf{x}, x \geq 0\big],$$

and the neglected terms don't matter much: $\mathbf{x}_k^-$ is at least the reciprocal of $(\mathbf{n}_{k+1})$ and $n_k(\mathbf{x}) = n_1(T^{k-1}\mathbf{x})$, so the error is not larger than

$$-\sum_0^{N-1} \log\big[n_1(T^k\mathbf{x}) + 1\big] \times \text{ the indicator of the event } (T^k\mathbf{x} < c)$$

$$-\sum_{N-\sqrt{N} < k \leq N} \log\big[n_1(T^k\mathbf{x}) + 1\big]$$

$$\simeq -N \times \mathbb{E}\big[\log(\mathbf{n}_1 + 1), \mathbf{x} < c\big].$$

Then all is well, c being small.

This proves a McMillan-type theorem, *viz.*

$$\mathbb{P}\left[\lim_{N\uparrow\infty} -\frac{1}{N} \log p(\mathbf{n}_1, \mathbf{n}_2, \ldots, \mathbf{n}_N) = \frac{\pi^2}{3\log 2}\right] = 1,$$

which is to say that the probability of *what you see*, namely $\mathbf{n}_1 = n_1(x)$,

$\mathbf{n}_2 = n_2(x), \ldots, \mathbf{n}_N = n_N(x)$, is typically $\exp(-N\pi^2/3\log 2)$, *the same for almost every x.*

Exercise 8.5.3   The information rate $h = \pi^2/3\log 2$ is $-2\,\mathbb{E}(\log x)$. Check it.
Hints: $\frac{1}{1+x} = 1 - x + x^2$ etc. helps; also you will want to know that $\sum_1^\infty 1/n^2 = \pi^2/6$.

Now the mean value H of $-\log p(\mathbf{n}_1, \ldots, \mathbf{n}_N)$ may be interpreted as the amount of information produced by the knowledge of $\mathbf{n}_1(x), \ldots, \mathbf{n}_N(x)$, relative to Gauss's measure $d\mathbb{P} = \frac{1}{\log 2}\,dx/(1+x)$. Then $\lim_{N\uparrow\infty} H/N$ is the "information production rate per letter", thinking of $\frac{1|}{|n_1} + \frac{1|}{|n_2} + \cdots$ as a message written in the alphabet $\mathbb{N}$ of whole numbers $n = 1, 2, 3$, etc. Its value is the number $h = \pi^2/6\log 2 \simeq 2.37$, as you would hope and may easily check, by looking over the proof just made with the help of Exercise 4.2.1. The analogue for the binary expansion $x = e_1/2 + e_2/4 + $etc., relative to Lebesgue's measure $d\mathbb{P} = dx$ is $h = \log 2$; similarly, for the decimal expansion, it's $h = \log 10$. Here, it is better to switch from the natural logarithm (log) to logarithms to the base 2 ($\log_2 x = \log x/\log 2$). The benchmark is now the binary rate $h = 1$, the decimal rate is $h \simeq 3.32$, and the rate for continued fractions is better: $h = 4.74/\log 2 \simeq 6.84$, roughly the decimal rate.

8.6 Geodesic flow

There are three standard 2-dimensional geometries:
(1) the "round" geometry of the unit sphere S^2, inherited from the ambient space $\mathbb{R}^3$ (curvature $+1$);
(2) the "flat" geometry of $\mathbb{R}^2$ itself (curvature 0); and
(3) the "hyperbolic" geometry of Poincaré's upper half-plane $\mathbb{H}^2 = \mathbb{R} \times (x_2 > 0)$.

(3) may not be familiar. I will describe it after some words about (1) and (2); see also Pogorelov [1967] and McKean–Moll [1997: §§1.9 and 4.3] for the unfamiliar bits.

8.6.1 Sphere

The shortest curve (geodesic) between two points of the unit sphere is the shorter arc of the great circle passing through them. The moving point $Q(t)$ traversing such a circle at speed 1 comes back to $Q(0)$ at time 2π and repeats itself so that

$$\overline{F} = \lim_{T\uparrow\infty} \frac{1}{T} \int_0^T F(Q)\,dt$$

exists, but as the outcome bears *no* resemblance to

$$\mathbb{E}(F) = \frac{1}{4\pi} \int F(x)\,d\,\text{area},$$

you may say that the geodesic flow has no law of large numbers.

8.6.2 Plane

Now the geodesics are ordinary straight lines $Q(t) = Q(0) + \omega t$, with direction ω, of unit length to keep speed 1, and LLN doesn't work at all. But wait. Let's make $\mathbb{R}^2$ look more like S^2 by factoring out the lattice $\mathbb{Z}^2$ to produce the (compact) torus $\mathbb{R}^2/\mathbb{Z}^2$. The straight lines are now taken mod $\mathbb{Z}^2$, winding through and around the hole in a complicated way, and H. Weyl's computation of §8.2 shows that

$$\lim_{T\uparrow\infty} \frac{1}{T} \int_0^T F(Q)\,dt = \int_{\mathbb{R}^2/\mathbb{Z}^2} F(x)\,d\,\text{area} \equiv \mathbb{E}(F)$$

for any smooth F, independently of $Q(0)$ provided $n \cdot \omega \neq 0$ for $n \in \mathbb{Z}^2 \setminus 0$, i.e. LLN is OK for geodesics issuing from any point in a "typical" direction. That's better.

8.6.3 Poincaré's half-plane

I am speaking of the open half-plane $\mathbb{H}^2 = \mathbb{R} \times (x_2 > 0)$ equipped with the line element $(x_2)^{-1}\sqrt{(\,dx_1)^2 + (\,dx_2)^2}$ giving the geometry its "hyperbolic" character. But first some words about the idea of curvature, itself, of which there are two aspects – inner and outer. The *inner* aspect has to do with what an ant crawling on a two-dimensional surface $S \subset \mathbb{R}^3$ perceives as to its local geometrical character. The *outer* aspect has to do with the presentation of the surface in $\mathbb{R}^3$ by, e.g., a US Geological Survey map showing the lines of equal height above sea level. The ant knows nothing of the sea, or of gravity, to indicate in what direction the water might lie, so these two aspects are not the same. Gauss (1820) was the first to understand this fully and to reconcile the two aspects in a satisfactory way. To understand what he did, look at Figure 8.6.1

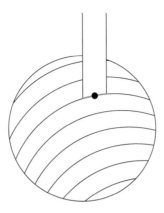

Figure 8.6.1

in which you see a patch of the surface intersected by an opaque pla-
nar patch meeting the surface at 90° and cutting out a green superficial
curve at a point of which, marked with a •, the conventional curvature
$\varkappa$ may be computed *in the ambient* $\mathbb{R}^3$. This may vary between a minimum $\varkappa_-$ and a maximum $\varkappa_+$ as the plane is rotated about the normal
to the surface at the marked point, and these numbers are generally of
the *outer* type, depending on the particular presentation of S in $\mathbb{R}^3$.
What Gauss discovered is the remarkable fact – his own words – that
their product $k = \varkappa_- \varkappa_+$ is *inner*, having nothing to do with how S is
presented in $\mathbb{R}^3$! It is this Gaussian curvature k that the ant perceives.

Example 8.6.1   Obviously, $\mathbb{R}^2$ is flat $(k = 0)$ and likewise $\mathbb{R}^2/\mathbb{Z}^2$. The
latter presents itself in $\mathbb{R}^3$ as a torus, by identification of opposite sides of
the unit square, and appears to be variously curved no matter what you
do by shrinking or stretching to make it look nice. But to the ant that
is all illusion. Coming, e.g., to the upper side, it pops up at the lower
side, walking straight on, not perceiving that anything has changed.

Example 8.6.2 S^2, aka $x_3 = \pm\sqrt{1 - (x_1)^2 - (x_2)^2}$, is different. Every
plane section, at, for example, the north pole, cuts out a great circle of
curvature $+1$, so Gauss's curvature k is $+1$, too.

Example 8.6.3 $\mathbb{H}^2$ is still different $(k = -1)$. I do not repeat its presentation in $\mathbb{R}^3$ or make the computation, for which see the lucid Needham [1997] and its instructive pictures, preferring to give you the flavor
of the thing by the simpler "mountain pass" presented by $x_3 = \frac{1}{2}(x_1^2 - x_2^2)$

in Figure 8.6.2. There you see two ridges rising on either hand, two val-

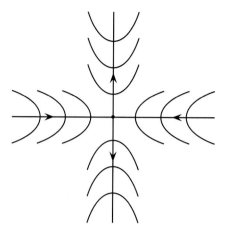

Figure 8.6.2

leys falling away, and arrows indicating the skier's "fall-line". Curvature $\varkappa_+ = 1$ comes from the plane $x_2 = 0$, curvature $\varkappa_- = -1$ from $x_1 = 0$, and Gauss's curvature k is now -1. Metrically, this surface is not $\mathbb{H}^2$, but it's similar at the saddle or height of land and gives an idea of what curvature -1 means. So much then for curvature.

Now the line $\mathbb{R} \times (x_2 = 0)$ bordering $\mathbb{H}^2$ cannot be seen from $\mathbb{R} \times (x_2 > 0)$ since the distance from, e.g. $\sqrt{-1}$ is not less than $\int_0^1 \mathrm{d}x_2/x_2 = +\infty$: x_2 must fall from 1 to 0 and any horizontal deviation from the vertical can only make the journey longer, and by the same remark, you will understand that any vertical line is a geodesic. What are the others? They are the semi-circles meeting $\mathbb{R} \times 0$ at $90°$, as can be seen from the following facts: $\mathbb{H}^2$ has a group of proper ($=$ orientation-preserving), rigid ($=$ distance-preserving) motions, like the (proper) rotations of S^2. These can be identified with the group $\mathrm{PSL}(2, \mathbb{R})$ of 2×2 real matrices $(a\,b/c\,d)$ of determinant $ad - bc = +1$, modulo its center ($\pm$ the identity), acting on $\mathbb{H}^2$ by the rule $z = x_1 + \sqrt{-1}\,x_2 \to (az+b)/(cz+d)$: such a motion preserves both orientation and the line element, as you may easily check, so it must map one geodesic to another, which is all that's needed here. You have only to verify that $\mathrm{PSL}(2, \mathbb{R})$ is capable of placing any two points of $\mathbb{H}^2$ on $V = 0 \times (x_2 > 0)$, of which the inverse image, if not a vertical line, can only be one of these semi-circles. In short, $\mathbb{H}^2$ has no other geodesics.

Now, in the large, LLN is hopeless here, just as for $\mathbb{R}^2$, but let's confine the geodesic flow by reducing $\mathbb{H}^2$ modulo the action of some subgroup of $PSL(2, \mathbb{R})$.

$\mathbb{H}^2$ is then reduced, just as $\mathbb{R}^2$ was reduced modulo $\mathbb{Z}^2$. A convenient choice is the "modular group" $\Gamma = PSL(2, \mathbb{Z})$ where now the *abcd* are integers, generated by the horizontal displacement $(1\,1/0\,1)$ mapping z to $z + 1$ and the inversion $(0 - 1/1\,0)$ mapping z to $-1/z$. The reduced half-plane $\mathbb{H}^2/\Gamma$ may be identified with the "fundamental cell" C seen in Figure 8.6.3, comprised of the part of the strip $-\frac{1}{2} \leq x_1 < \frac{1}{2}$ above the

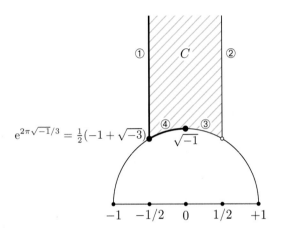

Figure 8.6.3

unit circle, with identification of the vertical sides ① and ② by translation $(z \to z + 1)$ and of the arcs ③ and ④ by inversion $(z \to -1/z)$, with a "cusp" running out to $\sqrt{-1}\,\infty$. The identifications appear to introduce bad creases to the "surface" $\mathbb{H}^2/\Gamma$ at $\sqrt{-1}$ and at $e^{2\pi\sqrt{-1}/3}$, but it is not really so. The correct attitude is to introduce complex local parameters $\mathfrak{z}$ as follows: at any interior point z_0, use $\mathfrak{z} = z - z_0$, and likewise on the open verticals ① and ② and on the open arcs ③ and ④ where two half discs are patched together; at $z_0 = \sqrt{-1}$, two quarter discs are patched together and may be opened up to a full disc by use of $\mathfrak{z} = (z - z_0)^2$; at $z_0 = e^{2\pi\sqrt{-1}/3}$, two 60° sectors are patched together, and $\mathfrak{z} = (z - z_0)^3$ does the job. Then the surface looks locally like a little disc at *every* point, and harmony is restored, with the result sketched in Figure 8.6.4. Note that the cusp really looks that way, all distances out near $\sqrt{-1}\,\infty$ being reduced by the factor $1/x_2$ in the line element.

Now $\mathbb{H}^2/\Gamma$ inherits the volume element $dx_1\,dx_2/x_2^2$. Happily, it is not

Figure 8.6.4

compact, unlike $\mathbb{R}^2/\mathbb{Z}^2$. Happily, it's not really very big since

$$\text{area } \mathbb{H}^2/\Gamma = \int_{-1/2}^{1/2} dx_1 \int_{\sqrt{1-x_1^2}}^{\infty} \frac{dx_2}{x_2^2} = 2 \int_0^{1/2} \frac{dx_1}{\sqrt{1-x_1^2}} = 2\pi/3$$

is finite, permitting the introduction of probabilities $d\mathbb{P} = \frac{3}{2\pi} \frac{dx_1 \, dx_2}{x_2^2}$, and that's good enough, as will be seen.

But first: how do the reduced geodesics look? The vertical lines are simple, but may be ignored, having only one degree of freedom, and likewise the semi-circles centered at (00) and of height 1 or more, which are periodic. The others are pretty complicated. Figure 8.6.5 shows such

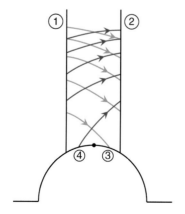

Figure 8.6.5

a geodesic starting upwards from side ④ (that's in red). By and by, it starts to fall (that's in green) until it meets the arc ③, whereupon it switches over to ④ by inversion, and starts to climb again. Figure 8.6.6 shows it in the format of Figure 8.6.4, spiraling out into the cusp, falling

Figure 8.6.6

back, fooling around at the bottom, then spiraling out again, on and on like that. Pretty complicated, and all that complicated behavior of reduced geodesics should lead to this: that if $F(Q)$ is any reasonable function and if $Q(t) : t \geq 0$ is any geodesic, departing from $Q(0)$ in a "typical" direction, then you should have

$$\lim_{T \uparrow \infty} \frac{1}{T} \int_0^T F(\mathbf{Q}) \, \mathrm{d}t = \mathbb{E}(F) = \int_C F \frac{\mathrm{d\,area}}{2\pi/3} \, .$$

8.6.4 $\mathbb{H}^2/\Gamma$: the circle bundle

Here a little detour is needed. The *place* Q is not adequate to tell where you are in geodesic flow. The *direction* P is also needed (it changes with time, unlike the case of $\mathbb{R}^2/\mathbb{Z}^2$), so the right setting is the circle bundle B, comprised of pairs $Q \in \mathbb{H}^2/\Gamma$ and directional vectors P of hyperbolic length $\sqrt{P_1^2 + P_2^2}/Q_2 = 1$.

Exercise 8.6.4   Show that the typical semi-circular geodesic centered at $(c, 0)$ and of height h, traversed at hyperbolic speed $\sqrt{\dot{Q}_1^2 + \dot{Q}_2^2}/Q_2 = 1$ may be written $Q(t) = (c, 0) + h(\sinh t / \cosh t, 1 / \cosh t)$.

Now introduce $P = \dot{Q}$ and think about the geodesic in this QP language. P is of length 1, so really it's only Q and the angle $\theta =$

$\tan^{-1}(P_2/P_1) = \tan^{-1}(\dot{Q}_2/\dot{Q}_1)$ that count, so you may use instead the variables $v_1 = Q_1$, $v_2 = 1/Q_2$, and $v_3 = \theta$. The reason for doing so is seen from the next exercise.

Exercise 8.6.5 Obviously, the natural volume element in $\mathbb{B}$ is $dQ_1\,dQ_2\,d\theta$ over $Q_2^2 = -dv_1\,dv_2\,d\theta$. Use Exercise 8.6.4, compute the vector field $\dot{v}$. Then check the divergence $\partial\dot{v}_1/\partial v_1 + \partial\dot{v}_2/\partial v_2 + \partial\dot{v}_3/\partial v_3 = 0$.

Now everything is ready: $\mathbb{H}^2/\Gamma$ is equipped with its circle bundle, for a more adequate description of things, and also with the probabilities $d\mathbb{P} = dQ_1\,dQ_2/2\pi Q_2^2 \times d\theta/2\pi$, invariant under the geodesic flow. Then you may hope to have

$$\lim_{T\uparrow\infty}\frac{1}{T}\int_0^T F(\mathbf{QP})\,dt = \mathbb{E}(F) = \int_\mathbb{B} F\,d\mathbb{P},$$

the *existence* of this average being guaranteed by Birkhoff's theorem. It is only its value that lies deeper (metric transitivity) which is proven next by a device of E. Artin [1924] going back to Gauss, involving, remarkably enough, continued fractions, just suited to this special problem! See also Hedlund [1934/37] and Hopf [1937] for a different take in more general circumstances where such special tricks are not available.

8.6.5 How continued fractions enter

Vertical geodesics being rare, I ignore them as before. The typical semi-circular geodesic is specified by its "feet" $\xi > \eta$ where it cuts the bordering line $x_2 = 0$. I take these to be irrational (anything else is also rare), and, as everything is done modulo Γ_1, I suppose $0 < \xi < 1$. Then η cannot be too close to ξ if the semi-circle is to cut the fundamental cell C_1: specifically, if it passes above the corner $\frac{1}{2}(-1+\sqrt{-3})$, then $\eta < -(\xi+2)/(2\xi+1) < -1$, as you may check. In short, $(0,1)\times(-\infty,-1)$ labels the active geodesics. Think next about Birkhoff's limit $\overline{F} = \overline{F}(\xi,\eta)$, recognizing by this notation that it depends only on the *locus* of the geodesic issuing (forwards and backwards) from its starting place, and write $0 < \xi < 1$ and $0 < -1/\eta < 1$ as ordinary continued fractions:

$$\xi = \frac{1|}{|n_0} + \frac{1|}{|n_1} + \frac{1|}{|n_2} \text{ etc.} \quad \text{and} \quad -\frac{1}{\eta} = \frac{1|}{|n_{-1}} + \frac{1|}{|n_{-2}} \text{ etc.}$$

When the motion, reduced modulo Γ, hits the boundary of C, the geodesic, and so also its feet, is changed: by the "diagonal" action (on both ξ and η) of $(1\pm1/0\,1)$: $x \to x\pm1$ if a vertical side is hit, or by the action of $(0\,1/-1\,0)$: $x \to -1/x$ if the circle is hit. Typically, encounters of both kinds occur infinitely often, and the corresponding substitutions may be combined as follows:

$$\xi = \cfrac{1}{n_0 + \cfrac{1}{n_1 + \cfrac{1}{n_2 \text{ etc.}}}} \qquad -\eta = n_{-1} + \cfrac{1}{n_{-2} + \cfrac{1}{n_{-3} \text{ etc.}}}$$

inversion $\downarrow$ $\qquad\qquad\qquad\qquad$ $\downarrow$

$$\cfrac{1}{n_{-1} + \cfrac{1}{n_{-2} + \cfrac{1}{n_{-3} \text{ etc.}}}} \qquad n_0 + \cfrac{1}{n_1 + \cfrac{1}{n_2} \text{ etc.}}$$

translation by n_0 $\downarrow$ $\qquad\qquad\qquad\qquad$ $\downarrow$

$$n_0 + \cfrac{1}{n_{-1} + \cfrac{1}{n_{-2} \text{ etc.}}} \qquad \cfrac{1}{n_1 + \cfrac{1}{n_2 \text{ etc.}}}$$

inversion $\downarrow$ $\qquad\qquad\qquad\qquad$ $\downarrow$

$$n_1 + \cfrac{1}{n_2 \text{ etc.}} \qquad \cfrac{1}{n_0 + \cfrac{1}{n_{-1} + \cfrac{1}{n_{-2} \text{ etc.}}}}$$

translation by $-n_1$ $\downarrow$ $\qquad\qquad\qquad\qquad$ $\downarrow$

$$\cfrac{1}{n_2 + \cfrac{1}{n_3 + \cfrac{1}{n_4 \text{ etc.}}}} \qquad n_1 + \cfrac{1}{n_0 + \cfrac{1}{n_{-1} \text{ etc.}}},$$

so as to produce, from the original pattern $-3-2-1012$, the new pattern -101234, which is the *double* shift of the *two-sided* continued fraction.

Exercise 8.6.6 Check that the probabilities

$$d\mathbb{P} = (\log 2)^{-1}(\xi - \eta)^{-2}\, d\xi\, d\eta$$

on the region $(0,1) \times (-\infty, -1)$ are invariant under the single shift of the two-sided fraction. *Hint*: The action of the shift is $\xi \to 1/\xi - n_0$, $-1/\eta \to 1/(n_0 - 1/\eta)$ if $1/(n_0 + 1) < \xi < 1/n_0$.

Now, like the single shift for one-sided continued fractions, the double shift is metrically transitive (just take N even in the proof of §8.5), and this is inherited by the double shift of two-sided continued fractions.

Exercise 8.6.7 Check it out.
Hint: The function $F(\xi, \eta)$ of the two-sided continued fraction is well approximated by $\mathbb{E}(F(\mathbf{n}_{-m}, \mathbf{n}_{-m+1}, \text{etc.}))$ if m is large. Why? What then?

The proof is nearly done. You have only to reflect that Birkhoff's function $\overline{F}(\xi, \eta)$ is insensitive to the (diagonal) action of Γ on ξ and η: as such, it must be constant, and what could the constant be but $\mathbb{E}(F)$? Mixing is inherited in the same way.

Exercise 8.6.8 The distance $d(x)$ from the place x to $\sqrt{-1}$ is

$$\cosh^{-1}\left[1 + \frac{1}{2x_2}(x_1^2 + x_2^2)\right],$$

for which see Exercise 8.6.9 below, so

$$\mathbb{P}\left[\lim_{T\uparrow\infty} \frac{1}{T} \int_0^T d(\mathbf{Q})\,dt = \mathbb{E}\left[d(\mathbf{Q})\right]\right] = 1 \quad \text{by Birkhoff.}$$

I invite you to estimate

$$\mathbb{E}(d) = \frac{6}{\pi} \int_0^{1/2} dx \int_{\sqrt{1-x^2}}^{\infty} \cosh^{-1}\left[(2y)^{-1}(x^2 + y^2 + 1)\right] \frac{dy}{y^2}.$$

Try to get $0.9 < \mathbb{E} < 1.2$ or better. *Hint*: With $\frac{1}{2}\left(y + \frac{1}{y}\right) = \cosh 2$, you get something between 2 and $\cosh^{-1}(1/4\sqrt{3}) + 7 = \log(2/\sqrt{3}) + 7$. I owe the numerical value $\mathbb{E}(d) = 1.0249830927$ to the kindness of A. Yilmaz.

Exercise 8.6.9 The formula for $d(x)$ in Exercise 8.6.8 is a special case of a general rule, *viz.*, the (hyperbolic) distance $d = d(x, y)$ between two points x and y of $\mathbb{H}^2$ is given by

$$\cosh(d) = 1 + \frac{1}{2}\frac{|x - y|^2}{x_2 y_2}.$$

Hint: The right side is unchanged by the rigid motions $\mathrm{PSL}(2, \mathbb{R})$. Yes? What then?

8.6.6 CLT

LLN is now proven in $\mathbb{H}^2/\Gamma_2$. What's new is that CLT is working, too:

$$\lim_{T\uparrow\infty} \mathbb{P}\left[a \le \frac{1}{\sqrt{T}} \int_0^T F(\mathbf{QP})\,\mathrm{d}t < b\right] = \int_a^b \frac{\mathrm{e}^{-x^2/2}}{\sqrt{2\pi}}\,\mathrm{d}x$$

for any function $F\colon \mathbb{B}/\Gamma_2 \to \mathbb{R}$ with $\mathbb{E}(F) = 0$ and $\mathbb{E}(F^2) = 1$. The proof is not simple, and I do not reproduce it, but see Sinai [1960], Ratner [1973], and/or Le Jan [1994] for proofs in various styles for surfaces of variable negative curvature and finite total area. I must also mention but do not prove a pretty fact, due to Sullivan [1982], *viz.*,

$$\mathbb{P}\left[\limsup_{T\uparrow\infty} \frac{d(\sqrt{-1}, \mathbf{Q}(T))}{\log T} = \frac{1}{2}\right] = 1,$$

descriptive of the way $\mathbf{Q}$ spirals out along the cusp.

These facts are remarkable. *Really, there is no probability here at all, only a perfectly predictable mechanical motion with a wild behavior, produced by the geometry alone.*

8.6.7 Back to $\mathbb{R}^2/\mathbb{Z}^2$

This is *not* the case for $\mathbb{R}^2/\mathbb{Z}^2$. LLN works, CLT doesn't. Let's see why. You would want $T^{-1/2}\int_0^T F(\mathbf{Q})\,\mathrm{d}t$ to be very nearly Gaussian in the case $n \cdot \omega \ne 0$ for $n \in \mathbb{Z}^2 \setminus 0$ if, for example, F is smooth, with $\hat{F}(0) = \int F(x)\,\mathrm{d}^2x = 0$, and T is large. But if $\mathbf{P} = \omega$ is uniformly distributed on the circle $|\omega| = 1$, then

$$\mathbb{P}\left[|n \cdot \omega|^{-1/2} > |n|^{5/2}\right] \le |n|^{-5/2} \times \frac{1}{2\pi}\int_0^{2\pi} \frac{\mathrm{d}\theta}{\sqrt{|n|\,|\cos\theta|}}$$

is comparable to $|n|^{-3}$, and as this is summable over $\mathbb{Z}^2 \setminus 0$, the first Borel–Cantelli lemma guarantees that

$$\mathbb{P}\left[|n \cdot \omega|^{-1} \le |n|^5 \text{ for large } n \in \mathbb{Z}^2\right] = 1.$$

Then, as in §8.2, with $Q(0) = 0$ as it plays no role, you find

$$\int_0^T F(n \cdot \omega\, t)\,\mathrm{d}t = \sum_{\mathbb{Z}^2\setminus 0} \hat{F}(n) \frac{\mathrm{e}^{2\pi\sqrt{-1}\,n\cdot\omega T} - 1}{2\pi\sqrt{-1}\,n\cdot\omega}$$

to be controlled by $\sum|\hat{F}(n)|\,|n|^5 < \infty$ if F is of class $\mathcal{C}^8$ or better. Not a hope for CLT.

8.6.8 Why $\mathbb{H}^2/\Gamma$ is better

The difference is caused by the way geodesics spread in $\mathbb{H}^2$.
$\mathbb{R}^2$ is simple: the two speed-one geodesics seen in Figure 8.6.7, issuing

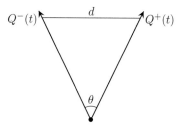

Figure 8.6.7

from 0 with an angle $\theta < \pi$ between them, are at distance $d = 2t\sin(\theta/2)$.

$\mathbb{H}^2$ is different: the figure shows two speed-one geodesics made from circles with centers, $\pm C$ and height h, issuing from the point $\sqrt{-1}$, Q^- being the reflection of Q^+ across $x_1 = 0$. Here, $Q^+(t) = (h\tanh(t) + c, h/\cosh(t))$, so by the formula of Exercise 8.6.9,

$$d(Q^+, Q^-) = \cosh^{-1}\left(1 + \frac{1}{2}\frac{|Q^+ - Q^-|^2}{h_+ h_-}\right) \simeq \cosh^{-1}(4\tanh^2 t \times \cosh^2 t)$$

$$\simeq 2t.$$

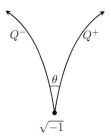

Figure 8.6.8

You see the difference: in $\mathbb{R}^2$, there is spreading at the rate $2\sin(\theta/2) < 2$; in $\mathbb{H}^2$, there is spreading *at rate* 2 for $t \uparrow \infty$, regardless of the initial angle θ, or as you may say, the two geodesics seem to be heading in *opposite* directions. It is this anomaly, combined with confinement by means of the modular group Γ, that produces in $\mathbb{H}^2/\Gamma$ a degree of chaotic behavior, not to say utter confusion, unknown in $\mathbb{R}^2/\mathbb{Z}^2$.

9

Communication over a Noisy Channel

Next comes a short account of C. Shannon's [1948] ideas on the subject, building upon Chapters 2, 7 and 8.

The general set-up is seen in the diagram.

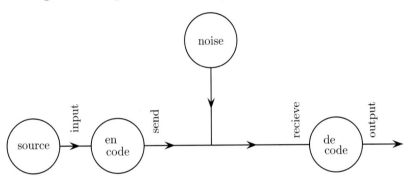

The input from the source is the spoken word or any sort of communication expressed by a stream of individual "letters" from some "alphabet". It is to be conveyed over the channel (a wire, the air, what have you, the long arrow in the picture), but this (the channel) is corrupted by noise, so what is received is only an imperfect copy of what was sent. What's wanted is to *en*code the input before it is sent and to *de*code the noisy communication upon reception so as to make the output as faithful as you can to the (typical) input, keeping the rate at which correct information is conveyed pretty high.

Obviously, there is a competition between rate and faithful communication. The Irish solution ("Do you know what I'm about to tell you?") is to talk charmingly, on and on, not worrying if anything is really understood. The English solution on the Continent is to say it (in English, never in French) over and over, in a louder and louder voice. Both ways can be pretty faithful if you have plenty of time, but the rates are very bad.

How to do better? How to measure the rate of information sent or of correct information received? And thinking of the channel as a pipe, How big is it? What, so to say, is its diameter? How fast can you pass (more or less correct) information through it? and how close to such an optimal rate can you get by clever encoding/decoding?

All these questions were posed and largely answered in two brilliant papers by C. Shannon [1948]. In my opinion, this stuff is mathematics (and electrical engineering) of the highest order, not sufficiently appreciated as such by the mathematical community, and I want to give you a brief account of it here in a simplified form.

The best popular account is the book of Shannon–Weaver [1949]. The best account with all the technical niceties is Wolfowitz [1978], but see also Ash [1965], Jellinek [1968] and/or Cover–Thomas [1991], for something easier.

9.1 Information/Uncertainty/Entropy

The first task is to explain how information is to be measured. Think of an experiment $\mathfrak{E}$ having a finite number of possible outcomes $e = 1, 2, \ldots, n$ with positive probabilities $p_1, \ldots, p_n$ – the cases $p = 0$ or 1 are not interesting – and let's ask two equivalent questions: How much information does the outcome provide once the trial is made? Or, what is the same: How much uncertainty prevails before? The common answer is to be a non-negative number $H(\mathfrak{E})$, subject to three natural rules:

Rule 1 $H(\mathfrak{E})$ is fully determined by the probabilities $p_1, \ldots, p_n$, without regard to order, i.e. it is a universal, symmetric function $H_n(p_1, \ldots, p_n)$ of these.

Rule 2   The information $H_2(p, 1 - p)$ for a single Bernoulli trial is to vanish when the outcome is certain ($p = 0$ or 1); it is to be biggest when the outcome is most *uncertain* ($p = 1/2$); and *that* number is to be 1, so setting the units in which information is to be measured. $H_2(1/2, 1/2) = 1$ is one "bit" of information, as supplied by the answer to a single, unbiased question: Yes or No?

Rule 3 is the really important one, requiring that information be *additive*. To explain what is meant, think of two experiments $\mathfrak{A}$ and $\mathfrak{B}$,

with outcomes a and b and probabilities p_a and p_b, performed simultaneously, the joint experiment $\mathfrak{A} \,\&\, \mathfrak{B}$ having outcomes ab with any probabilities p_{ab}, consistent with those of $\mathfrak{A}$ and $\mathfrak{B}$, i.e. with $\sum_a p_{ab} = p_b$
and $\sum_b p_{ab} = p_a$. Then it is required that

$$H(\mathfrak{A} \,\&\, \mathfrak{B}) = H(\mathfrak{A}) + H(\mathfrak{B}/\mathfrak{A}) = H(\mathfrak{B}) + H(\mathfrak{A}/\mathfrak{B}),$$

in which, for example, $H(\mathfrak{B}/\mathfrak{A})$ is the (mean) information, over and
above $H(\mathfrak{A})$, supplied by the outcome of $\mathfrak{B}$ once the outcome of $\mathfrak{A}$ is
known, construed as follows. Fix the outcome a of $\mathfrak{A}$; ascribe to $\mathfrak{B}$ the
conditional probabilities $p_{b/a} = p_{ab}/p_a$; take the information $H(\mathfrak{B}/a)$
supplied by *that* experiment, and average it as in $\sum_a p_a H(\mathfrak{B}/a) \equiv$
$H(\mathfrak{B}/\mathfrak{A})$.

The fact is that these rules allow just one form for the informa-
tion/uncertainty of the experiment $\mathfrak{C}$:[1]

$$H(\mathfrak{C}) = -\sum_e p_e \log_2 p_e.$$

This may look outlandish if you've not seen it before. Actually, it is a
rudimentary form of the "entropy" introduced into statistical mechanics
in the 1800s – but more of that below and in Chapter 10, postponed in
favor of the formula itself.

Proof. Take $n = 3$, $p_1 = p$, $p_2 = q$, and $p_3 = 1 - p - q$, and write
$H_2(p, 1 - p) = h(p)$. The information supplied by the joint experiment
depicted in Figure 9.1.1 is, variously,

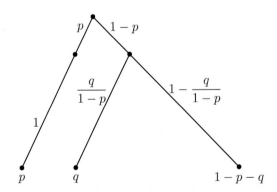

Figure 9.1.1

---

[1] $\log_2$ is the logarithm to the base 2, so that $H_2(1/2, 1/2) = 2 \times -\frac{1}{2} \log_2 \frac{1}{2} = 1$.

(1) $h(p) + (1 - p)h\left(\dfrac{q}{1 - p}\right)$

or

(2) $h(q) + (1 - q)h\left(\dfrac{p}{1 - q}\right)$

by Rules 1 and 3.

Exercise 9.1.1   Check that $h(p)$ is smooth for $0 < p < 1$ by judicious integration of (1) or (2).

Now differentiate (1) = (2), once by p and once by q, to obtain

(3′) $h''\left(\dfrac{q}{1 - p}\right)\dfrac{q}{(1 - p)^2} = h''\left(\dfrac{p}{1 - q}\right)\dfrac{p}{(1 - q)^2}$,

and re-express this in terms of $x = p/(1 - q)$ and $y = q/(1 - p)$, as in

(3″) $h''(x)x(1 - x) = h''(y)y(1 - y)$,

noting that the substitution $pq \to xy$ maps the triangle $[0 < p, 0 < q, p + q < 1]$ onto the square $[0 < x, y < 1]$, and vice versa. Evidently, $h''(x)$ is a constant multiple of $1/x(1 - x)$, so

$$h(x) = a + bx + c\left[-x\log_2 x - (1 - x)\log_2(1 - x)\right]$$

with $a = b = 0$ because of $h(0) = h(1) = 0$, and $c = 1$ because of $h(1/2) = 1$ – all this by Rule 2 which also dictates the choice of logarithms to the base 2. In short,

$$H_2(p, 1 - p) = -p\log_2 p - (1 - p)\log_2(1 - p), \quad 0 < p < 1.$$

Look now at Figure 9.1.2, depicting an experiment of n outcomes, and

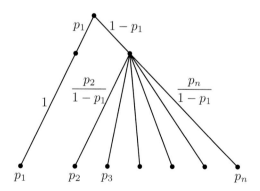

Figure 9.1.2

suppose the stated formula for $\#(\mathfrak{E})$ is correct for $n-1$ outcomes. By Rule 3,

$$H_n(p_1,\ldots,p_n) = H_2(p_1, 1-p_1) + (1-p_1)H_{n-1}\left(\frac{p_2}{1-p_1}, \frac{p_3}{1-p_1}, \text{etc.}\right)$$

$$= -p_1\log_2 p_1 - (1-p_1)\log_2(1-p_1)$$

$$- (1-p_1) \times \sum_2^n \frac{p_k}{1-p_1}\log_2\frac{p_k}{1-p_1}$$

$$= -\sum_1^n p_k\log_2 p_k$$

when you work it out. That does it.

Exercise 9.1.2 Here, Rule 3 was used only in a very special way. Check that $H(\mathfrak{E})$ satisfies the rule in its general form.

Exercise 9.1.3 $H(\mathfrak{E}) \leq \log_2 n$, with equality only if the outcomes of $\mathfrak{E}$ are equally likely.
Hint: Use Gibbs's lemma: $x\log x - x + 1 \geq 0$ with equality at $x=1$ only.

Exercise 9.1.4 $H(\mathfrak{A}|\mathfrak{B}) \leq H(\mathfrak{A})$, with equality only if $\mathfrak{A}$ and $\mathfrak{B}$ are independent, i.e. only if $p_{ab} = p_a p_b$; in particular, $H(\mathfrak{A}\,\&\,\mathfrak{B}) \leq H(\mathfrak{A}) + H(\mathfrak{B})$ with the same fine print.

Disclaimer I have to say that this derivation of the formula for H from the so-called rules for information/uncertainty is only housekeeping, to make things look neat after the real cooking is done. I could have picked it out of the air and promised that it would be useful now and then. Better still, I could (and now do) show you that it is a rudimentary form of the statistical-mechanical entropy, as I said before. Boltzmann's (1870) picture of the equilibration of a gas of like molecules makes the connection clear.

9.1.1 What Boltzmann said

Think of a gas of N indistinguishable molecules, confined to a large but not unlimited region in the phase space $\mathbb{R}^6$ of a single molecule with its 6 degrees of freedom: 3 for position + 3 for velocity. The occupied volume V_- is divided into a large but fixed number, M, of cells, each of volume 1 more or less, and it is desired to place the N molecules in the several cells: n_1 in the first cell, n_2 in the second, and so on, for a total of $n_1 + n_2 + \text{etc.} = N$.

Now there are

$$\binom{N}{n} = \frac{N!}{n_1! n_2! \cdots n_M!}$$

ways to do that, this number being, at the same time, the volume V_+ occupied by the totality of such configurations in the (joint) phase space $\mathbb{R}^{6N}$ of the N individual molecules – downstairs, the molecules are merely counted; upstairs, they are individually distinguished, one from another; that is why $V_+ = \binom{N}{n}$ is the volume they occupy in the big space. Now suppose, with Boltzmann, that all the occupation numbers are large so that Stirling's approximation may be used, and let's make a rough appraisal of V_+, thinking of that as indicating the degree of uncertainty as to the actual configuration:

$$\log_2 V_+ \simeq (N + \tfrac{1}{2}) \log_2 N - \sum_1^M (n_i + \tfrac{1}{2}) \log_2 n_i - (M - 1) \log_2 \sqrt{2\pi}$$

$$\simeq - \sum_1^M n_i \log_2 \frac{n_i}{N} \quad \text{with an error comparable to } \log_2 N$$

or, what is better,

$$\frac{1}{N} \log_2 V_+ \simeq - \sum_1^M \frac{n_i}{N} \log_2 \frac{n_i}{N} \equiv H,$$

which is to say $V_+ \simeq 2^{NH}$. Boltzmann perceived in this an interpretation of the Second Law of Thermodynamics: H, construed as entropy, increases with the passing time (that's the Second Law) because the gas wants to occupy a larger and larger volume in $\mathbb{R}^{6N}$, seeking out more and more uncertain configurations on its way to equilibrium when H is the biggest it can be. Then the occupation numbers are all about the same ($n \simeq N/M$ as per Exercise 9.1.3), and these 2^{NH} near-equilibrium configurations are evenly spread at over the volume V_+, with density 2^{-NH}, or very nearly so.[2]

More realistically, ascribe to each molecule an energy ε, typical of the cell in which it is found, and dropping the $\log_2$ in favor of the natural logarithm, bring the gas to equilibrium by maximizing

$$H = - \sum \frac{n_i}{N} \log \frac{n_i}{N},$$

subject to

---

[2] Here and below, I ignore the very likely nuisance that 2^{NH} is not a whole number.

(1) $\sum \frac{n_i}{N} = 1$ and

(2) $\sum \varepsilon_i \frac{n_i}{N} =$ some fixed number Q.

Thinking of n_1/N, n_2/N etc. as continuous variables, a formal application of Lagrange's rule at the maximum of H produces

$$\operatorname{grad} H = -\log \frac{n}{N} - 1 = \alpha \times \big[\operatorname{grad}(1) = 1\big] + \beta \times \big[\operatorname{grad}(2) = \varepsilon\big],$$

or, what is the same,

$$\frac{n_i}{N} = \frac{e^{-\varepsilon_i/T}}{Z},$$

where now $\alpha + 1 = \log Z$, $\beta = 1/T$, $Z = \sum e^{-\varepsilon_i/T}$ in accord with (1), and

$$H = \sum \frac{n_i}{N}\left(\frac{\varepsilon_i}{T} + \log Z\right) = \frac{Q}{T} + \log Z$$

in accord with (2). Here, T is interpreted as Kelvin's absolute temperature and Q (= mean energy per molecule) as heat per molecule in accord with Count Rumford's dictum that in an ideal gas of non-interacting molecules, heat is the equivalent of mechanical motion due to thermal agitation.

Now move the gas about a closed cycle of such (near) equilibrium configurations, permitting everything to change little by little. Then $dH = dQ/T - Q\,dT/T^2 + dZ/Z$, and this reduces to $dH = dQ/T$ by inspection of dZ/Z, so that $\oint dQ/T = 0$, expressing the *experimental* fact that dQ/T is an exact differential. Then with a change of attitude you could *define* H to be the indefinite integral $\int dQ/T$, determined up to an additive constant, which is how entropy is actually introduced in classical thermodynamics.

That's Boltzmann's picture; compare Fermi [1956:46–59] and, for a very different take, Flanders–Swann [1963:side 1, band 6]; see also §7.3.3 for McMillan's variant of the present theme.

I add a couple of illustrations of how information/uncertainty comes up in unexpected places.

9.1.2 Information as a guide to gambling[3]

Let e_1, e_2, etc. (= 0 or 1) be independent Bernoulli trials with success probability $p = \mathbb{P}(e = 1) > 1/2$ and play this game. Begin with capital $C_0 = 1$. At the nth play, bet a fraction $0 < f < 1$ of present capital

---

[3] After Kelly [1956].

C_{n-1} on heads ($\mathbf{e} = 1$). Your new capital will be $C_n = (1 + f)C_{n-1}$ if you win, or $(1 - f)C_{n-1}$ if you lose.

Exercise 9.1.5 $C_n \simeq 2^{nK}$ with probability 1 comes from LLN. Check this, compute $K = K(f)$, and maximize that over f to determine the best rule of play. *Answer:* $K = 1 - H_2(p, 1 - p)$.

This makes sense: with perfect information ($p = 1$), you bet everything ($f = 1$) and $C_n = 2^n$; for $p < 1$, H bits of information is the degree of uncertainty before the play is made, 1 is the biggest amount of information you could have, so $K = 1 - H$ is the amount of information you can count on and is the best choice. I think this clarifies the meaning of information in a specially nice way.

9.1.3 A dishonest coin[4]

To begin with, I must explain the notion of fractional dimension. Take compact $K \subset [0, 1]$, cover it, by a countable number of open intervals G of length $|G| \le h$, and define

$$\Lambda^\gamma(K) = \lim_{h \downarrow 0} \inf \sum |G|^\gamma \quad \text{for } 0 \le \gamma \le 1,$$

the infimum being taken over all such covers. When $\gamma = 0$ this is just the number of points in K; when $\gamma = 1$, it is the Lebesgue measure of K; and for γ in between it may take any value, $+\infty$ included. Now it's easy to see that if $\Lambda^\alpha(K) < \infty$ for some $\alpha < 1$, then $\Lambda^\beta(K) = 0$ for $\beta > \alpha$, and similarly, if $\Lambda^\alpha(K) > 0$ for some $\alpha > 0$, then $\Lambda^\beta(K) = \infty$ for $\beta < \alpha$. The critical exponent γ at which $\Lambda^\gamma(K)$ switches from ∞ to 0 is the fractional dimension of K.

Exercise 9.1.6 Let K be the familiar compact set obtained from $[0, 1]$ by removing the open middle third, the two remaining open middle ninths, the 4 remaining open middle twenty-sevenths, and so on. Prove that $\dim K = \log 2/\log 3$ and that $\Lambda^\gamma(K \cap [0, x])$ is the distribution function for $\mathbf{x} = \sum^\infty \mathbf{e}_n 3^{-n}$ in which the $\mathbf{e}$s are independent with common distribution $\mathbb{P}(\mathbf{e} = 2) = \mathbb{P}(\mathbf{e} = 0) = 1/2$.

Now take a dishonest coin with probabilities $p > 1/2$ for heads ($\mathbf{e} = 1$) and $q = 1 - p$ for tails ($\mathbf{e} = 0$). Then the sum $\mathbf{x} = \sum_1^\infty \mathbf{e}_n 2^{-n}$ occupies

---

[4] After Eggleston [1949].

a set $K \subset [0,1]$ with empirical frequencies

$$\lim_{n\uparrow\infty} \frac{1}{n}\#(k \leq n : \mathbf{e}_k = 1) = p \quad \text{and} \quad \lim_{n\uparrow\infty} \frac{1}{n}\#(k \leq n : \mathbf{e}_k = 0) = q.$$

K is not compact, but never mind. Let's still ask: What is its dimension? The pretty answer is $\dim K = H_2(p, 1-p)$. The *idea* of the proof is simple. For large n, the chance that $\mathbf{e}_1, \mathbf{e}_2, \ldots, \mathbf{e}_n$ is seen is

$$p^{\#\text{heads}}q^{\#\text{tails}} \simeq p^{np}q^{nq} = 2^{-nH} \quad \text{with } H = H_2(p,q) \text{ by LLN,}$$

and since each choice of the **es** cuts out an interval of length 2^{-n}, there should be about 2^{nH} such intervals covering K. This number cannot be much reduced. (Think it over.) The dimension is now read off from $\Lambda^\gamma(K) \simeq 2^{-n\gamma} \times 2^{nH}$ which breaks at $\gamma = H$.

9.1.4 Relative entropy

This notion made its appearance in connection with the large deviations of §2.4. I explain now what it means, imitating McMillan's theorem of §7.3.3 in a simpler setting.

Let $\mathbf{x}_1, \mathbf{x}_2$, etc. be independent copies of a variable $\mathbf{x}$, capable of a finite number of distinct values x with probabilities $\mathbb{P}(\mathbf{x} = x) = p_x$, and write $p(x_1 x_2 \ldots x_n) = \mathbb{P}[\mathbf{x}_1 = x_1, \mathbf{x}_2 = x_2, \ldots, \mathbf{x}_n = x_n]$. If you now ask: *What is the probability of what I just saw?* the answer will be roughly expressed by an easy application of LLN:

$$\frac{1}{n} \log p(\mathbf{x}_1 \mathbf{x}_2 \ldots \mathbf{x}_n) \simeq (-1) \times \sum p_x \log p_x$$

$$= \text{the entropy } H(p) \times \text{a nuisance factor } \log 2.$$

Now change the probabilities to $\mathbb{Q}(\mathbf{x} = x) = q_x$, keeping p_x as a "benchmark", and look at the "likelihood ratio"

$$\mathbf{r}_n = \frac{p(\mathbf{x}_1 \mathbf{x}_2 \ldots \mathbf{x}_n)}{q(\mathbf{x}_1 \mathbf{x}_2 \ldots \mathbf{x}_n)}.$$

Evidently,

$$\frac{1}{n} \log r_n \simeq \sum q_x \log \frac{p_x}{q_x} = \text{the "relative entropy"} - I(q) \text{ of §2.4.5,}$$

the meaning of which may now be clarified by the rough transcription

$$p(\mathbf{x}_1 \mathbf{x}_2 \ldots \mathbf{x}_n) \simeq q(\mathbf{x}_1 \mathbf{x}_2 \ldots \mathbf{x}_n) \times \mathrm{e}^{-nI(q)} \text{ for } n \uparrow \infty.$$

9.2 Noiseless coding

Shannon's story begins here with another interpretation of H. The channel is simplest when there is no noise at all. Then it is only a question of coding/decoding if, as it may be, the sender speaks one language and the receiver speaks another. B. McMillan [1952] found a pleasing interpretation of information for such a "noiseless channel".

The input is expressed in an alphabet $\mathfrak{A}$ of A letters a, these being emitted independently, with probabilities $0 < p_a < 1$, and the trouble is that the channel accepts only messages expressed in a foreign alphabet $\mathfrak{B}$ of B letters b, B being much smaller than A. Then you must make a "code book" which assigns, to each letter a, some code word $b_1 \ldots b_n$ of length $n = n(a)$. This the channel accepts and conveys uncorrupted by noise – and more is wanted, namely that no code word be the beginning of any other, so that the output can be unambiguously cut up into such words and decoded perfectly. This is "noiseless coding".

Now you may ask: Can this be done at all? And if so: How small can the mean word-length $\ell = \mathbb{E}(\mathbf{n}) = \sum n(a)p_a$ be made?

It is easily seen that the answer to the first question is "yes": indeed, if $\#(n)$ is the number of code words of length n, then you must have

$$\#(1) \leq B,$$
$$\#(2) \leq \big[B - \#(1) \big] B = B^2 - \#(1)B,$$
$$\#(3) \leq \Big[\big[B - \#(1) \big] B - \#(2) \Big] B = B^3 - \#(1)B^2 - \#(2)B,$$

and so on, i.e.

$$B^n \geq \#(1)B^{n-1} + \cdots + \#(n-1)B + \#(n),$$

which is to say

$$1 \geq \sum_{n=1}^{\infty} \#(n)B^{-n} = \sum_a B^{-n(a)};$$

and conversely, as you will check.

OK. Evidently, there are lots of codes if only the word lengths are big enough, but which code book is best or nearly so? Let's try

$$n(a) = \text{the smallest whole number} \geq -\log_2 p_a / \log_2 B,$$

ascribing short words to common letters. That's possible since

$$\sum_a B^{-n(a)} \leq \sum_a B^{\log_2 p_a / \log_2 B} = \sum_a p_a = 1,$$

and you find

$$\ell = \mathbb{E}(\mathbf{n}) \le \sum \left(1 - \frac{\log_2 p}{\log_2 B}\right) p = 1 + \frac{H}{\log_2 B}$$

with

$$H = -\sum p \log_2 p = \text{the information in a single output letter.}$$

Actually, this is not bad at all: in fact, ℓ is at least $H/\log_2 B$ for *any* code book of the type envisaged here. The proof is easy: $x \log x - x + 1 \ge 0$ for $x \ge 0$ (Gibbs's lemma), so $\log x \ge 1 - \frac{1}{x}$, and with $p_a = p$ and $n(a) = n$ for brevity, you find

$$\ell - \frac{H}{\log_2 B} = \sum pn + \sum \frac{p \log_2 p}{\log_2 B} = \frac{1}{\log_2 B} \sum p \log_2(B^n p)$$

$$\ge \frac{1}{\log_2 B} \sum p\left(1 - B^{-n}/p\right) = \frac{1}{\log_2 B}\left(1 - \sum B^{-n}\right) \ge 0.$$

Improvements can be made by "block coding". Think of inputs $a_1 \ldots a_m$ of fixed length m as a new kind of letter and encode *these*. The information in such a "letter" is mH, by independence, and with a near-optimal code, you have

$$\frac{mH}{\log_2 B} \le \mathbb{E}(\mathbf{n}) \le 1 + \frac{mH}{\log_2 B},$$

so that for large m, the mean word-length *per input letter* is as close to $H/\log_2 B$ as you want. That's McMillan's pretty interpretation of the information H.

9.3 The source

Now let's talk about the general set-up – input/channel/output – starting with the source which drives the whole thing.

Take an alphabet of a finite or infinite number of letters e and introduce the sample space of two-sided messages $\mathbf{e} = (\ldots \mathbf{e}_{-2}, \mathbf{e}_{-1}, \mathbf{e}_0, \mathbf{e}_1, \mathbf{e}_2, \mathbf{e}_3, \ldots)$ provided with probabilities $\mathbb{P}$ invariant under the shift

$$T \colon (\ldots \mathbf{e}_{-2}, \mathbf{e}_{-1} \overset{\downarrow}{,} \mathbf{e}_0, \mathbf{e}_1, \mathbf{e}_2, \mathbf{e}_3, \ldots) \to (\ldots \mathbf{e}_{-2}, \mathbf{e}_{-1}, \mathbf{e}_0 \overset{\downarrow}{,} \mathbf{e}_1, \mathbf{e}_2, \mathbf{e}_3, \ldots)$$

This is the source, denoted by $\mathfrak{E}$, representing the (uncoded) input to the channel.

The next examples illustrate the notion of memory, describing the dependence of the present ($\mathbf{e}_0$) upon the past $\mathbf{e}_n : n < 0$.

Example 9.3.1 The es are independent copies of e_0, and one speaks of memory $m = 0$.

Example 9.3.2 The es form a Markov chain, as in §7.1, i.e.

$$\mathbb{P}\big[e_0 = b \,|\, e_{-1}, e_{-2}, \text{etc.}\big] = p_{ab} > 0 \quad \text{with } a = e_{-1},$$

e_0 being distributed by the invariant distribution π, so that the two-sided chain is statistically invariant under the shift. Now the memory is $m = 1$.

Example 9.3.3 $\mathbb{P}\big[e_0 = b | e_1, e_2, \text{etc.}\big]$ depends not only upon e_{-1} but on e_{-2} as well. The Markov property is restored by doubling the state as in Chapter 7, and you speak of memory $m = 2$ – and so on, of $m = 3$ or more.

Example 9.3.4 Two-sided continued fractions $(\dots n_{-2}, n_{-1}, n_0, n_1, n_2, n_3 \dots)$ are shift-invariant under the distribution $(\log 2)^{-1}(\xi - \eta)^{-2}\, d\xi\, d\eta$ in which

$$\xi = [n_1, n_2, \text{etc.}] \in (0, 1] \quad \text{and} \quad \eta = -n_0 - [n_{-1}, n_{-2}, \text{etc.}] < -1;$$

see §8.5 for notation and background. Here, the alphabet $\mathbb{N} = 1, 2, 3, \text{etc.}$ is infinite, but never mind, and besides, there is no Markov property anymore $(m = \infty)$ – only the metric transitivity seen in §8.6.5. This is acceptable, too.

What is really wanted in a source is that it have (1) a short-term coherence, as natural languages do, and (2) some kind of long-term independence of widely separated letters/words, permitting the introduction of effective statistical methods, typified by LLN and its variants such as McMillan's theorem of §7.3.3 which plays a central role below. Finite memory m is enough (but not necessary) to satisfy (2), and as this number may be taken as large as you want, such a requirement is without practical significance.

9.3.1 The rate

The rate indicates the size of the source, measured in a natural way by the "information production rate":

$$h = \lim_{n \uparrow \infty} \frac{1}{n} H[e_0, e_1, \dots, e_{n-1}] = \text{entropy production per letter.}$$

But does this limit always exist? It's easy to see that it does. Indeed, with $H[e_0, \ldots, e_{n-1}] = H_n$, fixed $\ell > 0$, $n \uparrow \infty$, and $(k-1)\ell \le n < k\ell$, you have

$$H_n \le H_{k\ell} \le kH_\ell \quad \text{by Exercise 9.1.4 and shift-invariance}$$
$$= k\ell \times H_\ell/\ell$$
$$\le (n+\ell) \times H_\ell/\ell.$$

Then

$$\limsup_{n\uparrow\infty} H_n/n \le \inf_{\ell \ge 1} H_\ell/\ell \le \liminf_{n\uparrow\infty} H_n/n,$$

which is to say

$$\lim_{n\uparrow\infty} H_n/n = \inf_{n \ge 1} H_n/n.$$

Example 9.3.5   (Example 9.3.1 continued.) $H_n = nH[e_0]$ so h is simply $H[e_0]$.

Example 9.3.6   (Example 9.3.2 continued.)

$$H_n = H[e_0] + (n-1)H[e_1/e_0],$$

as you will check, so $h = H[e_1/e_0] = \sum_a \pi_a \left[-\sum_b p_{ab} \log_2 p_{ab}\right]$, as in §7.3.2.

9.3.2 McMillan's theorem (reprise)

This was proved in §7.5 for Markovian sources ($m = 1$) and also for the more complicated case of continued fractions ($m = \infty$) in §8.5.5. Now comes a variant, refining upon the mere existence of the information production rate

$$h = \lim_{n\uparrow\infty} \frac{1}{n} H[e_0 e_1 \ldots e_{n-1}] = \lim_{n\uparrow\infty} \mathbb{E}\left[-\frac{1}{n} \log_2 p(\mathbf{e}_0 \mathbf{e}_1 \ldots \mathbf{e}_{n-1})\right].$$

The statement is threefold:

(1) $\displaystyle\lim_{n\uparrow\infty} -\frac{1}{n} \log_2 p(\mathbf{e}_0, \mathbf{e}_1, \ldots, \mathbf{e}_{n-1}) = \overline{\mathbf{h}}(\mathbf{e})$ exists with probability 1,

(2) $\mathbb{E}(\overline{\mathbf{h}}) = h$,

and

(3) $\mathbb{P}(\overline{\mathbf{h}} = h) = 1$ if the shift is metrically transitive.

Here,

$$p(e_0, e_1, \ldots, e_{n-1}) = \mathbb{P}\big[\mathbf{e}_0 = e_0, \mathbf{e}_1 = e_1, \ldots, \mathbf{e}_{n-1} = e_{n-1}\big] \text{ as in } \S 7.5,$$

and you may think of these probabilities as positive since $e_0, e_1, \ldots, e_{n-1}$ is not seen otherwise. For simplicity, the number of letters e is assumed to be finite.

Proof after Breiman [1957]. Obviously,

$$
\begin{aligned}
p(\mathbf{e}_0, \ldots, \mathbf{e}_{n-1}) &= p(\mathbf{e}_{n-1}|\mathbf{e}_0, \ldots, \mathbf{e}_{n-2}) \times p(\mathbf{e}_{n-2}|\mathbf{e}_0, \ldots, \mathbf{e}_{n-3}) \\
&\quad \times \cdots \times p(\mathbf{e}_2|\mathbf{e}_0, \mathbf{e}_1) \times p(\mathbf{e}_1|\mathbf{e}_0) \times p(\mathbf{e}_0) \\
&= \prod_{k=0}^{n-1} T^k p(\mathbf{e}_0|\mathfrak{F}_{-k}),
\end{aligned}
$$

where T is the shift and $\mathfrak{F}_{-k}$ is the field of $\mathbf{e}_{-k}, \ldots, \mathbf{e}_{-1}$. In short,

$$-\frac{1}{n} \log_2 p(\mathbf{e}_0, \ldots, \mathbf{e}_{n-1}) = -\frac{1}{n} \sum_{k=0}^{n-1} T^k \log_2 p(\mathbf{e}_0|\mathfrak{F}_{-k}).$$

Here, two things are happening: for fixed $\mathbf{e}_0 = e_0$, $p(\mathbf{e}_0|\mathfrak{F}_{-n})$ has a limit $p(\mathbf{e}_0|\mathfrak{F}_{-\infty})$, by P. Lévy's Rule $4''$ from page 13 and so also for the actual $\mathbf{e}_0$ in hand:

$$\lim_{n \uparrow \infty} p(\mathbf{e}_0|\mathfrak{F}_{-n}) = p(\mathbf{e}_0|\mathfrak{F}_{-\infty}).$$

Besides, Birkhoff is working here, and you may hope that

$$\lim_{n \uparrow \infty} -\frac{1}{n} \log_2 p(\mathbf{e}_0, \ldots, \mathbf{e}_{n-1}) = \mathbb{E}\Big[-\log_2 p(\mathbf{e}_0|\mathfrak{F}_{-\infty})\big|\overline{\mathfrak{F}}\Big],$$

$\overline{\mathfrak{F}}$ being the field of shift-invariant events – and so it is. To begin with, for fixed e_0, $\mathbf{z}_n = -\log_2 p(\mathbf{e}_0|\mathfrak{F}_{-n})$ is a favorable game and also positive, as is $\mathbf{z}_n^2$, so

$$\mathbb{P}\Big[\max_{k \le n} \mathbf{z}_k > \lambda\Big] \le \frac{1}{\lambda^2} \mathbb{E}(\mathbf{z}_n^2) \quad \text{by Doob's inequality of } \S 1.8.$$

Now

$$\mathbb{E}(\mathbf{z}_n^2) = \sum_{j=0}^{\infty} \int_{j+1 \geq \mathbf{z}_n > j} \mathbf{z}_n^2 \, d\mathbb{P}$$

$$\leq \sum_{j=0}^{\infty} (j+1)^2 \, \mathbb{P}(\mathbf{z}_n > j)$$

$$= \sum_{j=0}^{\infty} (j+1)^2 \sideset{}{'}\sum p(e_0, e_{-1}, \ldots, e_{-n}),$$

in which the prime signifies that the sum is taken where

$$-\log_2 p(e_0|e_{-1}, \ldots, e_{-n}) > j.$$

There, $p(e_0, \ldots, e_{-n})$ is less than $2^{-j} p(e_{-1}, \ldots, e_{-n})$, so that the whole is not more than

$$\sum_{j=0}^{\infty} (j+1)^2 \sum_{e_0, \ldots, e_{-j}} 2^{-j} p(e_{-1}, \ldots, e_{-j})$$

$$= \sum_{j=0}^{\infty} (j+1)^2 2^{-j} \times \text{the number of letters } e_0.$$

That's Breiman's pretty estimate, from which $\mathbb{E}\left[\sup_{n \geq 1} \mathbf{z}_n^+\right] < \infty$ follows, justifying both the existence and the finiteness of $\lim_{n \uparrow \infty} \mathbf{z}_n$. But now, for any fixed $\ell \geq 1$ and $n \uparrow \infty$,

$$-\frac{1}{n} \log_2 p(\mathbf{e}_0, \ldots, \mathbf{e}_{n-1})$$

$$= \frac{1}{n} \sum_0^{n-1} T^k \left[\log_2 p(\mathbf{e}_0|\mathfrak{F}_{-k})\right]$$

$$\leq \frac{1}{n} \sum_0^{\ell-1} T^k \sup_{j \geq 0}\left[-\log_2 p(\mathbf{e}_0|\mathfrak{F}_{-j})\right]$$

$$+ \frac{1}{n} \sum_\ell^{n-1} T^k \left[-\log_2 p(\mathbf{e}_0|\mathfrak{F}_{-\infty})\right]$$

$$+ \frac{1}{n} \sum_\ell^{n-1} T^k \sup_{j \geq \ell}\left[-\log_2 p(\mathbf{e}_0|\mathfrak{F}_{-j}) + \log_2 p(\mathbf{e}_0|\mathfrak{F}_{-\infty})\right]$$

$$= (1') + (2') + (3'),$$

whence

$$\limsup_{n\uparrow\infty} -\frac{1}{n}\log_2 p(\mathbf{e}_0,\ldots,\mathbf{e}_{n-1})$$
$$\leq (1'') = 0$$
$$+(2'') = \mathbb{E}\big[-\log_2 p(\mathbf{e}_0|\mathfrak{F}_{-\infty})\big|\overline{\mathfrak{F}}\big] \quad \text{by Birkhoff}$$
$$+(3'') = \mathbb{E}\Big[\sup_{j\geq\ell}\big(\log_2 p(\mathbf{e}_0|\mathfrak{F}_{-j}) - \log_2 p(\mathbf{e}_0|\mathfrak{F}_{-\infty})\big)\big|\overline{\mathfrak{F}}\Big],$$

which is to say

$$\limsup_{n\uparrow\infty} -\frac{1}{n}\log_2 p(\mathbf{e}_0,\ldots,\mathbf{e}_{n-1}) \leq \mathbb{E}\big[-\log_2 p(\mathbf{e}_0|\mathfrak{F}_{-\infty})\big|\overline{\mathfrak{F}}\big] \equiv \overline{\mathbf{h}},$$

by making $\ell \uparrow \infty$ in $(3'')$. A similar bound from below will be obvious, with the stated conclusion:

$$\lim_{n\uparrow\infty} -\frac{1}{n}\log_2 p(\mathbf{e}_0,\ldots,\mathbf{e}_{n-1}) = \mathbb{E}\big[-\log_2 p(\mathbf{e}_0|\mathfrak{F}_{-\infty})\big|\overline{\mathfrak{F}}\big] = \overline{\mathbf{h}};$$

in particular

$$\overline{\mathbf{h}} = \mathbb{E}\big[-\log_2 p(\mathbf{e}_0|\mathfrak{F}_{-\infty})\big] = h$$

if the source is metrically transitive, and in *that* case, the "weak" form of the "individual" theorem just proved says that with overwhelming probability, $-\frac{1}{n}\log_2 p(\mathbf{e}_0,\ldots,\mathbf{e}_{n-1}) \simeq h$ for large n. Equivalently, if you look at long strings $\mathbf{e}_0\mathbf{e}_1\ldots\mathbf{e}_{n-1}$, then besides a little heap of small probability, you will see a (more or less) predictable number $N \simeq 2^{nh}$ of common or typical strings with (more or less) equal probabilities 2^{-nh}. It is this "equipartition" of probability that's the important thing, as you will see before the story's over.

Notice To make life simple, I suppose from now on that: (1) alphabets are finite; (2) the probabilities $p(e_0, e_1, \ldots, e_{n-1})$ are all positive; (3) the source is metrically transitive.

9.4 The noisy channel: capacity

The channel describes how communication is corrupted by noise. It is specified by the probabilities $p(b/a)$ that $b = (-b_{-2}, b_{-1}, b_0, b_1 -)$ is received, conditional on the knowledge that $a = (-a_{-2}, a_{-1}, a_0, a_1 -)$ was sent. These probabilities should be shift-invariant, i.e. $p(Tb/Ta) = p(b/a)$, and also non-anticipating, e.g. $p(b_0/a)$ should depend only upon the past $a_n : n \leq 0$, so that the channel works in, so to say, "real time".

Now hook up the channel to a source $\mathfrak{A}$ emitting messages a, coded or not, with probabilities $p(a)$; make up the joint probabilities $p(a \& b) = p(b/a)p(a)$; and sum (or should I say integrate) over a to produce shift-invariant probabilities $p(b)$ for the received communication b. Then you have three "sources" $\mathfrak{A}$ (in), $\mathfrak{B}$ (out), and the joint system $\mathfrak{A} \& \mathfrak{B}$, each with its private information production rate h, and from the addition Rule 3 of §9.1, you see that

$$h(\mathfrak{A}) + h(\mathfrak{B}/\mathfrak{A}) = h(\mathfrak{A} \& \mathfrak{B}) = h(\mathfrak{B}) + h(\mathfrak{A}/\mathfrak{B}),$$

or, what is now more to the point,

$$h(\mathfrak{A}) - h(\mathfrak{A}/\mathfrak{B}) = h(\mathfrak{A}) + h(\mathfrak{B}) - h(\mathfrak{A} \& \mathfrak{B}) = h(\mathfrak{B}) - h(\mathfrak{B}/\mathfrak{A}).$$

This little identity is indispensable to the understanding of what's going on.

What the *receiver* sees in the quantity $h(\mathfrak{A}) - h(\mathfrak{A}/\mathfrak{B})$ is the amount of information per sent letter *minus* the uncertainty or "equivocation" per letter, knowing what was received, i.e. it is *the* degree of certainty per letter that **a** was sent when **b** is received.

What the *sender* sees in the equivalent quantity $h(\mathfrak{B}) - h(\mathfrak{B}/\mathfrak{A})$ is the amount of information per received letter *minus* the uncertainty or equivocation per letter, knowing what was sent, i.e. it is the degree of certainty per letter that the reciept of **b** signifies that **a** was sent.

Either way, the number $H = h(\mathfrak{A}) - h(\mathfrak{A}/\mathfrak{B}) = h(\mathfrak{B}) - h(\mathfrak{B}/\mathfrak{A})$ is the rate per letter at which information is faithfully communicated by the channel, and if you believe all this, it is natural to declare the maximum (or, more precisely, the supremum) of this common number H, taken over some realistic class of sources, to be the "capacity" C of the channel. It's the "size of the pipe" vaguely introduced in §9.1.

Of course, at this point, "capacity" is just a word. It must be shown to have a practical significance, and it is the outstanding insight of Shannon [1948] that it does, *viz.*, for a wide variety of channels and sources, if you want faithfully to convey information over the channel at a rate $H < C$, you can do so by suitable coding so as to make the probability of error per letter as small as you like – but if you want to do it at a rate $H > C$, then the error cannot be so reduced no matter what mode of coding you may try. That's the meaning of C.

9.4.1 Simplest example

This is the "binary symmetric channel". The two alphabets (in and out) are the same, just ± 1; there is no memory, i.e. the letters are sent independently; and the letter by letter action of the channel is to transmit correctly with probabilities $p(+1/+1) = p(-1/-1) = p > 1/2$, and to mess up with the complementary probabilities $p(+1/-1) = p(-1/+1) = 1 - p < 1/2$. Obviously, if p is less than $1/2$, mostly garbage comes out, and the channel is useless.

To find the capacity, think about it this way. The noise may be represented by (two-sided) independent Bernoulli-type trials $\mathbf{e} = \pm 1$ with success probability $\mathbb{P}(\mathbf{e} = +1) = p$, acting letter by letter on the input $\mathbf{a}$ to produce the output $\mathbf{b}$, as per $\mathbf{b}_n = \mathbf{e}_n \mathbf{a}_n$. Now the conditional rate

$$h(\mathfrak{B}/\mathfrak{A}) = h(\mathbf{b}_0/\mathbf{a}_0) = H(\mathbf{e}_0) = H_2(p, 1 - p)$$

has to do with the channel only, not the source, and $h(\mathfrak{B}) \leq H(\mathbf{b}_0) \leq 1$, so C is not more than $1 - H_2(p, 1 - p)$.

But also, if the $\mathbf{a}$s are independent *honest* Bernoulli trials, then so are the $\mathbf{b}$s, as you will check, and $H(\mathfrak{B}) = H(\mathbf{b}_0) = 1$, i.e. $h(\mathfrak{B}) - h(\mathfrak{B}/\mathfrak{A}) = 1 - H_2(p, 1 - p)$ is the capacity C seen in Figure 9.4.1. This looks reason-

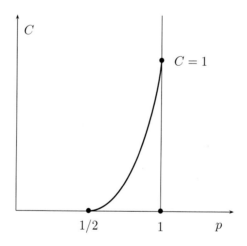

Figure 9.4.1

able: communication is perfect when $p = 1$, but when $p = 1/2$, you have $p(b_0/a_0) = 1/2$ independently of a_0, i.e. input and output are entirely independent, and the channel conveys no reliable information at all.

This simple channel will be reconsidered in §9.8, below, where the "strong converse" of Shannon [1948] and Elias [1961] is proved, *to wit*, with near optimal encoding of long messages of length n, the mean probability $f(n, H)$ of *wrong* decoding can be made $\leq 2^{-\alpha n}$ with a suitable

$\alpha > 0$ if $H < C$; contrariwise, if $H > C$, the mean probability $1 - f(n, H)$ of *correct* decoding is $\leq 2^{-\beta n}$ with some other number β, no matter what code books you employ. In short, the best performance is nearly perfect when $H < C$, but when $H > C$, only garbage comes out. That's pretty striking, don't you think? Here, the first part seems paradoxical: the longer the message, the better it works. That's because the encoding has to handle only 2^{nH} inputs more or less, and since $H < 1$, these are more and more rare compared to the 2^n possible strings of length n, so that they can be matched (and so faithfully decoded) to carefully chosen surrogates, so different from one another as to be easily recognized by the receiver even in a noisy disguise.

9.5 The noisy channel: coding

To make life still simpler from this place on, let the channel receive and emit the letters ± 1 only, and let the source emit letters e of any kind. Then it is necessary to *en*code the input $\mathbf{e}_0 \mathbf{e}_1 \ldots \mathbf{e}_{n-1}$ so the channel can receive it. This could be done by noiseless coding as in §9.1, but here's a better way. The source is metrically transitive, so by McMillan's theorem,

$$\mathbb{P}\left[\lim_{n\uparrow\infty} -\frac{1}{n}\log_2 p(\mathbf{e}_0, \mathbf{e}_1, \ldots, \mathbf{e}_{n-1}) = \text{the rate } h \text{ of the source}\right] = 1,$$

and if n is large, what you see in practice (i.e. with overwhelming probability) is about $N = 2^{nh}$ "common" strings of length n with more or less equal probabilities 2^{-nh}. These can be matched (*en*coding) to channel-adapted strings $\mathbf{a} = (\mathbf{a}_0, \mathbf{a}_1, \ldots, \mathbf{a}_{\ell-1})$ of length ℓ as soon as the count is right, i.e. as soon as $2^\ell > 2^{nh}$, and conveyed by the channel to the receiver, but imperfectly, as noisy strings $\mathbf{b} = (\mathbf{b}_0, \mathbf{b}_1, \ldots, \mathbf{b}_{\ell-1})$, and it is required to *de*code them, i.e. to decide, correctly one hopes, which string $\mathbf{a}$ was really sent. This is done by sorting out the 2^ℓ possible outcomes $\mathbf{b}$ into $2^{nh} < 2^\ell$ disjoint boxes B, one such to each input $\mathbf{a}$, declaring, rightly or wrongly, that $\mathbf{a}$ was sent if $\mathbf{b} \in B$.

This association of box to input is the "code book", and what is wanted, by judicious choice of this book, is faithfully to convey information as fast as you can, or, more realistically to keep the empirical rate $H = \log_2 N/\ell$ at which information is presented to the channel *high* and the mean probability $f = \sum_b p(b)[1 - p(a/b)]$ of mistakes *low*. This can be done for a wide variety of channels if $H < C$, but not otherwise:

when $H > C$ then f lies above the irreducible error $1 - C/H$, no matter what the channel or the code may be.

9.6 Communication when $H > C$

Now comes the pretty inequality $f > 1 - C/H$ of Fano [1961], just alluded to, expressing the impossibility of near error-free communication at empirical rates $H = \log_2 N/\ell$ above the capacity C.

Proof. The code book assigns to each received $b = (b_0, b_1, \dots, b_{\ell-1})$ an input $a = (a_0, a_1, \dots, a_{\ell-1})$. Let this be done in any way, and let f be the *mean* probability that a is *not* the string that was sent, i.e. $f = p(b)\big[1 - p(a/b)\big]$ summed over b. The key step is to bound $H(\mathbf{a}/\mathbf{b})$ in terms of f and N. Look at Figure 9.6.1: conditional on b, the experiment of deciding

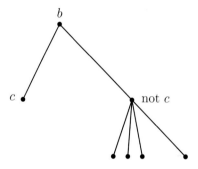

Figure 9.6.1

which a was sent is divided into two questions. Was it c? Or, if not, which $a \neq c$ was it? Now in computing $H(\mathbf{a}/\mathbf{b})$, it makes no difference how the as are listed for any particular b. Here, c has priority, so you my write

$$H(\mathbf{a}/\mathbf{b})$$
$$= \sum_b p(b) H_2\big[p(c/b), 1 - p(c/b)\big] \tag{1}$$
$$+ \sum_b p(b)\big[1 - p(c/b)\big] \times \Big[-\sum_{a \neq c} \frac{p(a/b)}{1 - p(c/b)} \log_2 \frac{p(a/b)}{1 - p(c/b)} \Big] \tag{2}$$

in which (1) is not more than $H_2(f, 1-f)$ by the concavity of $H_2(p, 1-p)$, and (2) is over-estimated by $\sum_b p(b)[1 - p(c/b)] = f$ multiplied by $\log_2 N$.

But also $H(\mathbf{a}) = H(\mathbf{a}_0, \ldots, \mathbf{a}_{n-1}) \simeq nh = \log_2 N$, so by definition of capacity, you see that for large n, and so also large ℓ,

$$\ell C \geq H(\mathbf{a}) - H(\mathbf{a}/\mathbf{b}) \geq \log_2 N - H_2(f, 1 - f) - f \log_2 N,$$

or very nearly so, i.e., neglecting various small errors,

$$H = \frac{\log_2 N}{\ell} \leq \frac{C}{1 - f} \text{ or what is the same, } f \geq 1 - \frac{C}{H}.$$

Summary

This is so important, that I spell it out once more: N is the number of more or less equally likely "common" inputs distinguished by McMillan's theorem, and $H = \log_2 N/\ell$ is the rate (per letter) at which information is presented to the channel. Fano's inequality, so interpreted, states that in the absence of any appreciable error, the rate cannot exceed the capacity, or contrariwise, that information cannot be conveyed at a rate $H > C$ without a probability of error at least $1 - C/H$.

9.7 Communication when $H < C$

It is still to be proved that near error-free communication is possible when the empirical rate $H = \log_2 N/\ell$ is less than the capacity C. *This cannot be true in general.*

Take a channel (no. 1) with probabilities $p(b/a)$ for inputs a and outputs b of length ℓ; make a second channel (no. 2) with probabilities $p(-b/a)$; and let some person unknown flip an honest coin to decide which channel to put into use. For channel no. 1, a good code book says (and it's mostly right) that $c(b)$ was sent when b is received. For channel no. 2, $c(-b)$ is the smart choice. But if you don't know which channel is in use, how do you choose: $c(b)$ or $c(-b)$? And if $c(-b) = -c(b)$ which is reasonable, won't you be wrong about half the time? That's the idea anyhow.

Clearly, this type of thing cannot occur if the output is metrically transitive, and that can be achieved by requiring of the channel an extra lack of memory. Roughly speaking, it is enough to suppose that if the number m is large enough then:

(1) the future $\mathbf{b}_n : n \geq 0$ be more or less independent of the *remote* past $\mathbf{a}_n : n < -m$, and

(2) two strings of output, separated by m letters or more, be independent, conditional on the input.

Neither (1) nor (2) has much to do with day-to-day communication: m

can be as large as you want, and we are talking, not about Sappho, say, so long ago, but about daily communication of information *now*, and what have (1) and (2) to do with that? What they *do* is to guarantee that the whole system $\mathfrak{A} \& \mathfrak{B}$ inherits the transitivity of the source. The proof is routine, so let's just assume it is so and pass on to more important things, *to wit*, to the construction of a good code book following Feinstein [1954].

Step 1. Hook up to the channel to a (transitive) source $\mathfrak{A}$ with empirical rate $H = \log_2 N/\ell$. By the transitivity of $\mathfrak{A} \& \mathfrak{B}$ and three applications of McMillan's theorem,

$$p(\mathbf{a}_0 \dots \mathbf{a}_{\ell-1}) \simeq 2^{-\ell h(\mathfrak{A})},$$

$$p(\mathbf{a}_0\mathbf{b}_0 \dots \mathbf{a}_{\ell-1}\mathbf{b}_{\ell-1}) \simeq 2^{-\ell h(\mathfrak{A} \& \mathfrak{B})},$$

$$\text{and} \quad p(\mathbf{b}_0 \dots \mathbf{b}_{\ell-1}) \simeq 2^{-\ell h(\mathfrak{B})},$$

with overwhelming probability if ℓ is large, and so also

$$\frac{p(\mathbf{a}_0\mathbf{b}_0 \dots \mathbf{a}_{\ell-1}\mathbf{b}_{\ell-1})}{p(\mathbf{a}_0 \dots \mathbf{a}_{\ell-1})p(\mathbf{b}_0 \dots \mathbf{b}_{\ell-1})} \simeq 2^{\ell[h(\mathfrak{A})+h(\mathfrak{B})-h(\mathfrak{A} \& \mathfrak{B})]}.$$

Note that any new source with rate $H' > H$ has $N' = 2^{\ell H'} > N = 2^{\ell H}$ common strings of length ℓ, so the old source can be matched to the new with something left over, i.e. you may suppose $h(\mathfrak{A})+h(\mathfrak{B})-h(\mathfrak{A} \& \mathfrak{B}) < C$ to be larger than some number $H' > H$, in which case

$$\frac{p(\mathbf{ab})}{p(\mathbf{a})p(\mathbf{b})} = \frac{p(\mathbf{a}_0\mathbf{b}_0 \dots \mathbf{a}_{\ell-1}\mathbf{b}_{\ell-1})}{p(\mathbf{a}_0 \dots \mathbf{a}_{\ell-1})p(\mathbf{b}_0 \dots \mathbf{b}_{\ell-1})} > 2^{\ell H'}$$

with probability $> 1 - e/2$, say.

Step 2. Now assign to each common input $a = (a_0 \dots a_{\ell-1})$, the box B of all outputs $b = (b_0 \dots b_{\ell-1})$ which make $\log_2 [p(ab)/p(a)p(b)] > \ell H'$. Leaving aside a little heap of probability $< e/2$, this produces what is most likely an *over*lapping cover of *some* of the 2^{ℓ} possible outputs b from which a non-overlapping cover is to be made.

Step 3. For brevity, I now write a and b for strings, not letters. Then the sum $\sum p(a)p(B/a)$ taken over the common strings a is the probability that $\log_2 [p(ab)/p(a)p(b)] > \ell H'$, and as this is more than $1 - e/2$, $p(B/a)$ is more than $1 - e$ for some input a. Pick such an input a_1 and call the associated box B_1. Look next at the leftover pairs aB, and if $p(B - B_1/a) \le 1 - e$ for them all, *stop*. But if $p(B - B_1/a) > 1 - e$ for some leftover a, call it a_2, put $B_2 = B - B_1$, and *repeat*, looking now at the pairs aB *still* leftover. And if $p(B - B_1 - B_2/a) \le 1 - e$ for them all, *stop*. But if $p(B - B_1 - B_2/a) > 1 - e$ for some still leftover a, call it a_3 and put $B_3 = B - B_1 - B_2$. You see the plan. The procedure stops after $N' \leq 2^{\ell H'}$ steps since there only so many common inputs a.

These non-overlapping boxes B_1, B_2, etc. and their associated inputs a_1, a_2, etc. constitute the code book, with the addendum that if B_0 is the class of outputs not so covered, then $\mathbf{b} \in B_0$ is registered as a mistake: *input* $\mathbf{a}$ *unknown*. Keep in mind for Step 5 that $p(B/a) > 1 - e$ for each pair aB in the code book.

Step 4. is to see how big this number N' really is. That's easy now: if the input $\mathbf{a}$ is left over, i.e. not distinguished in the code book, then $p(B_0/a)$ is $\leq 1 - e$. And besides, for any of the (very likely overlapping) boxes B of Step 2, you have

$$p(b/a) > p(b)2^{\ell H'} \text{ for } b \in B \text{ and so also } p(B) < 2^{-\ell H'}p(B/a) \leq 2^{-\ell H'},$$

with the result that

$$1 - e/2 < \text{ the probability that } p(ab)/p(a)p(b) > 2^{\ell H'}$$
$$= \sum_a p(a)p(B/a)$$
$$< p(B_0) + p(B_1) + p(B_2) + \text{etc.}$$
$$< 1 - e + N'2^{-\ell H'},$$

showing that $N' > 2^{\ell H'} \times (e/2)$ is *much* bigger than $N = 2^{\ell H}$ for large ℓ. Good.

Step 5. Now you can match the N common strings a from the source to the first N strings a distinguished in the code book, present these to the channel with equal probabilities $p(a) = N^{-1}$, and decode correctly with probabilities $p(B/a) > 1 - e$. The probability $1 - f$ of correct decoding is then the sum over a of $p(a)p(B/a) > 1 - e$, which is to say $f < e$, and that's as small as you like.

Summary The estimate $N' > 2^{\ell H}$ shows that if $N(\ell, e)$ is the maximal number of inputs of length ℓ with decoding error $\leq e$, then

$$\liminf_{e \downarrow 0} \liminf_{\ell \uparrow \infty} \frac{1}{\ell} \log_2 N(\ell, e) \geq C.$$

But also, by Fano's inequality,

$$\limsup_{e \downarrow 0} \limsup_{\ell \uparrow \infty} \frac{1}{\ell} \log_2 N(\ell, e) \leq C,$$

so

$$\lim_{e \downarrow 0} \lim_{\ell \uparrow \infty} \frac{1}{\ell} \log_2 N(\ell, e) = C,$$

exhibiting the capacity in new and more instructive way.

Reservation   Naturally, all this is provisional. The *existence* of good codes is proven without any practical suggestion for *how to make one*. That's a whole art in itself, for which see, e.g. Cover–Thomas [1991].

9.8 The binary symmetric channel

This simplest of channels was already seen in §9.4. It sends the input $\mathbf{a}_n = \pm 1$ to the output $\mathbf{b}_n = \mathbf{e}_n \mathbf{a}_n$, the noise $\mathbf{e}_n = \pm 1$ being modeled by independent Bernoulli trials with success probability $p > 1/2$. Here, $m = 0$, and for inputs $\mathbf{a} = (\mathbf{a}_0, \ldots, \mathbf{a}_{n-1})$ and outputs $\mathbf{b} = (\mathbf{b}_0, \ldots, \mathbf{b}_{n-1})$ of length n, $p(\mathbf{b}/\mathbf{a}) = p^{\#}(1-p)^{n-\#}$ with

$$\# = \#(0 \le k < n : \mathbf{e}_k = +1)$$
$$= \text{the number of matches } \mathbf{b}_k = \mathbf{a}_k \text{ for } 0 \le k < n.$$

The capacity $C = 1 - H_2(p, 1-p)$ was computed before. Now it is desired to send $N < 2^n$ equally likely strings $\mathbf{a}$ of length n through the channel at the empirical rate $H = \frac{1}{n} \log_2 N$, i.e. with $N = 2^{nH}$, and to decode the output $\mathbf{b}$ upon receipt with the smallest possible mean probability of error $f \equiv f(n, H)$. What type of decoding is to be used will be explained shortly, but first the facts: for large n and suitable positive numbers α and β, $f(n, H) < 2^{-\alpha n}$ if $H < C$, but if $H > C$, then $f(n, H) > 1 - 2^{-\beta n}$, making more precise what was just said in §9.7. Namely, that if $H < C$, nearly perfect communication can be achieved, but if $H > C$, mostly garbage comes out, more and worse the longer it goes on. This remarkable dichotomy is due to Shannon [1948] with refinements by Elias [1961]; see Wolfowitz [1978] for the full treatment. The present proof is slightly simplified by use of Cramér's nice estimate for Bernoulli trials from §2.4.1.

9.8.1 Shannon's idea for $H < C$

Instead of looking for a *single* good code, how about showing that the *average* of $f(n, H)$ over a restricted class of codes is small? Then there must be at least *one* good code, and indeed many. Fix $H < C$, pick $N = 2^{nH}$ inputs of length n, let the distance $d(a, b)$ be the number of mismatches $a_k \ne b_k$ for $0 \le k < n$, and decode according this rule: upon receipt of b declare that a was sent if $d(a, b) < d(a', b)$ for each of the other $N - 1$ chosen inputs $a' \ne a$, with the understanding that an error is recorded if the rule is ambiguous, i.e. if there are 2 or more inputs *a* closest to *b*. Now $H < C < 1$, so $N = 2^{nH}$ is much smaller than $2^n =$ the number of all possible strings of length n, so there are $\binom{2^n}{N}$ ways of selecting N equally likely inputs, and so also an equal number of codes of this type. The codes themselves are now regarded as equally likely, the average probability of correct decoding being

$$1 - f(n, H) = \binom{2^n}{N}^{-1} \sum_{\text{codes}} \sum_{a \in \text{code}} \sum_{b: d(a,b) < d(a',b)} [p(ab) = p(b/a)p(a)]$$

with $p(b/a) = p^{n-d}(1-p)^d$, $d = d(a, b)$ being the number of mismatches and $p(a) = N^{-1}$. In detail,

$$1 - f = \frac{1}{N}\binom{2^n}{N}^{-1} \sum_{\text{codes}} \sum_{a \in \text{code}} \sum_{d=0}^{n} \sum_{\substack{b: d = d(a,b) \\ < d(a',b)}} p^{n-d}(1-p)^d$$

$$= \frac{1}{N}\binom{2^n}{N}^{-1} \sum_{d=0}^{n} p^{n-d}(1-p)^d$$

$\times$ the sum, over a and b with $d(a, b) = d$, of the number of codes a in which b is unambiguous

$$= \frac{1}{N}\binom{2^n}{N}^{-1} \sum_{d=0}^{n} p^{n-d}(1-p)^d$$

$\times 2^n$ \quad choices for $a \in$ code

$\times \binom{n}{d}$ \quad choices for b at distance d from a

$\times \binom{2^n - V}{N - 1}$

where $V = V(d)$ is the number of strings a at distance $\le d$ from b, so that $2^n - V$ is the number of candidates from which the $N - 1$ strings $a' \ne a$ must be selected to complete the code book – and it is to be understood that the last factor (line 7) vanishes if $N - 1 > 2^n - V$. To spell it out,

$$1 - f = \sum_{d=0}^{n} \binom{n}{d} p^{n-d}(1-p)^d \frac{(2^n - N)!(2^n - V)!}{2^n - 1)!(2^n - V - N + 1)!}$$

$$= \sum_{d=0}^{n} \binom{n}{d} p^{n-d}(1-p)^d \prod_{M=1}^{N-1} \left(1 - \frac{V - 1}{2^n - M}\right),$$

or, what is the same,

$$f = \sum_{d=0}^{n} \binom{n}{d} p^{n-d}(1-p)^d \times \left[1 - \prod_{M=1}^{N-1} \left(1 - \frac{V-1}{2^n - M} \right) \right],$$

with the inherited convention that the product vanishes if $2^n - N + 1$ is $\leq V - 1$. It is this number f which ought to be small for large n.

To see that this is true, divide the sum in two, according as $d > d'$ or $d \leq d'$, with the following choice of d'. Pick H' a little bigger than $H < C$, keeping $1 - H' > 1 - C = H_2(p, 1 - p)$. Then you may write $1 - H' = H_2(p', 1 - p')$ with $1/2 < p' < p$ and take $d' = n(1 - p')$, ignoring what's more than likely, that this is not a whole number. Now the part of the sum for $d > d'$ is easy to estimate. It is not more than

$$\sum_{d>d'} \binom{n}{d} p^{n-d}(1-p)^d$$

$$= \left(\begin{array}{l} \text{the probability of more than } n(1 - p') \text{ failures in } n \\ \text{independent Bernoulli trials with failure probability } 1 - p \end{array} \right)$$

$$\simeq 2^{-nI} \quad \text{with } I = p' \log_2 \frac{p'}{p} + (1 - p') \log_2 \frac{1 - p'}{1 - p}$$

by Cramér's estimate of §2.4.1. That's the half of it. As to the rest,

$$\sum_{d \leq d'} \binom{n}{d} p^{n-d}(1-p)^d \times \left[1 - \prod_{M=1}^{N-1} \left(1 - \frac{V(d) - 1}{2^n - M} \right) \right]$$

$$< 1 - \left(1 - \max_{\substack{d \leq d' \\ M < N}} \frac{V(d) - 1}{2^n - M} \right)^N$$

$$< 1 - \left(1 - \frac{V(d')}{2^n - N} \right)^N$$

$$\simeq 1 - \left(1 - \frac{V(d')}{2^n} \right)^N \qquad \text{since } N = 2^{nH} \text{ is small compared to } 2^N$$

$$\leq N \times \frac{V(d')}{2^n} \qquad \begin{array}{l} \text{since } V(d') < 2^n \text{ and } 1 - (1 - x)^N \leq Nx \\ \text{for } 0 \leq x \leq 1 \end{array}$$

$$= 2^{nH} \times \begin{array}{l} \text{the probability of at least } n(1 - p') \text{ successes} \\ \text{in } n \text{ independent, honest Bernoulli trials} \end{array}$$

$$\simeq 2^{nH} \times 2^{-nI} \qquad \begin{array}{l} \text{with } I = p' \log_2 2p' + (1 - p') \log_2 2(1 - p') \\ \text{by Cramér's estimate} \end{array}$$

$$= 2^{-n(H'-H)} \qquad \text{since } I = 1 - H_2(p', 1 - p') = H'.$$

That does the trick.

9.8.2 Garbage out

Now take $H > C$ and think of any code you like, with $N = 2^{nH}$ equally likely inputs a of length n paired with disjoint boxes B covering the whole class of outputs $(\pm 1)^n$, the rule being "read a if $b \in B$". The mean probability $1 - f$ of correct decoding is now the sum, over the strings a distinguished in the code book, of

$$p(a \,\&\, B)$$

$$= p(a) \sum_{e: b = ea \in B} p(e)$$

$$= 2^{-nH}) \,\mathbb{P}(e \in aB)$$

$$\simeq 2^{-nH} \times 2^{-nH_2(p,1-p)} \times \text{the number of strings in } B,$$

by McMillan's theorem, and since the total number of strings b involved is just 2^n, you find

$$1 - f \simeq 2^{-nH} \times 2^{-nH_2(p,1-p)} \times 2^n = 2^{-n(H-C)}$$

in view of $C = 1 - H_2(p, 1 - p)$. In short, if you talk too fast, they won't understand *anything*!

10
Equilibrium Statistical Mechanics

J.C. Maxwell [1861] and J.W. Gibbs [1902] proposed a systematic recipe for how thermodynamics – descriptive of, e.g. a gas of like molecules in terms of pressure, volume, and temperature – may be deduced from the (over-detailed) Hamiltonian mechanics sketched in §8.1. Here, I will explain what Gibbs had in mind, illustrated by two fairly complicated examples of how it works in practice. The recipe is very general, and very successful, too, but Gibbs [1902] is modesty itself. I quote from his preface:

Here, there can be no mistake in regard to the agreement of the hypotheses with the facts of nature, for nothing is assumed in that respect. The only error into which one may fall is the want of agreement between the premises and the conclusions, and this, with care, one may hope, in the main, to avoid.

Fermi [1956] is recommended for an elementary introduction to the thermodynamics, beautifully done; then Thompson [1972] for the statistical mechanics; and, naturally, Feynman [1963/5] for either or both. Incidentally, there is a lovely biography of Gibbs by M. Rukeyser [1942] – highly recommended.

10.1 What Gibbs said

I derive the *canonical density* $\Gamma' = Z^{-1}e^{-H/kT}$ of §8.2 in Gibbs's style and indicate how thermodynamics comes out of that.

10.1.1 Phase space and energy

Gibbs begins with a large number (N) of indistinguishable, let us say hard, spherical molecules, confined in a vessel $V \subset \mathbb{R}^3$, with centers

317

Q_1, Q_2, etc. $\in V$ and momenta P_1, P_2, etc. $\in \mathbb{R}^3$, jointly specifying a point QP in the $(6N)$-dimensional phase space $V^N \times \mathbb{R}^{3N}$. The associated total *energy* or *Hamiltonian* is commonly the sum of kinetic energy (of motion) + potential energy (of position) of the form

$$H(Q, P) = \frac{1}{2m} \sum_{k=1}^{N} P_k^2 + \sum_{1 \leq i < j \leq N} U(Q_i - Q_j),$$

in which m is the mass of a single molecule and U is a radial function of the general shape seen in Figure 10.1.1; the corresponding force $-\operatorname{grad} U$ is partly repulsive ($U' < 0$) and partly attractive ($U' > 0$), and a hard core ($U' = +\infty$) is seen below the molecular diameter d. The

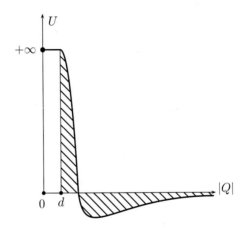

Figure 10.1.1

molecules move according to Hamilton's equations $\dot{Q} = \partial H / \partial P$ and $\dot{P} = -\partial H / \partial Q$ with, e.g. perfect reflection at the surface of the vessel. The conservation of energy ($\dot{H} = 0$) is immediate; also, the vector field $(QP) \rightarrow (\dot{Q}\dot{P}) = (\partial H / \partial P, -\partial H / \partial Q)$ is divergence-free, so the flow in phase space is incompressible, i.e. it is volume-preserving.

10.1.2 The microcanonical ensemble

Now it is natural to think of the energy as fixed, restricting the flow to the (invariant) submanifold or surface of constant energy where $H = h$, say, and to note, with Gibbs, that the volume element, $|\operatorname{grad} H|^{-1} \times \operatorname{d} \text{area}$, inherited from the invariant volume element $\operatorname{d}Q \operatorname{d}P$ in the ambient space is invariant, too.

The proof is easy. Take a little patch on the surface and look at the "pill box" seen in Figure 10.1.2, produced by moving the patch in the

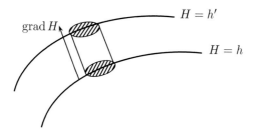

Figure 10.1.2

normal direction $\operatorname{grad} H$ up to the nearby surface $(H = h')$ with h' just a bit bigger than h. Then $h' - h \simeq |\operatorname{grad} H| \times$ the height of the pill box is invariant, and since volume = area $\times$ height is invariant, too, so also is their ratio: area divided by $|\operatorname{grad} H|$.

Gibbs now declares that, lacking any further information, the statistical equilibrium of the gas means that the phase QP is distributed over the surface as per the *microcanonical ensemble* with probabilities $d\Gamma = \dfrac{1}{Z} \dfrac{d\,\text{area}}{|\operatorname{grad} H|}$, Z being the necessary normalizer. Take it on faith, $Z < \infty$ included.

10.1.3 Heat bath and the canonical ensemble

Gibbs's next step is to explain what it means for the gas to be in equilibrium with a *heat bath* at *temperature* T. The bath is modeled by a large number N' of like molecules of mass 1, say, with no interaction whatsoever (ideal gas), i.e. with $H_{\text{bath}} = \frac{1}{2} \sum (P_k')^2$, and the joint system with Hamiltonian $H = H_{\text{bath}} + H_{\text{gas}}$ is placed in equilibrium in *its* microcanonical ensemble for fixed $H = h$. Gibbs now computes the reduced distribution for the gas alone by integrating out the variables $Q'P'$ pertaining to the bath as follows. For the big microcanonical ensemble, write

$$d\Gamma = \lim_{h' \downarrow h} \frac{dQ'\,dP'\,dQ\,dP \text{ restricted to the shell } h \leq H < h'}{\text{the volume of this shell}},$$

fix the variables QP pertaining to the gas, and integrate out the rest. Now $H_{\text{bath}} \simeq h - H_{\text{gas}}$ is effectively fixed so that P' is confined to the sphere of radius $\sqrt{2(h - H_{\text{gas}})}$ in $\mathbb{R}^{3N'}$, and up to factors of proportionality, you find

$$\int_{\text{bath only}} d\Gamma \simeq \int_{|P'|=\sqrt{2(h-H_{\text{gas}})}} dP' \times dQ\, dP$$

$$= \frac{(h - H_{\text{gas}})^{(3N'-1)/2}}{h^{(3N'-1)/2}} \times dQ\, dP$$

$$\simeq \exp\left[-H_{\text{gas}} \times \frac{3}{2}\frac{N'}{h}\right] dQ\, dP$$

if N' is large and h is more or less proportional to it, as it is perfectly reasonable to suppose. Indeed, if you take $h = \frac{3}{2}N' \times kT$ and make $N' \uparrow \infty$, then you find the reduced probabilities $d\Gamma = Z^{-1}e^{-H/kT}\, dQ\, dP$, in which $Z = \int e^{-H/kT}$ is the *partition function*, so-called, T is identified with Kelvin's absolute temperature, and $k \simeq 1.38 \times 10^{-16} \times (\text{cm})^2(\text{sec})^{-2}(\text{deg})^{-1}$ is Boltzmann's constant to get the numbers and the dimensions right; compare §5.9.4 on Maxwell's distribution. This is Gibbs' *canonical ensemble*, descriptive of the gas *in equilibrium with an infinite heat bath at temperature T*. This Gibbsian derivation is lovely, but see also Feynman [1964] for something short and sweet, using only the simplest ideas in his inimitable way.

Note 10.1.1 T is determined by the mean energy[1] $U = \int H\, d\Gamma$, in conformity with Count Rumford's dictum: that molecular motion is the mechanical equivalent of heat/temperature: in fact,

$$kT^2 \partial U/\partial T = \partial U/\partial(-1/kT) = \int H^2\, d\Gamma - \left(\int H\, d\Gamma\right)^2$$

is positive, which does the trick.

Note 10.1.2   Recall from §8.2 that the canonical density $\Gamma' = \frac{1}{Z}e^{-H/kT}$ maximizes entropy for fixed U and so also fixed T, i.e. it is the most random way of distributing the phase QP, absent further information. This confirmed Gibbs in his way of thinking.

10.1.4 Large volume

Gibbs now takes one step more. It is the *large volume limit* taken for fixed *density*[2] $\varrho = N/V$ as $N \uparrow \infty$ and V expands to fill the whole 3-dimensional space, in the hope that the particular shape of the vessel (a

---

[1] Forgive the temporary abuse of notation. The present letter U is conventional here.

[2] V now stands for the vessel or its volume, as the context requires.

cube, a sphere, or any reasonable figure) will not affect the outcome, and
that simplifications will obtain, suppressing unimportant details in favor
of a (thermodynamical) *bulk* description, as experience with LLN, CLT,
large deviations, etc. suggests. Here, an indispensable adjustment is
needed: the molecules being indistinguishable, to permute them changes
nothing. For this reason, Z is divided by an $N!$. Then $\lim_{N \uparrow \infty} \sqrt[N]{Z/N!} =$
$\mathfrak{Z}$ is taken – indeed, $\mathfrak{Z}$ does not exist without this adjustment, as will be
illustrated in §10.2. The whole thermodynamics of the gas is now hidden
in $\mathfrak{Z}$, epitomized by the *equation of state*, $p = kT\partial \log \mathfrak{Z}/\partial v$, expressing
$p =$ pressure as a function of the *temperature* T and the *specific volume*
$v = 1/\varrho$. This relation is the hoped for simplification that large numbers
produce. But why is $kT\partial \log \mathfrak{Z}/\partial v$ called pressure? Gibbs says only that
p is *analogous* to pressure, being, himself, too modest to say more. The
real justification is in the applications as will be seen presently, but first
a preliminary demonstration via the notion of free energy.

10.1.5 Thermodynamics: free energy

The idea is a commonplace, though you may not have thought about
it much. In Nature, to do real mechanical work requires a self-evident
amount of mechanical energy and *something more*. Any "engine" capable of converting energy into work is to some degree *in*efficient, losing
energy in the form of heat which is then *un*available. For this reason, the
conventional conservation of energy is not quite the right idea. What's
wanted is the *free* energy that *is* available.

A radically simplified illustration will serve to introduce the idea. A
mass ($\bullet$) moving in one dimension is attached to the origin ($\circ$) a Hooke's
spring, exerting a restoring force $(-k) \times$ the displacement Q. Gravity is

Figure 10.1.3

ignored, so Newton's $f = ma$ says that $m\,\mathrm{d}^2Q/\,\mathrm{d}t^2 = -kQ$. The mechanical energy $H = \frac{1}{2}m(\dot{Q})^2 + \frac{1}{2}kQ^2$ is then a constant of motion, and
$Q(t) = Q(0)\cos\omega t + \dot{Q}(0)\omega^{-1}\sin\omega t$ with $\omega = \sqrt{k/m}$. Now put in a
little dissipation, introducing an extra force $-f\,\mathrm{d}Q/\,\mathrm{d}t$, f for friction
if you will. Then $m\,\mathrm{d}^2Q/\,\mathrm{d}t^2 + f\,\mathrm{d}Q/\,\mathrm{d}t + kQ = 0$, the oscillation is
damped, dying out as $t \uparrow \infty$, and the energy H available at time $t > 0$

is the energy available at time $t = 0$ *less* the energy $f \int_0^t (\dot{Q})^2 \, dt'$ made *un*available due to dissipation, *aka* the *free* energy still available to do an honest day's work, as you will please check.

Naturally, the real thing is much more complicated. Gibbs writes $3 = e^{-F/kT}$. Roughly speaking, $Z/N! = e^{-NF/kT}$ if N is large, exhibiting

$$F = \frac{1}{N} \times -kT \log \int \frac{e^{-H/kT}}{N!} \, d\,\text{vol}$$

as a sort of *geometrical* mean value of H *per* molecule. It is the *Helmholtz free energy*, so-called, and plays a leading role in, e.g., the equation of state which may now be written $p = -\partial F/\partial v$. The reason for the adjective *free* is not far off. The system of molecules (call it a gas or what you will) is *mechanically* isolated from the rest of the world due to its confinement to the vessel, but it is not *thermally* isolated as it may exchange, or better to say *lose* heat, *aka* energy due to an incessant thermal agitation of the molecules. This is where entropy and the Second Law of Thermodynamics come in. The total entropy of the gas (S) is adjusted by subtracting the entropy $(\log N!)$ for the labeling of the *un*labelled molecules – after all, labeling is only an artefact of the description, not any thing in Nature. Then, still roughly speaking,

$$S = -\int \Gamma' \log \Gamma' \, d\,\text{vol} - \log N!$$

$$= \int \left(\frac{H}{kT} + \log Z \right) d\Gamma - \log N!$$

$$= \frac{NU}{kT} - \frac{NF}{kT},$$

U being the *arithmetical* mean value of H *per* molecule. Now replace S by $N \times (S =$ entropy per molecule) to obtain $F = U - kST$ in the infinite volume limit.[3] But what does *that* mean? S is determined only up an additive constant. There is no natural zero of entropy unless it be at Kelvin's absolute zero of temperature where all disorderly molecular agitation is stilled, so it is only the *change* of entropy, from one molecular configuration to another that has true meaning. Now the Second Law of Thermodynamics says that if Q is the amount of heat in the gas, then such a change (ΔS) is the integral of the differential dQ/T taken over any path in configuration space leading from the first configuration to

---

[3] Forgive the abuse of notation: U for mean energy and presently Q for heat. It is conventional and only for the moment.

the second, provided this is done *slowly*, little by little, to permit step-wise equilibration along the way, and it is self-evident that if this is done at constant temperature, then $\Delta S = \Delta Q/T$, which may be reconciled with $F = U - kST$ only if $\Delta F = \Delta U - k\Delta Q$ where Q is now heat per molecule. In short, counting *everything* per molecule, the change in free energy (ΔF) is the change in mechanical energy (ΔU) *less* the energy $(k\Delta Q)$ made *un*available for real work due to thermal agitation, correcting in this way the naive conception of the conservation of energy. That's the story. I say no more about it, but see Fermi [1956:77–79] for a better telling of it and for suggestions of a close affinity between this Gibbsian picture and the format for large deviations described in §§2.4 and 7.7 which you may wish to puzzle out.

10.1.6 Thermodynamics: pressure

Take N large but fixed. Then a small enlargement of the vessel $V \to V + \Delta V$, introduces a small change in the free energy, as in

$$\frac{-N\Delta F}{kT} \simeq \Delta \log Z \simeq \frac{\Delta \int e^{-H/kT}}{\int e^{-H/kT}} \simeq N \frac{\int_{\text{restricted}} e^{-H/kT}}{\int_{\text{unrestricted}} e^{-H/kT}},$$

in which the restricted integral is taken with any one molecule confined to the shell ΔV, i.e.

$$-\Delta F \simeq kT \int_{\text{restricted}} e^{-H/kT} \text{ divided by } \int_{\text{unrestricted}} e^{-H/kT}$$

$$= kT \times N^{-1} \times \text{ the expected number of molecules in the shell.}$$

This is now divided by the change $\Delta v = \Delta V/N$ in the specific volume to obtain

$$-\frac{\Delta F}{\Delta v} \simeq kT \times \frac{\text{the expected number of molecules in the shell}}{\text{the volume of the shell}},$$

and it is natural to say, with Gibbs, that the right-hand side is reminiscent of pressure so that, in the infinite volume limit where $\log \mathfrak{Z} = -F/kT$, you have

$$p = -\partial F/\partial v = kT \partial \log \mathfrak{Z}/\partial v.$$

In this way, everything is finally expressed in terms of p, T, $v = 1/\varrho$, and this equation of state which relates them, realizing the simplified (bulk) description that large numbers was supposed to produce.

Gibbs's recipe is then a systematic way of predicting the thermodynamics of the gas from the molecular picture. For example, it should now be possible to tell in what circumstances (of pressure, temperature and volume) the gas (e.g. steam) changes into a liquid (water), though to tell the truth, the computations required for any real gas are way beyond the capabilities of any numerical scheme, existing or projected. So for the *numbers*, observation is the only way, but for the *understanding*, Gibbs is *everything*.

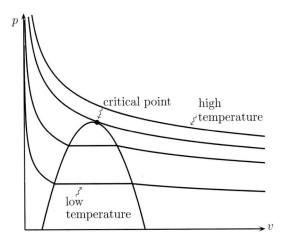

Figure 10.1.4

Figure 10.1.4 depicts several isotherms, i.e. $p(v) = p(v, T)$ plotted against v for a variety of fixed temperatures T. The change of phase from, e.g. steam to water, is indicated by a flat piece of the curve, appearing below some critical temperature T_0 = the boiling point. Above T_0, you see something like the ideal gas law $pv = \text{constant} \times T$; below T_0, you see condensation as the steam liquifies at constant pressure, as it does in Nature.

Part of this *empirical* diagram is already built into Gibbs' general recipe $p(v, T) = kT \partial \log 3/\partial v$, assuming that the particular shape of the vessel has no ultimate influence. For example, let V be a cube. Then Z increases with v for fixed N and T, so that $p(v) = kT \partial \log 3/\partial v$ is ≥ 0. It is also a general fact that $p(v)$ is decreasing in v. To see this, take 8 adjacent copies $V_1, \ldots, V_8$ of V, forming a big cube (= $8V$ by abuse of notation), place $\#(V_1) = n_1$ molecules in V_1, $\#(V_2) = n_2$ molecules in

V_2 etc., and write

$$Z(8N, 8V) = \frac{1}{(8N)!} \sum \frac{(8N)!}{n_1! n_2! \dots n_8!} \times$$

$$\times \int_{\substack{(8V)^{8N} \times \mathbb{R}^{24N} \\ \#(V_1)=n_1, \#(V_2)=n_2, \text{ etc.} \\ n_1+\dots+n_8=8N}} e^{-H/kT} \, dQ \, dP.$$

Here, $n_1! n_2! n_3! n_4!$ peaks when n_1, n_2, n_3, n_4 are all about $\frac{1}{4} \times (N_1 = n_1 + n_2 + n_3 + n_4)$, and similarly for $n_5! n_6! n_7! n_8!$, so

$$Z(8N, 8V) \gtrsim |Z(N_1, V)|^4 \times |Z(N_2, V)|^4 \text{ with } N_2 = n_5 + n_6 + n_7 + n_8,$$

from which you read off

$$\frac{\log Z(8N, 8V)}{8N} \gtrsim \frac{N_1}{2N} \frac{\log Z(N_1, V)}{N_1} + \frac{N_2}{2N} \log Z(N_2, V)$$

and so also

$$\log 3(\varrho) \geq \frac{1}{2} \frac{\varrho_1}{\varrho} \log 3(\varrho_1) + \frac{1}{2} \frac{\varrho_2}{\varrho} \log 3(\varrho_2)$$

by making $N \uparrow \infty$ with fixed densities

$$\varrho = \frac{8N}{8V} = \frac{N}{V} = \frac{1}{2} \frac{N_1}{V} + \frac{1}{2} \frac{N_2}{V} = \frac{1}{2} \varrho_1 + \frac{1}{2} \varrho_2.$$

This shows that $\varrho \log 3(\varrho)$ is a *concave* function of $\varrho = 1/v$, and switching back, from ϱ to v, you find

$$0 \geq \frac{\partial^2}{\partial \varrho^2} [\varrho \log 3(\varrho)] = v^2 \frac{\partial}{\partial v} v^2 \frac{\partial}{\partial v} \frac{1}{v} \log 3(v) = \frac{\partial^2}{\partial v^2} \log 3(v)$$

when you work it out: in short, $\partial p / \partial v = kT \partial^2 \log 3 / \partial v^2 \leq 0$. Note also for the record, that $F(v) = -kT \log 3(v)$ is *convex*.

OK so far. Now what? Mostly $p(v, T)$ is supremely difficult to compute though the recipe has a wide practical success. It *really works*, and I *can* show you a couple of beautiful examples of how it plays out, *to wit*, the van der Waals phase change, reminiscent of the change of steam to water seen in Figure 10.1.4 (§10.3), and the phase change occurring in the Ising model of a ferromagnet when disorder (no magnetization) gives way to order (positive magnetization) as the temperature T falls below a critical temperature T_0 (§10.4). But first it may be best to see what Gibbs is really saying in the very simplest cases.

Pause This notion, *change of phase*, is at the heart of the matter. The equation of state is the analogue of the law of large numbers, descriptive of the gas *in the bulk*, but what comes out is not a simple number any more. It describes, rather, the result of molecular *cooperation* (as opposed to independence) differing *in kind* at high and low temperature, and that's a whole new ball game.

10.2 Two simple examples

10.2.1 Ideal gas

The ideal gas is comprised of N indistinguishable molecules confined to a vessel $V \subset \mathbb{R}^3$. They do not interact at all (that's what *ideal* means) so the Hamiltonian is $H = P^2/2m$, and the partition function is simply[4]

$$Z = \frac{1}{N!}(V)^N \int_{\mathbb{R}^{3N}} e^{-P^2/2mkT}\, dP$$
$$= \frac{1}{N!}(V)^N (2\pi mkT)^{3N/2}.$$

Then, for fixed specific volume v, you find

$$\log Z \simeq -\left(N + \frac{1}{2}\right)\log N + N - \log\sqrt{2\pi} \quad \text{by Stirling}$$
$$+ \left[N\log V = N\log N + N\log v\right] \quad \text{since } V = Nv$$
$$+ \frac{3N}{2}\log(2\pi mkT),$$

showing, in lines 1 and 2, the indispensable role of $N!$, and providing, at infinite volume, the evaluation

$$\log 3 = 1 + \log v + \frac{3}{2}\log(2\pi kmT).$$

The pressure is now seen to be $p = kT/v$, i.e. $pv = kT$, or, what is more or less the same for large N, $pV = NkT$ – the conventional ideal gas law of Robert Boyle (1660), seen in Figure 10.2.1.

10.2.2 Hard balls

Now let the molecules be hard balls of diameter d, without interaction except that $|Q_i - Q_j|$ must be $\geq d$ for $i \neq j$. Now $H = P^2/2m$, plus ∞

---

[4] V stands for the vessel or its volume, as before.

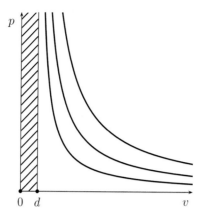

Figure 10.2.1

or 0 according as $|Q_i - Q_j| < d$ for some $i \neq j$ or no, and

$$Z = (2\pi mkT)^{3N/2} \times$$

$$\times V^N - \text{the excluded volume of } V^N \cap \left[\bigcup_{i<j} |Q_i - Q_j| < d \right].$$

Throw out the first factor which does not contribute to the pressure. The rest is independent of the temperature and looks simple but is *not*. In fact, it is not known how to compute the excluded volume for large N and fixed v except in the artificial case of dimension 1. Then the "molecules" are hard rods of length d, placed in a "vessel" $(Q : 0 \leq Q \leq V)$, and after removal of the irrelevant factor $(2\pi mkT)^{N/2}$, as for the ideal gas, you have

$$Z = \int_{\substack{0 \leq Q_1, \, Q_1 + d \leq Q_2, \\ Q_2 + d \leq Q_3, \dots, Q_N \leq V + d}} \mathrm{d}Q_1 \, \mathrm{d}Q_2 \cdots \mathrm{d}Q_N,$$

in which Q_k is the left end of the kth rod. Now change coordinates to $Q'_k = Q_k - (k-1)d$ so that

$$Z = \int_0^{V-Nd} \mathrm{d}Q'_N \int_0^{Q'_N} \mathrm{d}Q'_{N-1} \cdots \int_0^{Q'_2} \mathrm{d}Q'_1 = \frac{(V - Nd)^N}{N!} \quad \text{for } V > Nd.$$

Then with $v > d$, you find

$$\frac{1}{N} \log Z = \log(V - Nd) - \log N = \log(v - d) \quad \text{with } v = V/N$$

leading at once to the modified Boyle's law seen in Figure 10.2.1, *to wit* $p(v) \times (v - d) = kT$, which is to say $p(v) \times (V - Nd) \simeq NkT$ for large N.

Here, $v \simeq d$ signifies close packing, but no change of phase appears. In dimension 3, there could be a change of phase if the molecules got stuck in some rigid configuration that a little kick could suddenly reduce to close packing, but nobody really knows.

10.3 Van der Waals' gas law: dimension 1

The repulsive hard core – $U(Q) = +\infty$ for $|Q| \leq d$ – is now modified by an attractive tail $U(Q) = -ae^{-b|Q|}$ ($|Q| > d$) as in Figure 10.3.1.

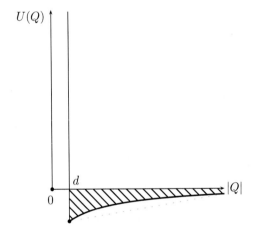

Figure 10.3.1

Hemmer–Kac–Uhlenbeck [1963/64] showed that, for fixed a and $b > 0$, nothing new happens at large volume: the phase diagram imitates Figure 10.2.1, and no phase change is seen. But when the attractive part is made long and weak by scaling a and b by a small common factor c, thereby keeping the total attraction $\int_0^\infty cae^{-cbx}\,dx \simeq a/b$ roughly fixed but making its "reach" $\int_0^\infty xcae^{-cbx}\,dx \simeq a/cb^2$ very large, then a phase change appears at $c = 0+$ as in Figure 10.1.4, recapitulated here as Figure 10.3.2 – a vapor/liquid change if you will. This was proven by a more or less explicit computation of $\mathfrak{Z}$, making heavy use of the precise shape of the tail: the large volume limit is taken first. Then c is reduced to 0, leading to the van der Waals' gas law (1873)

$$p(v, T) = \frac{kT}{v - d} - \frac{1}{2}\frac{a/b}{v^2},$$

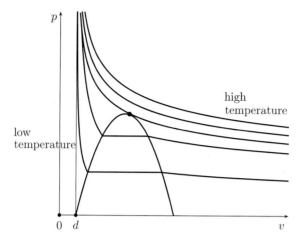

Figure 10.3.2

for $v > d$ when the temperature is high. But this could not to be the whole story. For low temperature,

$$\frac{\partial p}{\partial v} = -\frac{kT}{(v-d)^2} + \frac{a/b}{v^3}$$

vanishes at the roots of the cubic $v^3 - \frac{a}{b}(v-d)^2/kT$ which has always one irrelevant root to the left of d and two more (simple) roots $v_- < v_+$ to the right if the temperature is low. Then $p(v)$ looks like Figure 10.3.3, and this *must be wrong* since the pressure should decrease as v goes up, as noted in §10.1.6.

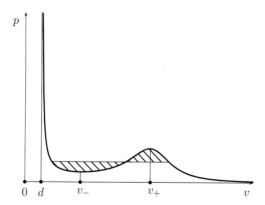

Figure 10.3.3

Exercise 10.3.1 Check that the extra roots appear at the critical temperature $T_0 = 4a/27bkd$ as a double root, splitting into a pair of simple roots at still lower temperatures.

What then? Well, the fact is that the van der Waals free energy

$$F_0(v) = kT \log \frac{1}{v-d} - \frac{1}{2}\frac{a/b}{v}$$

is only an upper bound to the true free energy $F = -kT \log \mathfrak{Z}$, and as the latter is convex, so $F \le F_0$ can be improved to $F \le$ the biggest convex minorant of F_0 with no further thought. Remarkably enough, F *is* the biggest convex minorant: at high temperatures $(T > T_0)$, there F_0 is already convex, $F = F_0$, and the van der Waals *pressure* $p(v,T)$ is correct as it stands. At low temperatures $(T < T_0)$ it is not so: F_0 looks as in Figure 10.3.4 with a (shaded) piece above the chord. This corre-

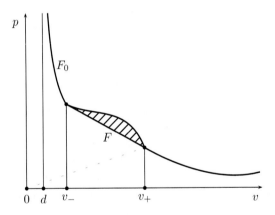

Figure 10.3.4

sponds to the flat piece in the true pressure $p(v) = -F'(v)$ indicated in Figure 10.3.3, and it is not hard to see where it's located if you observe that, with $p_0(v)$ denoting the unmodified $p(v,T)$,

$$
\begin{aligned}
-\int_{v_-}^{v_+} [p_0(v) - p_0(v_-)]\, dv &= \int_{v_-}^{v_+} [F_0'(v) - F_0'(v_-)]\, dv \\
&= F_0(v_+) - F_0(v_-) - (v_+ - v_-)F_0'(v_-) \\
&= F(v_+) - F(v_-) - (v_+ - v_-)F'(v_-) \\
&= 0,
\end{aligned}
$$

as is plain from Figure 10.3.4. In short, in the true phase diagram, the two shaded areas in Figure 10.3.3 must balance. The rule is due to Maxwell and bears his name.

10.4 Van der Waals: dimension 3

It is a happy fact, due to van Kampen [1964] and Lebowitz–Penrose [1966] that the complicated discussion of Hemmer–Kac–Uhlenbeck [1963/64] can be redone and much simplified in dimension 3 already, supposing only that the attractive tail $U(Q) < 0$ coming in at the molecular diameter $|Q| = d$ is a little smooth and of fairly slow variation, falling off like $-|Q|^{-(3+)}$ or better at $Q = \infty$. The large-volume limit $N \uparrow \infty$ with fixed density $\varrho = N/V$ is taken first. Then U is scaled as in $c^3U(cQ)$, and the van der Waals equation of state appears at $c = 0+$, as will be seen at the end.

10.4.1 Z bounded below

Let's forget the scaling for now. V is a big cube of side L. Divide it into a large number M of identical smaller cubes of side $\ell = L/\sqrt[3]{M}$, fixed but also large; shrink these symmetrically by a $d/2$ (right/left, backward/forward, and up/down) to produce identical cubical cells K at distance d apart. Place $n = N/M$ molecules in each of these, pretending for simplicity that this is a whole number, and noting that these reduced cells are big enough to accommodate so many molecules since their volume is still about ℓ^3 and the local density required is then $n/\ell^3 = N/M\ell^3 = N/V = \varrho$ which is, of course, to be fixed throughout. Now observe that this special configuration supplies a simple lower bound to the partition function. It is clear that

$$
Z(N,V) = \frac{1}{N!} \int e^{-H/kT}
$$

$$
\geq \frac{1}{N!} \sum_{n_1+\cdots+n_M=N} \frac{N!}{n_1!\cdots n_M!} \int_{(K_1)^{n_1} \times \cdots \times (K_M)^{n_M}} e^{-H/kT}
$$

$$
\geq \left(\frac{1}{n!}\right)^M \int_{(K_1)^n \times \cdots \times (K_M)^n} e^{-H/kT}.
$$

Here, the hard core comes in only for pairs of molecules in the same cell, so with $W = -U$ for $|Q| > d$ and vanishing before, you have

$$Z(N, V) \geq [Z(n, K)]^M \times$$

$$\times \text{ some lower bound to } \exp\left[\frac{1}{kT}\sum_{i<j} W(Q_i - Q_j)\right].$$

For the present purpose, the sum under the *exp* may be restricted to pairs $i < j$ corresponding to different cells with centers $k_i \neq k_j$, whereupon $W(Q_i - Q_j)$ is underestimated by $\underline{W}(k_i - k_j)$ with $\underline{W}(k) = \min_{|Q|\leq 2\ell} W(Q+k)$, as you will please check. The restricted sum is then underestimated by

$$\sum_{1 \leq i < j \leq M} \frac{1}{2} n(n-1)\underline{W}(k_i - k_j)$$

$$= \frac{1}{4} n(n-1)M \sum_{\mathbb{L}} \underline{W}(k) + \text{ some error,}$$

$\mathbb{L}$ being the lattice of centers $(L/\sqrt[3]{M} = \ell) \times \mathbb{Z}^3$ extended to the whole of $\mathbb{R}^3$. Here the error is controlled by

$$n^2 \sum_{k \in V} \sum_{k' \in V'} \underline{W}(k - k')$$

which may be estimated as follows: $\underline{W}(Q)$ inherits from $W(Q)$ a decay like $|Q|^{-3+}$ or better, so $\sum_{\mathbb{L}} \underline{W}(k)$ is comparable to $\int_1^\infty r^{-(1+)} \, dr < \infty$, and you may choose a number D so that $\sum_{|k|>D} \underline{W}(k)$ is small. The part of the error with $|k - k'| > D$ is then not more than $n^2 M \times$ that small sum. For the rest, when $|k - k'| \leq D$, k' lies *inside* a cube with side $L + 2D$ but *outside* the cube V with side L, so there are about $((L + 2D)^3 - L^3)/\ell^3 \simeq 6L^2 D/\ell^3$ such summands, contributing at most $n^2 6L^2 D/\ell^3 = n^2 M \times 6D/L$, which is also small compared to $n^2 M$, L being large. In short, the total error is small compared to N, n being fixed. Now back to the partition function:

$$\log Z(N, V) \geq M \log Z(n, K) + \frac{1}{4}\frac{n(n-1)}{kT}\sum_{\mathbb{L}} \underline{W}(k) + o(N)$$

is divided by $N \uparrow \infty$ to produce the large-volume underestimate:

$$\log 3 \geq \frac{1}{n} \log Z(n, K) + \frac{1}{4kT}\sum_{\mathbb{L}} \underline{W}(k) \times \varrho K$$

upon noting that $n = N/M = \varrho\ell^3 = \varrho K$ more or less.

10.4.2 Scaling and the van der Waals limit

$W(Q)$ is now replaced by $c^3 W(cQ)$ and c is taken down to $c \equiv 0+$, whereupon

(1) the whole of the attractive tail is removed from the $\exp(-H/kT)$ in $Z(n, K)$, leaving only the hard core;

(2) $\sum_{L-0} c^3 \underline{W}(cK)K$ tends to $\int_{\mathbb{R}^3} W(Q)\, d^3 Q \equiv a$; and

(3) you find that

$$\log 3 \geq \frac{1}{n} \log Z_{\text{hardcore}}(n, K) + \frac{\varrho a}{4kT} \,.$$

Then n is taken to $+\infty$ so that K expands to fill the whole of $\mathbb{R}^3$ at density $n/K \simeq \varrho$, producing the final underestimate:

$$\log 3(v) \geq \log 3_{\text{hardcore}}(v) + \frac{\varrho a}{4kT}$$

and so also the free-energy overestimate:

$$F(v) = -kT \log 3 \leq F_{\text{hardcore}}(v) - \frac{a}{4v} \equiv F_0(v),$$

much as in §10.3.

10.4.3 Finishing the proof

Now $F(v)$ is a convex function of v lying under $F_0(v)$; as such, it must be smaller than the biggest convex minorant of the latter, just as in §10.3: in fact, it *is* this biggest minorant, as may be proved by a more elaborate argument of the same type, for which see Lebowitz–Penrose [1966:104–107]. I do not reproduce it. Then the reasoning of §10.3 produces pictures like Figures 10.3.2–10.3.4, Maxwell's rule and all, assuming the hardcore free energy has some reasonable shape reminiscent of $-\log(v - v_3)$, where v_3 is the specific volume for close packing of molecules of diameter d in $\mathbb{R}^3$. The only real difference between dimensions 1 and 3 is that the hard-core free energy is now *unknown*; it could even produce a loose/close packing phase change as noted in §10.2, but as this is not seen in Nature, it may be supposed that it is absent here as well. The numbers v_3 *over* the molecular volume for dimensions 1, 2 and 3 may be recorded here. They are 1 in dimension 1, $\frac{1}{2}(1 + \sqrt{3})$ in dimension 2, and who knows what in dimension 3.

10.5 The Ising model

10.5.1 Overview

Now the scene changes from a gas of like molecules to a ferromagnet, but the interesting question is the same, *to wit*: Does cooperative behavior intervene at some critical temperature to produce a change of phase? Think of a box in $\mathbb{Z}^d$ ($d = 1, 2, 3$) with n points per edge and volume $N = n^d =$ the number of points enclosed, as in Figure 10.5.1 for $d = 2$ and $n = 5$. There you see $5 \times 5 = 25$ points or "sites" ($\bullet$) connected by

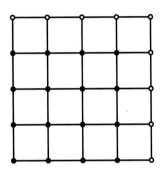

Figure 10.5.1

40 "bonds". Sometimes it will be convenient to identify the edges. That is what the open circles ($\circ$) indicate. Then there are only $4 \times 4 = 16$ sites and twice as many (32) bonds, which is more symmetrical. That won't make any difference in the end so I leave it vague for now.

Next, at each site ($\bullet$), place a "spin" $\sigma = \pm 1$. These are coupled along the bonds ($\bullet\!\!-\!\!\bullet$ and $\begin{smallmatrix}\bullet\\\bullet\end{smallmatrix}$) between nearest neighboring sites and also with an external (magnetic) field to produce the total energy or Hamiltonian

$$H = -J \sum_{\text{bonds}} \sigma'\sigma'' - K \sum_{\text{sites}} \sigma,$$

in which $J > 0$ is the (fixed) strength of the internal coupling and $K > 0$ is the (variable) strength of the external field. The (canonical) probability of such a configuration of spins is then $\mathrm{e}^{-H/kT}/Z$ with normalizer (partition function)

$$Z = \sum_{\text{spins}} \mathrm{e}^{-H/kT} = \sum_{\text{spins}} \exp\left[\frac{J}{kT} \sum_{\text{bonds}} \sigma'\sigma'' + \frac{K}{kT} \sum_{\text{sites}} \sigma\right] = \mathrm{e}^{-NF/kT},$$

F being the free energy per spin. Z now lacks the factor $1/N!$ it had in

§§10.1–10.4, and you may wonder why. The reason is simple. Previously, H was a symmetric function of the labelled molecular configuration, but there is nothing like that here. The configuration of the lattice and the insistence on nearest neighbor coupling are irreducible features of the description. Try it out for $n = 2$ with two +1s and two −1s and you will see what I mean. Note that $J > 0$ favors alignment of spins – mostly ↑ (+1) or mostly ↓ (−1) – while $K > 0$ favors spin ↑ (+1).

Now look at

$$\mathfrak{m}(k, T) = \frac{1}{N} \frac{\partial \log Z}{\partial K/kT} = \frac{1}{Z} \sum_{\text{spins}} \left(\frac{1}{N} \sum_{\text{sites}} \sigma \right) e^{-H/kT}$$

$$= \text{the mean value of } \sigma,$$

interpreted as the magnetization per spin induced by the external field at volume N, temperature T, and field strength K, and take the large-volume limit $N \uparrow \infty$. Then with $\mathfrak{Z} = \lim_{N\uparrow\infty} \sqrt[N]{Z}$, the magnetization per spin is

$$\mathfrak{m}(K, T) = \frac{\partial \log \mathfrak{Z}}{\partial K/kT}.$$

The important thing is now the "spontaneous" magnetization at fixed temperature T, *to wit,* $\lim_{K\downarrow 0} \mathfrak{m}(K, T) = \mathfrak{m}(0+, T)$, of which the meaning is this: The lattice caricatures a piece of iron, divided up into little cells, labelled by lattice sites as in Figure 10.5.2, each with a spin or

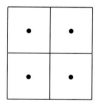

Figure 10.5.2

magnetic moment attached, the idea being that alignment of spins with preponderance of spin ↑ should be descriptive of magnetization of the iron, in alignment with the external field. Place the iron in the field and wait a bit for the spins to settle down. Now take N to $+\infty$ to simplify the description and switch off the field $(K \downarrow 0)$ *in that order.* If $\mathfrak{m}(0+, T)$ is positive, the iron itself is now a magnet; if not, it didn't work. For $d = 1$, it *doesn't* work: $\mathfrak{m}(0+, T) = 0$ at any temperature $0 < T < \infty$.

But, for $d = 2$ or 3, there is a "critical" temperature $0 < T_0 < \infty$ such that $\mathfrak{m}(0+, T) = 0$ at higher temperatures $(T > T_0)$ and $\mathfrak{m}(0+, T) > 0$ at lower temperatures $(T < T_0)$, signifying a phase change, from disorder above to order below. Obviously, the picture is far from Nature even for $d = 3$, but it is of great interest because it can be worked out more or less completely in the critical dimension 2 and much can be said in dimension 3 as well. Dimension 1 is too simple, as noted already, but I spell it out anyhow, as a warm-up; dimension 3 is hopeless for explicit computation, but more of that below. Thompson [1972: Chapters 4 and 5] is recommended for more information, attractively presented; see also Münster [1956: 568–603] for a full account in dimension 2 as of that date, and subsequent references cited here for aspects not then understood.

10.5.2 Dimension 1

Ising [1925] worked it out. I'm sure he thought: "This is something I can do in *any* dimension, even if it's pretty artificial". He was mistaken for $d = 2$ already, but $d = 1$ is really easy. The spins are arranged in a circle as in Figure 10.5.3 ($d = 1$, $n = 5$, with edge identification). The

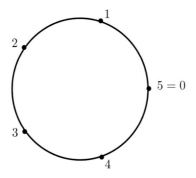

Figure 10.5.3

partition function is

$$Z = \sum_{\text{spins}} e^{-H}$$

$$= \sum_{\text{spins}} \exp\Big[J(\sigma_0\sigma_1 + \sigma_1\sigma_2 + \cdots + \sigma_{N-1}\sigma_0)$$

$$+ K(\sigma_0 + \sigma_1 + \cdots + \sigma_{N-1}\Big]$$

in which the $1/kT$ is suppressed for temporary brevity. Introducing the symmetric 2×2 matrix

$$\exp\left[J\sigma'\sigma'' + \frac{K}{2}(\sigma' + \sigma'')\right] = \begin{array}{c|c|c} & +1 & -1 \\ \hline +1 & e^{J+K} & e^{-J} \\ \hline -1 & e^{-J} & e^{J-K} \end{array} \, ,$$

you see that Z is just the trace of its Nth power and that $\mathfrak{Z} = \lim_{N\uparrow\infty} \sqrt[N]{Z}$ is its biggest eigenvalue:

$$\Lambda = \cosh K e^{J} + \sqrt{e^{2J}\sinh^2(K) - e^{-2J}},$$

as you will please check. The (spontaneous) magnetization per spin, with $1/kT$ restored to its place, is now found to be

$$\mathfrak{m}(K,T) = \frac{\partial \log \Lambda}{\partial K/kT} = \frac{\sinh(K/kT)}{\sqrt{\sinh^2(K/kT) + e^{-4J/kT}}} \, ;$$

in particular, $\mathfrak{m}(0+, T) = 0$ for any temperature T, as advertised. Figure 10.5.4 shows $\mathfrak{m} = \mathfrak{m}(K,T)$ in its dependence on K at various temperatures. With $x = e^{2J/kT} \times \sinh(K/kT)$, you have $\mathfrak{m} = x/\sqrt{x^2 + 1}$, $\mathfrak{m}' > 0$, and $\mathfrak{m}'' < 0$, so $\mathfrak{m}$ is concave, increasing in K, and rises as the temperature goes down.

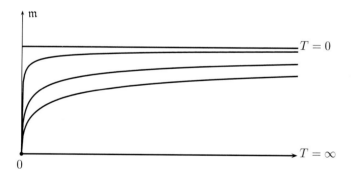

Figure 10.5.4

10.5.3 Dimension 2

Dimension 2 is *much* harder. The spins are placed on an $n \times n$ lattice, as in Figure 10.5.1, with or without edge identifications, and $\mathfrak{Z} =$

$\lim_{N \uparrow \infty} \sqrt[N]{Z(K,T)}$ is sought. Onsager [1944] found an explicit expression for $\mathfrak{Z}$ when $K = 0$, *viz.*, with $C = J/kT$ and $\varkappa = \frac{2 \sinh(2C)}{\cosh^2(2C)}$,

$$\log \mathfrak{Z}(0,T) = \log \sqrt{2} + \log \cosh(2C) + \frac{1}{\pi} \int_0^{\pi/2} \log \left(1 + \sqrt{1 - \varkappa^2 \sin^2 \theta} \right) d\theta.$$

The computation, simplified by Kaufmann and Onsager (1949), proceeds in the same style as for dimension 1, Z being the trace of the nth power of a suitable $2n \times 2n$ matrix, and $\mathfrak{Z}$ is its top eigenvalue in the ∞-volume limit. The details are not simple, involving particular knowledge of a double cover (spin group) of the $2n$-dimensional rotation group $SO(2n)$. Kac–Ward [1952] and others found other ways to do it, but none of these is really easy. As to $K > 0$, no explicit expression for $\mathfrak{Z}$ is known, though Onsager [1949] indicated, and Yang–Lee [1952] confirmed, the value of the spontaneous magnetization, *viz.*

$$\mathfrak{m}(0+, T) = 0 \qquad\qquad (T \geq T_2),$$

$$= \sqrt[8]{1 - \sinh^{-4}(2C)} \quad (T < T_2),$$

in which T_2 is the critical temperature $2J/k \log(1 + \sqrt{2})$ determined by $\sinh(2C) = 1$. I do not attempt to derive these remarkable formulae here, but I *will* explain (1) the existence of $\mathfrak{Z}$ (neglected in §10.1), (2) the presence of spontaneous magnetization in dimensions 2 and 3, and (3) how the critical temperature may be located.

10.6 Existence of $\mathfrak{Z}$

This is the easiest computation of its type, owing its simplicity to the short range of the coupling (nearest neighbors only); long range coupling, as for the gas of like molecules in §10.1, introduces difficulties, but these may be mostly overcome in a similar style; see, e.g. Yang–Lee [1952]. The factor $1/kT$ is suppressed as it plays no real role.

Note first, that edge identifications do not affect the value of $\mathfrak{Z}$: for a lattice of side n, they modify Z only by a factor $\exp(\pm 4nJ)$ at worst, and this disappears when the $N = n^2$th root is taken and $N \uparrow \infty$. Now take a lattice of side 2^{n+m} (n large) and divide it up into 2^{2n} small cells of side 2^m by means of the (red) dual lattice seen in Figure 10.6.1 for $n = m = 1$. The spins in two adjacent cells couple across their common border of dual bonds, and as these borders are of total length $l \leq 2^{2n} \times 4 \, 2^m$, the couplings $\sigma' \sigma''$ are additive over the small cells, up to

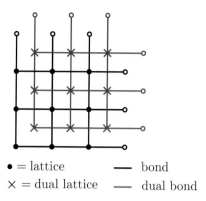

• = lattice   — bond
× = dual lattice   — dual bond

Figure 10.6.1

a total (absolute) error $\leq l$, so the several partition functions involved stand in the following approximate relation:

$$[Z_{2^m \times 2^m}]^{2^{2n}} \times e^{-J2^{2n} \cdot 4\, 2^m} \leq Z_{2^{n+m} \times 2^{n+m}}$$
$$\leq [Z_{2^m \times 2^m}]^{2^{2n}} \times e^{+J2^{2n} \cdot 4\, 2^m}.$$

But this is the same as saying

$$\frac{\log Z_{2^m \times 2^m}}{2^{2m}} - \frac{4J}{2^m} \leq \frac{\log Z_{2^{n+m} \times 2^{n+m}}}{2^{2n+2m}} \leq \frac{\log Z_{2^m \times 2^m}}{2^{2m}} + \frac{4J}{2^m},$$

from which the existence of

$$3' = \lim_{n \uparrow \infty} \sqrt[2^{2n}]{Z_{2^n \times 2^n}}$$

follows at once.

Obviously, the same type of thing can be done with 2^n replaced by[5] $[x^n]$ with a fixed number $x = (1+)$, producing a variant 3, depending perhaps on this choice. Now choose $[x^{m-1}] = a \leq c < [x^m] = b$ and compare the small $c \times c$ lattice to the big $b \times b$ lattice seen in Figure 10.6.2 with a (red) border of dual bonds separating the two, distinguishing the "inner" spins in the small lattice from the "outer" spins. The inner and outer spins couple only across the dual bonds of the border, and as these are only $4c$ in number, this may be neglected, which is to say that the

---

[5] $[x]$ is the biggest whole number $\leq x$.

inner and outer spins are, effectively, de-coupled. Then roughly speaking,

$$Z_{b \times b} = \sum_{\text{spins}} \exp\left[J \sum \sigma'\sigma'' + K \sum \sigma\right]$$

$$\leq \sum_{\text{spins}} \exp\left[J \underset{\substack{\text{inner} \\ \text{bonds}}}{\sum} \sigma'\sigma'' + J(b^2 - c^2) + K \underset{\substack{\text{inner} \\ \text{spins}}}{\sum} \sigma + K(b^2 - c^2)\right]$$

$$\leq 2^{b^2 - c^2} \qquad \text{from the sum over the outer spins}$$

$$\times Z_{c \times c} \times e^{(J+K)(b^2 - c^2)},$$

or what is more to the point,

$$\frac{1}{b^2} \log Z_{b \times b} \leq \left(1 - \frac{c^2}{b^2}\right) \log 2 + \frac{1}{c^2} \log Z_{c \times c} + (J + K)\left(1 - \frac{c^2}{b^2}\right).$$

Now make $c \uparrow \infty$ in such a way that $\frac{1}{c^2} \log Z_{c \times c}$ approximates $\liminf_{n \uparrow \infty} \frac{1}{n^2} \log Z_{n \times n}$ and also $n \uparrow \infty$ in response so that $a \leq c < b$. Then

$$\log \mathfrak{Z}(x) \leq \liminf_{n \uparrow \infty} \frac{1}{n^2} \log Z_{n \times n} + (\log 2 + J + K)\left(1 - \frac{1}{x^2}\right),$$

which is to say

$$\limsup_{x \downarrow 1} \log \mathfrak{Z}(x) \leq \liminf_{n \uparrow \infty} \frac{1}{n^2} \log Z_{n \times n},$$

with a similar bound from below, *viz.*

$$\liminf_{x \downarrow 1} \log \mathfrak{Z}(x) \geq \limsup_{n \uparrow \infty} \frac{1}{n^2} \log Z_{n \times n},$$

which I leave to you. That does it.

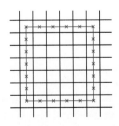

Figure 10.6.2

10.7 Magnetization per spin: dimension 2

The discussion is for dimension 2 but works just as well for dimension 3, as you will agree at the end.

10.7.1 Shape of $\mathfrak{m}$, T fixed

For brevity, suppress the factor $1/kT$ as before, write

$$Z = \sum_{\text{spins}} \exp(JX + KY)$$

with $X = \sum \sigma\sigma'$ and $Y = \sum \sigma$, and let the bracket $\langle \, \cdot \, \rangle$ signify the expectation relative to the canonical probabilities $Z^{-1}\mathrm{e}^{-H}$. Here is what you need to know for fixed T:

Fact 1 $\partial \log Z / \partial K = \langle Y \rangle \geq 0$.

Proof.

$$\frac{\partial \log Z}{\partial K} = \frac{1}{Z} \sum_{\text{spins}} Y \mathrm{e}^{JX+KY}$$

$$= \frac{1}{Z} \sum_{\text{spins}} -Y \mathrm{e}^{JX-KY}$$

$$= \frac{1}{Z} \sum_{\text{spins}} Y \mathrm{e}^{JX} \sinh(KY)$$

$$\geq 0,$$

in which line 2 follows from line 1 by simultaneous reversal of spins: $\sigma \to -\sigma$.

Fact 2 $\partial^2 \log Z / \partial K^2 = \langle Y^2 \rangle - \langle Y \rangle^2 \geq 0$.

Proof. Just compute.

Fact 3 $\partial^3 \log Z / \partial K^3 = \langle Y^3 \rangle - 3\langle Y \rangle\langle Y^2 \rangle + 2\langle Y \rangle^3 \leq 0$.

Now the signature lies deeper. I postpone the proof (indicated at the end) in favor of the consequences for $\mathfrak{m}$ = the magnetization per spin.

To begin with, by Facts 1, 2, and 3

$$\mathfrak{m}(K) = \lim_{n\uparrow\infty} \frac{1}{n^2} \partial \log Z / \partial K$$

lies between 0 and 1 and is an *increasing, concave* function of $K \geq 0$. Recall $\mathfrak{Z} = \lim_{n \uparrow \infty} \sqrt[n^2]{Z}$. By Fact 2,

$$\log Z \Big|_{K=A}^{K=B} \geq (B - A) \frac{\partial \log Z}{\partial K} \quad \text{taken at } K = A$$

so

$$\log \mathfrak{Z} \Big|_{K=A}^{K=B} \geq (B - A) \limsup_{n \uparrow \infty} \frac{1}{n^2} \frac{\partial \log Z}{\partial K} \quad \text{taken at } K = A,$$

and similarly,

$$\log \mathfrak{Z} \Big|_{K=A}^{K=B} \leq (B - A) \liminf_{n \uparrow \infty} \frac{1}{n^2} \frac{\partial \log Z}{\partial K} \quad \text{taken at } K = B.$$

Now $\log \mathfrak{Z}$ is convex by Facts 1 and 2; as such, it has right and left derivatives $\partial^+ \log \mathfrak{Z}/\partial K$ and $\partial^- \log \mathfrak{Z}/\partial K$ subject to

$$\frac{\partial^+ \log \mathfrak{Z}}{\partial K} \geq \limsup_{n \uparrow \infty} \frac{1}{n^2} \frac{\partial \log Z}{\partial K} \geq \liminf_{n \uparrow \infty} \frac{1}{n^2} \frac{\partial \log Z}{\partial K} \geq \frac{\partial^- \log \mathfrak{Z}}{\partial K}$$

by the last display. But $\partial^\pm \log \mathfrak{Z}/\partial K$ is not only increasing by Fact 2, but also concave by Fact 3, and cannot jump *up* so that

$$\partial^+ \log \mathfrak{Z}/\partial K = \partial^- \log \mathfrak{Z}/\partial K = \partial \log \mathfrak{Z}/\partial K, \text{ plain}:$$

in short, you have a proper (increasing, concave) magnetization per spin in the infinite-volume limit:

$$\frac{\partial \log \mathfrak{Z}}{\partial K} = \lim_{n \uparrow \infty} \frac{1}{n^2} \frac{\partial \log Z}{\partial K} = \mathfrak{m}(K).$$

10.7.2 Shape of $\mathfrak{m}$, K fixed

So far, T was fixed. Now let's prove that for fixed K, $m(K, T)$ and so also $\mathfrak{m}(K, T)$ decreases with increasing temperature, indicating an increase of disorder.

Proof. The true magnetization per spin for $n < \infty$ is found by restoring the factor $1/kT$ to J and K, and it is an easy exercise to check that $\partial m(J, K, T)/\partial T \leq 0$ is the same as saying

$$\frac{\partial m}{\partial J} + \frac{\partial m}{\partial K} = \frac{1}{n^2} \left[\frac{\partial^2 \log Z}{\partial J \partial K} + \frac{\partial^2 \log Z}{\partial K^2} \right] \geq 0$$

in the old format with $1/kT$ removed. Then by Fact 2, it suffices to check

Fact 4 $\partial^2 \log Z/\partial J \partial K = \langle XY \rangle - \langle X \rangle \langle Y \rangle \geq 0$.

In fact, it is just as easy to prove more in both dimensions 2 and 3, *to wit*, that for any three spins, located at the same, or at different, sites,

Fact 5 $\langle \sigma_1 \sigma_2 \sigma_3 \rangle - \langle \sigma_1 \rangle \langle \sigma_2 \sigma_3 \rangle \geq 0$.

The proof is very pretty, and this more general fact will be wanted presently, in dimension 3. Inequalities of this type are due to Griffiths [1967] and Kelly–Sherman [1968]. The argument follows Sylvester [1976].

Proof. Take two independent copies of the spins, α and β, corresponding to the double Hamiltonian

$$ H = -J \sum_{\text{bonds}} (\alpha'\alpha'' + \beta'\beta'') - K \sum_{\text{sites}} (\alpha + \beta) $$

and note that the 2×2 rotation

$$ \begin{array}{c} \alpha \\ \beta \end{array} \longrightarrow \begin{array}{c} \dfrac{\alpha + \beta}{\sqrt{2}} = a \\[2ex] \dfrac{\alpha - \beta}{\sqrt{2}} = b, \end{array} $$

carried out at each site separately, does not change the little quadratic forms so

$$ H = -J \sum_{\text{bonds}} (a'a'' + b'b'') - \sqrt{2}\, K \sum_{\text{sites}} a. $$

Now the as and bs are independent, by inspection of H, and

$$ \sqrt{2} \big[\langle \sigma_1 \sigma_2 \sigma_3 \rangle - \langle \sigma_1 \rangle \langle \sigma_2 \sigma_3 \rangle \big] $$
$$ = \sqrt{2} \big[\langle \alpha_1 \alpha_2 \alpha_3 \rangle - \langle \alpha_1 \beta_2 \beta_3 \rangle \big] $$
$$ = \frac{1}{2} \big\langle (a_1 + b_1) \big[(a_2 + b_2)(a_3 + b_3) - (a_2 - b_2)(a_3 - b_3) \big] \big\rangle $$
$$ = \big\langle (a_1 + b_1)(a_2 b_3 + a_3 b_2) \big\rangle $$
$$ = \langle a_1 a_2 \rangle \langle b_3 \rangle + \langle a_1 a_3 \rangle \langle b_2 \rangle + \langle a_2 \rangle \langle b_1 b_3 \rangle + \langle a_3 \rangle \langle b_1 b_2 \rangle $$

by the independence of the as and bs; also $\langle a \rangle = \sqrt{2}$ by Fact 1 and $\langle b \rangle = 0$, so it is enough to check that, e.g., $\langle b_1 b_2 \rangle \geq 0$. Now $\langle b_1 b_2 \rangle$ is a positive multiple of

$$ \sum_{\text{spins}} b_1 b_2 \exp \left[J \sum_{\text{bonds}} b'b'' \right] $$

which may expanded in powers of J to obtain a sum of terms like $b_1^{m_1} b_2^{m_2} b_3^{m_3}$ etc. $\times$ a positive constant $\times$ a power of J. Now at each site

(0), the total factor $b_0^m = 2^{-m/2}(\alpha_0 - \beta_0)^m$ may be collected together; and these collected factors may be summed over the values of α_0 and β_0, individually, site by site. If m is even, b_0^m is ≥ 0 anyhow, while if m is odd you get

$$(-1-1)^m + (+1+1)^m + (-1+1)^m + (+1-1)^m = -2^m + 2^m = 0.$$

That does it.

A proof of this type may be made for Fact 3 as well. It is based upon the fact that for any three spins,

Fact 6 $\langle\sigma_1\sigma_2\sigma_3\rangle - \langle\sigma_1\rangle\langle\sigma_2\sigma_3\rangle - \langle\sigma_2\rangle\langle\sigma_3\sigma_1\rangle - \langle\sigma_3\rangle\langle\sigma_1\sigma_2\rangle + 2\langle\sigma_1\rangle\langle\sigma_2\rangle\langle\sigma_3\rangle \leq 0$.

The proof now employs *four* independent copies of the spins, and as it is a little tedious, I do not reproduce it, but see Ellis [1985:142–146] if you want to see how it's done.

Preview

The magnetization per spin $\mathfrak{m}(K,T)$ is sketched in Figure 10.7.1. $T = T_2$ is the critical temperature at which spontaneous magnetization sets in:

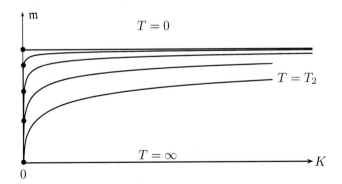

Figure 10.7.1

for $T > T_2$, $\mathfrak{m}(0+,T) = 0$; while for $T < T_2$, $\mathfrak{m}(0+,T)$ increases as the temperature drops. The *existence* of T_2 is not yet proved; that is the subject of §10.8.

Aside on dimension 3

The 3-dimensional lattice of side n may be viewed as a stack of $n \times n$ 2-dimensional lattices, of height n, coupled by vertical bonds which I propose to switch off, one by one. Let J_{12} be a *variable* coupling constant for a particular one of these vertical bonds. Then for the 3-dimensional partition function,

$$\partial^2 \log Z_3 / \partial J_{12} \partial K = \left\langle \sigma_1 \sigma_2 \sum \sigma \right\rangle - \langle \sigma_1 \sigma_2 \rangle \left\langle \sum \sigma \right\rangle \geq 0,$$

by Fact 5. Now when all the vertical bonds are switched off, the n levels of the stack are statistically independent copies of the $n \times n$ lattice with its little partition function Z_2, so that $\partial \log Z_3 / \partial K$ is reduced to $n \times \partial \log Z_2 / \partial K$, and

$$\mathfrak{m}_3 = \partial \log \mathfrak{Z}_3 / \partial K \geq \mathfrak{m}_2 = \partial \log \mathfrak{Z}_2 / \partial K.$$

The moral is this: that if, in dimension 2, spontaneous magnetization sets in at a critical temperature T_2 (and it does), then the same is true in dimension 3 at some temperature $T_3 \geq T_2$.

10.8 Change of phase: dimension 2

The argument comes in two parts, *to wit*, that $\mathfrak{m}(0+, T) = 0$ when the temperature is high and that $\mathfrak{m}(0+, T) > 0$ when it's low.

10.8.1 High temperature

A so-called "mean-field" approximation is used. Make the edge identifications in the $(n+1) \times (n+1)$ lattice so that the n^2 *unidentified* sites all look alike. Ising has it that each spin couples with its four neighbors, each with coupling J; in the approximation, *all* the spins interact with the *same* coupling $4J/n^2$ so that the total coupling per spin is the same as before, and as the new coupling favors cooperation of spins, you may hope that the new magnetization $(\overline{\mathfrak{m}})$ is bigger than the old $(\mathfrak{m})$. If then $\overline{\mathfrak{m}}(0+, T)$ vanishes at high temperatures (and it does), $\mathfrak{m}(0+, T)$ will vanish, too.

First compute $\overline{\mathfrak{Z}}$ and $\overline{\mathfrak{m}}$. Think of $\sum \sigma \equiv S$ as the terminus, after n^2 steps, of the standard random walk $RW(1)$, starting at the origin, so that

$$2^{-n^2} \overline{Z} = \mathbb{E} \exp \left[n^2 \left(4J\mathbf{x}^2 + K\mathbf{x} \right) \right] \text{ with } \mathbf{x} = S/n^2,$$

as you will please check. Here, you may use the large-deviations estimate of §2.4 to the following effect: for $n \uparrow \infty$ and any closed interval
$-1 \leq a \leq x \leq b < 1$, excluding $x = 0$,

$$\mathbb{P}\left[a \leq S/n^2 \leq b\right] \simeq \exp\left[-n^2 \min_{ab_-} I(x)\right]$$

with

$$I(x) = \max_{y \in \mathbb{R}}\left[xy - \log \mathbb{E}(e^{y\sigma})\right] \text{ for } |x| \leq 1.$$

The computation of I is trivial: the little expectation is $\cosh y$, $\tanh y = x$ at the maximum, so $I(x)$ is the function $\frac{1}{2}(1+x)\log(1+x) + \frac{1}{2}(1-x)\log(1-x)$. Then

$$2^{-n^2}\overline{Z} \simeq \exp\left[n^2 \max_{|x| \leq 1}\left(4Jx^2 + Kx - I(x)\right)\right],$$

up to a multiplicative error not worse than $\exp\left[o(n^2)\right]$, which is to say

$$\log \overline{3} = \max_{|x| \leq 1}\left[4Jx^2 + Kx - \frac{1}{2}(1+x)\log(1+x) - \frac{1}{2}(1-x)\log(1-x)\right] + \log 2.$$

Now remember that J is really J/kT which is small since the temperature is high, and K likewise, look at Figure 10.8.1, which you will please

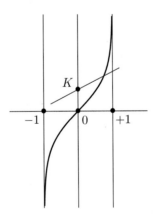

Figure 10.8.1

explain, and conclude that the maximum is found at the place $x = x(K)$
where the line $8Jx + K$ crosses the curve $I'(x) = \frac{1}{2}\log\frac{1+x}{1-x}$. The magnetization may now be found: $\overline{m}(K, T) = x(K)$, as you will check, and the
picture shows that $x(K = 0+) = 0$ since $\frac{1}{2}\log(x+1)/(x-1)$ has slope
1 at $x = 0$.

It remains to prove that $\overline{\mathfrak{m}} \geq \mathfrak{m}$, as will now appear. In fact, since four bonds meet at each site,

$$\sum_{\text{bonds}} (x - \sigma')(x - \sigma'') = n^2 x^2 - 8x \sum_{\text{sites}} \sigma + \sum_{\text{bonds}} \sigma'\sigma'',$$

so with $x = x(K) = \overline{\mathfrak{m}}$ and large n so that $\mathfrak{m}$ is roughly $\langle \sigma \rangle$, you find

$$\overline{\mathfrak{m}} - \mathfrak{m}$$

$$\simeq x - \langle \sigma \rangle$$

$$= \frac{1}{Z} \sum_{\text{spins}} (x - w) e^{J \sum_{\text{bonds}} (x - \sigma')(x - \sigma'')} e^{(8Jx + K) \sum_{\text{sites}} \sigma} \times e^{-Jn^2 x^2}$$

for any particular spin w. To see that this is non-negative, $1/Z$ and $e^{-Jn^2 x^2}$ are removed, $8Jx + K$ is identified with $\frac{1}{2}\log(1 + x)/(1 - x)$ as above, and the rest is developed in powers of J to produce a sum of terms like

$$\prod \left[\sum_{w = \pm 1} (x - w)^\ell \left(\frac{1 + x}{1 - x} \right)^{w/2} \times \text{a positive constant} \right],$$

in which the product is taken over the several sites and $\ell \geq 1$ may depend on the site in hand. But

$$\sum_{w = \pm 1} (x - w)^\ell \left(\frac{1 + x}{1 - x} \right)^{w/2} = \sqrt{1 - x^2} \left[(1 + x)^{\ell-1} + (-1)^\ell (1 - x)^{\ell-1} \right],$$

when you work it out, and that's non-negative for any $\ell \geq 1$ since $0 < x < 1$. Neat, no?

The same thing works in dimension 3 to show that $\mathfrak{m}(0+, T) = 0$ when the temperature is high.

Aside. The following exercises, adapted from Thompson [1972], provide an alternative to large deviations for the evaluation of $\overline{\mathfrak{z}}$.

Exercise 10.8.1 Check the evaluation

$$2^{-n^2} \overline{Z} = \int_{-\infty}^{+\infty} \frac{e^{-x^2/2}}{\sqrt{2\pi}} \left[\cosh\left(\sqrt{8J}\frac{x}{n} + K \right) \right]^{n^2} dx.$$

Deduce

$$-\log 2 + \log \overline{\mathfrak{z}} = \max_{x \in \mathbb{R}} \left[-\frac{x^2}{2} + \log \cosh(\sqrt{8J}x + K) \right].$$

Exercise 10.8.2 Reconcile that with the previous evaluation of $\overline{\mathfrak{z}}$.

10.8.2 Low temperature

The onset of spontaneous magnetization at low temperature will now be verified by a beautiful argument of Peierls [1936], corrected in part by Griffiths [1964] and Dobrushin [1965].

ℨ is insensitive to edge identifications, so take your $n \times n$ lattice plain, enclose it in an artificial border with all spins $\sigma = +1$, and let ω be any particular spin inside. It is to be proved that for $K = 0$, $\mathbb{P}(\omega = -1) \le \vartheta$ with a fixed number $\vartheta < 1/2$, independently of $n \uparrow \infty$ if the temperature is low. The factor $1/kT$ is suppressed as before.

Proof of Spontaneous Magnetization. If $\omega = -1$, there is a simple (i.e. non-self-intersecting) closed path γ of dual bonds enclosing ω, with $\sigma = -1$ inside and $\sigma = +1$ immediately outside. To see this, start at ω; step towards the outer boundary along any path of lattice bonds; stop when you hit a spin $+1$ (as you must do); and draw a dual bond separating it from the spin -1 just before. These dual (red) bonds form the path γ seen in Figure 10.8.2.

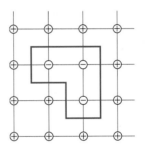

Figure 10.8.2

Now consider the effect of changing the spins inside γ from $\sigma = -1$ to $\sigma^* = +1$, leaving fixed the spins outside. Only the couples $\sigma'\sigma''$ connected by bonds crossing γ are changed so that $-H(\sigma) = -H(\sigma^*) - 2J \times$ the length ℓ of γ, and

$$\mathbb{P}(\gamma) = \frac{\text{the sum of } e^{-H(\sigma)} \text{ over configurations}}{\text{of spins with } \gamma \text{ enclosing } \omega \text{ as above}} \Big/ \frac{\text{the unrestricted}}{\text{sum of } e^{-H(\sigma)}}$$

$$= e^{-2J\ell} \times \text{the same sum of } e^{-H(\sigma^*)} \Big/ \frac{\text{the unrestricted}}{\text{sum of } e^{-H(\sigma)}}$$

$$\le e^{-2J\ell}$$

since each choice of spins σ^* on top appears also as a choice of spins σ

at the bottom. Obviously, many different γs may enclose ω in the style described above. That depends on the rest of the spins, and since a simple path of length ℓ (≥ 4) may be translated at most ℓ^2 ways and has, besides, only 3^ℓ more degrees of freedom owing to its simple character, you have

$$\mathbb{P}(\omega = -1) = \sum_\gamma \mathbb{P}(\gamma) \leq \sum_{\ell=4}^\infty e^{-2J\ell} \times \ell^2 3^\ell \equiv \vartheta < 1/2,$$

provided J (alias J/kT) is big, as indeed it is when the temperature is low. That's the proof.

The estimate $\langle \sigma \rangle = 1 - 2\,\mathbb{P}(\omega = -1) \geq 1 - 2\vartheta$ is now rewritten:

$$\sum_{\substack{\text{outer} \\ \text{border} \\ \text{included}}} \sigma'\sigma'' = \sum_{n\times n} \sigma'\sigma'' \quad \text{from the } n\times n \text{ lattice}$$

$$+\, 4n + 4 \quad \text{from the self-coupling of the border}$$

$$+ \sum_{\partial n\times n} \sigma \quad \begin{array}{l} \text{from the coupling of the} \\ n\times n \text{ lattice to the border} \end{array}$$

with a self-evident notation for the border in line 3, from which you see that $1 - 2\vartheta$ is overestimated by

$$\frac{\sum_{\substack{n\times n \\ \text{spins}}} \frac{1}{n^2}\left(\sum_{n\times n} \sigma\right) \times \exp\left[J\sum_{n\times n} \sigma'\sigma'' + J\sum_{\partial n\times n} \sigma\right]}{\sum_{\substack{n\times n \\ \text{spins}}} \exp\left[J\sum_{n\times n} \sigma'\sigma'' + J\sum_{\partial n\times n} \sigma\right]}$$

$$= \frac{1}{n^2}\frac{\partial \log Z}{\partial K} \quad \text{taken at } K = 0$$

with the amended partition function

$$Z = \sum_{\substack{n\times n \\ \text{spins}}} \exp\left[J\sum_{n\times n} \sigma'\sigma'' + J\sum_{\partial n\times n} \sigma + K\sum_{n\times n} \sigma\right].$$

This is not quite correct since the spins on $\partial n\times n$ are on a different statistical footing from those inside, but the discrepancy cannot change the mean of $\sum_{n\times n} \sigma$ by more than a $\pm 4n$, and this may be neglected upon division by n^2. The rest is easy: with the amended partition function Z, $\lim_{n\uparrow\infty} \sqrt[n^2]{Z}$ is the true ∞-volume partition function $\mathbf{3}$ since edge effects do not matter. And besides, $\partial \log Z/\partial K$ increases with K (yes?), with the result that $0 < 1 - 2\vartheta \leq \partial \log \mathbf{3}/\partial K = \mathfrak{m}(K, T)$ for any $K \geq 0$. Then $\mathfrak{m}(0+, T) \geq 1 - 2\vartheta > 0$ at low temperatures, and as it must increase as the temperature drops, so there is a specific (critical) temperature T_2 at which spontaneous magnetization sets in.

10.9 Duality and the critical temperature

In the absence of the external field ($K = 0$), if $C = J/kT$, and if high and low temperatures are exchanged by the rule $\sinh(2C_*)\sinh(2C^*) = 1$, then

$$\mathfrak{Z}(T_*) = \sqrt{\sinh(2C_*)}\,\mathfrak{Z}(T^*).$$

This is the duality of Kramers–Wannier [1941]. The fixed point $T = T^*$ determined by $2C = \sinh^{-1}(1) = \log(1 + \sqrt{2})$ is then special, and it is an educated guess that this should be the critical temperature T_2 at which spontaneous magnetization comes in. It is not obvious that this should be so, but in fact, it agrees with Yang–Lee's [1952] evaluation of $\mathfrak{m}(0+, T)$ reported in §10.5.3. No such duality is known in dimension 3 or in dimension 2 if $K > 0$; indeed, it is thought not to exist.

Proof of the duality after van der Waerden [1941]. Fix $T > 0$ and define the dual temperature T^* by

$$\sinh(2C)\sinh(2C^*) = 1.$$

A little manipulation produces

$$e^{\pm C} = \sqrt{\frac{1}{2}\sinh(2C)} \times \left(e^{C^*} \pm e^{-C^*}\right),$$

so that, in the $n \times n$ lattice without edge identifications,

$$Z = \sum_{\text{spins}} \prod_{\text{bonds}} e^{C\sigma'\sigma''}$$

$$= \left(\frac{1}{2}\sinh(2C)\right)^{B/2} \times \sum_{\text{spins}} \prod_{\text{bonds}} \left(e^{C^*} + \sigma'\sigma''e^{-C^*}\right)$$

with $B = 2n(n - 1) =$ the total number of lattice bonds. Denote by γ any collection of bonds of "length" $=$ their number $\ell = \ell(\gamma)$. Then you may write

$$\prod_{\text{bonds}} \left(e^{C^*} + \sigma'\sigma''e^{-C^*}\right) = \sum_{\gamma} e^{(B-\ell)C^* - \ell C^*} \prod_{\substack{\text{bonds} \\ \text{in } \gamma}} \sigma'\sigma''.$$

Here you may observe that if an odd number of bonds in γ meet at any particular site, then the sum over spins ±1 at that place kills the product under the right-hand sum, i.e. γ survives in the sum for Z only if it is a collection of simple, closed paths, such as are seen in black in Figure 10.9.1. For paths of this type the product reduces to $+1$ and the sum over spins to 2^{n^2}. The letter γ is now reserved for them.

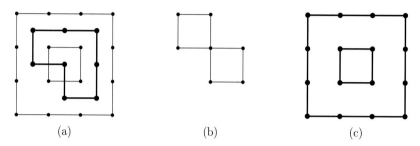

(a) (b) (c)

Figure 10.9.1

Next, take spins σ^* in the (red) dual lattice of side $n+1$ with $\sigma^* = +1$ inside γ and $\sigma^* = -1$ outside. Obviously, there are $\ell = \ell(\gamma)$ dual bonds crossing γ with $\sigma^{*\prime}\sigma^{*\prime\prime} = -1$. Otherwise $\sigma^{*\prime}\sigma^{*\prime\prime} = +1$, so you may now write

$$Z = \left(\frac{1}{2}\sinh(2C)\right)^{B/2} \times \sum_{\gamma} e^{(B-\ell)C^* - \ell C^*} \times 2^{n^2}$$

$$= \left(\frac{1}{2}\sinh(2C)\right)^{B/2} \times \frac{1}{2} \times e^{-4n} \times 2^{n^2/2} \sum_{*\text{spins}} \exp\left[C^* \sum_{\substack{\text{dual}\\\text{bonds}}} \sigma^{*\prime}\sigma^{*\prime\prime}\right]$$

with extra factors: $1/2$ due to the ambiguity between inside and outside ($\sigma^* \to -\sigma^*$) and an e^{-4n} due to the discrepancy between the bond numbers $B = 2n^2 - 2n$ and $B^* = 2n^2 + 2n$. Now take the root $\sqrt[n^2]{Z}$ and make $n \uparrow \infty$, keeping in mind that the outer boundary of the dual lattice does not matter. This produces the duality of partition functions: $\mathfrak{Z}(T) = \sqrt{\sinh(2C)}\,\mathfrak{Z}(T^*)$.

McKean [1975b] derives the duality in a very different style via the Poisson summation formula for paths in the $n \times n$ lattice with edge identifications, viewed as the (commutative) group $G =$ the product $\mathbb{Z}_2 \times \cdots \times \mathbb{Z}_2$ of $2(n-1)^2$ copies of $\mathbb{Z}_2 = \mathbb{Z} \mod 2$, one such for each bond of the lattice. The present method is more direct, so I do not reproduce this other way.

11

★ Statistical Mechanics Out of Equilibrium

This is *much* harder. I quote from Feynman [1964] at some length[1]

With this chapter we begin a new subject which will occupy us for some time. It is the first part of the analysis of the properties of matter from the physical point of view, in which, recognizing that matter is made out of a great many atoms, or elementary parts, which interact electrically and obey the laws of mechanics, we try to understand why various aggregates of atoms behave the way they do.

It is obvious that this is a difficult subject, and we emphasize at the beginning that it is in fact an *extremely* difficult subject, and that we have to deal with it differently than we have dealt with the other subjects so far. In the case of mechanics and in the case of light, we were able to begin with a precise statement of some laws, like Newton's laws, or the formula for the field produced by an accelerating charge, from which a whole host of phenomena could be essentially understood, and which would produce a basis for our understanding of mechanics and light from that time on. That is, we may learn more later, but we do not learn different physics, we only learn better methods of mathematical analysis to deal with the situation.

We cannot use this approach effectively in studying the properties of matter. We can discuss matter only in a most elementary way; it is much too complicated a subject to analyze directly from its specific basic laws, which are none other than the laws of mechanics and electricity. But these are a bit too far away from the properties we wish to study; it takes too many steps to get from Newton's laws to the properties of matter, and these steps are, in themselves, fairly complicated. We will now start to take some of these steps, but while many of our analyses will be quite accurate, they will eventually get less and less accurate. We will have only a rough understanding of the properties of matter.

One of the reasons that we have to perform the analysis so imperfectly is that the mathematics of it requires a deep understanding of the theory of probability; we are not going to want to know where every atom is actually moving,

---

[1] Reproduced with permission from *The Feynman Lectures on Physics*, New Millennium Edition, (Feynman, Leighton and Sands), Volume 1, Chapter 39.

but rather, how many move here and there on the average, and what the odds are for different effects. So this subject involves a knowledge of the theory of probability, and our mathematics is not yet quite ready and we do not want to strain it too hard.

... Anyone who wants to analyze the properties of matter in a real problem might want to start by writing down the fundamental equations and then try to solve them mathematically. Although there are people who try to use such an approach, these people are the failures in this field; the real successes come to those who start from a *physical* point of view, people who have a rough idea where they are going and then begin by making the right kind of approximations, knowing what is big and what is small in a given complicated situation. These problems are so complicated that even an elementary understanding, although inaccurate and incomplete, is worth-while having, ...

Think now, as in §10.1, of a gas comprised of a realistically large number $N \simeq 10^{23}$ of like spherical molecules, of diameter d and mass $m = 1$ say, with centers Q confined to a vessel V, and momenta P, *aka* velocities, together specifying the gaseous phase. The total energy is

$$H = \frac{1}{2} \sum_k P_k^2 + \sum_{i<j} U(Q_i - Q_j),$$

$U(Q)$ being a radial function of the type seen in Figure 10.1.1, exhibiting a hard core ($U = +\infty$) at distances $d' \le d$, followed by a short, steep repulsive part ($U' < 0$) and a long attractive tail ($U' > 0$) beyond. The Qs and Ps obey Hamilton's equations $\dot{Q} = \partial H/\partial P$ and $\dot{P} = -\partial H/\partial Q$, and as there are some 6×10^{23} of these, coupled together, it is not possible to understand much of anything without the introduction of statistical methods, as I said before. This means that you have to think, not of a single *individual* gas, but of an *ensemble* of copies of it, with phase distributed in some plausible way, and try to devise some simpler (but not too simple) approximate, statistical description of that which may be followed in its response to *streaming* ($\dot{Q} = P$) and the *scattering* due to collisions and/or close encounters of the molecules, hoping for simplifications as N is taken to $+\infty$ and V expands in proportion to fill up the whole of $\mathbb{R}^3$. That is the subject of Boltzmann's *Vorlesungen über Gastheorie* [1912], describing work dating from around 1870, and of Maxwell's *Dynamical Theory of Gases* [1861], elaborated since that time into a long, and if I may say so, not entirely satisfactory story, with, it must also be said, a seemingly miraculous practical success, so much of the reasoning being a type of *Desperazions Physik* to borrow M. Kac's nice phrase. I cannot go into all that, referring you instead to O. Lanford [1975] for the best mathematical introduction I know, to

Grad [1958] for the practical aspects, and to Yau [1998] for more recent news if you want more than the rudimentary *Guide to the Perplexed* presented next. Please skim it to see what the issues are and read the final bit about the two-speed gas to see if that sounds attractive. Byron said: "Your easy writing is your damned hard reading". Hoping that the rule operates backwards, I begin.

11.1 What Boltzmann said and what came after

The story unfolds in stages, successively simplifying the description of the gas.

Stage 1.   The molecules being indistinguishable, you don't need to know their names. It should be enough to divide the one-molecule phase space $V \times \mathbb{R}^3$ into a large number of small cells $A \times B$, to count the number of molecules with $Q \in A$ and $P \in B$, and to describe how these counts change with time. Clearly, this description is still much too complicated. But suppose these "small" cells are so "large" as each to accommodate very many molecules. Then as N is taken to infinity and V expands in proportion to fill up the whole of $\mathbb{R}^3$, you may hope to see some *bulk effect*, reminiscent of the weak law of large numbers, whereby for *most* cells, the count $\#(A \times B)$ is well approximated by the integral $\int_{A \times B} F \, dQ \, dP$ of a nice function F of $t \geq 0$ and the phase $QP \in \mathbb{R}^6$. Note that $\#(A) = \#(A \times \mathbb{R}^3) = \int_A n(Q) \, dQ$ with $n(Q) = \int_{\mathbb{R}^3} F \, dP$, from which you see that $n(Q) \simeq \#(A)/\operatorname{vol} A$ if A is not too big, exhibiting $n(Q)$ as a local *number* density or *concentration* and $F(QP)/n(Q)$ as the local *probability* density for velocities.

This is a very big simplification of the description, not at all easy to justify, but it should be more or less reliable, and one can always hope.

Stage 2.   Now the question comes: How does F change with time? The influence of *streaming* is simple enough. Absent collisions and/or close encounters, the phase $QP \in \mathbb{R}^6$ is in free flight $Q \to Q + tP$ with fixed velocity P, and the associated infinitesimal operator is $\mathfrak{G} = P \cdot \partial/\partial Q$, to which F responds as in $\partial F/\partial t = \mathfrak{G}^\dagger F = -(P \cdot \partial/\partial Q)F$. The rest of $\partial F/\partial t$ comes, of course, from the *scattering*, in respect of which it is assumed that the gas is *dilute*, meaning that the density $n(Q)$ is generally so small that only pair-wise collisions and/or close encounters need be counted, anything else being so rare or of so little consequence.

At this point, it is best to think only of hard spheres with $d = 1$

say, $U(Q) = \infty$ at distances $d' \leq d$, and $U(Q) = 0$ beyond, so that
the molecules are in free flight except for actual collisions with instanta-
neous, perfect reflection. The picture shows such a collision of molecules
nos. 1 and 2, in free flight before and again after collision, referred to
such coordinates as make molecule no.1 seem to be at rest. Nothing else
is seen. Molecule no.2 comes in from the left at constant relative veloc-

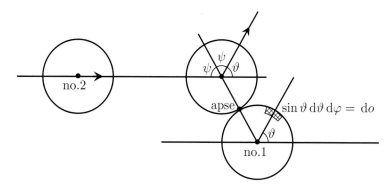

Figure 11.1.1

ity $P = P_2 - P_1$, scatters off molecule no.1 in the direction of the little
shaded spherical patch at co-latitude ϑ and longitude φ, and flies away
at relative velocity $P' = P'_2 - P'_1$. The "apse" is the point of contact, e
is the apse direction, ψ is the apse angle, ϑ is the scattering angle, and
$\vartheta + 2\psi = \pi$ as you see by looking at the symmetrical, reversed collision.

To determine how many such collisions take place in a short time
dt, think about the center $Q_2(t)$ of molecule no.2, projected on the
horizontal line through the apse along which molecule no.2 is coming in. The collision takes place by time $t + dt$, but not before time
t, only if the projection of $Q_2(t)$ is the left of the apse, and that of
$Q_2(t) + P\,dt$ is to the right, which is the same as saying $Q_2(t)$ lies
in the little cylinder seen in Figure 11.1.2 with ends congruent to the
patch of Figure 11.1.1 rotated counter-clockwise by ψ, cross-section
$\cos\psi\,do = \cos\psi \sin\vartheta\,d\vartheta\,d\varphi$, length $|P|\,dt$, and so of volume $dQ_2 = |P|\cos\psi\,do \times dt$, as you will see most easily by looking once more at
the symmetrical, reversed collision in Figure 11.1.1. Now the number
of molecules no.2 of this type with velocity in a little patch dP_2 is
roughly $F(Q_2 P_2)\,dQ_2\,dP_2$ while the number of molecules no.1 in a little patch $dQ_1 \times dP_1$ is roughly $F(Q_1 P_1)\,dQ_1\,dP_1$, so such collisions
account for a *loss* to $F(Q_1 P_1)\,dQ_1\,dP_1$ in the amount $F(Q_1 P_1)\,dP_1 \times$

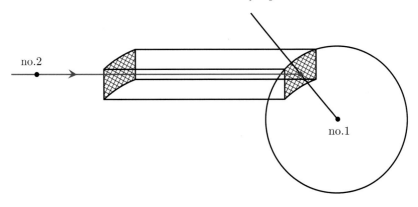

Figure 11.1.2

the sum of $F(Q_2 P_2)\, \mathrm{d}Q_2\, \mathrm{d}P_2$, taken over all the little cylinders of Figure 11.1.2. This loss is compensated by a *gain* due to the symmetrical, reverse collisions $P'_1 P'_2 \to P_1 P_2$ in the amount $F(Q'_1 P'_1)\, \mathrm{d}Q'_1\, \mathrm{d}P'_1 \times$ a similar sum of $F(Q'_2 P'_2)\, \mathrm{d}Q'_2\, \mathrm{d}P'_2$; and since the collision is Hamiltonian, preserving the 12-dimensional volume element $\mathrm{d}Q_1\, \mathrm{d}P_1\, \mathrm{d}Q_2\, \mathrm{d}P_2$, you may write the net *rate* of gain/loss to $F(Q_1 P_1)$ as

$$\int_{S \times \mathbb{R}^3} \left[F(Q'_1 P'_1) F(Q'_2 P'_2) - F(Q_1 P_1) F(Q_2 P_2) \right] |P| \cos \psi\, \mathrm{d}o\, \mathrm{d}P_2$$

upon recalling that $\mathrm{d}Q_2 = \cos \psi\, \mathrm{d}o |P|\, \mathrm{d}t$ and cancelling a common factor $\mathrm{d}Q_1\, \mathrm{d}P_1$. Now you may (though perhaps you shouldn't) identify all the Qs – at collision, they differ by d, and if d is not too big and if F varies slowly in respect to Q, it should be OK. Never mind. Just do it to produce Boltzmann's final equation in which I suppress both time and the common Q, so as to show only the essentials: *viz.*

$$\frac{\partial F}{\partial t} + P \cdot \frac{\partial F}{\partial Q} = \int_{S \times \mathbb{R}^3} \left[F(P'_1) F(P'_2) - F(P_1) F(P_2) \right] |P| \cos \psi\, \mathrm{d}o\, \mathrm{d}P_2,$$

in which S is the unit sphere centered at molecule no.1, and where $\mathrm{d}o$ lives, and the P_1 to the right is identified with the P to the left.

That's pretty much what Boltzmann said, hoping for the best. Obviously, it may be criticized in several respects. To begin with the identification of all the Qs is a bit much, but that is not the worst of it. Surely, the description of the *two*-molecule collision should involve the *two*-molecule concentration function, but is here replaced by the product of two *one*-molecule functions, indicative of a *chaotic, aka* statistically-independent character of the velocities of molecules nos. 1 and 2. It is this Stosszahlansatz, so-called by Boltzmann, that makes the whole thing click. I suppose he thought: Why not? Molecule no.2 comes in from far away to the left, not knowing that no.1 is in the way, and why then should their velocities *not* be identically and independently distributed at collision? Indeed, the whole thing has been justified for moderate times by O. Lanford [1975] in the limit as the occupied volume $N \times \frac{4}{3}\pi(d/2)^3$ tends to zero (that's diluteness) and the total cross-section $N \times \pi(d/2)^2$ tends to 1, say (that means lots of collisions still).

The same type of equation may be so to say "derived", using the same Stosszahlansatz and additional vague talk about the rapidity of collisions and so forth. The only change is that the $\cos\psi$ seen before is replaced by a more complicated so-called scattering cross-section J, of which a brief explanation here. Figure 11.1.3 shows the close encounter of two molecules, pictured here as mere points, with the addition of an invisible artificial unit sphere S centered at molecule no.1 on which

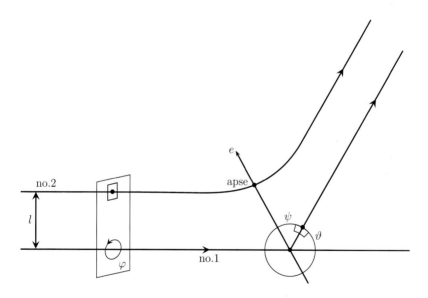

Figure 11.1.3

the scattering direction (•) is marked. The apse is the point of closest approach, e is the (unit) apse direction, ψ is the apse angle, ϑ the scattering angle, and $\vartheta + 2\psi = \pi$, as before. Molecule no.2 is seen far at the left, travelling in more or less free flight at velocity P, passing through a little patch of area $\ell \, d\ell \, d\varphi$ in the plane perpendicular to the horizontal line through molecule no.1, scattering swiftly off no.1, and flying away symmetrically via the reflection of its incoming trajectory across the apse line. Here, the two-body mechanics is described by Hamilton's equations for $Q = Q_2 - Q_1$ and $P = P_2 - P_1$, *viz.* $\dot{Q} = \partial H/\partial P = P$ and $\dot{P} = -\partial H/\partial Q = -\operatorname{grad} U(Q)$ with $H = \frac{1}{2}P^2 + U(Q)$, and these may be integrated most conveniently in polar coordinates r and θ in the plane of the paper, centered at molecule no.1, to show that

$$\psi = \int_{r_0}^{\infty} \ell \left(1 - \frac{2U(r)}{v^2} - \frac{\ell^2}{r^2} \right)^{-1/2} \frac{dr}{r^2},$$

in which r_0 is the apse distance (from the center of molecule no.1) and v is the relative speed $|P|$ far to the left. In this way, $\vartheta = \pi - 2\psi$ is determined as a function of ℓ and v, r_0 being so determined too, whereupon the patch in the perpendicular plane may be put in correspondence with the little spherical patch after matching the azimuth φ in the plane to the longitude on the sphere. The scattering cross-section is then the Jacobian $J(v, \vartheta) = \ell \, d\ell \, d\varphi / \sin \vartheta \, d\vartheta \, d\varphi$ and enters into the loss/gain balance for hard spheres, relating the volume dQ_2 of the slant cylinder of Figure 11.1.2 to $do = \sin \vartheta \, d\vartheta \, d\varphi$ in the same way that $\cos \psi$ did before.

Exercise 11.1.1 The computation of φ is not difficult. $Q = r(\cos \vartheta, \sin \vartheta)$ so $P^2 = 2H - 2U(Q) = |\dot{Q}|^2 = (dr/dt)^2 + r^2(d\vartheta/dt)^2$. Here H is a constant of motion and so is the rate $A = \frac{1}{2}r^2 \, d\vartheta/dt$ of area production, just as for Kepler's problem of §8.1. These may be evaluated by looking far to the left where Q is in more or less free flight at speed $v = |P|$ and distance ℓ from the horizontal line through molecule no.1, where they reduce to $H = v^2/2$, and $A = \ell v/2$.

Conclude that $\dfrac{d\vartheta}{dr} = \dfrac{\ell/r^2}{\sqrt{\Delta}}$ where $\Delta = \left(1 - \dfrac{2b}{v^2} - \dfrac{\ell^2}{r^2} \right)$ and integrate *that* along the trajectory, from the apse $(r_0, \vartheta + \varphi)$ out to (∞, ϑ) explaining also how the apse distance is determined.

Coming back to business, it is clear that all this is more than a little suspect. No apology is offered, nor could any be accepted. Quite frankly, I do not know of any derivation which is not either confusing, or confused, or else plainly wrong. Such reasoning cannot be expected to have any practical validity at all. But it *does*, much to one's astonishment, for which see, e.g. Grad [1958]. I leave the matter there, passing on to the next installment of the story.

Stage 3 concerns the tendency of the gas to some sort of *local* statistical equilibration of velocities. Now it is important to understand that there are two time-scales at work, descriptive of *fast* collisions and *slow* streaming. The suggestion is that these rapid collisions have such an equilibrating effect, reminiscent of CLT, complementing the LLN-type bulk effect of Stage 1. Evidence in favor is provided by Boltzmann's celebrated H-theorem, stating that if streaming be ignored (in which case F may be taken as a function of P alone) then the entropy[2] $H = -\int F \log F \, dP$ increases steadily with time unless F is already a local Maxwellian of the familiar form

$$F(Q, P) = n(2\pi kT)^{-3/2} \exp\left[-(P - \overline{P})^2/2kT\right]$$

- with local density

$$n(Q) = \int F(Q, P) \, dP,$$

- mean velocity

$$\overline{P}(Q) = \frac{1}{n(Q)} \int P F(Q, P) \, dP,$$

- temperature

$$T(Q) = \frac{1}{3kn(Q)} \int F(Q, P)(P - \overline{P})^2 \, dP.$$

A little mechanics will set the stage. Go back to Figure 11.1.3 depicting the close encounter of molecules nos. 1 and 2, travelling in more or less free flight, at velocities $P_1 P_2$ and $P_1' P_2'$ before, resp. after, collision. These obey

(1) $P_1' + P_2' = P_1 + P_2$ (conservation of momentum),

(2) $(P_1')^2 + (P_2')^2 = P_1^2 + P_2^2$) (conservation of energy), and so also their consequence

(3) $|P_2' - P_1'| = |P_2 - P_1|$.

(1) permits us to write $P_1' = P_2 + ce$ and $P_2' = P_1 - ce$ with some constant c and unit direction e. Then (2) leads easily to

(4) $P_1' = e \otimes e \, P_1 + (1 - e \otimes e) P_2$ and $P_2' = e \otimes e \, P_2 + (1 - e \otimes e) P_1$,

$e \otimes e$ being the projection on e, and inspection of Figure 11.1.3 identifies e as the apse direction pictured there.

---

[2] Too many Hs. Forgive it please.

Exercise 11.1.2 Please check all that, noting further that the scattering $P_1 P_2 \to P_1' P_2'$ is a rotation of $\mathbb{R}^6$, preserving the volume $dP_1\, dP_2$ if the scattering direction ($\bullet$), and so also the apse direction (e), is fixed.

Now comes Boltzmann's lovely proof: Q being ignored, the entropy production rate may be written

$$\frac{dH}{dt} = -\int_{S \times \mathbb{R}^6} \log F(P_1) \big[F(P_1')F(P_2') - F(P_1)F(P_2) \big] \, |P|\, J\, do\, dP_1\, dP_2$$
$$- \int_{S \times \mathbb{R}^6} \big[F(P_1')F(P_2') - F(P_1)F(P_2) \big] \, |P|\, J\, do\, dP_1\, dP_2,$$

$|P|$ being the relative speed $|P_2 - P_1|$ and J the scattering cross-section of Stage 2, depending only upon $|P|$ and the scattering angle ϑ. Here, line 2 of the display vanishes since $P_1 P_2 \to P_1' P_2'$ is a rotation of $\mathbb{R}^6$ for fixed $e \in S$ and $|P|J$ is the same before collision and after, and for the same reason you may exchange $P_1 P_2$ in line 1 in favor of $P_2 P_1$ or of $P_1' P_2'$ or else in favor of $P_2' P_1'$, producing four variants of it which may be averaged to obtain

$$\frac{dH}{dt} = \frac{1}{4} \int_{S \times \mathbb{R}^6} \big[F(P_1')F(P_2') - F(P_1)F(P_2) \big] \times$$
$$\times \log \frac{F(P_1')F(P_2')}{F(P_1)F(P_2)} \times |P|\, J\, do\, dP_1\, dP_2.$$

That's non-negative, by inspection, and cannot vanish unless $F(P_1')F(P_2') = F(P_1)F(P_2)$ for any collision whatsoever.

Exercise 11.1.3 completes the proof. Write $F = \exp(G)$. Then

(5) $G(P_1') + G(P_2') = G(P_1) + G(P_2)$,

and it is enough to show that $G(P) = A + B \cdot P + C P^2$ with constants A, B, C.

Hints: Use (4) to differentiate (5) twice in respect to P_1 to obtain

(6) $e \otimes e\, G''(P_1')\, e \otimes e + (1 - e \otimes e)G''(P_2')(1 - e \otimes e) = G''(P_1)$,

G'' being the (symmetric) Hessian $\big[\partial^2 G / \partial P_i \partial P_j : 1 \le i, j \le 3 \big]$. Use (6) to show that G'' is a diagonal matrix $\Lambda(P)$. Go back to (4) to infer that $\Lambda(P)$ does not depend on P and integrate back.

Aside 1 For the heat equation $\partial F / \partial t = \Delta F$ and $H = \int h(F)\, dP$,

$$\frac{dH}{dt} = \int h'(F)\, \text{div grad}\, F\, dP = -\int h''(F)|\text{grad}\, F|^2\, dP$$

is non-negative for *any* concave function h, provided the computation makes sense. Contrariwise, for Boltzmann's equation, it is a striking fact that up to trivialities, *only* $h(F) = -F \log F$ *will do!* See §11.7.2 for a simplified case and Vedenyapin [1988] for, e.g. hard spheres.

Aside 2 Note that the increase of H in response to collisions seems to contradict the reversibility of the *individual* Hamiltonian flow. The culprit can only be the Stosszahlansatz, but that's correct in the limit as the occupied volume $N \times \frac{4}{3}\pi(d/2)^3$ tends to zero and the cross-section $N \times \pi(d/2)^2$ tends to 1, as noted in Stage 2 – an old riddle of statistical mechanics, still not fully understood. It seems to me that the answer is otherwise and simpler – that the real reason is the simplification of the description by passing from labelled to *un*labelled molecules in Stage 1. Is it not so that the mechanics is already lost, or if not really *lost*, then *obscured*? Evidence on favor of this view is provided by Boltzmann's pretty calculation of §9.1.1, showing that the increase in occupied volume caused by passing to N named molecules from cell-by-cell counts n_1, n_2, etc. is expressed by a factor $\exp(H)$ with $H = -\sum(n_i/N) \log(n_i/N)$. Surely, that tells *something* since H signifies *dis*order. I do not find this elementary observation in any literature known to me.

Stage 4. Boltzmann's H-theorem just explained suggests, but surely does not prove, that F comes rapidly (?) close to a local Maxwellian with density $n(Q)$, mean velocity $\overline{P}(Q)$, and temperature $T(Q)$, changing with time and the place Q, so now you must ask: How *do* they change? How *close* is that approximation anyhow? In response to these questions, Hilbert [1912] proposed a formal, and, if I may say so, ill-judged scheme of successive approximations to F, on which I will comment both *pro* and *con* in §11.6. In Boltzmann's equation $\partial F/\partial t + P \cdot \partial F/\partial Q \equiv A(F) = B(F \otimes F)$, the collisions are speeded up by a big factor $1/\varepsilon$ to the right, with a view to enhancing the statistical equilibration of velocities, F is written as a formal power series $F_0 + \varepsilon F_1 + \varepsilon^2 F_2$ etc., and the equation is integrated recursively, starting at level -1 so to speak, with terms in ε^{-1}. There, you have $0 = B(F_0 \otimes F_0)$, forcing F_0 to be a local Maxwellian. Next, at level 0, $A(F_0) = B(F_0 \otimes F_1) + B(F_1 \otimes F_0)$, which may be integrated only if $\int P^n A(F_0) \, dP = 0$ for $n = 0, 1$, and 2, reflecting the structure of B and the character of collisions, preserving as they do *number, momentum* and *energy*. That's five, coupled, partial differential equations, of hydrodynamical type so called, for the five "hydrodynamical moments" $m = (n, \overline{P}, T)$ of the local Maxwellian, expressing $\partial m/\partial t$ in terms of m and $\partial m/\partial Q$. This is already very complicated. In the conventional hydrodynamical format with $Q = x$ and $\overline{P} = v$, and in the special inviscid case of incompressible, irrotational flow ($\text{div } v = 0$ and grad $\times v = 0$), it reduces at constant temperature to Euler's equation:

$$\frac{\partial v}{\partial t} + (v \cdot \text{grad})v + \text{grad}\, p = 0$$

with pressure

$$p = \frac{1}{4\pi} \int_{\mathbb{R}^3} \frac{1}{|x - y|} \text{ trace} \left[\left(\frac{\partial v}{\partial x} \right)^\dagger \left(\frac{\partial v}{\partial x} \right) \right] dy,$$

descriptive of Feynman's "dry water". "Wet water" (with viscosity) enters at level 1, and things get still more complicated, as you can imagine. Once more, a condition of integrability is required to reconcile the left- and right-hand sides of the equation, and so it goes. And there is worse to come, for now you say, without any apology or visible shame: "Oh. I didn't really mean that. The epsilon was just for book-keeping. Put it back to 1 and let's see what comes out". A very impudent thing to do. And besides nobody really knows, in any generality, whether such hydro- dynamical equations as appear have nice solutions or whether the formal power series for F makes any show of convergence. But never mind. Sup- pose it works somehow. Then the recipe distinguishes a *tiny* submanifold K of concentration functions $F(Q, P)$, depending in some very compli- cated way, only upon the five hydrodynamical moments $m(Q)$, with the suggestion that the general solution $F(t, Q, P)$ of Boltzmann's equation comes rapidly close to K on its way to some final (Maxwellian?) equi- librium. K is then an "attractor" for the Boltzmann flow in which it reduces to a flow of $m(Q)$ alone, simplifying the description from *one* function (F) of $QP \in \mathbb{R}^6$ to *five* functions (m) of $Q \in \mathbb{R}^3$. Doubtless, this so-called "Hilbert paradox" is simply wrong, but see §11.6 for commen- tary both *pro* and *con*. Better to follow Yau [1998] starting in Gibbs's canonical ensemble, speeding up the collisions and scaling both space and time, *once* to elicit a *bulk effect* (dry water) and *once more* to elicit fluctuations (wet water). Then the hydrodynamical equations enter on a better footing, but really it's all too complicated for words.

Stage 5. OK. Let it go, supposing that pretty soon F begins to stabi- lize in respect to velocities so as to approximate the local Maxwellian, and that the slow streaming is now having its effect, *viz* to kill the mean velocity, or so one may hope, and to bring density and tempera- ture to some final, constant resting values *in the large*. Along the way, the pressure may be computed in Maxwell's style in §5.9.4 for a final, thermodynamical reduction to the three *numbers* – density, *aka* specific volume, pressure, and temperature, related by an equation of state, as in Chapter 10. End of story, and a disappointing end it is. Only the ideal gas law is found, and this for three, fairly obvious reasons.

Reason 1 is too much diluteness, with the implication that a molecule (no.2) may hit a tagged molecule (no.1) but then flies off and is not seen again.

Reason 2 is the Stosszahlansatz, encouraging molecule no.1 to think still more that it is immersed in a statistically independent bath of like molecules no.2.

Reason 3 is that $U(Q)$ is seen, but only darkly in the scattering cross-section J. So while things may go well enough for moderate times, as they seem to do, Gibbs's beautiful thermodynamics is lost. Too bad.

The rest of this chapter is devoted to what I call the two-speed gas, on which most of the issues raised above may be illustrated and more or less resolved. Radically simplified though it may be, it is still instructive, and, besides, it has nice connections to a variety of seemingly foreign topics introduced before: to the Markov chains in §7.5; to Sparre–Andersen's combinatorial method in §5.8; to the space-time harmonic functions in §4.9; to Itô's stuff in §6.9; and even to our old friend $S^\infty(\sqrt{\infty})$ from §5.10. So even though this whole chapter is marked with a black star, indicating its difficulty and/or *un*necessity, I commend this two-speed gas to your attention.

11.2 The two-speed gas: chaos and the law of large numbers

Maxwell (1870) noticed that Boltzmann's collision integral

$$B(F \otimes F) = \int \left[F(P_1')F(P_2') - F(P_1)F(P_2)\right] |P| \, J(|P|, \vartheta) \, \mathrm{d}o \, \mathrm{d}P_2$$

is particularly simple when $U(Q)$ is proportional to $|Q|^{-4}$. Then and only then $|P| \times J(|P|, \vartheta)$ is a function of the scattering angle ϑ alone. Forces such as $\mathrm{grad}|Q|^{-4}$, proportional to $|Q|^{-5}$, are not seen in Nature, but computations are facilitated, and something may be learned. That was Maxwell's idea. The two-speed gas is a terminally artificial variant of Maxwell's gas, imitating the intermediate gas of Kac [1956] treated in §11.7, but still more simple. I follow McKean [1975b] with corrections and additional information.

The two-speed gas is comprised of n "molecules" with velocities $v \in Z = (\pm 1)$, moving on the line. The streaming is temporarily ignored, so it is only "collisions" that count. These do not have *any mechanical meaning*. I take, instead, a *statistical* rule which plays the role of scattering:

$$v_1 \atop v_2 \to {v_1' \atop v_2'} = {v_1 v_2 \atop v_2} \text{ or } {v_1 \atop v_1 v_2} \text{ with probabilities } 1/2 \text{ each.}$$

Here, it is permitted that one or the other of two molecules moving left suddenly decides to move right: ridiculous but never mind. The full rule is this. Begin with velocities $v = (v_1, \ldots, v_n) \in Z^n$; choose an exponential holding time T with mean $1/n$, i.e., with $\mathbb{P}(T > t) = e^{-nT}$; at this moment, pick a pair $1 \le i < j \le n$ at random, i.e. with equal probabilities $\binom{n}{2}^{-1}$; let v_i and v_j "collide" as above; and repeat, over and over, independently of what went before. This produces a Markov chain $\mathbf{v}(t) : t \ge 0$ on Z^n with rate n and jump probabilities[3]

$$\mathbb{P}(v_{ij} \to v_{ij}') = \frac{1}{2} \bigg/ \binom{n}{2},$$

so that the distribution $p(t, v) = \mathbb{P}[\mathbf{v}(t) = v]$ is regulated by

$$\frac{\partial p}{\partial t} = \left[n \times \binom{n}{2}^{-1} = \frac{2}{n-1} \right] \times \sum_{i<j} \text{ave} \left[p(v_{ij}') - p(v_{ij}) \right] \equiv \mathfrak{G}^\dagger p,$$

where † denotes transpose, $p(0+, \cdot) = f \in D(\mathbb{Z}^n)$ = the class of probability densities on $\mathbb{Z}^n$, and ave is the average over the two possible outcomes of the collision. This looks pretty complicated, especially if n is large, but is not: in fact, large n helps via a law of large numbers to be explained presently.

11.2.1 The empirical distribution

A few simple remarks prepare the way. $\mathfrak{G}^\dagger$ commutes with permutations of the molecules, reflecting the fact that these are indistinguishable, so if, as it is natural to assume, f is a symmetric function, then the same is true of p at any time $t \ge 0$ and – what is the important point – the empirical density

$$\mathbf{p}_{n/1}(t, e) = \frac{1}{n} \#(k \le n : \mathbf{v}_k = e), \qquad e = \pm 1,$$

is itself a Markov chain with rate n on the simpler space $[k/n : 0 \le k \le n]$ after reduction to $\mathbf{p} = \mathbf{p}_{n/1}(t, +1)$ permitted by $\mathbf{p}_{n/1}(-1) = 1 - \mathbf{p}_{n/1}(+1)$. This is all you really want, not caring about the names of the molecules but only the proportion of $(+1)$s to (-1)s.

---

[3] v_{ij}' is v with v_i, resp. v_j replaced by v_i', resp. v_j'. And v_{ij} is just v.

Exercise 11.2.1 Check the Markovian character of **p**.

Table 1 shows the jump probabilities for the several types of collision when $\mathbf{p} = k/n$ just before, from which you read off the jump probabilities for the reduced empirical chain seen in Table 2:

Table 1

collision	probability
$\begin{matrix} -1 & +1 \\ -1 & \to -1 \end{matrix}$ or $\begin{matrix} -1 \\ +1 \end{matrix}$	$\frac{(n-k)(n-k-1)}{n(n-1)}$
$\begin{matrix} +1 & -1 \\ -1 & \to -1 \end{matrix}$ or $\begin{matrix} +1 \\ -1 \end{matrix}$	$\frac{1}{2}\frac{k(n-k)}{n(n-1)}$ apiece
$\begin{matrix} -1 & -1 \\ +1 & \to +1 \end{matrix}$ or $\begin{matrix} -1 \\ -1 \end{matrix}$	$\frac{1}{2}\frac{k(n-k)}{n(n-1)}$ apiece
$\begin{matrix} +1 & +1 \\ +1 & \to +1 \end{matrix}$	$\frac{k(k-1)}{n(n-1)}$

Table 2

$p \to$	probability
$p + \frac{1}{n}$	$\frac{n}{n-1}(1-p)(1-p-\frac{1}{n})$
$p - \frac{1}{n}$	$\frac{n}{n-1}(1-p)p$
$p + \frac{0}{n}$	$p - 1$

Exercise 11.2.2 Check the tables.

11.2.2 Chaos and the law of large numbers

By "chaos" I mean that for any $2 \le n < \infty$, the initial (symmetric) n-molecule distribution $f_n \in D(Z^n)$ splits as the product $f_{\infty/1} \otimes \cdots \otimes f_{\infty/1}$ of a fixed 1-molecule distribution $f_{\infty/1} \in D(Z)$, expressing the statistical independence of the individual velocities at time 0. This independence is immediately broken by collisions but is happily re-established when $n \uparrow \infty$: indeed, with a self-evident notation,

$$p_{\infty/m}(t, v_1, \ldots, v_m) = \lim_{n \uparrow \infty} p_{n/m}(t, v_1, \ldots, v_m)$$

$$= p_{\infty/1}(t, v_1) \times \cdots \times p_{\infty/1}(t, v_m)$$

for any fixed $1 \le m < \infty$, a fact discovered by Kac [1956] in his somewhat more complicated setting. This *propagation* of chaos may be taken as a statistical vindication of Boltzmann's *Stosszahlansatz* in a simpler world. I postpone the proof in favor of the computation of $p_{\infty/1}$ and the simpler law of large numbers

$$\text{LLN:} \quad \mathbb{P}\left[\lim_{n \uparrow \infty} \mathbf{p}_{n/1}(t, e) = p_{\infty/1}(t, e), e = \pm 1\right] = 1$$

which leads to it for chaotic f.

To begin with, $p_{n/1}(t_1 + 1) \equiv p$ is the mean of $\mathbf{p}_{n/1}(t, e) \equiv \mathbf{p}$, and from the rate n and the jump probabilities of Table 2, you read off

$$\frac{\mathrm{d}}{\mathrm{d}t}\mathbb{E}(\mathbf{p}) = n \times \mathbb{E}\left[\frac{n}{n-1}(1-\mathbf{p})\left(1-\mathbf{p}-\frac{1}{n}\right)\left(\mathbf{p}+\frac{1}{n}\right)\right.$$

$$\left.+\frac{n}{n-1}(1-\mathbf{p})\mathbf{p}\left(\mathbf{p}-\frac{1}{n}\right)+\mathbf{p}^2-\mathbf{p}\right]$$

$$= \mathbb{E}[(1-\mathbf{p})(1-2\mathbf{p})] \quad \begin{array}{l}\text{errors under control of } 1/n \\ \text{when you work it out,}\end{array}$$

and, in the same vein,

$$\frac{\mathrm{d}}{\mathrm{d}t}\mathbb{E}(\mathbf{p}^2) = n \times \mathbb{E}\left[\frac{n}{n-1}(1-\mathbf{p})\left(1-\mathbf{p}-\frac{1}{n}\right)\left(\mathbf{p}+\frac{1}{n}\right)^2\right.$$

$$\left.+\frac{n}{n-1}(1-\mathbf{p})\mathbf{p}\left(\mathbf{p}-\frac{1}{n}\right)^2+\mathbf{p}^3-\mathbf{p}^2\right]$$

$$= \mathbb{E}[(1-\mathbf{p})(1-2\mathbf{p})2\mathbf{p}] \quad + \quad \text{a like error.}$$

Now introduce the solution of $\dot{x}=(1-x)(1-2x)$ with $x(0)=f_{\infty/1}(+1)$ $=p(0)$. Then, up to errors comparable to $1/n$,

$$\frac{\mathrm{d}}{\mathrm{d}t}\mathbb{E}(\mathbf{p}-x)^2$$

$$= \mathbb{E}\left[(1-\mathbf{p})(1-2\mathbf{p})2\mathbf{p}-2x(1-x)(1-2x)\mathbf{p}-2x(1-\mathbf{p})(1-2\mathbf{p})\right.$$

$$\left.+2x(1-x)(1-2x)\right]$$

$$= \mathbb{E}\left[2(\mathbf{p}-x)^2(2\mathbf{p}+x-3)\right]$$

$$\leq 2\mathbb{E}(\mathbf{p}-x)^2.$$

Here, $2p+x-3$ is negative and $n\mathbf{p}(0)$ is binomially distributed with success probability $x(0)$, so

$$\mathbb{E}[\mathbf{p}(0)-x(0)]^2 = \frac{1}{n}x(0)[1-x(0)] \leq \frac{1}{4n},$$

with the result that

(1) $\mathbb{E}[\mathbf{p}(t)-x(t)]^2$ is not more than $1/4n$,

(2) $x(t)=\lim_{n\uparrow\infty}p_{n/1}(t,+1)\equiv p_{\infty/1}(t,+1)$,

and

(3) $\mathbb{P}\left[\lim_{n\uparrow\infty}\mathbf{p}_{n/1}(t,+1)=p_{\infty/1}(t,+1)\right]=1.$

(1) is trivial and likewise (2). (3) is the law of large numbers and is also easy to prove: just replace n by n^2 in (1), use Borel–Cantelli to conclude that $\lim\limits_{n \uparrow \infty} \mathbf{p}_{n^2/1} = p_{\infty/1}$, and observe that

$$\frac{(n-1)^2}{m} \mathbf{P}_{(n-1)^2/1} \le \mathbf{p}_{m/1} < \frac{n^2}{m} \mathbf{P}_{n^2/1} \text{ if } (n-1)^2 \le m < n^2.$$

Exercise 11.2.3   Write out what is or ought to be Boltzmann's equation for the 2-speed gas as follows. Sum the n-molecule equation $\partial p/\partial t = \mathfrak{G}^\dagger p$ over v_2, v_3, etc. to produce

$$\frac{\partial}{\partial t} p_{n/1}(t, v_1) = 2 \operatorname{ave} \left[p_{n/2}(t, v_1', v_2') - p_{n/2}(t, v_1, v_2) \right] \qquad \begin{array}{c} \text{summed over} \\ v_2 = \pm 1. \end{array}$$

Then take n to $+\infty$ to establish chaos, reducing the display, first to

$$\frac{\partial}{\partial t} p(v) = \sum_{v_2 = \pm 1} 2 \operatorname{ave} \left[p(v_1') p(v_2') - p(v_1) p(v_2) \right]$$

with $p = p_{\infty/1}(t, v)$ and v_1 identified with v, and then to

$$\frac{\partial}{\partial t} p(t, v) = p(t, -1) \left[p(t, -v) - p(t, v) \right].$$

In short, $\dot{x} = (1 - x)(1 - 2x)$ is just Boltzmann's equation in disguise.

Exercise 11.2.4   Check that $\dot{x} = (1 - x)(1 - 2x)$ is exactly what is produced by the n-molecule propagation, $\partial p_n / \partial t = \mathfrak{G}^\dagger p_n$, when $n \uparrow \infty$ *if chaos propagates.*

Exercise 11.2.5   Compute

$$p_{\infty/1}(t, +1) = x(t) = \frac{1 - x(0) + (2x(0) - 1)e^{-t}}{2(1 - x(0)) + (2x(0) - 1)e^{-t}}.$$

11.2.3 Why chaos propagates

The fact that for fixed $m < \infty$ and chaotic f,

$$\lim_{n \uparrow \infty} p_{n/m} = p_{\infty/1} \otimes \cdots \otimes p_{\infty/1} \ \ m\text{-fold}$$

may be reduced to the simpler law of large numbers obtained just by a trick of Grünbaum [1971] adapted from Sparre-Andersen [1953/54], for which see §5.8.

Proof. What's wanted is the binomial distribution

$$\lim_{n\uparrow\infty} p_{n/m}(+1 \text{ repeated } \ell \text{ times and } -1 \text{ repeated } m-\ell \text{ times})$$

$$= \binom{m}{\ell} x^\ell (1-x)^{m-\ell} \quad \text{with } x = p_{\infty/1}(\;\cdot\;, +1) \text{ and } 0 \le \ell \le m.$$

Let $\mathbf{e} = 1$ or 0 according as $+1$ is repeated by $v_1, \ldots, v_m$ exactly ℓ times or not. Then $\mathbb{E}(\mathbf{e}) = p_{n/m}(+1 \text{ so repeated})$ is insensitive to permutations $v \in Z^m \to v^*$, so with a self-evident notation, $\mathbf{e}$ may be replaced by $\frac{1}{n!}\sum \mathbf{e}^*$, and the latter may be evaluated as follows. Let k be the number of times $+1$ is repeated, by $v_1, \ldots, v_n$, i.e. let $\mathbf{p}_{n/1}(+1) = k/n$. If $k < \ell$, then $\mathbf{e}$ is always 0, while if $k \ge \ell$, then

$$\frac{1}{n!}\sum \mathbf{e}^* = \binom{n}{m}^{-1} \binom{k}{\ell}\binom{n-k}{m-\ell} \qquad \text{as you will check please}$$

$$= \binom{m}{\ell} \times \frac{k(k-1)\ldots(k-\ell+1)}{n(n-1)\ldots(n-\ell+1)}$$

$$\times \frac{(n-k)(n-k-1)\ldots(n-k-m+\ell+1)}{(n-\ell)(n-\ell-1)\ldots(n-m+1)}$$

$$\simeq \binom{m}{\ell}\left(\frac{k}{n}\right)^\ell \left(1-\frac{k}{n}\right)^{m-\ell} \qquad \text{for large } n$$

$$= \binom{m}{\ell}\left[\mathbf{p}_{n/1}(+1)\right]^\ell \times \left[1-\mathbf{p}_{n/1}(+1)\right]^{m-\ell}$$

$$\simeq \binom{m}{\ell} x^\ell (1-x)^{m-\ell} \qquad \text{by the law of large numbers.}$$

That does it.

Exercise 11.2.6 Think it over: why this is the same as saying $p_{\infty/m} = p_{\infty/1} \otimes \cdots \otimes p_{\infty/1}$ m-fold.

Exercise 11.2.7 Speaking of chaos, the symmetrical distributions, f_∞ on Z^∞, may be viewed as a section of a convex cone with a self-evident compactness, of which the general member must be a center of gravity of the vertices, as in §4.9: indeed, the vertices are the chaotic fs with factor $f_\theta(+1) = \theta$ and $f_\theta(-1) = 1-\theta$, indexed by $0 \le \theta \le 1$, and every f_∞ is a center of gravity of these.
Hint: Change from $v = \pm 1$ to Bernoulli-type variables $v' = \frac{1}{2}(v+1)$ and look at

$$\mathbb{P}\left[\#(j \le n : v_j' = +1) = k\right] = \binom{n}{k}\mathbb{P}\left[v_1', \ldots, v_k' = 1 \text{ and } v_{k+1}', \text{ etc. } = 0\right]$$

with help from §4.9.4.

The idea that chaos propagates is due to Kac [1956] who confirmed it for his somewhat more complicated, but still artificial type of collision; see also Grünbaum [1971] for something closer to the real thing and Mischler–Mouhot [2013] for the latest news on this subject.

11.3 The two-speed gas: fluctuations

The law of large numbers may be refined by a description of how the empirical distribution $\mathbf{p}_{n/1}(t, e)$ fluctuates about its mean $p_{n/1}(t, e)$. It is to be proved that as $n \uparrow \infty$, the scaled fluctuations

$$\mathbf{x}_n(t) = \sqrt{n}\left[\mathbf{p}_{n/1}(t, +1) - p_{n/1}(t, +1)\right] \quad (t \geq 0)$$

approximate a (Gaussian) modification of BM(1) with drift proportional to displacement, mostly pulling it back towards the origin ($\mathbf{x} = 0$), reminiscent of the OU process of §6.9. This is the analogue of CLT for the 2-speed gas. I follow McKean [1975b] as before[4].

Aside on Centering. $\mathbf{x}_n$ is centered about $p_{n/1}$, but it is more convenient to use $p_{\infty/1}$ instead. This makes no difference to the outcome for large n: in fact, with $\mathbf{p}_{n/1}(t, +1) = \mathbf{p}(t, +1)$ as before and a change of notation better suited to the present task, from $p_{\infty/1}(t, +1) = x$ to $p_{\infty/1}(t, +1) = p$, you find

$$k\frac{\mathrm{d}}{\mathrm{d}t}\left(p_{n/1} - p_{\infty/1}\right) = \frac{\mathrm{d}}{\mathrm{d}t}\mathbb{E}(\mathbf{p} - p)$$
$$= \mathbb{E}\left[(1 - \mathbf{p})(1 - 2\mathbf{p}) - (1 - p)(1 - 2p)\right] + \mathrm{O}(1/n)$$
$$= (4p - 3)\mathbb{E}(\mathbf{p} - p) + 2\mathbb{E}(\mathbf{p} - p)^2 + \mathrm{O}(1/n)$$

as in §11.2, from which $p_{n/1} = p_{\infty/1} + \mathrm{O}(1/n)$ follows since $\mathbb{E}(\mathbf{p} - p)^2$ is, itself, comparable to $1/n$.

Exercise 11.3.1 Please check this.

Now for $n \uparrow \infty$, $\mathbf{x}_n$ approximates a Brownian motion $\mathbf{x}_\infty$ with restoring drift, variously described by

(1) $\mathrm{d}\mathbf{x}_\infty(t) = \sqrt{1 - p}\,\mathrm{d}\mathbf{b}(t) + (4p - 3)\mathbf{x}_\infty\,\mathrm{d}t$ with a standard Brownian motion $\mathbf{b}$;

---

[4] If you do look at this paper, you will find a silly mistake halfway down p. 447 which spoils pp. 448–9. What is wanted here is done correctly.

or, in integrated form, by

(2) $\mathbf{x}_\infty(t) = e^{-k(t)}\mathbf{x}_\infty(0) + e^{-k(t)} \int_0^t e^k \sqrt{1-p}\,d\mathbf{b}(t')$ with $k(t) = \int_0^t (3-4p)\,dt'$;

or again, via Brownian scaling, by

(3) $\mathbf{x}_\infty(t) = e^{-k(t)}\mathbf{x}_\infty(0) + e^{-k(t)}\mathbf{b}^*\left(\int_0^t e^{2k}(1-p)\,dt'\right)$ with a new Brownian motion $\mathbf{b}^*$.

Exercise 11.3.2   Check (2) and (3) from (1).

Here, $\mathbf{x}_\infty$ is "guided" by p which controls the amplitude of the Brownian part $\sqrt{1-p}\,d\mathbf{b}$ and also the drift $(4p-3)\mathbf{x}_\infty\,dt$. Note that $\mathbf{x}_\infty$ is Markovian if you don't mind the fact that its transition mechanism changes with time; if you *do* mind, this can be corrected by keeping track of both $\mathbf{x}_\infty$ and p. Then (1) is supplemented by $dp = (1-p)(1-2p)\,dt$; the transition mechanism is the same at all times; and the (backward) infinitesimal operator of the pair $\mathbf{x}_\infty$ and p is

$$\mathfrak{G}_\infty = \frac{1}{2}(1-p)\frac{\partial^2}{\partial x^2} + (4p-3)\frac{\partial}{\partial x} + (1-p)(1-2p)\frac{\partial}{\partial p}.$$

Proof.   The way to proceed is to look at the infinitesimal operator $\mathfrak{G}_n$ for the pair $\mathbf{x}_n$ and p. That's not hard to compute. For $\mathbf{p}(t) = k/n$ and $p(t) = p$, $\mathbf{x}(t) = \sqrt{n}(k/n - p)$ jumps at rate n:

- *up* by $+\dfrac{1}{\sqrt{n}}$ with probability $\dfrac{(n-k)(n-k-1)}{n(n-1)}$;
- *down* by $-\dfrac{1}{\sqrt{n}}$ with probability $\dfrac{(n-k)k}{n(n-1)}$;
- or *not at all* with probability $\dfrac{k}{n}$,

moving also in response to p at rate $-\sqrt{n}(1-p)(1-2p)$, from which you read off

$$\mathfrak{G}_n f(x,p) = n \times \frac{(n-k)(n-k-1)}{n(n-1)} \times \left[f\left(x + \frac{1}{\sqrt{n}}\right) - f(x)\right]$$
$$+ n \times \frac{(n-k)k}{n(n-1)} \times \left[f\left(x - \frac{1}{\sqrt{n}}\right) - f(x)\right]$$
$$- \sqrt{n}(1-p)(1-2p)\frac{\partial f}{\partial x}$$
$$+ (1-p)(1-2p)\frac{\partial f}{\partial p} \quad \text{with } \frac{k}{n} = \mathbf{p} \text{ and } \mathbf{x} = \sqrt{n}(\mathbf{p}-p) = x.$$

This may be spelled out for large n and[5] $f \in C^{31}(\mathbb{R}, [0,1])$ as follows.

---

[5] $C^{31}(\mathbb{R}, [0,1])$ means $C^3(\mathbb{R})$ for fixed p and $C^1[0,1]$ for fixed x.

Up to negligible errors,

$$
\mathfrak{G}_n f(x,p) = \frac{(n-k)(n-k-1)}{n-1}\left[\frac{1}{\sqrt{n}}\frac{\partial f}{\partial x} + \frac{1}{2n}\frac{\partial^2 f}{\partial x^2}\right]
$$
$$
+ \frac{(n-k)k}{n-1}\left[-\frac{1}{\sqrt{n}}\frac{\partial f}{\partial x} + \frac{1}{2n}\frac{\partial^2 f}{\partial x^2}\right]
$$
$$
- \sqrt{n}(1-p)(1-2p)\frac{\partial f}{\partial x} + (1-p)(1-2p)\frac{\partial f}{\partial p}
$$
$$
= \frac{1}{2}\frac{\partial^2 f}{\partial x^2}\times\left[\frac{n-k}{n} = 1-\mathbf{p}\right]
$$
$$
+ \frac{\partial f}{\partial x}\times\left[\begin{array}{c}\frac{(n-k)(n-k-1)}{\sqrt{n}(n-1)} - \sqrt{n}(1-p)(1-2p)\\ ((1-\mathbf{p})(1-2\mathbf{p}) - (1-\mathbf{p})(1-2\mathbf{p}))\sqrt{n}\\ (2(\mathbf{p}+p)-3)\sqrt{n}(\mathbf{p}-p)\end{array}\right]
$$
$$
+ (1-p)(1-2p)\frac{\partial f}{\partial p}
$$
$$
\simeq \frac{1}{2}(1-p)\frac{\partial^2 f}{\partial x^2} + (4p-3)\,x\,\frac{\partial f}{\partial x} + (1-p)(1-2p)\frac{\partial f}{\partial p}
$$
$$
= \mathfrak{G}_\infty f(x,p)
$$

by the law of large numbers!

Now under mild technical conditions, amply satisfied here, the tendency of $\mathfrak{G}_n$ to $\mathfrak{G}_\infty$ implies the (statistical) tendency of $\mathbf{x}_n$ to $\mathbf{x}_\infty$. That does the trick. I do not say more, so this is only a demonstration, but see H. Trotter [1958] for the precise statement and a proof.

★**Exercise 11.3.3** Use (3) and the Brownian scaling to show that for $T\uparrow\infty$, the shifted fluctuations $\mathbf{x}_\infty(\,\cdot\,+T)$ settle down to follow the shift-invariant, OU-type motion $\mathbf{x}(t) = e^{-t}\mathbf{b}^{**}(e^{2t}/4)$ with yet another Brownian motion and infinitesimal operator $\mathfrak{G} = \frac{1}{4}\partial^2/\partial x^2 - x\,\partial/\partial x$.

11.4 More about Boltzmann's equation

The little Boltzmann equation of Exercise 11.2.3, rewritten here as

$$
\partial p(v)/\partial t = \varkappa[p(-v) - p(v)]
$$

with solution $p_{\infty/1}(t,v)$ and rate $\varkappa = p_{\infty/1}(t,-1)$, calls for a bit more talk. It is descriptive of a "tagged" molecule in an infinite "bath" of like molecules and of how its velocity $\mathbf{v}(t)$ changes, its displacement being ignored as before. In equilibrium (i.e., $\varkappa = 1/2$), $\mathbf{v}(t) : t \geq 0$ is just the simplest 2-state Markov chain with states $e = \pm 1$ and identical rates $1/2$, represented by $\mathbf{v}(t) = \mathbf{v}(0)(-1)^{\#(t/2)}$ with a standard, rate-1 Poisson process $\#$. Out of equilibrium, $\mathbf{v}(t) = \mathbf{v}(0)(-1)^{\#(k)}$ is regulated by the "clock"

$$ k(t) = \int_0^t \varkappa \, dt' = \frac{1}{2} \log \left[\varkappa(0)(e^t - 1) + 1 \right], $$

running faster or slower in response to the updated "concentration" $\varkappa = p_{\infty/1}(t, -1)$ of velocities $v = -1$: what I call a "nonlinear" Markov process, guided by, i.e. with transition mechanism changing in response to, its updated distribution, just like the fluctuations $\mathbf{x}_\infty$ of §11.3.[6]

Now it is interesting to look also at the displacement $\mathbf{x}(t) = \mathbf{x}(0) + \int_0^t \mathbf{v}(t') \, dt'$, i.e. to take account of streaming as well. The pair $\mathbf{x}$ and $\mathbf{v}$ is Markovian in the same nonlinear sense, the analogue of Boltzmann's equation being $\partial p / \partial t + v \partial p / \partial x = \varkappa [p(-v) - p(v)]$. S. Goldstein [1951] showed that in equilibrium (i.e., $\varkappa = 1/2$), $\mathbf{x}(t) : t \geq 0$ approximates the standard Brownian motion $BM(1)$ when the collisions are speeded up by a factor $2/\varepsilon$, the displacement scaled up by a $1/\sqrt{\varepsilon}$, and ε is taken down to $0+$. This is easy to check with the help of CLT.

The underlying (scaled) displacement is now

$$ \mathbf{x}(t) = \mathbf{x}(0) + \frac{\mathbf{v}(0)}{\sqrt{\varepsilon}} \int_0^t (-1)^{\#(t'/\varepsilon)} \, dt' = \mathbf{x}(0) + \sqrt{\varepsilon} \, \mathbf{v}(0) \int_0^{t/\varepsilon} (-1)^{\#(t')} \, dt' $$

$$ \equiv \mathbf{x}(0) + \mathbf{z}(t) $$

which may be spelled out as follows. Let $0 < T_1 < T_2 <$ etc. be the jump times of $\#$. Then for $T_n \leq t/\varepsilon < T_{n+1}$ and $\mathbf{x}(0) = 0$,

$$ \mathbf{z}(t) = \sqrt{\varepsilon} \, \mathbf{v}(0) \int_0^{t/\varepsilon} (-1)^{\#(t')} \, dt' $$

$$ = \sqrt{\varepsilon} \, \mathbf{v}(0) \left[T_1 - (T_2 - T_1) + (T_3 - T_2) - \text{etc.} \pm (t - T_n) \right]. $$

Obviously, $t/\varepsilon \simeq T_n \simeq n$ by the standard LLN, so replacing n with $2n$ for simplicity, two applications of CLT show that, for $n \uparrow \infty$ and

---

[6] See McKean [1969] for a general account of such things.

regardless of $\mathbf{v}(0) = \pm 1$,

$\mathbf{z}(t)$

$$\simeq \sqrt{t}\,\mathbf{v}(0) \times \frac{1}{2\sqrt{n}}\big[(T_1 - 1) + (T_3 - T_2 - 1) + \cdots + (T_{2n-1} - T_{2n-2} - 1)$$
$$- (T_2 - T_1 - 1) - (T_4 - T_3 - 1) - \cdots - (T_{2n} - T_{2n-1} - 1)\big]$$

approximates $\sqrt{t} \times 1/\sqrt{2} \times$ the difference of two independent, standard Gaussian variables. In short,

$$\lim_{\varepsilon \downarrow 0} \mathbb{P}\,[\mathbf{z}(t) \le x] = \int_{-\infty}^x \frac{e^{-y^2/2t}}{\sqrt{2\pi t}}\,dy,$$

imitating BM(1).

Exercise 11.4.1 As a check, compute $\mathbb{E}(\mathbf{z})$ and $\mathbb{E}(\mathbf{z}^2)$ at $\varepsilon = 0+$ directly.

It remains to show that for fixed T and small ε, the future $\mathbf{x}^+(t) = \mathbf{x}(t + T) - \mathbf{x}(T)$ wants to be independent of the past $\mathbf{x}(t')$: $t' \le T$, for which you have only to observe that $\mathbf{x}^+(t)$ looks like $\mathbf{v}(t) \times \frac{1}{\sqrt{\varepsilon}}\int_0^t (-1)^{\#(t'/\varepsilon)}\,dt'$ with a new Poisson process $\#$, independent of the past. Here, $\mathbf{v}(T) = \dot{\mathbf{x}}(T)$ *does* depend upon the past, but that does not matter since the (Brownian) distribution of the rest at $\varepsilon = 0+$ does not care about a signature.

Exercise 11.4.2 Check that out of equilibrium, the scaled displacement with constant $\varkappa \ne 1/2$ approximates the modified Brownian motion $\varkappa^{-1}\mathbf{b}(\varkappa t)$, *aka* $\varkappa^{-1/2}\mathbf{b}(t)$, by Brownian scaling. This will be wanted shortly.

Exercise 11.4.3 Check that the scaled Boltzmann equation

(1) $\partial p/\partial t + \frac{v}{\sqrt{\varepsilon}}\,\partial p/\partial x = \frac{1}{\varepsilon}[p(-v) - p(v)]$

is equivalent to the telegrapher's equation

(2) $\partial^2 p/\partial t^2 + \frac{2}{\varepsilon}\,\partial p/\partial t = \frac{1}{\varepsilon}\,\partial^2 p/\partial x^2$,

subject to the special condition (1) imposed at $t = 0+$. Goldstein [1951] based his proof on this remark. The present, more efficient method stems from Kac [1974] who emphasized the curious fact that up to trivialities, $\partial p/\partial t + v\,\partial p/\partial x = p(-v) - p(v)$ is the only known "hyperbolic system" with a statistical interpretation, all attempts to do something similar in, e.g. dimension 3, having failed. Note, in particular, that (1) preserves positivity, but (2) in general does not.

11.5 The two-speed gas with streaming

The introduction of streaming brings with it serious difficulties. The n-molecule displacement $\mathbf{x}(t) = \mathbf{x}(0) + \int_0^t \mathbf{v}(t')\, dt'$ must now be tracked and some mechanism is wanted to rule out collisions unless the participating molecules are close together. The n-molecule concentration $p(t, x, v)$ is now a symmetric function of the pairs (x_1, v_1), (x_2, v_2) etc., for which I propose the forward equation

$$\frac{\partial p}{\partial t} + v \cdot \frac{\partial p}{\partial x} = \frac{2}{n-1} \sum_{i<j} \mathbb{E}\left[p(v'_{ij}) - p(v_{ij})\right] \times$$
$$\times \frac{n}{2} \text{ times the indicator of } |x_i - x_j| < \frac{1}{n}$$

in the notation of §11.2, with the understanding that x_i and x_j do not change when molecules i and j collide. If, then, chaos were to propagate and other miracles be seen, you would come to Boltzmann's equation for the one-molecule concentration function $p(t, x, v)$ of $t \geq 0$, $x \in \mathbb{R}$, and $v = \pm 1$ in the form

$$\frac{\partial p(v)}{\partial t} + v\frac{\partial p(v)}{\partial x} = p(-1)\left[p(-v) - p(v)\right]$$

with the same t and x on both sides. I have *no idea* how to prove this, so take it on faith and let's use it as a guide to what may be going on in general. What happens here is complicated enough, as you will see.

The underlying velocity is now $\mathbf{v}(t) = \mathbf{v}(0)(-1)^{\#(k)}$, with a standard Poisson process $\#$ run by the clock

$$k(t) = \int_0^t p(t', \mathbf{x}(t'), -1)\, dt',$$

displaying the apparent complication in comparison to §11.4, that the clock is now coupled to the displacement, but this is not so bad as you might think: for short times, $\mathbf{v}(t) = \mathbf{v}(0)$ and $\mathbf{x}(t) = \mathbf{x}(0) + t\mathbf{v}(0)$, up to the time $\mathbf{t}_1$, when

$$k(t) = \int_0^t p\big(t', \mathbf{x}(0) + t'\mathbf{v}(0), -1\big)\, dt' = \text{the first jump of } \#;$$

then $\mathbf{v}(t)$ changes sign and $\mathbf{x}(t) = \mathbf{x}(0) + \mathbf{t}_1\mathbf{v}(0) - (t - \mathbf{t}_1)\mathbf{v}(0)$, up to the time $\mathbf{t}_2$ when

$$k(t) = \text{the first jump of } \# + \int_{\mathbf{t}_1}^t p\big(t', \mathbf{x}(0) + \mathbf{t}_1\mathbf{v}(0) - (t - \mathbf{t}_1)\mathbf{v}(0), -1\big)\, dt'$$
$$= \text{the second jump of } \#;$$

and so on – in short, the motion is only a bit more complicated: once $p(t, x, v)$, $\mathbf{x}(t)$, and $\mathbf{v}(t)$ are known, you may compute $p(t', x, v)$ for $t' \geq t$, and the motion continues by the same recipe, updated. In short, once Boltzmann's equation is solved for the rate function $p(t, x, -1)$, you know what the velocity is doing.

But what about the displacement $\mathbf{x}(t) = \mathbf{x}(0) + \int_0^t \mathbf{v}(t')\, dt'$? To study that, the collisions are speeded up by a factor $1/\varepsilon$ and the displacement scaled up by a $1/\sqrt{\varepsilon}$, as in §11.4, so that

$$\frac{\partial p(v)}{\partial t} + \frac{x}{\sqrt{\varepsilon}}\frac{\partial p(v)}{\partial x} = \frac{p(-1)}{\varepsilon}\left[p(-v) - p(v)\right],$$

and ε is taken to 0+. At $\varepsilon = 0+$, the local mean velocity $\bar{v} = \frac{p(+1)-p(-1)}{p(+1)+p(-1)}$ vanishes like $\sqrt{\varepsilon}$, the local concentration $n = p(+1) + p(-1)$ solves the surprising equation $\partial n/\partial t = \partial^2 \log n/\partial x^2$, and the displacement reduces to a modified Brownian motion guided by n, with (backward) infinitesimal operator

$$\mathfrak{G} = \frac{\partial}{\partial x}\frac{1}{n}\frac{\partial}{\partial x} = \frac{1}{n}\frac{\partial^2}{\partial x^2} - \frac{n'}{n^2}\frac{\partial}{\partial x},$$

or so it should have if all goes well. You see I wasn't kidding – it's *not simple at all*, and the demonstration offered here is *not* a proof. Quite candidly, I do not know how to do much better.

11.5.1 Solving Boltzmann

Boltzmann's equation is readily converted into

$$p(t, x, v) = e^{-k(t)}p(0+, x, v) + e^{-k(t)}\int_0^t p\left(t', x + \frac{v}{\sqrt{\varepsilon}}t', -v\right) de^{k(t')}$$

with

$$k(t) = \frac{1}{\varepsilon}\int_0^t p\left(t', x + \frac{vt'}{\sqrt{\varepsilon}}, -1\right) dt'$$

and x replaced by $x - \frac{v}{\sqrt{\varepsilon}}t$ throughout the right side, as you will please check. This may be solved by routine iteration, starting from $f(x, v) = p(0+, x, v)$ as a first approximation, and it's easy to see that p inherits many nice properties of f. For example, any initial bounds such as $0 < A \leq f \leq B$ are inherited by p since the right side is an averaging in view of $e^{-k}(1 + e^k - 1) = 1$; also $p(t, \pm\infty, v = 1/2)$, signifying a Maxwellian equilibrium at infinity, if $f(\pm\infty, v) = 1/2$; and a little more work (tedious to do) shows that p is smooth on $[0, \infty) \times \mathbb{R} \times (\pm 1)$ if f is smooth on $\mathbb{R} \times (\pm 1)$. I take all that for granted without further notice, choosing $A = 1/3$ and $B = 2/3$ for definiteness.

11.5.2 Carleman's gas

Things would be much simpler if the rate $p(-1)$ to the right in Boltzmann's equation could be replaced by half the concentration n, as would be automatic if the mean velocity $\bar{v}$ were to vanish, and that's what it wants to do as $\varepsilon \downarrow 0$. To see this, write

$$I(t) = \int_{-\infty}^{\infty} \left| p(+1) - \frac{1}{2} \right|^2 dx + \int_{-\infty}^{\infty} \left| p(-1) - \frac{1}{2} \right|^2 dx$$

and compute

$$\frac{dI}{dt} = 2 \int \left[p(+1) - \frac{1}{2} \right] \times \left[-\frac{p'(+1)}{\sqrt{\varepsilon}} + \frac{p(-1)}{\varepsilon} [p(-1) - p(+1)] \right]$$

$$+ 2 \int \left[p(-1) - \frac{1}{2} \right] \times \left[\frac{p'(-1)}{\sqrt{\varepsilon}} + \frac{p(-1)}{\varepsilon} [p(+1) - p(-1)] \right]$$

$$= -\frac{2}{\varepsilon} \int p(-1)[p(+1) - p(-1)]^2,$$

with the implication that I decreases and that

$$\int_0^{\infty} dt \int_{-\infty}^{\infty} |p(+1) - p(-1)|^2 \le \frac{3\varepsilon}{2} [I(0) - I(\infty)] \quad \text{since } p(-1) \ge 1/3,$$

indicating, though it does not prove, that the local mean velocity $\bar{v}$ imitates $\sqrt{\varepsilon}$. I take this at face value, as permission to replace the rate $p(-1)$ by $n/2 =$ half the local concentration, as in the equation of Carleman [1957]:

$$\frac{\partial p}{\partial t} + \frac{v}{\sqrt{\varepsilon}} \frac{\partial p}{\partial x} = \frac{n}{2\varepsilon} [p(-v) - p(v)] = \frac{1}{2\varepsilon} [p^2(-v) - p^2(v)].$$

This is *wholly artificial* in that the rule of collision is now physically ridiculous, *to wit*, two molecules at the same place and with like velocity may of a sudden turn around and both go backwards. But it is not, I think, mathematically ridiculous and has this advantage over the two-speed gas: that $\partial p/\partial x$ can now be controlled, independently of $t \ge 0$, $x \in \mathbb{R}$, and $\varepsilon \downarrow 0$, and likewise $\partial p/\partial t$ after a harmless modification of its initial datum. In detail, if $f(x,v) = e + vo$ with e (for even) $= \frac{1}{2}[f(+1) + f(-1)]$ and o (for odd) $= \frac{1}{2}[f(+1) - f(-1)]$, then

$$-\frac{\partial p}{\partial t}(0+, x, v) = +\frac{1}{\sqrt{\varepsilon}} v(e' + vo') + \frac{e}{\varepsilon} \times 2vo = \frac{o'}{\sqrt{\varepsilon}} + v \left(\frac{e'}{\sqrt{\varepsilon}} + \frac{2oe}{\varepsilon} \right)$$

is controlled only if $o = -\frac{1}{2}\sqrt{\varepsilon}\, e'/e$, or very nearly so, in which case

$$\frac{\partial n}{\partial t} = -\frac{2}{\sqrt{\varepsilon}}\frac{\partial o}{\partial x} = \frac{\partial^2 \log n}{\partial x^2} \quad \text{at } t = 0+,$$

in accord with the "surprising equation" announced before, and

$$p(0+, x, v) = e - \frac{v}{2}\sqrt{\varepsilon}\,\frac{e'}{e}$$

is very nearly in its local Maxwellian equilibrium in accord with $p(t, \pm\infty, v) = 1/2$. McKean [1975a:84–85] gives the details for $\partial p/\partial x$ and $\partial p/\partial t$, together with a cautionary note, that the type of estimate employed there *cannot* be obtained for the two-speed gas. Something else is needed, but I don't know what.

11.5.3 The surprising equation

Now back to Boltzmann with the rate $p(-1)$ replaced by $n/2$ and $p = e + vo$ with

$$e \text{ (for even)} = \frac{1}{2}[p(+1) + p(-1)] \text{ and } o \text{ (for odd)} = \frac{1}{2}[p(+1) - p(-1)]$$

as before. In this language,

(1) $\dfrac{\partial e}{\partial t} = -\dfrac{1}{\sqrt{\varepsilon}}\dfrac{\partial o}{\partial x}$

(2) $\dfrac{\partial o}{\partial t} = -\dfrac{1}{\sqrt{\varepsilon}}\dfrac{\partial e}{\partial x} - \dfrac{no}{\varepsilon},$

so with

$$k(t) = \frac{1}{\varepsilon}\int_0^t n(t', x)\, dt',$$

you find, from (2),

$$o(t, x) = e^{-k(t)}o(0+, x) - \frac{1}{\sqrt{\varepsilon}}\frac{\partial}{\partial x}\, e^{-k(t)}\int_0^t \frac{\frac{\partial e}{\partial x}(t', x)}{n(t', x)/\varepsilon}\, de^{k(t')},$$

and then, from (1),

$$\frac{\partial n}{\partial t} = -\frac{2}{\sqrt{\varepsilon}}\frac{\partial}{\partial x}\, e^{-k(t)}o(0+, x) + \frac{\partial}{\partial x}\int_0^t \frac{\partial}{\partial x}\log n(t', x)e^{-k(t)}\, de^{k(t')}.$$

Presumably, the first piece to the right is negligible for $\varepsilon \downarrow 0$ since $k(t) \geq 2t/3\varepsilon$, n being $\geq 2/3$, and since, for the same reason, $e^{-k(t)}\, de^{k(t')}$ concentrates its weight at $t' = t$, you ought to have $\partial n/\partial t = \partial^2 \log n/\partial x^2$ at $\varepsilon = 0+$, but that requires more control of p than was elicited so far.

Fortunately, there is a more sophisticated machinery that does the trick. I refer to "Weyl's lemma", for which see, e.g. McKean [2005:85–90].

The story begins this way. Take a smooth "test function" φ of t and x, vanishing outside an open box such as $(0, T) \times (-L, L) \subset [0, \infty) \times \mathbb{R}$. If, then, $\partial n / \partial t = \partial^2 \log n / \partial x^2$, you would have

$$\int_0^\infty dt \int_{-\infty}^{+\infty} dx\, \varphi \left[\frac{\partial^2 \log n}{\partial x^2} - \frac{\partial n}{\partial t} \right]$$
$$= \int_0^\infty dt \int_{-\infty}^{+\infty} dx \left[\frac{\partial^2 \varphi}{\partial x^2} \log n + \frac{\partial \varphi}{\partial t} n \right] = 0.$$

Weyl's lemma turns the thing around to guarantee that if $n \in C[(0, \infty) \times \mathbb{R}]$ is positive so that $\log n$ makes sense, and if line 2 of the display vanishes for every test function, then in fact $n \in C^\infty[(0, \infty) \times \mathbb{R}]$ and solves $\partial n / \partial t = \partial^2 \log n / \partial x^2$ in the naive sense. The plan is to check that the concentration $n_\varepsilon(t, x)$ computed from Boltzmann's equation for fixed $\varepsilon > 0$ fits the bill at $\varepsilon = (0+)$.

Here's the computation for which I write mostly $n_\varepsilon = n$ plain to keep the notation as simple as I can.

Step 1 is nothing much. Ignoring the $-\frac{2}{\sqrt{\varepsilon}} \frac{\partial}{\partial x} e^{-k} o(0+)$ in $\frac{\partial n}{\partial t}$, you may write

$$\int_0^\infty dt \int_{-\infty}^{+\infty} \varphi \frac{\partial n}{\partial t}\, dx$$
$$= \int_0^\infty dt \int_{-\infty}^{+\infty} dx\, \varphi(t, x) \frac{\partial}{\partial x} \int_0^t \frac{1}{\varepsilon} \frac{\partial n}{\partial x}(t', x)\, e^{- \int_{t'}^t \frac{n}{\varepsilon}(t'', x)\, dt''}\, dt'$$
$$= - \int_{-\infty}^{+\infty} \int_0^\infty \frac{\partial \varphi}{\partial x}(t, x)\, dt\, dx \int_0^t \frac{1}{\varepsilon} \frac{\partial n}{\partial x}(t', x) e^{- \int_{t'}^t \frac{n}{\varepsilon}(t'', x)\, dt''}\, dt'.$$

Step 2 is to move the $\frac{\partial \varphi(t, x)}{\partial x}$ under the next integral to a $\frac{\partial \varphi(t', x)}{\partial x}$. This is harmless since the error so committed is controlled by $\frac{\partial n}{\partial x}$ and

$$\int_{-L}^L \int_0^T dt\, dx \int_0^t \frac{1}{\varepsilon}(t - t') e^{-(t - t')2/3\varepsilon} < 2LT \times \int_0^\infty s e^{-2s/3}\, ds \times \varepsilon,$$

so throw the error away and write

$$\int_0^\infty \mathrm{d}t \int_{-\infty}^{+\infty} \varphi \frac{\partial n}{\partial t} \, \mathrm{d}x$$

$$= -\int_{-\infty}^{+\infty} \mathrm{d}x \int_0^\infty \mathrm{d}t \int_0^t \frac{\partial \varphi}{\partial x} \frac{1}{\varepsilon} \frac{\partial n}{\partial x} \, \mathrm{e}^{-\int_{t'}^t \frac{n}{\varepsilon} (t'',x) \, \mathrm{d}t''} \, \mathrm{d}t'$$

$$= -\int_{-\infty}^{+\infty} \mathrm{d}x \int_0^\infty \mathrm{d}t' \frac{\partial \varphi}{\partial x} (t',x) \frac{\partial}{\partial x} \log n(t',x) \times$$

$$\times \frac{n(t',x)}{\varepsilon} \int_{t'}^\infty \mathrm{e}^{-\int_{t'}^t \frac{n}{\varepsilon} (t'',x) \, \mathrm{d}t''} \, \mathrm{d}t.$$

Step 3 is to take $n(t',x)/\varepsilon$ in the line just above and put it under the last integral as an $n(t,x)/\varepsilon$, committing an additional error controlled as in Step 2. This likewise may be thrown away, and the last integral computed to produce *unity*, whereupon you have

$$\int_0^\infty \mathrm{d}t \int_{-\infty}^{+\infty} \varphi \frac{\partial n}{\partial t} \, \mathrm{d}x = -\int_{-\infty}^{+\infty} \mathrm{d}x \int_0^\infty \frac{\partial \varphi}{\partial x} \frac{\partial \log n}{\partial x} \, \mathrm{d}t'$$

$$= \int_0^\infty \int_{-\infty}^{+\infty} \frac{\partial^2 \varphi}{\partial x^2} \log n \, \mathrm{d}x,$$

up to errors vanishing at $\varepsilon = 0+$.

Step 4. $\partial n/\partial x$ and $\partial n/\partial t$ being controlled, you may now take $\varepsilon \downarrow 0$ in such a way that $n = n_\varepsilon$ tends to a positive fraction $n_{0+} \in C[(0,\infty) \times \mathbb{R}]$ satisying the premises of Weyl's lemma.[7] Then $n_{0+} \in C^\infty[(0,\infty) \times \mathbb{R}]$ solves $\partial n/\partial t = \partial^2 \log n/\partial x^2$ with the fixed initial data determined at $\varepsilon = 0+$ by the recipe of §11.5.2, and as there can be only *one* such function, the *full* limit n_{0+} exists and solves the "surprising equation".

Don't worry too much about all that. Rather, I would have you realize that $\bar{v} = 0$ and $\partial n/\partial t = \partial^2 \log n/\partial x^2$ are the two-speed hydrodynamical equations, as in "Stage 4" of §11.1.

But what about the thermodynamical "Stage 5", describing what happens when t is very large?

Define

$$I_1 = \int_{-\infty}^{+\infty} (n-1)^2, \quad I_2 = \int_{-\infty}^{+\infty} (n')^2/n, \quad \text{and} \quad I_3 = \int_{-\infty}^{+\infty} |(\log n)''|^2.$$

Keeping in mind that $2/3 \le n \le 4/3$ and that $\partial n/\partial x$, $\partial n/\partial t$, and so also $\partial^2 \log n/\partial x^2$ are controlled, it is not too difficult to show that if $I_1(0)$

---

[7] See Exercise 1.7.4 for this.

is finite, then $I_1 \le -2I_2$ and $I_2 \le -2I_3$, whence $\int_0^\infty I_2 \, dt \le I_1(0) < \infty$ and $I_2(\infty) = 0$, from which follows

$$(n-1)^4 = \left[2 \times \int_{-\infty}^x (n-1)n' \right]^2 \le 4 \int_{-\infty}^{+\infty} n(n-1)^2 \times \int_{-\infty}^{+\infty} \frac{(n')^2}{n}$$

$$\le \frac{16}{3} I_1 I_2 \downarrow 0,$$

so that's the thermodynamical stage: $\overline{v} \equiv 0$ and $n \equiv 1$. And don't ask what happened to the temperature: it is merely $\frac{1}{2}(1+\overline{v})^2 + \frac{1}{2}(1-\overline{v})^2 = 1$, so it is pointless to speak about it.

Exercise 11.5.1   Do the estimates of I_1, I_2 and I_3. It's a little tricky if done carefully, but fun, and the entrance of Fisher's information is nice.

Exercise 11.5.2   I do not know any explicit solution of $\partial n / \partial t = \partial^2 \log n / \partial x^2$ with $n(t, \infty) = 1$ other than $n \equiv 1$. There *are* plenty of these. Despite its singular appearance, any non-negative datum $n(0+, \cdot)$, not identically zero, gives rise to a smooth, positive solution, inheriting the value of $n(0+, \infty)$, for which see Mckean [1975a]. At the opposite extreme, with $n(0+, \infty) = 0$, is the solution $n(t, x) = 2t(c^2 t^2 + x^2)^{-1}$, reducing at $t = 0+$ to the unit mass when $c = 2\pi$? Please check that. Perhaps you can find other solutions not known to me.

11.5.4 Velocity and displacement

I will be brief. Write $k(t)$ for the clock $\int_0^t (n/2)(t', \mathbf{x}(t')) \, dt'$. Then much as in §11.4,

$$\mathbf{v}(t) = \mathbf{v}(0)(-1)^{\#(k/\varepsilon)},$$

and the scaled displacement is

$$\mathbf{x}(t) = \mathbf{x}(0) + \frac{\mathbf{v}(0)}{\sqrt{\varepsilon}} \int_0^t (-1)^{\#(k'/\varepsilon)} \, dt' \quad \text{with } k' = k(t').$$

Now Exercise 11.4.2 shows that if $n/2$ is constant ($\varkappa$), then $\mathbf{x}(t) - \mathbf{x}(0)$ approximates the modified Brownian motion $\varkappa^{-1}\mathbf{b}(\varkappa t)$, *aka* $\varkappa^{-1/2}\mathbf{b}(t)$, by Brownian scaling, indicating what should happen here, *to wit*: that at $\varepsilon = 0+$, you should see

$$\mathbf{x}(t) = \mathbf{x}(0) + \int_0^t \frac{d\mathbf{b}(t')}{\sqrt{n(t', \mathbf{x}(t'))/2}},$$

where n is now the solution of $\partial n/\partial t = \mathfrak{G}^\dagger n = \partial^2 \log n/\partial x^2$. But what does this really mean? I suppose it means

$$\mathbf{x}(t) = \mathbf{x}(0) + \lim_{l\uparrow\infty} \sum_{k/2^l \leq t} \frac{\mathbf{b}((k+1)/2^l) - \mathbf{b}(k/2^l)}{\sqrt{n(k'/2^l, \mathbf{x}(k'/2^l))/2}}$$

for some choice of k' between k and $k+1$. The customary choice would be $k' = k$, making $\mathbf{b}((k+1)/2^l) - \mathbf{b}(k/2^l)$ stick out into the future in the style of §6.9, leading easily to the backward infinitesimal operator $\mathfrak{G} = \frac{1}{n}(\partial^2/\partial x^2)$. Then the forward operator would be $\mathfrak{G}^\dagger = (\partial^2/\partial x^2)\frac{1}{n}$, and you would have $\mathfrak{G}^\dagger n = 0$. Oh dear. But let's try $k' = k+1$ instead. Now to a good approximation,

$$\frac{1}{\sqrt{n(\frac{k+1}{2^l}, \mathbf{x}(\frac{k+1}{2^l}))/2}} =$$

$$\frac{1}{\sqrt{n/2}} + \sqrt{2}\left(-\frac{1}{2}\frac{n'}{n^{3/2}}\right) \times \left[\mathbf{x}\left(\frac{k+1}{2^l}\right) - \mathbf{x}\left(\frac{k}{2^l}\right) = \frac{\mathbf{b}(\frac{k+1}{2^l}) - \mathbf{b}(\frac{k}{2^l})}{\sqrt{n/2}}\right],$$

with the understanding that all the ns are taken at $t = k/2^l$ and $x = \mathbf{x}(k/2^l)$, and this reduces, after multiplication by the Brownian increment, to

$$\frac{b((k+1)/2^l) - b(k/2^l)}{\sqrt{n/2}}, \text{ with the corrective drift } -\frac{n'}{n^2} \times 2^{-l},$$

providing the correct, self-dual infinitesimal operator

$$\mathfrak{G} = \frac{1}{n}\frac{\partial^2}{\partial x^2} - \frac{n'}{n^2}\frac{\partial}{\partial x} = \frac{\partial}{\partial x}\frac{1}{n}\frac{\partial}{\partial x} = \mathfrak{G}^\dagger,$$

in accord with $\partial n/\partial t = \mathfrak{G}^\dagger n = \partial^2 \log n/\partial x^2$.

11.6 Chapman–Enskog–Hilbert

This recipe, cited in §11.1, was proposed by Hilbert [1912], subsequently elaborated by Chapman [1916] and Enskog [1917], and now bears all three names: CEH for short. Here's how it goes for the two-speed gas in local equilibrium ($\varkappa = 1/2$): the collisions are speeded up by a big factor $2/\varepsilon$ as per

(1) $\quad \dfrac{\partial p}{\partial t} = -v\dfrac{\partial p}{\partial x} + \dfrac{1}{\varepsilon}[p(-v) - p(v)] = \mathfrak{G}^\dagger p;$

p is written as a formal power series $p_0 + \varepsilon p_1 + \varepsilon^2 p_2 +$ etc. with the epsilon right side up, and then you say: Oh. I didn't really mean that: the epsilon was just for book keeping. Put it back to 1 and let's see what comes out. Quite an odd thing to do. I follow McKean [1967] where more can be found.

11.6.1 First pass

Write $p(-v) - p(v) = Dp(v)$ for brevity, noting for future use that $Dv + vD = -2v$ and $D^2 = -2D$. Evidently, p is killed by

$$\left(\frac{\partial}{\partial t} - v\frac{\partial}{\partial x} + \frac{1}{\varepsilon}D\right)\left(\frac{\partial}{\partial t} + v\frac{\partial}{\partial x} - \frac{1}{\varepsilon}D\right)$$
$$= \frac{\partial^2}{\partial t^2} - \frac{\partial^2}{\partial x^2} + \frac{2}{\varepsilon}\left(-v\frac{\partial}{\partial x} + \frac{D}{\varepsilon}\right) = \mathfrak{G}^\dagger,$$

which is to say that p solves the telegrapher's equation

(2) $\quad \dfrac{\partial^2 p}{\partial t^2} + \dfrac{2}{\varepsilon}\dfrac{\partial p}{\partial t} - \dfrac{1}{\varepsilon}\dfrac{\partial^2 p}{\partial x^2} = 0,$

subject to $p(0+) = f$ and $\partial p(0+)/\partial t = \mathfrak{G}^\dagger f$ as in (1), recapitulating Exercise 11.4.3 without the scaling of x by $\sqrt{\varepsilon}$. The way into CEH is now to observe that $\partial^2/\partial t^2 + \frac{2}{\varepsilon}\partial/\partial t - \frac{1}{\varepsilon}\partial^2/\partial x^2$ can be factored in another way with the help of the formal power series[8]

$$\varepsilon^{-1}\left(1 - \sqrt{1 + \varepsilon^2\partial^2}\right) = \sum_1^\infty (-1)^n \binom{2n}{n} 2^{-2n} \frac{\varepsilon^{2n-1}}{2n-1}\partial^{2n},$$

in which ∂ is short for $\partial/\partial x$: in fact,

$$\left(\frac{\partial}{\partial t} + \frac{1 + \sqrt{1 + \varepsilon^2\partial^2}}{\varepsilon}\right)\left(\frac{\partial}{\partial t} + \frac{1 - \sqrt{1 + \varepsilon^2\partial^2}}{\varepsilon}\right) = \frac{\partial^2}{\partial t^2} + \frac{2}{\varepsilon}\frac{\partial}{\partial t} - \frac{1}{\varepsilon}\frac{\partial^2}{\partial x^2}$$

by inspection, so if the formal power series $p = p_0 + \varepsilon p_1 +$ etc., solves (1), then $[\partial/\partial t + \varepsilon^{-1}(1 - \sqrt{1 + \varepsilon^2\partial^2})]p \equiv q$, which is itself a formal power series $q_0 + \varepsilon q_1 +$ etc., is killed by $\partial/\partial t + \varepsilon^{-1}(1 + \sqrt{1 + \varepsilon^2\partial^2})$. But this operator is $2/\varepsilon$ to a first approximation, from which you see that q_0 can only be 0, and indeed, that the whole of q must vanish, i.e.

(3) $\quad \dfrac{\partial p}{\partial t} = (\sqrt{1 + \partial^2} - 1)p$

after putting ε back to 1 as the recipe CEH says to do.

---

[8] Recall the sum $\sum_1^\infty \binom{2n}{n} 2^{-2n}(2n-1)^{-1}\gamma^{2n-1} = \frac{1 - \sqrt{1-\gamma^2}}{\gamma}$ from §3.2 on RW(1).

11.6.2 Second pass

Now let's forget this very queer procedure and try to see what (3) says in comparison to (1) to which it purports to be equivalent. Evidently,

$$\left(\sqrt{1+\partial^2}-1\right)p = \left(-v\frac{\partial}{\partial x}+D\right)p$$

imposes upon p a special form. Split p into its odd and even parts relative to $v = \pm 1$, as in $p = e + v\,o$. You have

$$\left(\sqrt{1+\partial^2}-1\right)e + v\left(\sqrt{1+\partial^2}-1\right)o = -ve' - o' - 2vo,$$

i.e.

$$\left(\sqrt{1+\partial^2}-1\right)e = -o'$$

and also

$$\left(\sqrt{1+\partial^2}+1\right)o = -e',$$

of which the first says that $o = -\partial^{-1}\left(\sqrt{1+\partial^2}-1\right)e$, and the second says the same since $\partial\left(\sqrt{1+\partial^2}+1\right)^{-1} = \partial^{-1}\left(\sqrt{1+\partial^2}-1\right)$. In short,

$$p(x,v) = \left[1 + v\partial^{-1}\left(\sqrt{1+\partial^2}-1\right)\right]e(x) \quad\text{with}\quad e = \frac{1}{2}\left[p(+1) + p(-1)\right].$$

Obviously, the class of functions of this special type is preserved by the equivalent formal flows (1) and (3). I call it K. There, the whole of p is fully specified by the initial concentration $n = 2e$ since (3) propagates e independently of everything else, this being the appropriate statement of the dubious "Hilbert's paradox" cited at Stage 4 of §11.1.

Note in passing, that the map

$$\Gamma : p \to \frac{1}{2}\left[1 + \frac{1 - v\partial + D}{\sqrt{1+\partial^2}}\right],$$

applied to the general function p, is the projection onto K: Indeed,

$$4(1+\partial^2)\Gamma(1-\Gamma) = \left(\sqrt{1+\partial^2}+1-v\partial+D\right)\left(\sqrt{1+\partial^2}-1+v\partial-D\right) = 0,$$

as you will check please, so that $\Gamma = \Gamma^2$ is a projection, and besides, it acts as the identity precisely on K since $\sqrt{1+\partial^2}-1$ and $-v\partial + D$ are equivalent in that class. Γ will be useful presently.

11.6.3 Making better sense of all that

It is conventional in this type of business to ask that the formal power series $p = p_0 + \varepsilon p_1 + $ etc. make naive sense, but that is somewhat wrong-headed, asking too much; in particular, it requires that p be band-limited to frequencies $-1 \le k \le 1$, i.e. that its Fourier transform $\hat{p}(k)$ vanish for $|k| \ge 1$, and what has that to do with anything? Nothing one would hope.

Note also that any naive reduction of the operator $\sqrt{1 + \partial^2} - 1 = \partial^2/2 - \partial^4/8 + \partial^6/16 - $ etc., beyond $\partial^2/2$ which is sensible enough, leads to nonsense: for example, if the density $n = p(+1) + p(-1) = x^4$ near $x = 0$, then $\partial n/\partial t = -3$ *at* 0, which is ridiculous, n being a density and so non-negative – and it is the same for the Boltzmann equation of §11.1. CEH begins well enough but soon produces similar nonsense in the form of so-called corrections which may well be *numerically* valuable in restricted circumstances but are, in reality, *wrong*, the moral being that once you have a law of large numbers and some account of the Gaussian fluctuations about that bulk effect **stop**. Beyond that, abandon hope, or else **go all the way**. This could have been said more plainly in Chapter 2 already, but I did not have a satisfactory example then. Now I do.

To continue, the best way to proceed is what the Swedish mathematician Lars Gårding used to call "the fearless Fourier transform", meaning just *do* it and hope for the best, that something helpful will come out. Take the transform as in $\hat{f}(k) = (2\pi)^{-1} \int e^{-\sqrt{-1}\,kx} f(x)\,\mathrm{d}x$. Then $\hat{f}' = \sqrt{-1}\,k\hat{f}$, i.e. $\sqrt{-1}\,k$ is the "symbol" of $\partial = \partial/\partial x$ and $\sqrt{1 - k^2}$ the symbol of $\sqrt{1 + \partial^2}$. A little care must be taken with the latter: the reality of f is reflected by the fact that $\hat{f}(-k)$ is the complex conjugate of $\hat{f}(k)$, and you need $\sqrt{1 + \partial^2}$ to respect reality, the best way to guarantee this being to extend $\sqrt{1 - k^2}$, from low frequencies $|k| < 1$ where it is taken even and positive, to high frequencies $|k| \ge 1$, by analytic continuation in the upper half-plane as in Figure 11.6.1, so that $\sqrt{1 - k^2}$ is odd and

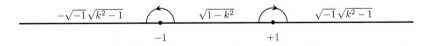

Figure 11.6.1 Analytic continuation of $\sqrt{1 - k^2}$

imaginary on the bordering line. OK. Keep it in mind. (1) is now easily solved for general $p(0+) = f$ and $\partial p/\partial t(0+) = \mathfrak{G}^{\dagger} f = (-v\,\partial/\partial x + D)f$. You have $\partial^2 \hat{p}/\partial t^2 + 2\partial \hat{p}/\partial t = -k^2 \hat{p}$, leading easily to a pleasant surprise when you work it out:

$$p = e^{-t} \int e^{\sqrt{-1}\, kx} (\Gamma f)^{\wedge} e^{\sqrt{1-k^2}\, t}\, dk$$

$$+ e^{-t} \int e^{\sqrt{-1}\, kx} ((1 - \Gamma) f)^{\wedge} e^{-\sqrt{1-k^2}\, t}\, dk,$$

Γ being the projection onto K just described, and this makes perfect sense if f is smooth and decays nicely at $x = \pm\infty$.

11.6.4 Focusing

Now something nice may already be seen: p decays as $t \uparrow +\infty$, but only slowly. The high frequency part plays only a minor role since $\sqrt{1 - k^2}$ is imaginary there and e^{-t} carries the day. Contrariwise, the low frequency part cannot be expected to decay much better than

$$e^{-t} \int_{-1}^{+1} e^{\sqrt{1-k^2}\, t}\, dk = \int_{-\pi/2}^{\pi/2} e^{(\cos\theta - 1)t} \cos\theta\, d\theta$$

$$\simeq \int_{-\pi/2}^{\pi/2} e^{-\theta^2 t/2}\, d\theta \simeq \sqrt{2\pi}/t.$$

But if now you project onto K, the error $(1-\Gamma)p$ decays like e^{-t} and even a little better owing to the fast oscillation of $\exp(\sqrt{1-k^2}\, t)$ for $|k| \geq 1$, with the moral that p comes very rapidly close to its projection, which is what I mean by "focusing". It represents a clear success for CEH.

Of course, this is not yet the whole story: p should be not only real but also positive if it is to make "philosophical" sense, and you must ask if projection preserves that. The answer is Yes and No, as simple examples show. The exact class of functions $f \geq 0$ with non-negative projection is not known to me. Perhaps there is something natural, i.e. "philosophically" correct about it. It would be nice to know.

11.7 Kac's gas

This gas of Kac [1956] is intermediate between Maxwell's gas and its two-speed caricature. The molecules move on the line as before, but now the velocities are arbitrary and the collision is a rotation

$$\begin{matrix} v_1 \\ v_2 \end{matrix} \longrightarrow \begin{matrix} v_1' \\ v_2' \end{matrix} = \begin{matrix} \cos\theta v_1 - \sin\theta v_2 \\ \sin\theta v_1 + \cos\theta v_2 \end{matrix} \equiv v_1 \circ v_2$$

with "scattering angle" θ uniformly distributed on the circle $S : 0 \leq \theta < 2\pi$. Obviously, the energy $\frac{1}{2}(v_1^2 + v_2^2)$ is preserved but the momentum $v_1 + v_2$ is not, as for the two-speed gas. The n-molecule probability density $p(t, v) : t \geq 0, v \in \mathbb{R}^n$ obeys

$$\frac{\partial p}{\partial t} = \frac{2}{n-1} \sum_{i<j} \text{ave} \left[p(v'_{ij}) - p(v_{ij}) \right] = \mathfrak{G}^\dagger p$$

with the old two-speed notation: $v_{ij} = (v_1, \ldots, v_n) \in \mathbb{R}^n$, v_k being fixed for $k \neq i, j$; v'_{ij} signifies the collision of v_i and v_j; and ave is the expectation with respect to the associated scattering angle. $\mathfrak{G}^\dagger$ is then the forward infinitesimal operator of a Markov chain $\mathbf{v}(t) : t \geq 0$ with rate n and equally likely choice of the colliding pair, preserving the total energy $\frac{1}{2}\mathbf{v}^2 = \frac{1}{2}(\mathbf{v}_1^2 + \mathbf{v}_2^2 + \cdots + \mathbf{v}_n^2)$. Streaming is ignored. The two-speed gas with streaming was trouble enough!

11.7.1 Boltzmann's equation and Wild's sum

The propagation of chaos is proven much as before, for which see Grünbaum [1971]: with the notation of §11.2.2,

$$p_{\infty/m} = \lim_{n \uparrow \infty} p_{n/m} = p_{\infty/1} \otimes \cdots \otimes p_{\infty/1} \ m\text{-fold}$$

if $f_\infty = p_\infty(0+, \cdot)$ is chaotic in the sense that $f_\infty = f_{\infty/1} \otimes f_{\infty/1} \otimes$ etc., and

$$\frac{\partial p_{\infty/1}(v)}{\partial t} = \int_{S \times \mathbb{R}} \left[p_{\infty/1}(v'_1) p_{\infty/1}(v'_2) - p_{\infty/1}(v_1) p_{\infty/1}(v_2) \right] \, \mathrm{d}o \, \mathrm{d}v_2$$

where the time is supressed, $\mathrm{d}o$ is the uniform measure $\mathrm{d}\theta/2\pi$ on S, and the v_1 to the right is identified with the v to the left in the customary way. That's Boltzmann's equation and a very interesting equation it is.

Write p *plain* in place of the cumbersome $p_{\infty/1}$, and likewise f in place of $f_{\infty/1}$. Then $\partial p/\partial t$ may be written $p \circ p - p$, the operation $\circ$ being a sort of convolution:

$$f_1 \circ f_2(v) = \int_{S \times \mathbb{R}} f(v'_1) f(v'_2) \, \mathrm{d}o \, \mathrm{d}v_2$$

$$= \int_{S \times \mathbb{R}} [f_1(\cos \theta v_1 - \sin \theta v_2) f_2(\sin \theta v_1 + \cos \theta v_2)] \, \mathrm{d}o \, \mathrm{d}v_2$$

with the same identification of v and v_1. Obviously $\circ$ is commutative – just change θ into $\pi/2 - \theta$ and v_2 into $-v_2$; but *it does not associate*, as the next exercise shows.

Exercise 11.7.1   Take $\int vf = 0$, $\int v^2f = 1$, and $I[f] = \int v^4 f < \infty$. Then

$$I\left[(f \circ f) \circ (f \circ f)\right] = \frac{9}{16} I[f] + \frac{21}{16}$$

but

$$I\left[(f \circ (f \circ (f \circ f)))\right] = \frac{159}{256} I[f] + \frac{291}{256},$$

with the equality only if $I[f] = 3$ as for $f = (2\pi)^{-1/2}e^{-v^2/2}$.

Exercise 11.7.2 $\int v^2 p = \int v^2 f \equiv \sigma^2$ since

$$\frac{\mathrm{d}}{\mathrm{d}t} \int v^2 p = \int v^2 \left[p(v_1')p(v_2') - p(v_1)p(v_2)\right] \mathrm{d}o\,\mathrm{d}^2 v$$

$$= \int \left[\int (v_1')^2 \,\mathrm{d}o - v_1^2\right] p(v_1)p(v_2)\,\mathrm{d}^2 v = 0.$$

Compute $I[p] = e^{-t/4}I[f] + 3(1 - e^{-t/4})\sigma^4$ in the same style. This little $e^{-t/4}$ plays a central role later, as you will see. The fact is that all the moments of p tend rapidly to the moments of the Maxwellian $(2\pi\sigma^2)^{-1/2}e^{-v^2/2\sigma^2}$ as $t \uparrow \infty$, $I[p] = \int v^4 p$ being the slowest of the lot.

Exercise 11.7.3   Note in passing that for $\varphi \in C(\mathbb{R})$, say,

$$\int \varphi f_1 \circ f_2 = \int_{\mathbb{R}^2} f_1(v_1)f_2(v_2)\,\mathrm{d}^2 v \int_S \varphi\left(\sqrt{v_1^2 + v_2^2}\,\sin\theta\right)\mathrm{d}o,$$

so that $f_1 \circ f_2$ is indifferent to the odd parts of f_1 and f_2 and is itself, an even function of $v = v_1$. This be helpful later on.

If $\circ$ had been associative, the solution of Boltzmann's equation would have been

$$p = \sum_1^\infty e^{-t}(1 - e^{-t})^{n-1} f \circ \cdots \circ f \quad n\text{-fold},$$

as you will please check. But what to do in the present non-associative case? E. Wild [1951] discovered the pretty answer: *just put parentheses into $f \circ \cdots \circ f$ in all possible ways and average*. The meaning is best explained by a picture such as Figure 11.7.1, showing a *partition* π of 5 into $3+2 = (2+1)+2 = ((1+1)+1)+(1+1) = 1+1+1+1+1$. The first splitting could have been $4+1$, or the present $3+2$, or else $2+3$ or $1+4$. These are given equal weights, $1/4$ apiece. Subsequent splittings

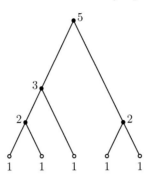

Figure 11.7.1

are given similar equal weights, independently of what was done higher up, the total weight of this particular partition being

$$|\pi| = \frac{1}{4} \text{ at level 1 down } (5 = 3{+}2) \times \frac{1}{2} \text{ at level 2 } (3 = 2{+}1 \text{ or } 1{+}2) = \frac{1}{8}.$$

Then a copy of $p(0+, \cdot) = f$ is placed at each node ($\circ$) at the bottom, and these are multiplied up in accordance with the picture in Figure 11.7.2, to produce the five-fold product $\pi(f) = (((f \circ f) \circ f) \circ (f \circ f)$

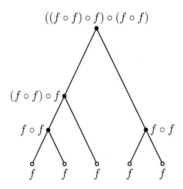

Figure 11.7.2

at the top. Wild's sum for the solution of Boltzmann's equation is then

$$p = \sum_{1}^{\infty} e^{-t}(1 - e^{-t})^{n-1} \sum_{\pi/n} |\pi| \pi(f),$$

the inner sum being taken over all partitions π of n, with weights $|\pi|$ and products $\pi(f)$ determined in the manner of Figure 11.7.2.

Exercise 11.7.4 Check that Wild's sum really does the trick.
Hint: If $n \geq 2$ splits into $n_1 + n_2$ at level 1 and if π_1 and π_2 are
the subsequent partitions of n_1 and n_2, then $|\pi| = \dfrac{|\pi_1| \times |\pi_2|}{n-1}$ and
$\pi(f) = \pi_1(f) \circ \pi_2(f)$.

But what is the meaning of $e^{-t}(1 - e^{-t})^{n-1}$? The next exercise tells.

Exercise 11.7.5 Extend Figure 11.7.2 indefinitely down, adding also
a little vertical line at the top as in Figure 11.7.3, and ascribe to each
leg an independent holding time T, distributed as in $\mathbb{P}(T > t) = e^{-t}$,
to be interpreted as its length or duration. Then draw the horizontal
line at level t down, seen also in Figure 11.7.3, and count the num-
ber $\mathbf{n}(t)$ of intersections ($\circ$). It is to be shown that $\mathbb{P}[\mathbf{n}(t) = n]$ is just
$e^{-t}(1 - e^{-t})^{n-1}$.
Hint: $\mathbf{n}(t_1 + t_2)$ is the sum of $\mathbf{n}(t_1)$ independent copies of $\mathbf{n}(t_2)$.

Figure 11.7.2 is now augmented as in Figure 11.7.3, by placing in-
dependent copies $\mathbf{v}_1, \mathbf{v}_2, \ldots, \mathbf{v}_5$ of the variable $\mathbf{v}_0$ underlying f at the
nodes, and colliding these with independent scattering angles, producing
a variable $v(t) = \big((v_1 \circ v_2) \circ v_3\big) \circ (v_4 \circ v_5)$ at the top, of which $p(t, v)$
is the probability density when all partitions of $n = \mathbf{n}(t)$ are taken into
account. Here and below, $v_1 \circ v_2$ signifies either $v_1' = \cos\theta v_1 - \sin\theta v_2$ or

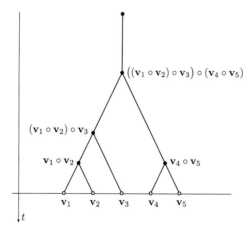

Figure 11.7.3

$v_2' = \sin\theta v_1 + \cos\theta v_2$ – it does not matter which as they have the same
distribution.

Aside. So far so good. But note that this is not yet a progressive, path-wise description of the velocity of a tagged molecule in an ∞-molecule, chaotic gas since Figure 11.7.3 must be continually revised as the level t goes down. That comes later in §11.7.6.

11.7.2 Entropy and the tendency to equilibrium

Boltzmann's H-theorem is proved in the usual way. The entropy production rate

$$\frac{\mathrm{d}H}{\mathrm{d}t} = \frac{1}{4} \int_{S \times \mathbb{R}^2} [p(v_1')p(v_2') - p(v_1)p(v_2)] \log \frac{p(v_1')p(v_2')}{p(v_1)p(v_2)} \,\mathrm{d}o \,\mathrm{d}^2 v$$

is visibly non-negative and cannot vanish unless $p(v_1')p(v_2') = p(v_1)p(v_2)$ for every collision, in which case $p(v)$ is the Maxwellian $(2\pi\sigma^2)^{-\frac{1}{2}}\mathrm{e}^{-v^2/2\sigma^2}$. This indicates, but as before does not prove, that $p(t,v)$ tends to the Maxwellian as $t \uparrow \infty$. The proof, under various conditions and with various results, of this Maxwellian variant of CLT, is postponed a bit to take up the statement of Aside 1, Stage 3, §11.1, that up to trivialities, $-\int p \log p$ is the *only* quantity of the form $H[p] = \int h(p)\,\mathrm{d}v$ which increases under Boltzmann's flow, totally *unlike*, e.g. heat flow $\partial p/\partial t = \partial^2 p/\partial x^2$ for which *any* concave function h will do. The two-speed gas is too simple to illustrate this fact, but Kac's gas will do. McKean [1963] proves it with all the necessary technical flourishes when $\sigma^2 = \int v^2 f = 1$. Here, a simple demonstration can be given if σ is allowed to vary, starting from the reasonable hope that $H[p]$ is always maximal at the Maxwellian $p = (2\pi\sigma^2)^{-1/2}\mathrm{e}^{-v^2/2\sigma^2}$.

To begin with, I remind you that the gradient of $F \in C(\mathbb{R}^n)$ may be defined by $F(x + \varepsilon y) = F(x) + \varepsilon \operatorname{grad} F(x) \cdot y + \mathrm{o}(\varepsilon)$ at $\varepsilon = 0+$, with the suggestion that the same rule should be used in function space for reasonable $F \colon f \to \mathbb{R}$. Before, $\operatorname{grad} F$ was a *vector* – now it will be a *function* – and the inner product $(\,\cdot\,)$ will be taken in $L^2(\mathbb{R})$. For example, $\operatorname{grad} \int f^2 = 2f$ and $\operatorname{grad} \int (f')^2 = -2f''$, as you will check. If then $H[p] = \int h(p)$, subject to $\int p = 1$, $\int vp = 0$, and $\int v^2 p = \sigma^2$, is maximal at the Maxwellian, you ought to have

$$\operatorname{grad} H = h'(p) = A \operatorname{grad} \int p + B \operatorname{grad} \int vp + C \operatorname{grad} \int v^2 p$$

$$= A + Bv + Cv^2$$

at that place, with coefficients A, B, C depending upon σ^2. That's just Lagrange's rule. Obviously, $B = 0$ since $h'(p)$ is an even function of v. Also $v^2 = -2\sigma^2 \log p + \text{a constant}$, so you may write $h'(p) = A + B \log p$

with new A and B, and this may be multiplied by p' and integrated back to obtain

$$h(p) = A(\sigma^2)p + B(\sigma^2)(p \log p - p)$$

without any constant of integration since $\int h(p)$ has to make sense. It remains to show that A and B do not really depend upon σ^2 at all. That's easy. Just differentiate by σ^2, producing

$$\frac{1}{2}h'(p)p'' = \frac{1}{2}A(\sigma^2)p'' + \frac{1}{2}B(\sigma^2)p'' \log p + A'(\sigma^2)p + B'(\sigma^2)(p \log p - p)$$

and compare to $h'(p) = A + B \log p$ multiplied by p''. Evidently, $A'p + B'(p \log p - p)$ must vanish and so also A' and B', and since p takes all values between 0 and $1/\sigma\sqrt{2\pi}$ by choice of α, so $h(p) = Ap + Bp \log p$ with fixed A and B, not just for the Maxwellian p but for any p whatsoever, with $B \leq 0$ to guarantee the increase of $H[p]$. That's all there is to it, technicalities aside.

Actually, there are other quantities that are increasing/decreasing, but not of the form $H[p] = \int h(p)$. Fisher's information $F[p] = \int p^{-1}(p')^2$ is decreasing, as will now be proved, suggesting that it is the only *new*, decreasing quantity, beyond $\int p$ and $\int p \log p$, of the form $\int h(p, p')$, but I could never find a proof or even a convincing demonstration of that. Perhaps you can.

11.7.3 Fisher's information

I remind you, from §5.5.2, that Fisher's information is also an entropy production rate: if w is the solution of $\partial w / \partial t = \partial^2 w / \partial x^2$, then

$$\frac{d}{dt} H[w] = - \int w'' \log w = \int \frac{w'^2}{w}$$

if all goes well. The proof, adapted from McKean [1967] that $F[p]$ is decreasing depends upon this fact.

Take $f \geq 0$ with $\int f = 1$, $\int vf = 0$, and $\int v^2 f = 1$ say, let p be the corresponding solution of Boltzmann's equation, and introduce the solution $q(s, v) = e_1 * p$ of the heat equation $\partial q / \partial s = \partial^2 q / \partial v^2$ with $q(0+, v) = p(t, v)$, $*$ being the ordinary convolution and $e_1 = (4\pi s)^{-1/2} e^{-v^2/4s}$ the corresponding elementary solution of that equation. Then for fixed $s \geq 0$, q solves Boltzmann's equation – not at all apparent from

$$\frac{\partial q}{\partial t} = e_1 * (p \circ p - p) = e_1 * p \circ p - q,$$

but it's true. Look at it this way: $e_2 * p \otimes p = q \otimes q$ where e_2 is the elementary solution $e_1 \otimes e_1$ of the two-dimensional heat equation $\partial w / \partial s = \Delta w$, and since two-dimensional heat flow commutes with rotations, you have $e_2 * p' \otimes p' = q' \otimes q'$, as well, with an indefensible but temporary abuse of notation *to wit*, $p' \otimes p' = p(v'_1) \times p(v'_2)$. Then

$$
\begin{aligned}
q \circ q &= \int q' \otimes q' \, \mathrm{do} \, \mathrm{d}v_2 \\
&= \int e_2 * p' \otimes p' \, \mathrm{do} \, \mathrm{d}v_2 \\
&= \int \mathrm{do} \, \mathrm{d}v_2 \int e_1(v_1 - \xi) p'(\xi) \int e_2(v_2 - \eta) p'(\eta) \, \mathrm{d}\xi \, \mathrm{d}\eta \\
&= \int e_1(v_1 - \xi) \, \mathrm{d}\xi \int e_2(v_2 - \eta) \, \mathrm{d}v_2 \int p'(\xi) p'(\eta) \, \mathrm{do} \, \mathrm{d}\eta \\
&= e_1 * (p \circ p).
\end{aligned}
$$

That does the trick. Now look at the entropy production for q written in a peculiar way:

$$
\frac{\mathrm{d}H}{\mathrm{d}t}[q] = \frac{1}{4} \int_{S \times \mathbb{R}^2} \left[\frac{q' \otimes q'}{q \otimes q} - 1 \right] \log \frac{q' \otimes q'}{q \otimes q} \times q \otimes q \, \mathrm{do} \, \mathrm{d}^2 v.
$$

Here,

$$
\frac{q' \otimes q'}{q \otimes q} = \frac{e_2 * \left[\frac{p' \otimes p'}{p \otimes p} \times p \otimes p \right]}{q \otimes q} \equiv A \left[\frac{p' \otimes p'}{p \otimes p} \right]
$$

is an averaging since $A[1] = 1$, so

$$
\begin{aligned}
\frac{\mathrm{d}H}{\mathrm{d}t}[q] &= \frac{1}{4} \int \left(A \left[\frac{p' \otimes p'}{p \otimes p} \right] - 1 \right) \log A \left[\frac{p' \otimes p'}{p \otimes p} \right] \times q \otimes q \\
&\leq \frac{1}{4} \int A \left[\left(\frac{p' \otimes p'}{q \otimes p} - 1 \right) \log \frac{p' \otimes p'}{p \otimes p} \right] \times p \otimes q, \\
&\quad \text{since } (x - 1) \log x \text{ is convex} \\
&= \frac{1}{4} \int e_2 * \left[p' \otimes p' - p \otimes p \right] \log \frac{p' \otimes p'}{p \otimes p} \\
&= \frac{\mathrm{d}H}{\mathrm{d}t}[p].
\end{aligned}
$$

That's the point. Now all you need to do is to convert the resulting

$$
0 \geq \frac{\mathrm{d}}{\mathrm{d}s} \frac{\mathrm{d}}{\mathrm{d}t} H[q] \quad \text{into} \quad \frac{\mathrm{d}}{\mathrm{d}t} \frac{\mathrm{d}}{\mathrm{d}s} H[q] = \frac{\mathrm{d}}{\mathrm{d}t} F[p] \text{ at } s = 0 + .
$$

I trust the idea is clear. Obviously, the technicalities are skimped. Take the necessary repairs as exercise if you want, but don't worry about it.

11.7.4 CLT: Trotter's method

Evidence in favor of CLT comes from Wild's sum and the revised version of Figure 11.7.3, decorated with n independent copies $\mathbf{v}_1, \ldots, \mathbf{v}_n$ of $\mathbf{v}_0$ placed at the nodes (○) at level t down. Above each of these you see m ramifications (●) indicating the collision of the two variables next below, all the scattering angles involved being independent of everything else, $\mathbf{v}$s and holding times alike. Then the variable $\mathbf{v}(t)$ at the top appears as a sum $c_1\mathbf{v}_1 + c_2\mathbf{v}_2 + \cdots + c_n\mathbf{v}_n$, each coefficient c being the product of m factors, sine or cosine, of independent scattering angles, subject to $c_1^2 + c_2^2 + \cdots + c_n^2 = 1$. Now if t is large, the bulk of Wild's sum comes from large n, and for most partitions of such n, the numbers m are mostly large as well, suggesting that Trotter's [1959] proof of the classical CLT, described in §5.3, might be applied here to show that for such typical partitions, $\pi(f)$ is nearly Maxwellian. The idea is carried out in McKean [1967] which I follow next, with improvements, taking $\sigma^2 = \int v^2 p = 1$ as it plays no role.

The idea is simple. Let $\mathbf{u}_1, \mathbf{u}_2, \ldots, \mathbf{u}_n$ be Gaussian variables with common density $(2\pi)^{-1/2} e^{-v^2/2}$, independent of each other and of everything else, and write

$$\mathbf{w}_k = \sum_{i<k} c_i\mathbf{v}_i + \sum_{j>k} c_j\mathbf{u}_j \quad \text{for } 1 \leq k \leq n,$$

noting that $c_1\mathbf{u}_1 + \cdots + c_n\mathbf{u}_n$ is Gaussian, in view of $c_1^2 + \cdots + c_n^2 = 1$ and the independence of the cs and the $\mathbf{u}$s. Then

$$\int \varphi\pi(f) - \int \varphi \frac{e^{-v^2/2}}{\sqrt{2\pi}}$$

is nothing but the expectation of the telescoping sum

$$S = \sum_1^n \left[\varphi(\mathbf{w}_k + c_k\mathbf{v}_k) - \varphi(\mathbf{w}_k + c_k\mathbf{u}_k) \right],$$

which may be estimated termwise for $\varphi \in C^4(\mathbb{R})$ and $n \geq 2$ as follows. Use the MacLaurin expansion of φ, centered at $\mathbf{w}_k$, to develop the general summand of S as in

$$\varphi(\mathbf{w}_k) + \varphi'(\mathbf{w}_k)c_k\mathbf{v}_k + \frac{1}{2}\varphi''(\mathbf{w}_k)c_k^2\mathbf{v}_k^2 + \frac{1}{6}\varphi'''(\mathbf{w}_k)c_k^3\mathbf{v}_k^3 + \frac{1}{24}\varphi''''(\star)c_k^4\mathbf{v}_k^4$$

$$- \varphi(\mathbf{w}_k) - \varphi'(\mathbf{w}_k)c_k\mathbf{u}_k - \frac{1}{2}\varphi''(\mathbf{w}_k)c_k^2\mathbf{u}_k^2 - \frac{1}{6}\varphi'''(\mathbf{w}_k)c_k^3\mathbf{u}_k^3$$

$$- \frac{1}{24}\varphi''''(\star\star)c_k^4\mathbf{u}_k^4$$

for suitable choices of $\star$ and $\star\star$, and take the expectation for fixed scattering angles. Here, n is ≥ 2, so by Exercise 11.7.3, it is harmless to suppose that f is even, and since $\mathbf{w}_k$ is independent of both $\mathbf{v}_k$ and $\mathbf{u}_k$, everything is either killed or cancels out except the terms in φ''''. In short, the discrepancy

$$\left| \int \varphi \pi(f) - \int \varphi \frac{e^{-v^2/2}}{\sqrt{2\pi}} \right|$$

is controlled by a fixed multiple of

$$\max_{\mathbb{R}} |\varphi''''| \times \mathrm{D}(\pi) \equiv \mathbb{E}\left[\sum_1^n c_k^4 \right]$$

and

$$\left| \int \varphi p - \int \varphi \frac{e^{-v^2/2}}{\sqrt{2\pi}} \right|$$

by

$$e^{-t} \max|\varphi| \qquad \text{coming from } n = 1 \text{ in Wild's sum}$$

$$+ \max|\varphi''''| \times \sum_1^\infty e^{-t}(1 - e^{-t})^{n-1} \sum_{\pi/n} |\pi| \, \mathrm{D}(\pi),$$

provided, what I should have said, that $\int v^4 f < \infty$. The inner sum $\overline{\mathrm{D}}(n)$ just above is now estimated using the splitting rules from Exercise 11.7.4, *to wit*, for $n = n_1 + n_2$, $|\pi| = |\pi_1| \, |\pi_2|/(n-1)$ and $\pi(f) = \pi_1(f) \circ \pi_2(f)$. In detail, putting $\overline{\mathrm{D}}(1) = 1$, you have

$$\overline{\mathrm{D}}(n) = \sum_{\pi/n} |\pi| \, \mathrm{D}(\pi)$$

$$= \sum_{n_1+n_2=n} \sum_{\substack{\pi_1/n_1 \\ \pi_2/n_2}} [\mathrm{D}(\pi_1) + \mathrm{D}(\pi_2)] \frac{|\pi_1| \, |\pi_2|}{n-1} \times \frac{3}{8}$$

$$\text{since } \int \sin^4 \, do = \int \cos^4 \, do = \frac{3}{8}$$

$$= \frac{3}{8} \times \frac{1}{n-1} \sum_{n_1+n_2=n} [\overline{\mathrm{D}}(n_1) + \overline{\mathrm{D}}(n_2)]$$

$$= \frac{3}{4} \frac{1}{n-1} \sum_1^{n-1} \overline{\mathrm{D}}(k),$$

leading easily to $\sum_1^\infty \overline{D}(n)\gamma^{n-1} = (1-\gamma)^{-3/4}$ and hence to the final estimate:

$$\left| \int \varphi p - \int \varphi \, \frac{\mathrm{e}^{-v^2/2}}{\sqrt{2\pi}} \right| \le \mathrm{e}^{-t} \max|\varphi| + \text{a fixed multiple of } \mathrm{e}^{-t/4} \max|\varphi''''|$$

in which you see the $\mathrm{e}^{-t/4}$ predicted in Exercise 11.7.2.

Exercise 11.7.6 Check the evaluation of $\sum_1^\infty \overline{D}(n)\gamma^{n-1}$.

Note The estimate may be used as in McKean [1960] to obtain

$$\int_{-\infty}^{+\infty} \left| p - \frac{\mathrm{e}^{-v^2/2}}{\sqrt{2\pi}} \right| \mathrm{d}v \le C(f) \times \mathrm{e}^{-\gamma t}$$

provided $\int v^4 f < \infty$ as here and also $\int f^{-1}(f')^2 < \infty$, but the γ is *much* smaller than $1/4$. Better luck next time.

11.7.5 CLT: Grünbaum's method

Grünbaum [1972] made a decisive improvement to §11.7.4, bringing γ up to the desired $\gamma = 1/4$. The idea stems from Grad [1958]. Write $p = (1+h)g$ with $g = (2\pi)^{-1/2}\mathrm{e}^{-v^2/2}$. Then $\int hg = 0$ since $\int p = 1$, so $\partial p/\partial t = p \circ p - p$ reads

$$\frac{\partial h(v)}{\partial t} = \int_{S \times \mathbb{R}} [h(v_1') + h(v_2') + h(v_1')h(v_2')] \, \mathrm{d}o \, g(v_2) \, \mathrm{d}v_2 - h(v),$$

as you will check. Here,

$$\int_{\mathbb{R}} g(v_1) \, \mathrm{d}v_1 \left| \int_{\mathbb{R}} \int_S h(v_1')h(v_2') \, \mathrm{d}o \, g(v_2) \, \mathrm{d}v_2 \right|^2$$

$$\le \int g(v_1)g(v_2) \, \mathrm{d}^2v \int_S h^2(v_1')h^2(v_2') \, \mathrm{d}o$$

$$= \left| \int h^2 g \right|^2,$$

and if this $\int h^2 g$ should be small, and smaller still as $t \uparrow \infty$, you could hope that near the equilibrium ($h = 0$), the full, nonlinear flow of $p = (1+h)g$ would be well approximated by the response of p to the simpler, linear flow of h regulated by

$$\frac{\partial h}{\partial t} = \int_{S \times \mathbb{R}} [h(v_1') + h(v_2')] \, \mathrm{d}o \, g(v_2) \, \mathrm{d}v_2 - h.$$

In fact, this simplified flow can be integrated explicitly in $L^2(\mathbb{R}, g)$, leading to the natural, optimal estimate

$$\int \left| p - \frac{e^{-v^2/2}}{\sqrt{2\pi}} \right| dv \leq C(f)e^{-t/4}$$

provided that $\int v^4 f < \infty$ as before and that $\int h^2 g$ is small enough at the start ($1/25$ is Grünbaum's magic number).

The integration of the simplified flow employs the Hermite polynomials $H_n(v) = (-1)^n e^{v^2/2} D^n e^{-v^2/2}$, viz. $H_0 = 1$, $H_1 = v$, $H_2 = v^2 - 1$, etc. Introduce

$$G_\gamma(v) = \sum_0^\infty H_n(v) \frac{\gamma^n}{n!} = e^{\gamma v - \gamma^2/2}.$$

Obviously

$$\int G_\alpha G_\beta \, g = \int e^{(\alpha+\beta)v} g \, dv \, e^{-\alpha^2/2} e^{-\beta^2/2} = e^{\alpha\beta}$$

from which you see that the Hs are mutually perpendicular in $L^2(\mathbb{R}, g)$ and span that space. Also,

$$G_\gamma(\cos\theta \, v_1 + \sin\theta \, v_2) = e^{\gamma\cos\theta \, v_1} e^{\gamma\sin\theta \, v_2} e^{-\cos^2\theta \, \gamma^2/2} e^{-\sin^2\theta \, \gamma^2/2}$$
$$= G_{\gamma\cos\theta}(v_1) \times G_{\gamma\sin\theta}(v_2),$$

whence

$$\frac{H_n(\cos\theta \, v_1 + \sin\theta \, v_2)}{n!} = \sum_{n_1+n_2=n} \frac{H_{n_1}(v_1)}{n_1!} \cos^{n_1}\theta \times \frac{H_{n_2}(v_2)}{n_2!} \sin^{n_2}\theta,$$

from which you read off the response of the Hs to the simplified flow

$$\frac{\partial h}{\partial t} = 2 \int_{S\times\mathbb{R}} h(\cos\theta \, v_1 + \sin\theta \, v_2) \, d\circ g(v_2) \, dv_2 - h,$$

to wit, $H_n \to e^{-\gamma_n t} H_n$ with $\gamma_n = 1$ if n is odd and $\gamma_n = 1 - 2\int \sin^n\theta \, d\circ$ if n is even, the smallest of these even rates, beyond $\gamma_2 = 0$, being $\gamma_4 = 1/4$. So that's where $\gamma = 1/4$ comes from, and what Grünbaum proves by comparison of flows is that $\sqrt{\int h^2 g} \leq C(f)e^{-t/4}$ if $\int h^2 g \leq 1/25$ to start with, and so also

$$\int \left| p - \frac{e^{-v^2/2}}{\sqrt{2\pi}} \right| \leq \sqrt{\int h^2 g} \leq C(f)e^{-t/4},$$

as advertised, than which nothing better could be hoped for. There is even a smooth change of (Hermite) coordinates in the vicinity of the equilibrium, intertwining the linear and the nonlinear flows. I omit the proof – it's not entirely easy – but the picture is so nice I could not leave it out.

Remark Carleman [1957] proved the tendency to the Maxwellian for a gas of hard spheres by showing *directly* that $\lim_{t \uparrow \infty} H^\bullet[p] = 0$ radial $p(0+, \cdot)$, a remarkable *tour de force*. See also Carlen–Carvalho–Gabetta [2000] and Mischler–Mouhot [2013] for Maxwell's gas, treated in the style of §§11.7.2–11.7.5.

Exercise 11.7.7 The old figures of merit (1) $\log \sqrt{2\pi e} - H$ and (2) $\frac{1}{2}(F - 1)$ of §5.5.2 are to be compared with Grünbaum's (3) $\int h^2 g$. (1) $\leq$ (2) comes from the logarithmic Sobolev inequality of §5.5.3. Check that (1) $\leq$ (3) in the large and that (3) $\simeq$ (1) in the small near $h = 0$. Presumably (2) is the most delicate. Show that it is roughly $(3') = \int (h')^2 g$ for small h and that $(3') \geq (3)$ with lots to spare. Hints for $(3') \geq (3)$: $H'_n = nH_{n-1}$ and $\int H_n^2 g = n!$.

11.7.6 A tagged molecule

Now back to the start and to the neglected question of what Boltzmann's equation really says about the chaotic ∞-molecule gas with initial density $f_\infty = f_{\infty/1} \otimes f_{\infty/1} \otimes$ etc. I follow Tanaka [1978] with simplifications, Kac's gas being simpler than Maxwell's which is treated there.

The velocities $\mathbf{v}(t) : t \geq 0$ of a distinguished molecule (no. 0) are to be pictured as a sort of rate-1 Poisson process with jumps $\mathbf{v} \to \mathbf{v} \circ \mathbf{w}$ due to collisions with the molecules of an independent, chaotic gas of like molecules in such a way that $\mathbf{v}$ is "guided" in the style of §11.3 by the solution of Boltzmann's $\partial p/\partial t = p \circ p - p$ with $p(0+, \cdot) = f_{0/1} \equiv f$. The one-step transition density[9] $\mathbb{P}_f[\mathbf{v}(t) = v | \mathbf{v}(0) = v_0]$ is declared to be the solution $q(t, v)$ of $\partial q/\partial t = p \circ q - q$, reducing at $t = 0+$ to the unit mass at v_0; for two steps, it is a little more complicated, *viz.*

$$\mathbb{P}_f[\mathbf{v}(t_1) = v_1, \mathbf{v}(t_2) = v_2 | \mathbf{v}(0) = v_0]$$
$$= \mathbb{P}_f[\mathbf{v}(t_1) = v_1 | \mathbf{v}(0) = v_0]$$
$$\times \mathbb{P}_{p(t_1, \cdot)}[\mathbf{v}(t_2 - t_1) = v_2 | \mathbf{v}(0) = v_1];$$

---

[9] Here is my usual sloppy notation ($\mathbf{v} = v$) for densities and a convenient disregard of the distinction between *density* and *distribution*.

for three steps, it is

$$\mathbb{P}_f[\mathbf{v}(t_1) = v_1, \mathbf{v}(t_2) = v_2, \mathbf{v}(t_3) = v_3 | \mathbf{v}(0) = v_0]$$
$$= \mathbb{P}_f[\mathbf{v}(t_1) = v_1 | \mathbf{v}(0) = v_0]$$
$$\times \mathbb{P}_{p(t_1, \cdot)}[\mathbf{v}(t_2 - t_1) = v_2 | \mathbf{v}(0) = v_1]$$
$$\times \mathbb{P}_{p(t_2, \cdot)}[\mathbf{v}(t_3 - t_2) = v_3 | \mathbf{v}(0) = v_2].$$

You see the pattern. The pair $\mathbf{v}$ and p is Markovian: p does what it does, and the guiding f is updated at each step in response, with the suggestion from $\partial q / \partial t = p \circ q - q$ that $\mathbf{v}(t) : t \geq 0$ is comparable to the standard, rate-1 Poisson process, but with complicated jumps, from $\mathbf{v}$ to $\mathbf{v} \circ \mathbf{w}$, the colliding velocity $\mathbf{w}$ being distributed by p. Figure 11.7.4 helps to explain what is meant.

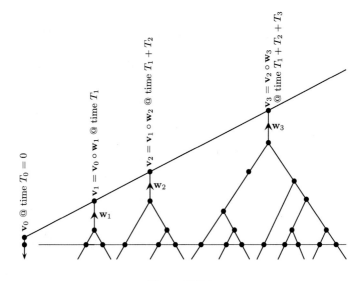

Figure 11.7.4

It purports to show the velocity $\mathbf{v}(t) : t \geq 0$ of the tagged molecule (no. 0) in an infinite, chaotic, statistically independent bath of like molecules. Its initial velocity $\mathbf{v}(0)$ is distributed by f: the Ts are independent holding times with common distribution $\mathbb{P}(T > t) = e^{-t}$. The little suspended diagrams are copies of Figure 11.7.3, independent of the Ts and of each other. These are cut off at levels T_1, $T_1 + T_2$, etc. to determine, at the top, the velocities $\mathbf{w}_1, \mathbf{w}_2$, etc. of molecules nos. $1, 2, 3$ etc., incoming from the bath. These collide with molecule no. 0 to produce, successively,

$$
\begin{aligned}
\mathbf{v}(t) &= \mathbf{v}(0) && \text{for } 0 \leq t < T_1 \\
&= \mathbf{v}(0) \circ \mathbf{w}_1 && \text{for } T_1 \leq t < T_1 + T_2 \\
&= \mathbf{v}(0) \circ \mathbf{w}_1 \circ \mathbf{w}_2 && \text{for } T_1 + T_2 \leq t < T_1 + T_2 + T_3
\end{aligned}
$$

and so on,

the claim being that this recipe recapitulates Wild's sum. Obviously, the number of *direct* collisions up to time t is Poisson distributed, as per $\mathbb{P}[\#(t) = n] = e^{-t} t^n / n!$, but what about the number of *indirect* collisions? That should be distributed *à la* Wild, as in $\mathbb{P}[\#(t) = n] = e^{-t}(1 - e^{-t})^{n-1}$. The count, artificially commenced at $\#(t) = 1$ for $t < T_1$, may be expressed as

$$
\#(t) = 1 + \mathbf{n}(T_1) + \mathbf{n}(T_1 + T_2) + \cdots + \mathbf{n}(T_1 + \cdots + T_n)
$$
$$
\text{for } T_1 + \cdots + T_n \leq t < T_1 + \cdots + T_{n+1}
$$

each $\mathbf{n}$ being an independent copy of the count of Exercise 11.7.2. Then with $0 \leq \gamma < 1$ and the understanding that $T_1 + \cdots + T_n = 0$ if $n = 0$,

$$
\mathbb{E}(\gamma^{\#})
$$
$$
= \sum_0^{\infty} \mathbb{E}\left[\gamma^{1 + \mathbf{n}(T_1) + \mathbf{n}(T_1 + T_2) + \cdots + \mathbf{n}(T_1 + \cdots + T_n)}, \right.
$$
$$
\left. T_1 + \cdots + T_n \leq t < T_1 + \cdots + T_{n+1} \right]
$$
$$
= \gamma \sum_0^{\infty} \int_{t_1 < \cdots < t_n \leq t} e^{-t_1} e^{-t_2} \cdots e^{-t_n} \, \mathrm{d}^n t \, \mathbb{E}\, \gamma^{\mathbf{n}(t_1)} \, \mathbb{E}\, \gamma^{\mathbf{n}(t_1+t_2)} \text{etc.} \times
$$
$$
\times \int_{t-(t_1 + \cdots + t_n)}^{\infty} e^{-t_{n+1}} \, \mathrm{d}t_{n+1}
$$
$$
= \gamma e^{-t} \sum_0^{\infty} \int_{0 \leq t_1' < t_2' < \cdots < t_n'} \frac{\gamma e^{-t_1'}}{1 - \gamma(1 - e^{-t_1'})} \frac{\gamma e^{-t_2'}}{1 - \gamma(1 - e^{-t_2'})} \text{etc. } \mathrm{d}^n t'
$$

after the substitution $t_1 = t_1', t_1 + t_2 = t_2'$, and so on

$$
= \gamma e^{-t} \sum_0^{\infty} \frac{1}{n!} \left[\log \frac{1}{1 - \gamma(1 - e^{-t})} \right]^n
$$
$$
= \frac{\gamma e^{-t}}{1 - \gamma(1 - e^{-t})}
$$
$$
= \sum_1^{\infty} \gamma^n e^{-t}(1 - e^{-t})^{n-1},
$$

as Exercise 11.7.2 says it ought to be.

Exercise 11.7.8 The verification of $\mathbb{P}[\mathbf{v}(t) = v] =$ Wild's sum is only a little more complicated. That's the exercise.

The recipe shows that in the chaotic, infinite-molecule gas, the molecules incoming from the bath fly away and are not seen again, neither they themselves, nor any of the molecules involved in *their* prior collisions, so that the next incoming molecule is statistically independent of what went before. That's because there are *so* many molecules, and why should one just seen be chosen twice out of a dilute bath? That's the picture.

11.7.7 CLT: $S^\infty(\sqrt{\infty})$ revisited

Let's go back, as a finale, to the idea of chaos. The general density f_∞ belongs to the section of the obvious convex cone specified by symmetry and the normalization $\int f_\infty \, \mathrm{d}^\infty v = 1$. Now with $n \geq 1$ and $v_{n+1} = v_1'$, $v_{n+2} = v_2'$, etc., you have

$$f_\infty(v_1', v_2', \text{etc.}) = \int_{\mathbb{R}^n} \frac{f_\infty(v_1, \ldots, v_n, v_1', v_2', \text{etc.})}{f_{\infty/n}(v_1, \ldots, v_n)} f_{\infty/n}(v_1, \ldots, v_n) \, \mathrm{d}^n v$$

expressing the f_∞ to the left as a center of gravity of the conditional densities to the right. It follows that f_∞ is a vertex of the section only if, e.g. $f_\infty = f_{\infty/1} \otimes f_\infty$, which is to say that the vs are independent, or what is the same, that $f_\infty = f_{\infty/1} \otimes f_{\infty/1} \otimes$ etc. is chaotic. Then the general f_∞ is, or ought to be a center of gravityy of these, as in

$$f_\infty = \int f_{\infty/1} \otimes f_{\infty/1} \otimes \text{etc.} \, \mathrm{d}\pi(f_{\infty/1}),$$

whatever that means, with apologies for too many πs – partitions before and now a distribution of factors $f_{\infty/1}$ – and a disregard of technicalities, notably some compactness of the section, for which see Hewitt–Savage [1955]. The corresponding ∞-molecule density p_∞ now appears likewise, as a center of gravity:

$$p_\infty = \int p_{\infty/1} \otimes p_{\infty/1} \otimes \text{etc.} \, \mathrm{d}\pi(f_{\infty/1}),$$

the factor $p_{\infty/1}$ being the solution of Boltzmann's equation with $p_{\infty/1}(0+, \cdot) = f_{\infty/1}$, so that the general, non-chaotic, ∞-molecule gas appears as a chaotic gas with $f_{\infty/1}$ chosen at random. Now in the n-molecule chaotic gas, $\mathbf{v}_1^2 + \cdots + \mathbf{v}_n^2$ is a constant of motion, and if $\int v^4 f_{\infty/1}$

is finite, then $(\mathbf{v}_1^2-1+\cdots+\mathbf{v}_n^2-1)/\sqrt{n}$ is nearly Gaussian by the classical CLT. Then with $\sigma^2 = \int (v^2-1)^2 f_{\infty/1}$,

$$\mathbb{P}\left[\left|\mathbf{v}_1^2 + \cdots + \mathbf{v}_n^2 - n\right| > c\sigma\sqrt{n\log n}\right] \simeq 2 \int_{c\sqrt{\log n}} \frac{e^{-x^2/2}}{\sqrt{2\pi}} < n^{-c^2/2}$$

is the general term of a convergent sum if $c > \sqrt{2}$, showing that

$$\mathbb{P}\left[\mathbf{v}_1^2 + \cdots + \mathbf{v}_n^2 = n + \mathrm{O}\sqrt{n\log n} \text{ as } n \uparrow \infty\right] = 1,$$

and this is not only for chaotic f_∞, but in general provided $\int (\int v^4 f_{\infty/1})\, d\pi$ is finite. Then it is clear that the general non-chaotic, ∞-molecule gas is, as you may say, *attracted* to $S^\infty(\sqrt{\infty})$ and that p_∞ is or ought to be tending to the natural round density

$$\frac{1}{\sqrt{2\pi}}e^{-v^2/2} \otimes \frac{1}{\sqrt{2\pi}}e^{-v^2/2} \otimes \text{etc.}$$

so that chaos, *absent* at the start, is *established* as $t \uparrow \infty$, a point of view emphasized in a slightly different way of speaking, by Kac [1956].

Coda

Statistical mechanics is such a vast and fascinating subject, I could go on and on, but I stop here, hoping to have given you some idea of it and some encouragement to look into it further, with one special recommendation: Max Dresden [1956], published by The Magnolia Petroleum Co. if you please, and probably impossible to find, but what I found more helpful than anything else when I started to learn these things.

12

Random Matrices

This subject is not so classical as what's gone before. It has its roots in practical statistics (1930 or so), quantum mechanics (1958), and, very recently, in an astonishing variety of other questions, touched upon below. §12.7 et seq. are technically more demanding than any thing we've done before, employing for the most part radically new methods, but it's fascinating and worth the effort. Mehta [1967] is *the* best general reference.

12.1 The Gaussian orthogonal ensemble (GOE)

Quantum-mechanically speaking, the energy levels E of a large collection of n like atoms, without internal degrees of freedom such as angular momentum or spin, may be found by solving $H\psi \equiv -\Delta\psi + V\psi = E\psi$, in which Δ is Laplace's operator in $\mathbb{R}^{3n} = \mathbb{R}^3 \times \cdots \times \mathbb{R}^3$, one copy for each atom to record its location, and V is the (classical) total potential energy of such a configuration. But, if n is big, the computation is already way out of reach of the biggest machine. What to do? Well, you can stop there, or you can look for something simpler, something you *can* compute, to see if that helps. Wigner [1958, 1967] proposed the simplest such caricature in connection with the scattering of neutrinos off heavy nuclei, replacing H by a typical, real symmetric, $n \times n$ matrix $\mathbf{x} = (\mathbf{x}_{ij} : 1 \leq i, j \leq n)$ and looking at its spectrum as $n \uparrow \infty$; far away from Nature perhaps, but never mind. The idea is to provide $\mathbf{x}$ with some natural probabilities $d\mathbb{P}$ and to look in this "ensemble" for typical features of its necessarily real spectrum, such as occur, for $n \uparrow \infty$, with overwhelming probability (as in LLN) or have a stable statistical description (as in CLT). For example, you may want to know

- How does the empirical distribution of $\operatorname{spec}\mathbf{x}$ look when properly scaled? Or

• How is it spaced in the *bulk*? Or again

• How does the biggest eigenvalue behave, at the *edge*, so to say?

Here, to make life simple, I stick to Gaussian probabilities:[1]

$$d\mathbb{P} = \frac{1}{Z} e^{-\operatorname{tr}(\mathbf{x}^2)/2} \, d\operatorname{vol} \quad \text{with normalizer } Z$$

$$= \prod_{j>i} \frac{e^{-x_{ij}^2}}{\sqrt{\pi}} \, dx_{ij} \prod_k \frac{e^{-x_{kk}^2/2}}{\sqrt{2\pi}} \, dx_{kk}.$$

This set up is the "Gaussian orthogonal ensemble", GOE for short.

Why this choice? For three reasons. First, why not take $\mathbb{E}(\mathbf{x}_{ij}) = 0$ for every $i \leq j$ to remove any bias? That's obvious. The second is the fact that the group $O(n)$ of rotations $o = (o_{ij} : 1 \leq i, j \leq n)$ with $o^{-1} = o^\dagger$, with the dagger signifying "transposed", acts on $\mathbf{x}$ as in $\mathbf{x} \to o\mathbf{x}o^\dagger$, and it's natural to suppose that the probabilities are not changed by this action – after all, it reflects only a rigid rotation of the coordinate frame which is an artefact of the description, having nothing to do with the geometrical action of $\mathbf{x}$ on $\mathbb{R}^n$. The third reason is the assumption that the x_{ij} are independent. Why not? Then, up to trivialities, no other *Gaussian* probabilities are possible.

Exercise 12.1.1 The point is that the ensemble of matrices $\mathbf{x}$ is a copy of $\mathbb{R}^{n(n+1)/2}$, endowed with its natural metric $\sqrt{\operatorname{tr}(\mathbf{x}^2)}$, and that $\mathbf{x} \to o\mathbf{x}o^\dagger$ acts as a rotation there. Think it over.

The first question above, about the empirical distribution of $\operatorname{spec} \mathbf{x}$, may now be answered. Write $\operatorname{spec} \mathbf{x} = (\lambda_1 \leq \lambda_2 \leq \cdots \leq \lambda_n)$. Then

$$\mathbb{P}\left[\lim_{n \uparrow \infty} \frac{1}{n} \#\left(k \leq n : \lambda_k \leq \sqrt{2n}\, x \right) = \frac{2}{\pi} \int_{-1}^x \sqrt{1-y^2} \, dy \text{ for } |x| x \leq 1) \right]$$

$$= 1,$$

in which you see the semi-circle $z = \sqrt{1-y^2}$, $-1 \leq y \leq 1$, whence the name Wigner's "semi-circle law". §12.2 provides a *demonstration* – why it should be so – and §12.3 a *proof* in a completely different style; see also Trotter [1984] for yet another way. §12.4 explains the beautiful connection between $\operatorname{spec} \mathbf{x}$ and $\mathrm{BM}(n)$ found by Dyson [1962], to be elaborated in §§12.5 and 12.6. The second and third questions about bulk and edge require a wholly new machinery, explained in §12.7 for a technically simpler variant of GOE, to which the rest of the chapter is devoted.

---

[1] tr is "trace".

12.2 Why a semi-circle?

This is not a proof but only a demonstration to convince a reasonable, not to say naive person[2].

12.2.1 Reduction to spec $\mathbf{x}$

The probabilities $d\mathbb{P} = Z^{-1}\exp\left[-\operatorname{tr}(\mathbf{x}^2/2)\right]d\operatorname{vol}$ are invariant under the action $\mathbf{x} \to o\mathbf{x}o^\dagger$ of $O(n)$ on the ensemble $\mathbb{R}^{n(n+1)/2}$, so for the reduction of $d\mathbb{P}$ to spec $\mathbf{x}$, it is enough to express the reduced volume element in spectral terms. The spectrum is of course real, and may be taken in increasing order: $\lambda_1 \leq \lambda_2 \leq \cdots \leq \lambda_n$; it is also simple with probability 1 since any coincidence such as $\lambda_1 = \lambda_2$ reduces the $n(n+1)/2$-dimensional ensemble to a region of co-dimension 1 or more, of volume 0. In short, $\lambda_1 < \lambda_2 < \cdots < \lambda_n$ for present purposes. Then the rotation o which brings $\mathbf{x}$ to diagonal form $o\mathbf{x}o^\dagger = \Lambda$ is unambiguous, up to a rotation o' commuting with Λ, and these are very special: $o'_{ij} = \pm 1$ on diagonal, vanishing off diagonal. This $2^n : 1$ ambiguity may be removed as follows. The volume element being invariant under the action $\mathbf{x} \to o\mathbf{x}o^\dagger$, it suffices to make the reduction in the vicinity of some diagonal $\mathbf{x} = \Lambda$ with simple spectrum. There you may take o to be the identity and extend it *un*ambiguously (and so also smoothly) to $\mathbf{x}$ in the vicinity of Λ. Now take the differential of $\mathbf{x} = o^\dagger\Lambda o$ *at* $\mathbf{x} = \Lambda$:

$$d\mathbf{x} = do^\dagger\Lambda o + o^\dagger\, d\Lambda o + o^\dagger\Lambda\, do = do^\dagger\Lambda + d\Lambda + \Lambda\, do;$$

note that do is skew since

$$0 = d(oo^\dagger) = (do)o^\dagger + o(do^\dagger) = do + do^\dagger;$$

and wedge the individual differentials

$$dx_{ij} = \begin{array}{ll} do_{ij}(\lambda_i - \lambda_j) & \text{off diagonal} \\ d\lambda_k & \text{on diagonal} \end{array}$$

up to top degree $n(n+1)/2$ to obtain

$$d\operatorname{vol} = \prod_{j>i}(\lambda_j - \lambda_i)\prod_k d\lambda_k \times \text{some volume element } \omega \text{ on } O(n).$$

[2] T. Suidan kindly introduced me to the line of thought followed here.

The latter may now be integrated out to obtain the final form[3] of the reduced probabilities on spec $\mathbf{x}$:

$$\frac{1}{Z} \prod_{j>i} (\lambda_j - \lambda_i) \times (2\pi)^{-n/2} e^{-\lambda^2/2}$$

with a new normalizer Z.

Exercise 12.2.1 Compute Z for $n = 2$. *Answer*: $1/\sqrt{\pi}$. The evaluation for $n \geq 3$ is complicated. Fortunately, it is not needed.

Exercise 12.2.2 If it is not familiar, check that $V = \prod_{j>i}(\lambda_j - \lambda_i)$ is the so-called Vandermonde determinant of $[\lambda_i^{j-1} : 1 \leq i, j \leq n]$. This will be needed later on.
Hint: V is a polynomial of total degree $n(n-1)/2$, vanishing when $\lambda_1 = \lambda_2$ or $\lambda_2 = \lambda_3$, etc.

The reduced density with its factor $\prod_{j>i}(\lambda_j - \lambda_i)$ is hard to handle for actual computation – in fact, the honest proof of the semi-circle law in §12.3 does not require the reduction at all – but it may be used very simply to get a rough idea of how spec $\mathbf{x}$ looks for large dimension $n \uparrow \infty$. That's what comes next.

12.2.2 Steepest descent

The trivial evaluation $\mathbb{E}(\text{spec } \mathbf{x}^2) = n(n+1)/2$ says that the expectation of the empirical mean $\frac{1}{n} \text{tr} \mathbf{x}^2 = \frac{1}{n}(\lambda_1^2 + \cdots + \lambda_n^2)$ is $(n+1)/2$, indicating that spec $\mathbf{x}$ should be viewed on the scale $\sqrt{n}$; in fact, the scale $\sqrt{2n}$ is just right, as will be confirmed in §12.3. Believe it for now, write the reduced density in the form

$$\frac{1}{Z} \exp\left[-\sum_k \lambda_k^2/2 + \sum_{j>i} \log(\lambda_j - \lambda_i) \right],$$

---

[3] This type of distribution is nothing new. It originates with the English statistician Wishart [1928].

and leaving the normalizer Z aside, put it on the proposed scale by taking $\lambda = \sqrt{2n}\,x$. The exponent is then changed to

$$-n\sum_k x_k^2 + \frac{1}{2}\sum_{j\neq i}\log|x_j - x_i| + \text{an irrelevant constant absorbed into } Z$$

$$= \frac{n^2}{2}\left[-\frac{1}{n}\sum_{i\&j}(x_i^2 + x_j^2) + \frac{1}{n^2}\sum_{j\neq i}\log|x_j - x_i|\right]$$

$$\equiv \frac{n^2}{2}\times A,$$

and if you will believe that some kind of large numbers effect is taking over as $n\uparrow\infty$, then you will want to know where the quadratic form is largest and how it looks near by, as in the much simpler derivation in §1.5 of Stirling's formula by Kelvin's method of steepest descent. That's where the action must be taking place. Obviously, the present case is much more complicated, but let's hope for the best, supposing that

$$\lim_{n\uparrow\infty}\frac{1}{n}\#(k\leq n : \mathbf{x}_k \leq x) = \int_{-1}^{x} f(y)\,dy \ \text{ for } -1 \leq x \leq 1$$

with a fixed density f, the restriction to $-1 \leq x \leq 1$ being in accord with the belief that most of spec $\mathbf{x}$ lives between $\pm\sqrt{2n}$. Of course, it's a big jump, but be patient and hope, expecting f to imitate Figure 12.2.1, with

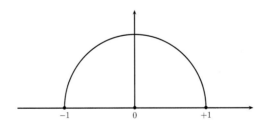

Figure 12.2.1

its peak at $x = 0$ and vanishing at $x = \pm 1$ since very large eigenvalues are surely rare. A look at the form A may now suggest that you replace it by

$$B = \int_{-1}^{+1}\int_{-1}^{+1}\left[-(x^2 + y^2) + \log|x - y|\right]f(x)f(y)\,dx\,dy$$

and try to determine f by maximizing *that*, subject to the sole constraint $C = \int_{-1}^{1} f = 1$. The more serious-looking constraint, that f be non-negative, is ignored, hoping it will come out by itself. The function space analogue of Lagrange's finite-dimensional recipe: grad B = constant×the gradient of the constraint function C, is applied for this purpose, producing the identity

(1)
$$\begin{aligned}\text{``grad } B\text{''} &= -2\int_{-1}^{+1}(x^2+y^2)f(y)\,dy + 2\int_{-1}^{+1}\log|x-y|f(y)\,dy\\ &= \text{a constant multiple of ``grad } C\text{''} = 1.\end{aligned}$$

Exercise 12.2.3 Think over why that's the right way to go.

The constant at the end is not known but can be removed by (careful) differentiation in respect to $-1 < x < +1$, assuming f is smooth to make life easy. The result is

$$(2)\quad \int_{-1}^{+1}\frac{1}{x-y}f(y)\,dy = 2x,$$

in which the singular-looking integral is taken for $|x-y| \geq h > 0$, and h is then reduced to $0+$.

Exercise 12.2.4 Justify (2) from (1).
Hint:
$$\int_{-1}^{+1}\frac{1}{x-y}f(y)\,dy = f(-1)\log(1+x) - f(+1)\log(1-x)$$

$$quad + \int_{-1}^{+1}\log|x-y|f'(y)\,dy,$$

as you will check. Do it first for $f \equiv 1$ to get the idea.

Now solve (2) for f. In the integral at the left of (2), replace x by $z = x + \sqrt{-1}\,h$ off the cut $-1 \leq x \leq 1$ seen in the figure. This defines an analytic function $G(z)$, and careful computation in the style of Exercise 12.2.4 shows that, for $-1 < x < +1$,

(3) $\lim\limits_{h\downarrow 0}$ real part $G(x + \sqrt{-1}\,h)$

$$= \lim_{h\downarrow 0}\int_{-1}^{+1}\frac{x-y}{(x-y)^2+h^2}f(y)\,dy$$

$$= \int_{-1}^{+1}\frac{1}{x-y}f(y)\,dy \text{ taken as in (2)}$$

$$= 2x.$$

More simply, you find

(4) $-\lim\limits_{h\downarrow 0}$ imaginary part $G(x + \sqrt{-1}\,h)$

$= +\lim\limits_{h\uparrow 0}$ imaginary part $G(x + \sqrt{-1}\,h)$

$= \lim\limits_{h\downarrow 0} h \int_{-1}^{+1} \dfrac{1}{(x-y)^2 + h^2} f(y)\,\mathrm{d}y$

$= \pi f(x),$

and from these two – (3) and (4) – it is evident that (2) has at most one solution: otherwise, the function G formed for the difference of two solutions would be purely imaginary on the *upper* bank of the cut, by (3), and so extend analytically *across* the cut, by the reflection principle[4], preventing its imaginary part from jumping as (4) predicts. It remains only to make an educated guess as to what G should be. Look at Figure 12.2.2, the function $\sqrt{1-z^2} = \sqrt{1-z} \times \sqrt{1+z}$ off the cut. Ordinarily, such a radical is double-valued, but this one is not once the

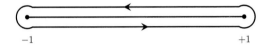

-1 $+1$

Figure 12.2.2

signature is fixed (as -1, say) on the upper bank, the point being that the radical changes sign *once* as you pass about the branch point -1 and *once more* as you pass about $+1$, for *two*, i.e. for *no*, changes at all. The signature is now chosen in conformity with the fact that the imaginary part of G is negative in the upper half-plane, and the proposal is that $G = 2z + 2\sqrt{z^2 - 1}$. In fact, it satisfies (3), by inspection, and (4) reduces to $f(x) = \frac{2}{\pi}\sqrt{1-x^2}$, as advertised.

Well, that was pretty swift, and sloppy, too. §12.3 does it properly in an entirely different way.

12.3 The semi-circle: a hands-on proof

The spectral density $Z^{-1}V(2\pi)^{-n/2}\mathrm{e}^{-\lambda^2/2}$ with its unknown Z and its awkward restriction to $\lambda_1 < \lambda_2 <$ etc. is not pleasant, so here comes an honest proof by elementary means: basically Markov's method of moments, introduced in §4.4, combined with a recipe of Wiener [1948:87] for computing the expectation of a product of Gaussian variables.

[4] Ahlfors [1979: 172–174]

12.3.1 Wiener's recipe

Everything here is Gaussian with mean 0. The recipe emerges in three steps.

Step 1 For a single variable $\mathbf{x}$ and a nice function $f(\mathbf{x})$,

$$\mathbb{E}[\mathbf{x}f(\mathbf{x})] = \mathbb{E}(\mathbf{x}^2)\,\mathbb{E}[f'(\mathbf{x})].$$

Proof. Just write it out and you will see.

Step 2 For $\mathbf{x} = (\mathbf{x}_1, \ldots, \mathbf{x}_n)$ with correlation[5] $\mathbb{E}[\mathbf{x} \otimes \mathbf{x}] = K$,

$$\mathbb{E}(\mathbf{x}_1\mathbf{x}_2\ldots\mathbf{x}_n) = \sum_{j>1} K_{1j}\,\mathbb{E}(\mathbf{x}_2\ldots\mathbf{x}_n \text{ with } \mathbf{x}_j \text{ left out}).$$

Proof. $\mathbf{x}$ may be written $\sqrt{K}\,\mathbf{y}$ with $\mathbb{E}[\mathbf{y}\otimes\mathbf{y}] = $ the identity, $\sqrt{K}$ being any symmetric root of K, so by Step 1, you have

$$\mathbb{E}(\mathbf{x}_1\mathbf{x}_2\ldots\mathbf{x}_n) = \sum_{1 \leq i \leq n} (\sqrt{K})_{1i}\,\mathbb{E}(\mathbf{y}_i\mathbf{x}_2\ldots\mathbf{x}_n)$$

$$= \sum_{\substack{1 \leq i \leq n \\ j>1}} (\sqrt{K})_{1i}\,\mathbb{E}\left(\mathbf{x}_2\ldots\frac{\partial\mathbf{x}_j}{\partial\mathbf{y}_i}\ldots\mathbf{x}_n\right)$$

$$= \sum_{\substack{1 \leq i \leq n \\ j>1}} (\sqrt{K})_{1i}(\sqrt{K})_{ji}\,\mathbb{E}(\mathbf{x}_2\ldots\mathbf{x}_n \text{ with } \mathbf{x}_j \text{ left out})$$

$$= \sum_{j>1} K_{1j}\,\mathbb{E}(\mathbf{x}_2\ldots\mathbf{x}_n \text{ with } \mathbf{x}_j \text{ left out}).$$

Step 3 $\mathbb{E}(\mathbf{x}_1\mathbf{x}_2\ldots\mathbf{x}_n) = 0$ or $\sum\prod K$ according as n is odd or even, the sum being taken over all pairings of the indices 1, 2 up to n, and the product of the correlations $K_{ij} = \mathbb{E}(\mathbf{x}_i\mathbf{x}_j)$ so distinguished.

Proof for even n. By Step 2, repeated until everything is paired.

12.3.2 Traces

Now take Gaussian $\mathbf{x} = [\mathbf{x}_{ij} : 1 \leq i, j \leq n]$ as before, with mean 0, mean-square 1, resp. $1/2$ on-, resp. off-diagonal, and independence for distinct $i \leq j$. The main task is to compute the "moments" $\mathbb{E}[\text{tr}(\mathbf{x}^Q)]$ for

---

[5] $\mathbf{x} \otimes \mathbf{x}$ is short-hand for $[\mathbf{x}_i\mathbf{x}_j : 1 \leq i, j \leq n]$.

fixed Q and $n \uparrow \infty$. Obviously, this vanishes for odd Q, while if $Q = 2q$ is even, then

$$\mathbb{E}[\text{tr}(\mathbf{x}^Q)] \simeq C(Q)n^{q+1} \text{ with } C(Q) = \binom{2q}{q}2^{-q}\frac{1}{q+1},$$

as will be seen below.

12.3.3 Samples

Let's do a couple of these as a warm-up. $\mathbb{E}[\text{tr}(\mathbf{x}^0)] = n$, by definition, and

$$\mathbb{E}[\text{tr}(\mathbf{x}^2)] = \sum_{i \& j} \mathbb{E}(\mathbf{x}_{ij}^2) = n + n(n-1) \times \frac{1}{2} \simeq \frac{n^2}{2} \equiv C(2)n^2.$$

OK. Now for $\mathbb{E}[\text{tr}(\mathbf{x}^4)]$, write

$$\mathbf{x}_{i_1 i_2} \mathbf{x}_{i_2 i_3} \mathbf{x}_{i_3 i_4} \mathbf{x}_{i_4 i_1} \text{ as } (12 \mid 23 \mid 34 \mid 41)$$

for economy, and use Step 3 to check, with tears, that

$$\mathbb{E}[\text{tr}(\mathbf{x}^4)] = \sum_{1234} \mathbb{E}(12 \mid 23 \mid 34 \mid 41)$$

$$= \sum_{1234} \big[\, \mathbb{E}(12 \mid 23)\,\mathbb{E}(34 \mid 41) + \mathbb{E}(12 \mid 34)\,\mathbb{E}(23 \mid 41) +$$

$$+ \mathbb{E}(12 \mid 41)\,\mathbb{E}(23 \mid 34)\big]$$

$$= \sum_{\substack{1=2 \\ 4}} \mathbb{E}(14 \mid 41) + \sum_{1=2} 1 + \sum_{\substack{1=2 \\ 3}} \mathbb{E}(23 \mid 32)$$

$$+ \sum_{\substack{1 \neq 2 \\ 4}} \frac{1}{2}\mathbb{E}(14 \mid 41) + \sum_{1 \neq 2} \frac{1}{4} + \sum_{\substack{1 \neq 2 \\ 3}} \frac{1}{2}\mathbb{E}(23 \mid 32)$$

$$= \mathbb{E}[\text{tr}(\mathbf{x}^2)] + n + \mathbb{E}[\text{tr}(\mathbf{x}^2)]$$

$$+ \frac{n-1}{2}\mathbb{E}[\text{tr}(\mathbf{x}^2)] + \frac{n(n-1)}{4} + \frac{n-1}{2}\mathbb{E}[\text{tr}(\mathbf{x}^2)]$$

$$\simeq (n+1)\,\mathbb{E}[\text{tr}(\mathbf{x}^2)]$$

$$\simeq \frac{n^3}{2} \equiv C(4)n^3,$$

as it is supposed to be.

12.3.4 A better way

Obviously, this is *not* the way to do it for large Q. Try this, using Step 1 in its self-evident conditional form and the standard basis $e_1 = (1, 0, \ldots, 0)$, e_2, etc. of $\mathbb{R}^n$:

$$
\begin{aligned}
\mathbb{E}[\mathrm{tr}(\mathbf{x}^{Q+2})] &= \sum_{i,j} \mathbb{E}[\mathbf{x}_{ij}(\mathbf{x}^{Q+1})_{ji}] \\
&= \sum_{i,j} \mathbb{E}(\mathbf{x}_{ij}^2)\,\mathbb{E}\left[\frac{\partial}{\partial \mathbf{x}_{ij}}(\mathbf{x}^{Q+1})_{ji}\right] \\
&= \sum_{i,j} \mathbb{E}(\mathbf{x}_{ij}^2) \sum_{A+B=Q} \mathbb{E}\left[\left(\mathbf{x}^A \frac{\partial \mathbf{x}}{\partial \mathbf{x}_{ij}} \mathbf{x}^B\right)_{ji}\right] \\
&= \sum_{i=j} \sum_{A+B=Q} \mathbb{E}\left[\left(\mathbf{x}^A e_i \otimes e_i\, \mathbf{x}^B\right)_{ji}\right] \\
&\quad + \sum_{i \neq j} \frac{1}{2} \sum_{A+B=Q} \mathbb{E}\left[\left(\mathbf{x}^A (e_i \otimes e_j + e_j \otimes e_i)\mathbf{x}^B\right)_{ji}\right] \\
&= \sum_{i=j} \sum_{A+B=Q} \mathbb{E}\left[(\mathbf{x}^A)_{ii}(\mathbf{x}^B)_{ii}\right] \\
&\quad + \frac{1}{2} \sum_{i \neq j} \sum_{A+B=Q} \mathbb{E}\left[(\mathbf{x}^A)_{ji}(\mathbf{x}^B)_{ji} + (\mathbf{x}^A)_{jj}(\mathbf{x}^B)_{ii}\right] \\
&= \frac{1}{2} \sum_{i,j} \sum_{A+B=Q} \mathbb{E}\left[(\mathbf{x}^A)_{ji}(\mathbf{x}^B)_{ij} + (\mathbf{x}^A)_{jj}(\mathbf{x}^B)_{ii}\right] \\
&= \frac{1}{2} \sum_{A+B=Q} \mathbb{E}\left[\mathrm{tr}(\mathbf{x}^{A+B}) + \mathrm{tr}(\mathbf{x}^A)\,\mathrm{tr}(\mathbf{x}^B)\right] \\
&= \frac{1}{2}(Q+1)\,\mathbb{E}\,\mathrm{tr}(\mathbf{x}^Q) + \frac{1}{2} \sum_{A+B=Q} \mathbb{E}\left[\mathrm{tr}(\mathbf{x}^A)\,\mathrm{tr}(\mathbf{x}^B)\right].
\end{aligned}
$$

The rule may be simplified by a reduction of

$$
\mathbb{E}\left[\mathrm{tr}(\mathbf{x}^A)\,\mathrm{tr}(\mathbf{x}^B)\right] = \sum_{\substack{A \text{ indices } 12\cdots* \\ B \text{ indices } 1'2'\cdots*'}} \mathbb{E}(12 \mid 23 \mid \cdots \mid *1 \mid 1'2' \mid 2'3' \mid \cdots \mid *'1'),
$$

written here in the style of the sample computation of $\mathbb{E}\,\mathrm{tr}(\mathbf{x}^4)$. By Step 3, the expectation under the sum is of the form $\sum \prod K$ taken over all pairings of the indices. Pairing only internal variables, primed or unprimed, and summing over $12\cdots*1'2'\cdots*'$ produces $\mathbb{E}\,\mathrm{tr}(\mathbf{x}^A) \times \mathbb{E}\,\mathrm{tr}(\mathbf{x}^B)$. Any other pairings involve some ij paired to some $i'j'$, and as it makes no difference after summing over $12\cdots* \, 1'2'\cdots*'$, you may as well pair

12 to $1'2'$. Then Step 2 produces

$$\sum_{12\cdots*'} \mathbb{E}(12 \mid 1'2') \, \mathbb{E}(23 \mid \cdots \mid *1 \mid 2'3' \mid \cdots \mid *'1')$$

$$= \sum_{1=2\,\text{etc.}} \mathbb{E}(13 \mid \cdots \mid *1 \mid 13' \mid \cdots \mid *'1) \qquad \text{on diagonal}$$

$$+ \sum_{1\neq2\,\text{etc.}} \frac{1}{2} \mathbb{E}(23 \mid \cdots \mid *1 \mid 23' \mid \cdots \mid *'1) \qquad \begin{array}{l}\text{off diagonal}\\ \text{with } 1'2' = 12\end{array}$$

$$+ \sum_{1\neq2\,\text{etc.}} \frac{1}{2} \mathbb{E}(23 \mid \cdots \mid *1 \mid 13' \mid \cdots \mid *'2) \qquad \begin{array}{l}\text{off diagonal}\\ \text{with } 1'2' = 21\end{array}$$

$$= \sum_{1\,\&\,2\,\text{etc.}} \mathbb{E}(23 \mid \cdots \mid *1 \mid 13' \mid \cdots \mid *'2) \qquad \begin{array}{l}\text{by reversal of } 3'\cdots*'\\ \text{in the summands}\\ \text{with } 1 \neq 2\end{array}$$

$$= \sum_{1\,\&\,2} \mathbb{E}\left[(\mathbf{x}^{A-1})_{21} (\mathbf{x}^{B-1})_{12} \right]$$

$$= \mathbb{E}\,\mathrm{tr}(\mathbf{x}^{A+B-2})$$

so that

$$\mathbb{E}\left[\mathrm{tr}(\mathbf{x}^A)\,\mathrm{tr}(\mathbf{x}^B) \right] = \mathbb{E}\,\mathrm{tr}(\mathbf{x}^A) \times \mathbb{E}\,\mathrm{tr}(\mathbf{x}^B)$$
$$+ \text{ some number } \#(Q) \text{ of copies of } \mathbb{E}\,\mathrm{tr}(\mathbf{x}^{A+B-2}),$$

with the final result

$$\mathbb{E}\,\mathrm{tr}(\mathbf{x}^{Q+2}) = \frac{1}{2}(Q+1)\,\mathbb{E}\,\mathrm{tr}(\mathbf{x}^Q)$$
$$+ \frac{1}{2} \sum_{A+B=Q} \mathbb{E}\,\mathrm{tr}(\mathbf{x}^A) \times \mathbb{E}\,\mathrm{tr}(\mathbf{x}^B)$$
$$+ \#(Q) \text{ copies of } \mathbb{E}\,\mathrm{tr}(\mathbf{x}^{Q-2}),$$

in which both A and B may be taken to be even.

12.3.5 Leading order

Now start from $\mathbb{E}\,\mathrm{tr}(\mathbf{x}^0) = n$ and check, by induction, that

$$\mathbb{E}\,\mathrm{tr}(\mathbf{x}^Q) = \text{some constant } C(Q) \times n^{Q/2+1} + \mathrm{O}(n^{Q/2}).$$

This reduces the appraisal of $\mathbb{E}\,\mathrm{tr}(\mathbf{x}^Q)$ for $n \uparrow \infty$ to the determination of $C(Q)$ with the aid of

$$C(Q+2) = \frac{1}{2} \sum_{A+B=Q} C(A)C(B) \text{ starting from } C(0) = 1.$$

Here, $C(Q) \leq 2^q$ with $q = Q/2$ by a self-evident induction, A and B now being even, so you may write $C(Q) = 2^{q+1}C'(Q)$ with $C'(Q) \leq 1/2$, form the generating function

$$f(\gamma) = \sum_0^{\infty} C'(Q)\gamma^{Q+1} \text{ for } 0 < \gamma < 1,$$

and deduce from

$$f(\gamma) = \frac{\gamma}{2} + \sum_0^{\infty} \frac{1}{2} \sum_{A+B=Q} C'(A)C'(B) = \frac{\gamma}{2} + \frac{\gamma}{2} f^2(\gamma)$$

that

$$f(\gamma) = \frac{1}{\gamma}\left[1 - \sqrt{1-\gamma^2}\right].$$

This function is familiar from §3.2 in connection with the passage time from 0 to 1 for RW(1)[6]: indeed,

$$C'(Q) = \mathbb{P}_0(T_1 = Q+1) = \binom{2q}{q}2^{-2q}\frac{1}{2q+2},$$

from which you read off

$$C(Q) = \binom{2q}{q}2^{-q}\frac{1}{q+1} \quad \text{for even } Q.$$

Exercise 12.3.1 Show that $\mathbb{P}_0(T_1 = Q+1)$ solves the recursion for $C'(Q)$ directly from RW(1).

12.3.6 Convergence of traces

Next fix Q and look at the normalized trace

$$Z_n = \frac{1}{n}\operatorname{tr}\left(\frac{\mathbf{x}}{\sqrt{2n}}\right)^Q$$

for large n. Now $\mathbb{E}(Z_n) = 0$ if Q is odd, while it is approximately $2C'(Q)$ if Q is even. Then also

$$n^{Q+2}\,\mathbb{E}\,|Z_n - \mathbb{E}(Z_n)|^2$$
$$= \mathbb{E}\left[\operatorname{tr}(\mathbf{x}^Q)\right]^2 - \left[\mathbb{E}\operatorname{tr}(\mathbf{x}^Q)\right]$$
$$= \text{a fixed multiple of } \mathbb{E}\operatorname{tr}(\mathbf{x}^{Q-2}) = \mathrm{O}(n^Q),$$

---

[6] This seems to be a fluke. I see no reason for it.

from which you see that

$$\mathbb{P}\left[|Z_n - \mathbb{E}(Z_n)| > n^{-1/3}\right] \le n^{2/3} \times O(n^{-2}) = O(n^{-4/3})$$

and

$$\mathbb{P}\left[\lim_{n\uparrow\infty} \frac{1}{n}\,\mathrm{tr}\left(\frac{\mathbf{x}}{\sqrt{2n}}\right)^Q = 2C'(Q)\right] = 1.$$

This was the goal of the whole discussion of traces.

12.3.7 The semi-circle

Now introduce the empirical distribution of scaled eigenvalues:

$$\mathbf{F}_n(\lambda) = \frac{1}{n}\,\#\left(k \le n : \boldsymbol{\lambda}_k/\sqrt{2n} \le \lambda\right)$$

and write

$$\frac{1}{n}\,\mathrm{tr}\left(\frac{\mathbf{x}}{\sqrt{2n}}\right)^Q = \int \lambda^Q\,\mathrm{d}\mathbf{F}_n(\lambda),$$

noting that the tendency of the left-hand side to $2C'(Q) \le 1$ means that $\mathbf{F}_n(\lambda)$ does not grow much outside $-1 \le \lambda \le +1$. Next, recall from §2.1.3, that polynomials are dense in, e.g. $C[-2, 2]$, and conclude from

$$\lim_{n\uparrow\infty} \int \lambda^Q\,\mathrm{d}\mathbf{F}_n(\lambda) = 2C'(Q)$$

that $\mathbf{F}_n(\lambda)$ tends to a (sure) distribution function $F(\lambda)$, increasing from 0 to 1 between -1 and $+1$ and fully specified by its non-trivial (even) moments $2C'(Q)$. It remains to show that $\mathrm{d}F(\lambda) = \frac{2}{\pi}\sqrt{1 - \lambda^2}\,\mathrm{d}\lambda$, and there are two ways to do it. You can start from

$$\gamma \int_{-1}^{+1} \frac{1}{1 - \gamma\lambda}\,\mathrm{d}F(\lambda) = \sum_0^\infty \int_{-1}^{+1} \lambda^Q\,\mathrm{d}F(\lambda)\,\gamma^{Q+1}$$

$$= 2\sum_0^\infty C'(Q)\gamma^{Q+1}$$

$$= \frac{2}{\gamma}\left[1 - \sqrt{1 - \gamma^2}\right] \quad \text{for } |\gamma| < 1,$$

or, what is the same with $x = 1/\gamma$,

$$\int_{-1}^{+1} \frac{1}{x - \lambda}\,\mathrm{d}F(\lambda) = 2x - 2\sqrt{x^2 - 1} \quad \text{for } |x| > 1,$$

and proceed as in §12.2, or more simply but less obviously, you may take advantage of the particular form of

$$2C'(Q)$$

$$= \binom{2q}{q} 2^{-2q} \frac{1}{2q+2}$$

$$= 2 \times \left[\binom{2q}{q} 2^{-2q} - \binom{2q+2}{q+1} 2^{-2q-2} \right]$$

$$= 2\, \mathbb{P}_0 \left[\mathbf{x}(Q) = 0 \right] - 2\, \mathbb{P}_0 \left[\mathbf{x}(Q+2) = 0 \right] \qquad \text{for RW}(1) \text{ with apologies for the } \mathbf{x}$$

$$= \frac{1}{\pi} \int_{-\pi}^{\pi} \cos^Q \theta \sin^2 \theta \, d\theta \qquad \text{by Pólya's formula of §4.1, adapted to dimension 1}$$

$$= \frac{2}{\pi} \int_{-1}^{+1} \lambda^Q \sqrt{1 - \lambda^2} \, d\lambda \qquad \text{by the substitution } \cos \theta = \lambda.$$

I like that better though it is a bit sneaky.

Amplification The fact is that the Gaussian character of $\mathbf{x}$ and the strict independence imposed have little to do with the semi-circle: Most of what is known about $\operatorname{spec} \mathbf{x}$ in high dimension has a certain universality, reminiscent of LLN in respect of the semi-circle, and of CLT in respect of bulk and edge. The proofs are not easy, but see Tracy–Widom [1993:105] for a first pass at the semi-circle, and Deift et al. [1999] for the whole story.

12.4 Dyson's Coulomb gas

Dyson [1962] discovered a beautiful interpretation of the spectral density in which $\operatorname{spec} \mathbf{x}$ is pictured as n copies of BM(1), moving independently on the line, modified by mutual repulsions to keep them from crossing.

12.4.1 2 × 2 hands-on

Let's take this simplest case to see what's going on. Introduce the 2×2 symmetric matrix

$$\mathbf{x} = \begin{pmatrix} \mathbf{x}_{11} & \mathbf{x}_{12} \\ \mathbf{x}_{12} & \mathbf{x}_{22} \end{pmatrix}$$

with three independent copies of BM(1), $\mathbf{x}_{11}$, $\mathbf{x}_{22}$, and $\sqrt{2}\,\mathbf{x}_{12}$, all starting at the origin. The eigenvalues are

$$\boldsymbol{\lambda}_\pm = \frac{\mathbf{x}_{11} + \mathbf{x}_{22}}{2} \pm \sqrt{\tfrac{1}{4}(\mathbf{x}_{11} - \mathbf{x}_{22})^2 + \mathbf{x}_{12}^2}$$

$$= \frac{1}{\sqrt{2}}\,\mathbf{b}_0 \pm \frac{1}{\sqrt{2}}\sqrt{\mathbf{b}_1^2 + \mathbf{b}_2^2}$$

with three new independent copies of BM(1):

$$\mathbf{b}_0 = \frac{\mathbf{x}_{11} + \mathbf{x}_{22}}{\sqrt{2}}, \quad \mathbf{b}_1 = \frac{\mathbf{x}_{11} - \mathbf{x}_{22}}{\sqrt{2}}, \quad \text{and} \quad \mathbf{b}_2 = \mathbf{x}_{12}\sqrt{2},$$

in which you see the process BES(2) $= \sqrt{\mathbf{b}_1^2 + \mathbf{b}_2^2} \equiv \mathbf{r}$ of §6.10.2, and you know that

$$\mathbf{r}(t) = \mathbf{b}_3(t) + \frac{1}{2}\int_0^t \frac{dt'}{\mathbf{r}}$$

with still another copy of BM(1). In short,

$$\boldsymbol{\lambda}_\pm = \mathbf{b}_\pm(t) \pm \frac{1}{2}\int_0^t \frac{dt'}{\boldsymbol{\lambda}_+ - \boldsymbol{\lambda}_-}$$

with still other independent Brownian motions, $\mathbf{b}_\pm = \frac{1}{\sqrt{2}}(\mathbf{b}_0 \pm \mathbf{b}_3)$ and $\boldsymbol{\lambda}_+ - \boldsymbol{\lambda}_- = \sqrt{2}\,\mathbf{r}$, and you may write down the corresponding backward infinitesimal operator by analogy with BES(2):

$$\mathfrak{G} = \frac{1}{2}\left(\frac{\partial^2}{\partial\lambda_-^2} + \frac{\partial^2}{\partial\lambda_+^2}\right) + \frac{1}{2}\frac{1}{\lambda_- - \lambda_+}\frac{\partial}{\partial\lambda_-} + \frac{1}{2}\frac{1}{\lambda_+ - \lambda_-}\frac{\partial}{\partial\lambda_+}\,.$$

That's Dyson's 2×2 Coulomb gas, so-called in reference to Coulomb's law for the repulsion f between like charges. Here, the repulsive drifts $\pm\frac{1}{2}(\lambda_+ - \lambda_-)^{-1}$, pushing λ_+/λ_- to the right/left, keep the spectrum simple, just like the drift $1/2r$ keeps BES(2) away from the origin.

12.4.2 $n \times n$ in general

The eigenvalues cannot be written out in any simple way if $n \geq 3$, but Itô's lemma from §6.9 makes the job easy anyhow. $\mathbf{x}$ is now the $n \times n$ symmetric matrix formed by independent copies of BM(1), modified by a $1/\sqrt{2}$ off diagonal. It is legitimate to take $\mathbf{x}(0)$ to be diagonal since the action of $O(n)$ will make it so without change of spec $\mathbf{x}$, and to assume also, what is typical, that spec $\mathbf{x}$ is simple. Then the initial eigenvectors are just $e_1 = (1, 0, 0, \ldots, 0)$, $e_2 = (0, 1, 0, \ldots, 0)$, and so on, and as these, together with their eigenvalues, move smoothly in response to the motion of $\mathbf{x}$, you may compute the differential of the updated spectral problem $\mathbf{x}\mathbf{e} = \boldsymbol{\lambda}\mathbf{e}$ *at* $t = 0$ in Itô's style:

(1) $d\mathbf{x}\,\mathbf{e} + \mathbf{x}\,d\mathbf{e} + d\mathbf{x}\,d\mathbf{e} = d\boldsymbol{\lambda}\,\mathbf{e} + \boldsymbol{\lambda}\,d\mathbf{e} + d\boldsymbol{\lambda}\,d\mathbf{e}.$

Now fix $1 \le i \le n$ and take the inner product of (1) for $\mathbf{e} = \mathbf{e}_i$ dotted with $\mathbf{e}_j$ to produce, when $i = j$,

(2′) $d\mathbf{x}_{ii} + \mathbf{e}_i \cdot \mathbf{x}\,d\mathbf{e}_i = d\boldsymbol{\lambda}_i + \boldsymbol{\lambda}_i\,\mathbf{e}_i \cdot d\mathbf{e}_i + d\boldsymbol{\lambda}_i\,\mathbf{e}_i \cdot d\mathbf{e}_i,$

in which the second and fifth terms cancel since $\mathbf{e}\mathbf{x} = \mathbf{x}^\dagger\mathbf{e} = \boldsymbol{\lambda}\mathbf{e}$, and the final term may be dropped if the normalization $\mathbf{e} \cdot \mathbf{e} = 1$ is maintained since $2\mathbf{e} \cdot d\mathbf{e} + (\,d\mathbf{e})^2 = 0$ shows that $\mathbf{e} \cdot d\mathbf{e}$ has no Brownian part, reducing (2′) to the simplified

(2″) $d\mathbf{x}_{ii} = +\mathbf{e}_i\,d\mathbf{x}\,d\mathbf{e}_i = d\boldsymbol{\lambda}_i$

and likewise, when $i \ne j$,

(2‴) $d\mathbf{x}_{ij} + \boldsymbol{\lambda}_j\mathbf{e}_j \cdot d\mathbf{e}_i + \mathbf{e}_j \cdot d\mathbf{x}\,d\mathbf{e}_i = \boldsymbol{\lambda}_i\mathbf{e}_j \cdot d\mathbf{e}_i + d\boldsymbol{\lambda}_i\,\mathbf{e}_j \cdot d\mathbf{e}_i.$

(2‴) now is multiplied by $d\mathbf{x}_{ij}$ to obtain

(3) $dt/2 = (\boldsymbol{\lambda}_i - \boldsymbol{\lambda}_j) \times d\mathbf{x}_{ij}\,\mathbf{e}_j \cdot d\mathbf{e}_i,$

the product of any three differentials being 0, from which you find that for fixed $1 \le i \le n$,

(4) $\dfrac{1}{2} \displaystyle\sum_{j \ne i} \dfrac{dt}{\boldsymbol{\lambda}_i - \boldsymbol{\lambda}_j}$

$= \displaystyle\sum_{j \ne i} d\mathbf{x}_{ij}\,\mathbf{e}_j \cdot d\mathbf{e}_i$ by (3)

$= \displaystyle\sum_{1 \le j \le n} d\mathbf{x}_{ij}\,\mathbf{e}_j \cdot d\mathbf{e}_i$ since $\mathbf{e}_i \cdot d\mathbf{e}_i$ has no Brownian part

$= \mathbf{e}_i \cdot d\mathbf{x}\,d\mathbf{e}_i$ as you will check, keeping $t = 0$ in mind

$= d\boldsymbol{\lambda}_i - d\mathbf{x}_{ii}$ by (2″).

In short,

$$(5) \quad d\lambda_i = dx_{ii} + \frac{1}{2} \sum_{j \neq i} \frac{dt}{\lambda_i - \lambda_j},$$

in which you see the Coulomb-like drifts $\frac{1}{2}(\lambda_i - \lambda_j)^{-1}$, keeping the eigenvalues from meeting as for $n = 2$. The backward infinitesimal operator is now read off:

$$\mathfrak{G} = \frac{1}{2}\left(\Delta = \frac{\partial^2}{\partial\lambda_1^2} + \cdots + \frac{\partial^2}{\partial\lambda_n^2}\right) + \frac{1}{2}\sum_{i=1}^{n}\left(\sum_{j \neq i}\frac{1}{\lambda_i - \lambda_j}\right)\frac{\partial}{\partial\lambda_i}.$$

12.4.3 Coda

The elementary solution $e(t, x, y)$ of $\partial w/\partial t = \mathfrak{G}w$ is not readily expressible in an elementary way. Let's look, instead, only at the simpler density for spec $\mathbf{x}$ itself. By the reduction to spec $\mathbf{x}$ in §12.2 and Brownian scaling, this density has to be

$$\frac{1}{Z}\frac{\prod_{j>i}(\lambda_j - \lambda_i)}{t^{n(n-1)/4}}\frac{e^{-\lambda^2/2t}}{(2\pi t)^{n/2}},$$

with the same normalizer Z as before, and must agree with the elementary solution of $\partial w/\partial t = \mathfrak{G}w$ specialized to paths starting at spec $\mathbf{x} = 0$. Now, in general, if $\mathbf{x}(t) : t \geq 0$ is some Brownian motion modified by drift as here, if $\mathfrak{G}$ is its infinitesimal operator, and if $e = e(t, x, y)$ is the elementary solution of $\partial w/\partial t = \mathfrak{G}w$, then $w(t, x) = \int ef\,dy$ is *the* solution with $w(0+, \cdot) = f$, and you must have $\partial e/\partial t = \mathfrak{G}e$ where $\mathfrak{G}$ is acting upon $\mathbf{x}$. But also, if f is nice, $\dot{w} = \mathfrak{G}w$ will be the solution of $\partial\dot{w}/\partial t = \mathfrak{G}\dot{w}$ with $\dot{w}(0+, \cdot) = \mathfrak{G}f$, so that $\partial w/\partial t = \int e\mathfrak{G}f$ and, by partial integration, you must have $\partial e/\partial t = \mathfrak{G}^{\dagger}e$ where the forward operator $\mathfrak{G}^{\dagger}$, the transpose of $\mathfrak{G}$, is acting now upon y. For BM(n), itself, $\mathfrak{G} = \frac{1}{2}\Delta = \mathfrak{G}^{\dagger}$, and it makes no difference, but for the OU process of Note 6.9.4, with $k = 1$ for simplicity, $\mathfrak{G} = \frac{1}{2}\partial^2/\partial x^2 - x\,\partial/\partial x$ is *not* the same as $\mathfrak{G}^{\dagger} = \frac{1}{2}\partial^2/\partial x^2 + (\partial/\partial x)x$, reflecting the fact that the corresponding elementary solution

$$e(t, x, y) = (2\pi\sigma^2)^{-1/2}\exp\left[-(y - e^{-t}x)^2/2\sigma^2\right] \quad \text{with } \sigma^2 = \frac{1}{2}(1 - e^{-2t})$$

is not symmetric in x and y. Accordingly, the spectral density w for spec $\mathbf{x}$ displayed above, will be solving $\partial w/\partial t = $ not $\mathfrak{G}w$ but $\mathfrak{G}^{\dagger}w$.

Exercise 12.4.1   Check that with $V = \prod_{j>i}(\lambda_j - \lambda_i)$ as before,

$$\mathfrak{G} = \frac{1}{2}V^{-1}\,\mathrm{div}\,V\,\mathrm{grad} \quad \text{and} \quad \mathfrak{G}^\dagger = \frac{1}{2}\,\mathrm{div}\,V\,\mathrm{grad}\,V^{-1}.$$

Exercise 12.4.2   Now check by hand that, indeed, $\partial w/\partial t = \mathfrak{G}^\dagger w$.

12.5 Brownian motion without crossing

A very attractive picture of the Coulomb gas may be obtained in terms of n independent copies of BM(1) conditioned so as not to meet.

12.5.1 Crossing times

Take independent copies $\mathbf{x}_1, \mathbf{x}_2, \ldots, \mathbf{x}_n$ of BM(1), starting at

$$\mathbf{x}_1(0) = x_1 < \mathbf{x}_2(0) = x_2 < \text{etc.}$$

These are bound to cross since, e.g. $(\mathbf{x}_2 - x_1)/\sqrt{2}$ is a copy of BM(1), so you may introduce the "crossing time":

$$0 < T = \min\left(t : \mathbf{x}_i(t) = \mathbf{x}_j(t) \text{ for some } i < j\right) < \infty :$$

actually, at time T, two (and only two) paths meet; for example, if $\mathbf{x}_1(T) = \mathbf{x}_2(T) = \mathbf{x}_3(T)$, then T is a stopping time for the Brownian motion $(\mathbf{x}_1 - \mathbf{x}_2)/\sqrt{2}$ and $\mathbf{x}_3$ is independent of that.

Now what is the probability (or better the density) for $\mathbf{x} = \mathbf{x}_1 < \cdots < \mathbf{x}_n$ to arrive at $y = y_1 < \cdots < y_n$ at time t without crossing, i.e. with $T > t$? The answer, expressed in my customary, sloppy but convenient notation, is

$$\mathbb{P}_x\left[\mathbf{x}(t) = y,\, T > t\right] = \det\left[p(t, x_i, y_j) : 1 \le i, j \le n\right]$$

with the Brownian transition density

$$p(t, x, y) = (2\pi t)^{-n/2} \times \exp\left[-(x - y)^2/2t\right].$$

This elegant formula is due to Karlin–McGregor [1959].

Proof.   Take a function f on the figure $F = (x : x_1 < \cdots < x_n)$ and extend it anti-symmetrically to the whole space, so that $f(\pi x) = \mathrm{sign}\,\pi\, f(x)$ for any permutation of the n "letters" $x_1, \ldots, x_n$. Then[7]

$$\mathbb{E}_x\left[f \circ \mathbf{x}(t), T > t\right] = \mathbb{E}_x\left[f \circ \mathbf{x}(t)\right] - \mathbb{E}_x\left[f \circ \mathbf{x}(t), T < t\right],$$

[7] $\mathbb{P}(T = t) = 0$ for any fixed $t \ge 0$, as you will easily see.

in which the third piece vanishes since T is a stopping time, and if, for example, it's $\mathbf{x}_1$ and $\mathbf{x}_2$ that meet, you may exchange $\mathbf{x}_1(t') : t' \geq T$ and $\mathbf{x}_2(t') : t' \geq T$ without changing anything, except that their termini are exchanged, changing thereby the signature of $f \circ \mathbf{x}(t)$. But then

$$\mathbb{E}_x \left[f \circ \mathbf{x}(t), T > t \right] = \mathbb{E}_x \left[f \circ \mathbf{x}(t) \right]$$

$$= \int_{\mathbb{R}^n} \frac{e^{-(x-y)^2/2t}}{(2\pi t)^{n/2}} \, f(y) \, dy$$

$$= \text{the sum over all permutations } \pi \text{ of}$$

$$\int_{\pi F} \prod_1^n \frac{e^{-(x_i - y_i)^2/2t}}{\sqrt{2\pi t}} \, f(y) \, dy$$

$$= \text{the sum of}$$

$$\text{sign}(\pi) \times \int_F \prod_1^n \frac{e^{-(x_i - \pi y_i)^2/2t}}{\sqrt{2\pi t}} \, f(y) \, dy$$

$$= \int_F \det \left[\frac{e^{-(x_i - y_j)^2/2t}}{\sqrt{2\pi t}} \right] f(y) \, dy.$$

Now specialize f.

12.5.2 Connection to $\text{spec}\, \mathbf{x}$

$\mathbb{P}_0[\mathbf{x}(t) = y \mid T > t]$ makes no sense as it stands: for example, the Brownian motion $(\mathbf{x}_1(t) - \mathbf{x}_2(t))/\sqrt{2} = 0$ has infinitely many small roots t – think about Brownian scaling and you'll see – but take it in this sense:

$$\lim_{\substack{x \in F \\ \to 0}} \mathbb{P}_x[\mathbf{x}(t) = y \mid T > t] \quad \text{for fixed } y \in F.$$

That's kosher and something nice comes out, namely the spectral density of §12.4: To begin with,

$$\mathbb{P}_x[\mathbf{x}(t) = y \mid T > t]$$

$$= \frac{\mathbb{P}_x[\mathbf{x}(t) = y, T > t]}{\mathbb{P}_x(T > t)}$$

$$= \frac{1}{Z} \det \left[\frac{e^{-(x_i - y_j)^2/2t}}{\sqrt{2\pi t}} \right] \quad \text{with normalizer } Z = \mathbb{P}_x(T > t) = \int_F \det \, dy$$

$$= \frac{1}{Z} (2\pi t)^{-n/2} e^{-x^2/2t} \det \left[e^{x_i y_j/t} \right] e^{-y^2/2t}.$$

Now, for $x \in F$ approaching 0 and fixed $y \in F$,

$$\det \left[e^{x_i y_j / t} \right] \simeq t^{-n} V(x) V(y) \times \frac{1}{2!} \frac{1}{3!} \cdots \frac{1}{(n-1)!} \,,$$

for which see Exercise 12.5.1 below, so

$$\mathbb{P}_x \left[\mathbf{x}(t) = y \mid T > t \right]$$

$$= e^{-x^2 / 2t} t^{-n} V(x) V(y) \frac{e^{-y^2 / 2t}}{(2\pi t)^{n/2}} \bigg/ e^{-x^2 / 2t} t^{-n} V(x) \int_F V(y) \frac{e^{-y^2 / 2t}}{(2\pi t)^{n/2}} \, \mathrm{d}y$$

(Don't worry; there's plenty of domination to justify the denominator)

$$= Z \frac{V(y)}{t^{n(n-1)/2}} \frac{e^{-y^2 / 2t}}{(2\pi t)^{n/2}} \qquad \begin{array}{l} \text{with the old normalizer} \\[4pt] Z = \int_F V(y) \frac{e^{-y^2 / 2}}{(2\pi)^{n/2}} \text{ of §§12.1 and 12.4;} \end{array}$$

in short, the spectral density of §12.4 is just the density for n independent copies of BM(1), starting at $\mathbf{x}_1(0) = \cdots = \mathbf{x}_n(0) = 0$ in the present sense, to come to $y_1 < y_2 < \cdots < y_n$ at time $t > 0$, *conditioned* not to have crossed before. Quite beautiful, no?

Exercise 12.5.1 Prove that $\det \left[e^{x_i y_j} : 1 \le i, j \le n \right] = V(x) V(y) \times [1 + o(1)]$ for fixed y and x approaching 0 via F.
Hints: Do $n = 2$ by hand. For $n = 3$, think about the ratio of

$$\det \left[e^{x_i y_j} : 1 \le i, j \le 3 \right] \quad \text{to} \quad V(x) = \det \begin{bmatrix} 1 & x_1 & x_1^2 \\ 1 & x_2 & x_2^2 \\ 1 & x_3 & x_3^2 \end{bmatrix}$$

making $0 < x_1 < x_2 < x_3$ decrease to 0 in that order, keeping L'Hôpital's rule in mind.

Aside. This is *not* to say that $\mathbb{P}_x [\mathbf{x}(t) = y \mid T > t]$ is Dyson's transition density, but it's amusing to note that $\mathbb{P}_x [\mathbf{x}(t) = y \mid T = \infty]$ is almost the same: By Exercise 12.5.1, $\mathbb{P}_x (T > N) \simeq V(x) \times$ a fixed multiple of $N^{-n(n-1)/4}$ for $N \uparrow \infty$, so

$$\mathbb{P}_x [\mathbf{x}(t) = y \mid T > N]$$

$$= \mathbb{P}_x [\mathbf{x}(t) = y \mid T > t] \times \frac{\mathbb{P}_y (T > N - t)}{\mathbb{P}_x (T > N)}$$

$$\simeq \frac{1}{V(x)} \mathbb{P}_x [\mathbf{x}(t) = y \mid T > y] V(y) \quad \text{as } N \uparrow \infty :$$

obviously, a transition density. It follows that the Brownian motion, conditioned *not to cross ever*, is Markovian with backward infinitesimal operator

$$\mathfrak{G} = \frac{1}{2}V^{-1}\Delta V = \frac{1}{2}\Delta + \operatorname{grad}\log V \cdot \operatorname{grad} + \frac{1}{2}V^{-1}(\Delta V),$$

which is easily spelled out: $\Delta V = 0$ since ($\mathfrak{G}1$) must vanish and $\operatorname{grad}\log V$ is twice Dyson's drift. Too bad, but more of this in the next section.

12.6 The Gaussian unitary ensemble

GOE is now abandoned in favor of the Gaussian *unitary* ensemble (GUE) of complex symmetric matrices $x = [x_{ij} : 1 \leq i, j \leq n] = x^\dagger$, where the dagger now signifies the *conjugate* transposed, and the probabilities are taken to be

$$d\mathbb{P} = \frac{1}{Z}\,e^{-\operatorname{tr}(xx^\dagger)/2}\,d\operatorname{vol}$$

$$= \prod_{j>i} \frac{e^{-(\operatorname{real} x_{ij})^2}}{\sqrt{\pi}}\,d\operatorname{real} x_{ij}$$

$$\times \prod_{j>i} \frac{e^{-(\operatorname{imag} x_{ij})^2}}{\sqrt{\pi}}\,d\operatorname{imag} x_{ij}$$

$$\times \prod_{k} \frac{e^{-x_{kk}^2/2}}{\sqrt{2\pi}}\,dx_{kk}.$$

The analogue of $O(n)$ acting in $\mathbb{R}^n$ is now the unitary group $U(n)$ of $n \times n$ complex matrices u with $uu^\dagger = $ the identity, acting in $\mathbb{C}^n$, and $d\mathbb{P}$ is the simplest choice of probabilities to make the whole ensemble (statistically) invariant under the action $\mathbf{x} \to \mathbf{x}' = u\mathbf{x}u^\dagger$. But why this switch? It does not make much difference in the end: the semi-circle is seen once more with a slightly different scaling, and a more pleasing variant of Dyson's gas, a little modified, appears as well. It is only in the bulk and at the edge (still postponed) that there is some advantage. There, the results are scarcely different, but the way to compute is simpler, and I did not want to bother you with anything more.

12.6.1 Reduction to spec x

The transcription of the volume element into spectral language now changes in a technically advantageous way, the reduced volume element being $\frac{1}{Z} V^2 \, d^n \lambda$ with $V = \prod_{j>i} (\lambda_j - \lambda_i)$ as before, but *squared* so as to be a symmetric rather than an anti-symmetric function. This makes the spectral density $\frac{1}{Z} V^2 (2\pi)^{-n/2} e^{-\lambda^2/2}$ *much* easier to handle, as will be seen in §§12.7 and 12.8. The plan of reduction is effectively the same as in §12.2, only now $x_{ij} = x'_{ij} + \sqrt{-1}\, x''_{ij}$ is complex off diagonal and both dx'_{ij} and dx''_{ij} enter into the volume element; the factor $\lambda_j - \lambda_i$ then enters *twice*, producing the V^2 announced above.

12.6.2 Scaling and the semi-circle

The variable $\mathbf{x}$ is now bigger: the trace

$$\operatorname{tr}(\mathbf{x}\mathbf{x}^\dagger) = \sum_k \mathbf{x}_{kk}^2 + 2\sum_{i<j} \left[(\mathbf{x}'_{ij})^2 + (\mathbf{x}''_{ij})^2 \right] \quad \text{for GUE}$$

is roughly double the trace

$$\operatorname{tr}(\mathbf{x}^2) = \sum_k \mathbf{x}_{kk}^2 + 2\sum_{i<j} (\mathbf{x}_{ij})^2 \quad \text{for GOE,}$$

and since the diagonal does not count for much, you may expect the scaling factor $\sqrt{2n}$ for GOE to change to $\sqrt{2} \times \sqrt{2n} = 2\sqrt{n}$. This is, in fact, correct, and the semi-circle is found once more:

$$\mathbb{P}\left[\lim_{n\uparrow\infty} \frac{1}{n} \#(k \le n : \lambda_k \le 2\sqrt{n}\, x) = \frac{2}{\pi} \int_{-1}^{x} \sqrt{1 - y^2}\, dy \text{ for } |x| \le 1 \right] = 1.$$

The "demonstration" of §12.2 and/or the proof of §12.3 may be modified to do the work – in fact, as for CLT, it does not matter very much what you do: for example, if all the x_{ij} are independent, with common (uniform) distribution $\mathbb{P}(\mathbf{x} \le x) = \frac{1}{2}(x+1)$ for $|x| \le 1$, then the semi-circle still comes out with modified scaling $3/2\sqrt{n}$; see Tracy–Widom [1993:105] for a picture.

12.6.3 Dyson's gas

Itô's lemma is applied as in §12.4 to produce

$$d\lambda_i = dx_{ii} + \sum_{j>i} (\lambda_i - \lambda_j)^{-1} |dx_{ij}|^2$$

where now the old $(dx_{ij})^2 = dt/2$ is replaced by $|dx_{ij}|^2 = (dx'_{ij})^2 + (dx''_{ij})^2 = dt$, leading at once to the infinitesimal operator

$$\mathfrak{G} = \frac{1}{2}\left(\Delta = \frac{\partial^2}{\partial\lambda_1^2} + \cdots + \frac{\partial^2}{\partial\lambda_n^2} \right) + \sum_{i=1}^{n} \sum_{j \ne i} \frac{1}{\lambda_i - \lambda_j} \frac{\partial}{\partial\lambda_i}$$

of the Brownian motion conditioned never to cross as per §12.5: $\mathfrak{G}$ is no longer $\frac{1}{2}V^{-1}\operatorname{div}V\operatorname{grad}$ as in §12.4, but $\frac{1}{2}V^{-1}\Delta V$. In short, Dyson's gas and the Brownian motion without crossing are now one and the same – a very pretty coincidence, I think.

Exercise 12.6.1   This is a bit beside the point, but it's amusing and also instructive as to the form of $\mathfrak{G} = \frac{1}{2}V^{-1}\Delta V$. I ask you: What is BM(1) starting at $\mathbf{x}(0) > 0$, conditioned never to hit $x = 0$? In particular: What is its infinitesimal operator?
Hint: Compute $\mathbb{P}_x\left[\mathbf{x}(t) = y \,|\, T_0 > N\right]$ for $N \uparrow \infty$. *Answer*: It is the radial part BES(3) of BM(3) with $\mathfrak{G} = (1/2x)(\partial^2/\partial x^2)x = \frac{1}{2}\partial^2/\partial x^2 + \frac{1}{x}\partial/\partial x$.

12.7 How to compute

To deal with the reduced density $(1/Z)V(2\pi)^{-n/2}e^{-\lambda^2/2}$ both in the bulk and at the edge, a whole new mode of computation comes into play, unlike anything before. I follow Tracy–Widom [1993]; see also Mehta [1967] who started it all.

12.7.1 Andréief's lemma

Let $P_1, \ldots, P_n$ and $Q_1, \ldots, Q_n$ be functions on the line with reasonable decay at $\pm\infty$. Andréief's [1883] lemma is

$$\int_{\mathbb{R}^n} \det\left[P_i(x_j)\right] \times \det\left[Q_i(x_j)\right] = n! \times \det\left[\int_{-\infty}^{+\infty} P_i Q_j \, dx\right]$$

Proof.   Just write out the left-hand side and manipulate:

$$\int_{\mathbb{R}^n} \sum_{\pi} \operatorname{sign}(\pi) P_{\pi 1}(x_1) \ldots P_{\pi n}(x_n) \times \sum_{\pi'} \operatorname{sign}(\pi') Q_{\pi' 1}(x_1) \ldots Q_{\pi' n}(x_n)$$

$$= \sum_{\pi} \operatorname{sign}(\pi) \sum_{\pi'} \operatorname{sign}(\pi'\pi) \int_{-\infty}^{+\infty} P_{\pi 1} Q_{\pi'\pi 1} \times$$

$$\times \int_{-\infty}^{+\infty} P_{\pi 2} Q_{\pi'\pi 2} \times \cdots \times \int_{-\infty}^{+\infty} P_{\pi n} Q_{\pi'\pi n}$$

$$= \sum_{\pi} \sum_{\pi'} \operatorname{sign}(\pi') \int_{-\infty}^{+\infty} P_1 Q_{\pi' 1} \times \cdots \times \int_{-\infty}^{+\infty} P_n Q_{\pi' n}$$

$$= n! \times \det\left[\int_{-\infty}^{+\infty} P_i Q_j\right]$$

since $\pi'1, \pi'2, \ldots, \pi'n = 1, 2, \ldots, n$ in some order.

12.7.2 Application to spec x: first pass

Take a reasonable function f on the line and look at the spectral expectation[8]

$$\mathbb{E}\prod_1^n [1 + f(\mathbf{x}_k)] = \frac{1}{Z} \int_{x_1 < \cdots < x_n} V^2(x) \frac{e^{-x^2/2}}{(2\pi)^{n/2}} \prod_1^n [1 + f(\mathbf{x}_k)]$$

Recall $V(x) = \prod_{j>i}(x_j - x_i) = \det\left[x_j^{i-1} : 1 \le i,j \le n\right]$ from Exercise 12.2.2 and apply Andréief's lemma to the functions

$$P_k = x^{k-1}(2\pi)^{-1/4}e^{-x^2/4} \quad \text{and} \quad Q_k = P_k \times [1 + f(x)].$$

Here

$$\det[P_i(x_j)] \times \det[Q_i(x_j)] = \left[\det x_j^{i-1}\right]^2 \frac{e^{-x^2/2}}{(2\pi)^{n/2}} \prod_1^n [1 + f(x_k)],$$

and since this is all symmetric in $x_1, \ldots, x_n$, you may divide by $n!$ and integrate over the whole of $\mathbb{R}^n$ to obtain

$$\mathbb{E}\prod_1^n [1 + f(\mathbf{x}_k)]$$

$$= \frac{1}{Z} \det\left[\int_{-\infty}^{+\infty} x^{i+j-2} \frac{e^{-x^2/2}}{\sqrt{2\pi}} [1 + f(x)] \, dx : 1 \le i,j \le n\right].$$

Now specialize f to be the indicator of an interval $J \subset \mathbb{R}$ with a factor $-\gamma$ in front. Then

$$\prod_1^n [1 - \gamma f(\mathbf{x}_k)] = (1 - \gamma)^{\#},$$

being the occupation number $\#(k \le n : \mathbf{x}_k \in J)$, so

$$\mathbb{E}(1-\gamma)^{\#} = \frac{1}{Z} \det\left[\int_{-\infty}^{+\infty} x^{i+j-2} \frac{e^{-x^2/2}}{\sqrt{2\pi}} [1 - \gamma f(x)] \, dx : 1 \le i,j \le n\right],$$

from which the probabilities $\mathbb{P}(\# = k)$, $1 \le k \le n$, could (in principle) be found.

Two more ingredients are needed before this recipe is of real, practical use, *viz.* Hermite polynomials and Fredholm determinants.

---

[8] I change spec x to $\mathbf{x}_1 < \cdots < \mathbf{x}_n$ for uniformity of notation.

12.7.3 Hermite polynomials

These are the powers x^n, $n \geq 0$, reorganized to make a unit perpendicular frame in $L^2(\mathbb{R}, (2\pi)^{-1/2}e^{-x^2/2})$ already seen in §§6.11.2 and 11.7.5, but I repeat with a bit of extra information. The recipe is[9]

$$H_n(x) = (-1)^n (n!)^{-1/2} e^{x^2/2} D^n e^{-x^2/2}.$$

Fact 1 $\displaystyle\int_{-\infty}^{+\infty} H_i H_j \frac{e^{-x^2/2}}{\sqrt{2\pi}}\,dx = 1$ or 0 according as $i = j$ or not.

Proof. Discard for a moment the normalizer $\sqrt{n!}$, form the function

$$G(\gamma) = \sum_0^\infty H_n \frac{\gamma^n}{n!} = e^{x^2/2} e^{-(x-\gamma)^2/2} = e^{x\gamma} e^{-\gamma^2/2},$$

and compute

$$\int G(\alpha)G(\beta) \frac{e^{-x^2/2}}{\sqrt{2\pi}} = \int e^{(\alpha+\beta)x} e^{-\alpha^2/2} e^{-\beta^2/2} \frac{e^{-x^2/2}}{\sqrt{2\pi}}$$
$$= e^{(\alpha+\beta)^2/2} e^{-\alpha^2/2} e^{-\beta^2/2} = e^{\alpha\beta}$$
$$= \sum_0^\infty \frac{(\alpha\beta)^n}{n!}.$$

The rest is obvious.

Fact 2

$$H_n' = xH_n - \sqrt{n+1}\,H_{n+1} \quad \text{(Part 1)}$$
$$= \sqrt{n}\,H_{n-1} \quad\quad\quad\quad \text{(Part 2)}.$$

Proof. Part 1 is immediate from the definition. Now H_n is a polynomial of precise degree n, perpendicular to x^k for $k < n$, by Fact 1, so for $k \leq n-2$,

$$\int H_k H_n' \frac{e^{-x^2/2}}{\sqrt{2\pi}} = \int H_k\, xH_n \frac{e^{-x^2/2}}{\sqrt{2\pi}} \quad \text{by Part 1,}$$

and this must vanish. H_n' is then a multiple of H_{n-1}, and the multiplier is read off from the top power of x. That's Part 2.

Fact 3 Denote $A(x)B(y)$ by $A \otimes B$. Then

$$\sum_0^{n-1} H_j \otimes H_j = \frac{H_n \otimes H_n' - H_n' \otimes H_n}{x - y}.$$

---

[9] D is differentiation with regard to x.

Proof. $xH_n = \sqrt{n}\,H_{n-1} + \sqrt{n+1}\,H_{n+1}$ by Fact 2, so

$$x\sum_0^{n-1} H_j \otimes H_j - \sum_0^{n-1} H_j \otimes H_j\, y$$

$$= \sum_0^{n-1}\left[\left(\sqrt{j}\,H_{j-1} + \sqrt{j+1}\,H_{j+1}\right) \otimes H_j\right.$$

$$\left. - H_j \otimes \left(\sqrt{j}\,H_{j-1} + \sqrt{j+1}\,H_{j+1}\right)\right]$$

$$= \sum_0^{n-1}\left[\sqrt{j+1}\,H_{j+1} \otimes H_j - \sqrt{j}\,H_j \otimes H_{j-1}\right]$$

$$+ \sum_0^{n-1}\left[\sqrt{j}\,H_{j-1} \otimes H_j - \sqrt{j+1}\,H_j \otimes H_{j+1}\right]$$

$$= \sqrt{n}\,H_n \otimes H_{n-1} - \sqrt{n}\,H_{n+1} \otimes H_n \qquad \text{by telescoping of sums}$$

$$= H_n \otimes H_n' - H_n' \otimes H_n \qquad\qquad \text{by Fact 2.}$$

12.7.4 Application to spec x: second pass

The use of Fact 3 is postponed, but already Fact 1 is helpful: H_n being a polynomial of precise degree n, you may write

$$V(x) = \det\left[x_i^{j-1}\right] = \det\left[H_{j-1}(x_i)\right]$$

up to a numerical factor, permitting the re-expression of the expectation

$$\mathbb{E}\prod_1^n [1 + f(\mathbf{x}_k)]$$

$$= \frac{1}{Z}\,\det\left[\int_{-\infty}^{+\infty} x^{i+j-2}\,\frac{e^{-x^2/2}}{\sqrt{2\pi}}\,[1 + f(x)]\,\mathrm{d}x : 1 \le i, j \le n\right]$$

as

$$\det\left[\int_{-\infty}^{+\infty} H_i H_j\,\frac{e^{-x^2/2}}{\sqrt{2\pi}}\,[1 + f(x)]\,\mathrm{d}x : 0 \le i, j \le n\right]$$

$$= \det\left[\text{the identity } I + \int_{-\infty}^{+\infty} H_i H_j f(x)\,\frac{e^{-x^2/2}}{\sqrt{2\pi}}\,\mathrm{d}x : 0 \le i, j \le n\right].$$

No numerical factor appears anymore, and that's correct as you see by putting $f = 0$, so the unknown normalizer, Z taking its cue, leaves the stage. That's welcome.

12.7.5 Fredholm determinants

Fredholm determinants are needed for the reduction of $\prod[1 + f(\mathbf{x}_k)]$ in the bulk and at the edge when $n \uparrow \infty$, and as these may be unfamiliar, I will sketch what you need to know: see Lax [2002] and/or McKean [2011] for any technical details skated over here or later in §12.9.

Take a matrix $K = [K_{ij} : 1 \le i, j \le n]$ and let $\mathbf{k}$ be an ordered set $k_1 < k_2 < \cdots < k_\ell$ of $\#(\mathbf{k}) = \ell$ indices. Then with $[e_{ij} : 1 \le i, j \le n] =$ the $n \times n$ identity I, you have

$$\det(I + K) = \sum_\pi \operatorname{sign} \pi \prod_{i=1}^n (e_{i\pi i} + K_{i\pi i}),$$

in which any partial product of es vanishes unless the permutation π fixes the affected indices. What survives is

$$\sum_{\ell=0}^n \sum_{\mathbf{k}:\#(\mathbf{k})=\ell} \sum_{\substack{\pi \text{ permuting } \mathbf{k}, \\ \text{fixing the rest}}} \operatorname{sign} \pi \prod_{i \in \mathbf{k}} K_{i\pi i}$$

$$= \sum_{\ell=0}^n \sum_{\mathbf{k}:\#(\mathbf{k})=\ell} \det\left[K_{ij} : i, j \in \mathbf{k}\right]$$

$$= 1$$

$$+ \sum_i K_{ii}$$

$$+ \sum_{i<j} \det \begin{bmatrix} K_{ii} & K_{ij} \\ K_{ji} & K_{jj} \end{bmatrix}$$

$$+ \sum_{i<j<k} \det \begin{bmatrix} K_{ii} & K_{ij} & K_{ik} \\ K_{ji} & K_{jj} & K_{jk} \\ K_{ki} & K_{kj} & K_{kk} \end{bmatrix}$$

$$+ \text{etc.}$$

$$+ \det\left[K_{ij} : 1 \le i, j \le n\right].$$

This is the key to Fredholm's idea of how to extend the customary determinant for $\mathbb{R}^n$ to the infinite-dimensional (function) space "$\mathbb{R}^\infty$". Take, for example, a well-behaved integral operator

$$K \colon f(x) \longrightarrow Kf(x) = \int_0^1 K(x, y) f(y) \, dy \quad \text{acting in } C[0, 1],$$

with $K(x, y) \in C[0, 1]^2$ or better. This will in some sense be small compared to the identity I, and it may be hoped that the determinant of $I +$ *the operator* K will have a sensible meaning. In fact, if you replace $f(x) : 0 \le x \le 1$ by $f(k/n) : 1 \le k \le n$ and approximate the action $K : f \to Kf$ by

$$Kf\left(\frac{i}{n}\right) = \sum_{j=1}^{n} K\left(\frac{i}{n}, \frac{j}{n}\right) \frac{1}{n} f\left(\frac{j}{n}\right), \qquad 1 \le i \le n$$

in $\mathbb{R}^n$, then Fredholm's formula for $\det\left[I + K\left(\frac{i}{n}, \frac{j}{n}\right)\frac{1}{n} : 1 \le i, j \le n\right]$ will tell you what to do in $C[0, 1]$: *in extenso*, you find

$$1$$

$$+ \sum_i K\left(\frac{i}{n}, \frac{i}{n}\right) \frac{1}{n}$$

$$+ \sum_{i<j} \det \begin{bmatrix} K\left(\frac{i}{n}, \frac{i}{n}\right) & K\left(\frac{i}{n}, \frac{j}{n}\right) \\ K\left(\frac{j}{n}, \frac{i}{n}\right) & K\left(\frac{j}{n}, \frac{j}{n}\right) \end{bmatrix} \frac{1}{n^2}$$

$$+ \sum_{i<j<k} \det \begin{bmatrix} K\left(\frac{i}{n}, \frac{i}{n}\right) & K\left(\frac{i}{n}, \frac{j}{n}\right) & K\left(\frac{i}{n}, \frac{k}{n}\right) \\ K\left(\frac{j}{n}, \frac{i}{n}\right) & K\left(\frac{j}{n}, \frac{j}{n}\right) & K\left(\frac{j}{n}, \frac{k}{n}\right) \\ K\left(\frac{k}{n}, \frac{i}{n}\right) & K\left(\frac{k}{n}, \frac{j}{n}\right) & K\left(\frac{k}{n}, \frac{k}{n}\right) \end{bmatrix} \frac{1}{n^3}$$

$$+ \text{etc.,}$$

suggesting that the determinant of $(I + K)$ should be

$$1$$

$$+ \int_0^1 K(x_1, x_1)\, dx_1$$

$$+ \int_{x_1 < x_2} \det \begin{bmatrix} K(x_1, x_1) & K(x_1, x_2) \\ K(x_2, x_1) & K(x_2, x_2) \end{bmatrix} dx_1\, dx_2$$

$$+ \int_{x_1 < x_2 < x_3} \det \begin{bmatrix} K(x_1, x_1) & K(x_1, x_2) & K(x_1, x_3) \\ K(x_2, x_1) & K(x_2, x_2) & K(x_2, x_3) \\ K(x_3, x_1) & K(x_3, x_2) & K(x_3, x_3) \end{bmatrix} dx_1\, dx_2\, dx_3$$

and so on, *ad infinitum*.

This is Fredholm's way (1903). Don't worry about the infinite sum: each term, both before and after taking n to $+\infty$, has a domination comparable to $(n!)^{-1/2}$ that takes care of everything.

I list five rules for these determinants. Check them for matrices. Then use them freely as if you were an old hand at it.

Rule 1 $\det(I + A)(I + B) = \det(I + A) \times \det(I + B)$.

Rule 2 $\det(I + K^\dagger) = \det(I + K)$, $K^\dagger$ being the transpose of K, as in $K^\dagger(x, y) = K(y, x)$.

Rule 3 $I - K$ has a nice inverse $(I - K)^{-1}$ of the same type if (and only if) $\det(I - K) \neq 0$. Then $\det(I - K)^{-1}$ is then the reciprocal of $\det(I - K)$, as per rule 1, and $(I - K)^{-1} = I + K + K^2 +$ etc. if all goes well, i.e. if K is not too big. Contrariwise, $(I - K)f = 0$ has a nice solution f if (and only if) $\det(I - K) = 0$.

Rule 4 $- \log \det(I - K) = \text{tr} \left[- \log(I - K) = K + K^2/2 + K^3/3 + \text{etc.} \right]$ if all goes well. (Recall tr is the trace, i.e. the integral of $K(x, y)$ down the diagonal $x = y$.) Naturally, $\det(I - K)$ had better be positive here, and if $(I - K)^{-1} = I + K + K^2 +$ etc. is OK, surely, $K + K^2/2 + K^3/3 +$ etc. will be better still.

Rule 5 $\det(I + AB) = \det(I + BA)$ may be the least familiar of these rules. Here, AB must be "square", mapping some space into itself, and likewise BA – otherwise the determinants won't make sense – but the individual A and B may be merely "rectangular": for example, you might have

$$\mathbb{R}^3 \xrightarrow{A} \mathbb{R}^2 \xrightarrow{B} \mathbb{R}^3 \xrightarrow{A} \mathbb{R}^2,$$

AB being a self-map of $\mathbb{R}^2$ and BA a self-map of $\mathbb{R}^3$. Think it over in such a case, using Rule 4 if you want, under the assumption that A and B are not too big. In any case, if A is rectangular, mapping space no. 1 into space no. 2, you'll want B to map space no. 2 back to no. 1. Once that's taken care of, the rest follows easily enough.

12.7.6 Application to spec x: third pass

Rule 5 is applied to

$$A \colon Q \in L^2 \left(\mathbb{R}, \frac{e^{-x^2/2}}{\sqrt{2\pi}} \right) \longrightarrow P = \left[\int_{-\infty}^{+\infty} H_i f Q \frac{e^{-x^2/2}}{\sqrt{2\pi}} : 0 \le i < n \right] \in \mathbb{R}^n$$

and

$$B \colon P \in \mathbb{R}^n \longrightarrow Q = \sum_{0}^{n-1} P_j H_j(x) \in L^2 \left(\mathbb{R}, \frac{e^{-x^2/2}}{\sqrt{2\pi}} \right).$$

AB is now the self-map of $\mathbb{R}^n$ implemented by

$$\left[\int_{-\infty}^{+\infty} H_i H_j f \frac{e^{-x^2/2}}{\sqrt{2\pi}} : 0 \le i, j < n \right]$$

and BA is the self-map of $L^2\left(\mathbb{R}, (2\pi)^{-1/2}\, e^{-x^2/2}\right)$ implemented by

$$\left[\sum_0^{n-1} H_j(x) \otimes H_j(y) f(y) \frac{e^{-y^2/2}}{\sqrt{2\pi}} : x, y \in \mathbb{R}\right],$$

the outcome being

$$\mathbb{E} \prod_1^n [1 + f(\mathbf{x}_k)] = \text{the ordinary } \det(I + AB)$$

$$= \text{Fredholm's } \det(I + BA) \qquad \text{by Rule 5}$$

$$= \det \left[I + \sum_0^{n-1} H_j \otimes H_j\, f\, \frac{e^{-y^2/2}}{\sqrt{2\pi}} : x, y \in \mathbb{R} \right]$$

$$= \det \left[I + \frac{e^{-x^2/4}}{\sqrt[4]{2\pi}} \frac{H_n \otimes H'_n - H'_n \otimes H_n}{x - y} \frac{e^{-y^2/4}}{\sqrt[4]{2\pi}}\, f : x, y \in \mathbb{R} \right]$$

$$\equiv \det \left[I + K f : x, y \in \mathbb{R} \right],$$

in which line 4 comes from Fact 3 under Hermite polynomials in §12.7.3. This may be specialized as before by taking $f = -\gamma \times$ the indicator of an interval $J \subset \mathbb{R}$ so that $\prod_1^n [1 - \gamma f(\mathbf{x}_k)] = (1 - \gamma)^{\#}$, $\#$ being the occupation number $\#(B \le n = \mathbf{x}_k \in J)$; in particular,

$$\mathbb{P}(\# = 0) = \det \left[I - K : x, y \in J \right].$$

Now everything is ready to deal with bulk and edge for $n \uparrow \infty$: the dimension of the determinant is fixed (it is already $+\infty$) and it's only

$$K(x, y) = \frac{e^{-x^2/4}}{\sqrt[4]{2\pi}} \frac{H_n \otimes H'_n - H'_n \otimes H_n}{x - y} \frac{e^{-y^2/4}}{\sqrt[4]{2\pi}}$$

which changes with the dimension of the ensemble. That feature simplifies life, as will be seen next.

12.8 In the bulk

For large n, the semi-circle predicts that nearly all the n eigenvalues of $\mathbf{x}$ are packed into the interval between $\pm 2\sqrt{n}$, spaced, away from the edge, more or less evenly on the scale $1/\sqrt{n}$, so if you want to see what's happening near the origin, say, you will have to rescale that way. A natural question is then: What is the probability of having a spectral gap between 0 and $1/\sqrt{n}$? Or again: What is the probability of such a gap, knowing that there is an eigenvalue *at* 0?

12.8.1 A single gap

The probability of a gap of length $L/\sqrt{n}$ is known from §12.7: it is Fredholm's determinant

$$\det\left[I - \sqrt{w(x)}\,\frac{H_n \otimes H_n' - H_n' \otimes H_n}{x - y}\,\sqrt{w(y)} : 0 \le x, y \le L/\sqrt{n}\right]$$

with $w = (2\pi)^{-1/2}e^{-x^2/2}$, so you need to know how $H_n(x)$ looks for small x and $n \uparrow \infty$. That's a delicate business. The fact is that $H_n(x)$ exhibits two modes of behavior: out beyond the edge at $x = 2\sqrt{n}$, it runs off to $\pm\infty$, while below $2\sqrt{n}$, it is a oscillatory. That's where you have to work now.

The details can be found in Bateman $[1953(2):199(2)]^{10}$. It is convenient to take the dimension even $(= 2n)$; it hardly matters, though a correction will be needed at the end. Then

$$H_{2n}(x) = (-1)^n \sqrt[4]{\pi}\left(\frac{x^2}{2}\right)^{1/4}\,J_{-1/2}(\sqrt{2n}\,x)$$

with the (happily elementary) Bessel function $J_{-1/2}(x) = (2/\pi x)^{\frac{1}{2}}\cos x$, and to get something useful, you need to rescale, replacing $\sqrt{2n}\,x$ by x', say, so that $H_{2n}(x) \simeq (-1)^n(\pi n)^{-1/4}\cos x'$. This may be reliably differentiated to obtain

$$H_{2n}'(x) \simeq (-1)^{n-1}(\pi n)^{-1/4}\sin x' \times \text{ the not to be forgotten } \sqrt{2n} = \frac{dx'}{dx},$$

the upshot being

$$\sqrt{w(x)}\,\frac{H_{2n} \otimes H_{2n}' - H_{2n}' \otimes H_{2n}}{x - y}\,\sqrt{w(y)}$$

$$\simeq \frac{1}{\sqrt[4]{2\pi}}\,\frac{\sqrt{2n}}{\sqrt[4]{\pi n}}\,\frac{\sin x' \cos y' - \cos x' \sin y'}{(x' - y')/\sqrt{2n}}\,\frac{1}{\sqrt[4]{2\pi}} \times dy' \times \left(\frac{1}{\sqrt{2n}} = \frac{dy}{dy'}\right)$$

$$= \frac{1}{\pi}\,\frac{\sin(x' - y')}{x' - y'}\,dy';$$

in short,

$$\lim_{n\uparrow\infty} \mathbb{P}\left[\operatorname{spec}\mathbf{x} \cap [0, L/\sqrt{n}] \text{ is void}\right] = \det[I - K : 0 \le x, y \le L] \equiv D(L),$$

where now $K(x, y) = \frac{1}{\pi}(x - y)^{-1}\sin(x - y)$, and the scaling has been corrected from $x = x'/\sqrt{2n}$ for even dimension $2n$ to $x = x'/\sqrt{n}$ for

---

^{10} *Warning*: If you look this up, keep in mind that Bateman's Hermite polynomials (BH) are related to those in §12.7 as $H_n(x) = 2^{-n/2} \times$ Bateman's $H_n(x/\sqrt{2})$. A nuisance but there it is.

the general dimension n. Naturally, some domination is needed here, in Fredholm's sum for the determinant, but don't worry. It's OK.

$D(L)$ will be fully investigated in §§12.9 and 12.10. For the moment, just observe that $D(0+) = 1$ as it should be, and let me tell you that $D(L)$ is roughly $C\exp(-L^2/8)$ at $L = +\infty$, for which see §12.10. In the meantime, here's a demonstration that $D(+\infty) = 0$, as it should be if the $\sqrt{n}$ scaling is correct. Let $Z(AB)$ be the event that $\mathrm{spec}\,\mathbf{x} \cap [A, B]$ is void. Then it is reasonable to think that

$$\mathbb{P}[Z(0B)] = \mathbb{P}[Z(0A)] \times \mathbb{P}[Z(AB) \mid Z(0A)] \le \mathbb{P}[Z(0A)]\,\mathbb{P}[Z(AB)]$$

since the absence of spectrum in $[0, A)$ should make it more likely to find some in $[A, B]$. Now make $n \uparrow \infty$ to obtain $D(B) \le D(A)D(B - A)$ after recognizing that $\mathbb{P}[Z(AB)] = \det[I - K : A \le x, y \le B]$ depends on $B - A$ alone. Note now $D(L) \simeq 1 - L/\pi$ for small L, and conclude from $D(L) \le [D(L/n)]^n$ that $D(L) \le \exp(-L/\pi)$. Rough, but good enough for now. Think it over.

12.8.2 Wigner's surmise

Now let's work out the conditional probability

$$\lim_{n \uparrow \infty} \mathbb{P}\left[\mathrm{spec}\,\mathbf{x} \cap [0, L/\sqrt{n}] \text{ is void} \,\Big|\, 0 \in \mathrm{spec}\,\mathbf{x}\right].$$

Wigner [1958] conjectured that this should be like $\exp(-\pi L^2/4)$ for large L, which is nearly right. That was his "surmise" as it is now called, to be corrected in §12.10.1 to something like $\exp(-L^2/8)$.

Here's how it goes. Take $f = -\alpha\,\mathbf{1}_{0A} - \beta\,\mathbf{1}_{AB}$ with $0 \le \alpha, \beta \le 1$ and $0 < A < B < \infty$ to produce[11]

$$\mathbb{E}\left[(1 - \alpha)^{\#(0A)}(1 - \beta)^{\#(AB)}\right]$$
$$= \det[I - \alpha K\,\mathbf{1}_{0A} - \beta K\,\mathbf{1}_{AB} : 0 \le x, y \le B],$$

in which $K(x, y) = \sqrt{w(x)}\,(H_n \otimes H_n' - H_n' \otimes H_n)\,\sqrt{w(y)}/(x - y)$ as before, and differentiate by α with the help of Rule 5, §12.7 to obtain[12]

$$\mathbb{E}\left[\#(0A)(1 - \alpha)^{\#(0A)-1}(1 - \beta)^{\#(AB)}\right]$$
$$= \det[I - \alpha K\,\mathbf{1}_{0A} - \beta K\,\mathbf{1}_{AB}] \times \mathrm{tr}\left[(I - \alpha K\,\mathbf{1}_{0A} - \beta K\,\mathbf{1}_{AB})^{-1}\,K\,\mathbf{1}_{0A}\right].$$

---

[11] # indicates the obvious occupation number.
[12] the existence of the inverse is assumed; see Fact 1§12.9 for reassurance if you want it.

Think of $K(x, y)$ as restricted to $0 \leq x, y \leq B$ throughout and read off

(1) $\mathbb{P}\left[\#(0A) = 1\right] = \det(I - K \mathbf{1}_{0A}) \operatorname{tr}\left[(I - K \mathbf{1}_{0A})^{-1} K \mathbf{1}_{0A}\right]$

 by choice of $\alpha = 1$ and $\beta = 0$,

(2) $\mathbb{P}\left[\#(0A) = 1 \ \& \ \#(AB) = 0\right] = \det(I - K) \operatorname{tr}\left[(I - K)^{-1} K \mathbf{1}_{0A}\right]$

 by choice of $\alpha = \beta = 1$,

and so also

(3)
$$\begin{aligned} &\mathbb{P}\left[\#(AB) = 0 \,\middle|\, \#(0A) = 1\right] \\ &= \frac{\det(I - K) \operatorname{tr}\left[(I - K)^{-1} K \mathbf{1}_{0A}\right]}{\det(I - K \mathbf{1}_{0A}) \operatorname{tr}\left[(I - K \mathbf{1}_{0A})^{-1} K \mathbf{1}_{0A}\right]} \end{aligned}$$

Taking A down to $0+$, you find

(1') $\mathbb{P}\left[\#(0A) \simeq 1\right] = \operatorname{tr}\left[K \mathbf{1}_{0A}\right] = K(0,0) \times A$,

(2') $\mathbb{P}\left[\#(0A) \simeq 1 \ \& \ \#(AB) = 0\right] = \det(I - K) \times (I - K)^{-1} K(0,0) \times A$,

and

(3')
$$\begin{aligned} &\mathbb{P}\left[\operatorname{spec} \mathbf{x} \cap (0, B] \text{ is void} \,\middle|\, 0 \in \operatorname{spec} \mathbf{x}\right] \\ &= \det(I - K) \times \frac{(I - K)^{-1} K(0,0)}{K(0,0)} \end{aligned}$$

Now put $h = L/\sqrt{n}$ and make $n \uparrow \infty$ as before to obtain the final formula

$$\lim_{n \uparrow \infty} \mathbb{P}\left[\operatorname{spec} \mathbf{x} \cap (0, L/\sqrt{n}] \text{ is void} \,\middle|\, 0 \in \operatorname{spec} \mathbf{x}\right]$$
$$= \det(I - K) \times \frac{(I - K)^{-1} K(0,0)}{K(0,0)}$$

where $K(x, y)$ is now $\frac{1}{\pi} \sin(x - y)/(x - y)$ and $\det(I - K) = D(L)$ as above. This, too, will be further explicated in §§12.9 and 12.10.

12.9 The ODE

Jimbo–Miwa–Môri–Sato [1980] discovered that $D(L)$ may be expressed pretty explicitly with the help of a simple 2×2 Hamiltonian system with an interesting history, not in any visible way connected with what we are thinking about. It is this type of coincidence that lends mathematics so much of its charm; compare §12.12. Tracy–Widom [1993] is shorter and easier to read than Jimbo et al. Here, I follow Tracy–Widom [2004] with minor simplifications.

 I collect some preliminary facts, skimmed over before, about $D(L) =$

$\det(I-K)$, thinking of $K(x,y)=\sin(x-y)/\pi(x-y)$ as a map of $L^2[0,L]$ into $C[0,L]$.

Fact 1 $\det(I-K)\neq 0$; in particular, $I-K$ has a nice inverse, by Rule 3, §12.7.

Proof.

$$\frac{\sin x}{\pi x}=\frac{1}{2\pi}\int_{-1}^{+1}e^{\sqrt{-1}\,kx}\,dk,$$

from which you read off[13] $(\sin x/\pi x)^{\wedge}=1$ or 0 according as $|k|\leq 1$ or not. Now take f vanishing *outside* $0\leq x\leq L$ and suppose $(I-K)f=0$ *inside*. Inside, you have

$$f(x)=Kf(x)=\frac{1}{2\pi}\int_{-1}^{+1}e^{\sqrt{-1}\,kx}\,dk\int_{0}^{L}e^{-\sqrt{-1}\,ky}f(y)\,dy$$
$$=\frac{1}{2\pi}\int_{-1}^{+1}e^{\sqrt{-1}\,kx}\hat{f}(k)\,dk$$

so

$$\frac{1}{2\pi}\int_{-\infty}^{+\infty}|\hat{f}|^2=\int_{-\infty}^{+\infty}f^2=\int_{0}^{L}f\,Kf=\frac{1}{2\pi}\int_{-1}^{+1}|\hat{f}|^2,$$

and this is not possible unless $f\equiv 0$: $\hat{f}$ must vanish outside $|k|\leq 1$, in which case, the prior display has to be correct not only inside $0\leq x\leq L$ but outside as well, and that won't do: for then $f(x)$ would be an analytic function of the *complex* variable x and could not vanish for, e.g. $x\leq 0$ without its vanishing everywhere. Rule 3, §12.7 does the rest.

Fact 2   Fredholm's Rules 4 and 5, §12.7 require that $(I-K)^{-1}=I+K+K^2$ etc. and $-\log(I-K)=K+K^2/2+K^3/3$ etc. make sense. Let's look into that. $K=K^\dagger$ is non-negative in the sense that the quadratic form $\int f\,Kf=\frac{1}{2\pi}\int_{-1}^{+1}|\hat{f}|^2$ is such; similarly, $\int f(I-K)f=\frac{1}{2\pi}\int_{|k|\geq 1}|\hat{f}|^2$ is non-negative, or, as you may say, $0\leq K\leq I$. The kernel K is also *compact* in the sense that the image of the unit ball $(f:\int f^2\leq 1)$ of $L^2[0,L]$ is compact in $C[0,L]$ because $|(Kf)'(x)|$ is overestimated by a fixed constant, independently of f. These facts guarantee that K has an infinite number of mutually perpendicular eigenfunctions, e_0,e_1,e_2, etc., normalized by $\int e_n^2=1$, with eigenvalues $\gamma_0\geq\gamma_1\geq\gamma_2\geq\gamma_3$ etc. $\downarrow 0$ and a rapidly converging expansion $K(x,y)=\sum_0^\infty\gamma_n e_n\otimes e_n$, much as for

---

[13] $^\wedge$ signifies the Fourier transform of §1.6 in a convenient variant:
$\hat{f}(k)=\int_{-\infty}^{+\infty}e^{-\sqrt{-1}\,kx}f(x)\,dx$ with $f(x)=(\hat{f})^\vee(x)=\frac{1}{2\pi}\int e^{\sqrt{-1}\,kx}\hat{f}(k)\,dk$ and $\int f^2\,dx=\frac{1}{2\pi}\int|\hat{f}|^2\,dk.$

matrices with a like symmetric, non-negative character[14]. Here, $\gamma_0 \leq 1$ since $\gamma_0 = \int e_0 K e_0 = \frac{1}{2\pi} \int_{-1}^{+1} |\hat{e}_0|^2 \leq \int e_0^2 = 1$; indeed, $\gamma_0 < 1$: otherwise, $K e_0 = e_0$ would violate Fact 1. The powers of K may now be estimated:

$$Ke = \gamma e \text{ and } |\gamma e| = |Ke| \leq \sqrt{\int_{-\infty}^{\infty} (\sin x/\pi x)^2} = \frac{1}{\sqrt{\pi}},$$

so $K^p(x,y) = \sum_0^\infty \gamma_n^p e_n \otimes e_n$ is majorized by a fixed multiple of $\sum_0^\infty \gamma_n^{p-2}$, independently of $p \geq 3$. Here,

$$\sum_{p=3}^{\infty} \sum_{n=0}^{\infty} \gamma_n^{p-2} = \sum_0^\infty \frac{\gamma_n}{1-\gamma_n} \leq \frac{1}{1-\gamma_0} \sum_0^\infty \gamma_n,$$

in which the final sum is nothing but $\operatorname{tr} K = L/\pi$; in short, the sums $I + K + K^2$ etc. and $K + K^2/2 + K^3/3$ etc. are just fine.

Fact 3 The non-negative character of K has still another advantage. This property is automatic for K^2, K^4 etc. and is inherited by K^3, K^5 etc. It follows that every power of K is non-negative on the diagonal so that the individual traces figuring in $-\log\det(I-K) = \operatorname{tr}(K + K^2/2 + K^3/3$ etc.$)$ contribute in a cooperative way to $D(L) = \exp\left[-\operatorname{tr}(K + K^2/2 \text{ etc.})\right]$; in particular, $D(L) \leq \exp(-\operatorname{tr} K) = \exp(-L/\pi)$, as demonstrated already in §12.8. No real improvement is got from $\operatorname{tr}(K^2/2)$ or $\operatorname{tr}(K^3/3)$. You have to go all the way before much more is seen, but that comes later, in §12.10.

Now to the real business.

Step 1 introduces a helpful expression for $(I-K)^{-1}K \equiv G$ in terms of the functions

$$P(y) = (I-K)^{-1}\sin(y-L/2) \quad \text{and} \quad Q(y) = (I-K)^{-1}\cos(y-L/2),$$

in which the operator $(I-K)^{-1}(x,y)$ acts on y, the resulting functions being evaluated, *to wit*[15]

$$G(x,y) = \frac{P \otimes Q - Q \otimes P}{\pi(x-y)}.$$

---

[14] Lax [2002] is the best place to find this if it is unfamiliar.
[15] Here and below, permit me to confuse the operator and its kernel; the context will make plain what's meant.

Proof. With M being the operation of multiplication by x, you find

$$
(x - y)G(x, y)
$$
$$
= MG - GM
$$
$$
= MK(I - K)^{-1} - (I - K)^{-1}KM
$$
$$
= (I - K)^{-1}\left[(I - K)MK - KM(I - K)\right](I - K)^{-1}
$$
$$
= (I - K)^{-1}(MK - KM)(I - K)^{-1}
$$
$$
= (I - K)^{-1}\frac{\sin(x - y)}{\pi}(I - K)^{-1}
$$
$$
= \frac{1}{\pi}(I - K)^{-1}\left[\sin(x - L/2)\cos(y - L/2)\right.
$$
$$
\left. - \cos(x - L/2)\sin(y - L/2)\right](I - K)^{-1}
$$
$$
= \frac{1}{\pi}[P \otimes Q - Q \otimes P],
$$

where the last line uses the fact that $(I - K)^{-1}$ is symmetric.

Step 2. The reflection of $0 \le x \le L$ about its midpoint commutes with K and reverses/preserves the signature of $\sin / \cos(\,\cdot\, - L/2)$, so $P(L - x) = -P(x)$ and $Q(L - x) = Q(x)$; in particular, $P(L) = -P(0)$, $Q(L) = Q(0)$, $P'(L) = P'(0)$, $Q'(L) = -Q'(0)$, and so on. These remarks facilitate the proof of two useful identities:

for Q: $\qquad Q'(x) = -G(x, L)Q(0) + G(x, 0)Q(0) - P(x),$

and similarly,

for P: $\qquad P'(x) = +G(x, L)P(0) + G(x, 0)P(0) + Q(x).$

Proof for P. With D being differentiation by x, you have

$$
D(I - K)^{-1} = D + DK(I - K)^{-1}
$$
$$
= D + (I - K)^{-1}\left[(I - K)DK - KD(I - K)\right](I - K)^{-1}
$$
$$
+ (I - K)^{-1}KD
$$
$$
= (I - K)^{-1}(DK - KD)(I - K)^{-1} + (I - K)^{-1}D.
$$

This identity is now applied to $\sin(\,\cdot\, - L/2)$ to produce

$$
P' = (I - K)^{-1}(DK - KD)P + Q,
$$

and $(DK - KD)P$ is worked out: $\partial K(x,y)/\partial x = -\partial K(x,y)/\partial y$, so

$$(DK - KD)P(x) = -\int_0^L \frac{\partial}{\partial y} K(x,y)P(y)\,\mathrm{d}y$$
$$= -K(x,L)P(L) + K(x,0)P(0).$$

$P(L) = -P(0)$ does the rest.

Step 3. $P(0)$ and $Q(0)$, considered as functions of L alone, solve the 2×2 Hamiltonian system:

$$\frac{\partial Q}{\partial L} = -2\frac{Q^2 P}{\pi L} + \frac{1}{2}P = \frac{\partial H}{\partial P}$$
$$\frac{\partial P}{\partial L} = +2\frac{P^2 Q}{\pi L} - \frac{1}{2}Q = -\frac{\partial H}{\partial Q}$$

with Hamiltonian $H = \frac{1}{4}(P^2 + Q^2) - P^2 Q^2/\pi L$ and initial values $Q(L = 0) = 1$ and $P(L = 0) = 0$. That's the ODE advertised in the section's title, upon which is based the large L approximation $D(L) \simeq \exp(-L^2/8)$ to come in §12.10.

Proof for Q.

$Q(0) = I + K + K^2 +$ etc. applied to $\cos(\cdot - L/2)$, taken at $x = 0$

$= \cos(L/2)$ (line 0)

$+ \displaystyle\int_0^L K(0,y_1)\cos(y_1 - L/2)\,\mathrm{d}^1 y$ (line 1)

$+ \displaystyle\int_0^L \int_0^L K(0,y_1)K(y_1,y_2)\cos(y_2 - L/2)\,\mathrm{d}^2 y$ (line 2)

$+ \displaystyle\int_0^L \int_0^L \int_0^L K(0,y_1)K(y_1,y_2)K(y_2,y_3)\cos(y_3 - L/2)\,\mathrm{d}^3 y$ (line 3)

and so on.

$\mathrm{d}Q/\,\mathrm{d}L$ is now computed by hand: the L figuring in $\cos(\cdot - L/2)$ contributes

$$\frac{1}{2}\sin\left(\frac{-L}{2}\right) + (K + K^2 + \text{etc.}) \text{ applied to } \frac{1}{2}\sin\left(\cdot - \frac{L}{2}\right)$$
$$\text{taken at } x = 0 = \frac{1}{2}P(0).$$

The upper limits of integration contribute, too. I spell that out in an abbreviated notation, line by line:

$K(0, L) \cos(L/2)$ from line 1

$K(0, L)(K \cos)(L) + K^2(0, L) \cos(L/2)$ from line 2

$K(0, L)(K^2 \cos)(L) + K^2(0, L)(K \cos)(L) + K^3(0, L) \cos(L/2)$

                                                                             from line 3

and so on,

which is now added up as follows:

$$\sum_{i=1}^{\infty} \sum_{j=0}^{\infty} K^i(0, L)(K^j \cos)(L) \qquad \text{with the understanding that } K^0 = I$$

$$= G(0, L) \times (I - K)^{-1} \cos(\cdot - L/2) \qquad \text{evaluated at } x = 1$$

$$= \frac{P(0)Q(L) - Q(0)P(L)}{\pi(0 - L)} \times Q(L) \qquad \text{by Step 1}$$

$$= -2\frac{Q^2(0)P(0)}{\pi L} \qquad \text{by Step 2.}$$

That does it.

Step 4. $-\,\mathrm{d}\log D/\,\mathrm{d}L = (4/\pi) \times$ the Hamiltonian H of Step 3.

Proof $-\log D = \mathrm{tr}(K + K^2/2 + K^3/3 \text{ etc.})$ is written out and differentiated by hand much as in Step 3 to produce

$$K(L, L) + K^2(L, L) + K^3(L, L) \text{ etc.} = G(L, L) = G(0, 0) \quad \text{by Step 2,}$$

and this may be identified as $(4/\pi) \times H$ with a little more help from Steps 1 and 2. I leave it to you as an exercise.

12.10★ The tail

The estimate $D(L) \simeq \exp(-L^2/8)$ has still to be confirmed. The literature conveys the idea that a heavy machinery is needed, but it is not so as you will see. The best way to proceed is via the little Hamiltonian system of Step 3, §12.9, repeated here for convenience:

$$\dot{Q} = -2\frac{Q^2 P}{\pi L} + \frac{1}{2}P = +\frac{\partial H}{\partial P} \quad \text{and} \quad \dot{P} = +2\frac{P^2 Q}{\pi L} - \frac{1}{2}Q = -\frac{\partial H}{\partial Q}$$

with

$$H = \frac{1}{4}(P^2 + Q^2) - \frac{P^2 Q^2}{\pi L},$$

both $P = P(x = 0)$ and $Q = Q(x = 0)$ being regarded as functions of L alone, starting at $P(0) = 0$ and $Q(0) = 1$, and the spot $(\dot{\ })$ signifies d/dL. Note for future use that $dH/dL = \partial H/\partial Q\, \dot{Q} + \partial H/\partial P\, \dot{P} + \partial H \partial L = 4P^2 Q^2/\pi L^2$ is positive.

Step 1 shows that $Q^2(L) > \pi L/4$ for $L \geq 0$. That's obvious early on, so let's suppose $Q^2(L) = \pi L/4$ at e.g. $L = 1$. Now

$$\left(Q^2 - \frac{\pi L}{4}\right)^{\bullet} = -\frac{4PQ}{\pi L}\left(Q^2 - \frac{\pi L}{4}\right) - \frac{\pi}{4} \quad \text{and}$$

$$\left(P^2 - \frac{\pi L}{4}\right)^{\bullet} = +\frac{4PQ}{\pi L}\left(P^2 - \frac{\pi L}{4}\right) - \frac{\pi}{4},$$

so

$$Q^2 - \frac{\pi L}{4} = -\frac{\pi}{4}Z^{-1}\int_1^L Z\, dL' \quad \text{and} \quad P^2 - \frac{\pi L}{4} = -\frac{\pi}{4}Z\int_0^L Z^{-1}\, dL'$$

where

$$Z = \exp\left[\int_0^L \frac{4PQ}{L'}\, dL'\right],$$

and the inequality

$$\left(P^2 - \frac{\pi L}{4}\right)\left(Q^2 - \frac{\pi L}{4}\right) = \frac{\pi^2}{16}\int_0^L Z^{-1}\int_1^L Z \geq \frac{\pi^2}{16}(L-1)^2$$

may be spelled out to produce $H(\infty) \leq \pi/8 < \infty$. The fact is $-\log D = \int_0^L \frac{4}{\pi}H$ tends to $+\infty$ faster than L in contradiction to $H(\infty) < \infty$.

Proof. The point is that K acting on the *whole* line is a projection $K : f \to \left[\hat{f} \times \mathbf{1}_{|k|\leq 1}\right]^{\vee}$ so that[16] K^n is just K itself, so that for large L, you ought to have $\mathrm{tr}(K^n) \simeq \mathrm{tr}(K) = L/\pi$, with the result that

$$\lim_{L\uparrow\infty} -\log D(L) \geq \sum_{n=1}^{\infty} \liminf_{L\uparrow\infty} \frac{\mathrm{tr}\, K^n}{n} = \frac{1}{\pi}\sum_{n=1}^{\infty}\frac{1}{n} = +\infty.$$

Exercise 12.10.1 That's only a "demonstration", but try the first few traces ($n = 1, 2, 3$). $\mathrm{tr}(K) = L/\pi$ is trivial and $\mathrm{tr}(K^2) \simeq L/\pi$ is easy too. Do it. Now justify

$$\mathrm{tr}(K^3) = \frac{1}{2\pi}\int_{-1}^{+1} dk \int_{-L/2}^{L/2} dx \left|\int_{-L/2}^{L/2} e^{\sqrt{-1}ky}\frac{\sin y}{\pi y}\, dy\right|^2$$

---

[16] Here and in the display below, K^n signifies the nth power of the *operator* or of the *kernel*, as the context requires.

and estimate. The rest is more subtle so let's leave it at that, though you could try $\operatorname{tr}(K^5)$, noting that the odd traces suffice to make the point.

Step 2. $Q^2 > \pi L/4$ now prevails for all $L \geq 0$ and also $P^2 < \pi L/4$, as seen in Step 1. The content of the present step is that $P^2 = \pi L/4 + O(\sqrt{L})$.

Proof. Note first that $\dot{P} = (\frac{2Q}{\pi L})(P^2 - \pi L/4)$ is negative so that P decreases from 0 to $-\infty$: if not, then $P(\infty) > -\infty$, and $\dot{P} < 2(P^2(\infty) - \pi L/4)$ drives P down to $-\infty$ any how. Then

$$0 \leq \frac{\pi L}{4} - P^2$$

$$= \frac{\pi}{4} Z \int_0^L Z^{-1}$$

$$= \frac{\pi}{4} \int_0^L \exp\left(\int_{L'}^L \frac{4PQ}{\pi L''} \, dL''\right) dL'$$

$$< L_0 + \int_{L_0}^L \exp\left(-\int_{L'}^L \frac{2L''}{\sqrt{L''}}\right) dL' \qquad \text{if } P < -\sqrt{\pi} \text{ for } L \geq L_0$$

$$= O(\sqrt{L})$$

when you work it out. This can be much improved, as will appear soon.

Step 3. P is now under pretty good control, but how about Q? Rewrite

$$\left(\frac{P}{Q}\right)^{\bullet} = \frac{Q\left(\frac{2P^2Q}{\pi L} - \frac{1}{2}Q\right) - P\left(-\frac{2Q^2P}{\pi L} + \frac{1}{2}P\right)}{Q^2} = \frac{4P^2}{\pi L} - \frac{1}{2} - \frac{1}{2}\left(\frac{P}{Q}\right)^2$$

in terms of $2\dot{\psi}/\psi = P/Q$, with $\psi(0) = 1$, $\psi > 0$ and $\dot{\psi} < 0$, to obtain $\ddot{\psi} = (2P^2/\pi L - \frac{1}{4})\psi$. Evidently, $\psi(\infty) = 0$ since $2P^2/\pi L \simeq 1/2$ by Step 2, $\ddot{\psi} = \frac{1}{4}\psi$, and $\psi(\infty) > 0$ makes $\dot{\psi}$ turn positive; similarly, $\dot{\psi}(\infty) = 0$ since $\ddot{\psi}(\infty) < 0$ makes ψ turn negative. Putting it all together, you find

$$-\ddot{Q}(L) = 2\int_L^\infty \dot{\psi}\ddot{\psi} = \frac{1}{2}\int_L^\infty \dot{\psi}\psi = -\frac{1}{4}\psi^2(L),$$

and so

$$\left(\frac{P}{Q}\right)^2 = 4\left(\frac{\dot{\psi}}{\psi}\right)^2 \simeq 1, \quad \text{i.e.,} \quad Q^2 \simeq \frac{\pi L}{4}.$$

This, too, can be improved but wait. It's not needed.

Step 4.   Go back to Step 1 and write

$$Q^2 - \frac{\pi L}{4} = Z^{-1}\left(1 - \frac{\pi}{4}\int_0^L Z\right) \text{ with } Z = \exp\int_0^L \frac{4PQ}{\pi L'}\, dL' \text{ as before.}$$

Here, $4PQ/\pi L \simeq -1$, so roughly speaking, $Z \simeq e^{-L}$, $Z^{-1} \simeq e^L$, and $1 - \frac{\pi}{4}\int_0^L Z = \mathrm{O}(Le^{-L})$ since Q^2 is comparable to $\pi L/4$. But now

$$\pi L \times \left(H - \frac{\pi L}{16}\right) = \left(\frac{\pi L}{4} - P^2\right)\left(Q^2 - \frac{\pi L}{4}\right)$$

$$= \frac{\pi}{4} \times Z\int_0^L Z^{-1} \times Z^{-1}\left(1 - \frac{\pi}{4}\int_0^L Z\right)$$

$$= \frac{\pi}{4}\left(1 - \frac{\pi}{4}\int_0^L Z\right) = \mathrm{O}\left(Le^{-L}\right),$$

with the result that $H = \pi L/16 +$ a summable function, i.e. $-\log D(L) \simeq L^2/8$ as promised, and a little more.

12.10.1 Wigner's surmise (corrected)

Recall from §12.8 the spacing distribution for the next eigenvalue *to the right*, conditional on the presence of a single eigenvalue *at* the origin, to wit, $\pi G(0,0)D(L) = 4H\exp\left(-\int_0^L 4H/\pi\right)$. This is now seen to look like a constant multiple of $L\exp(-L^2/8)$ at $L = \infty$. What it does at $L = 0+$ is also interesting. There, the corresponding density looks like $L^2/3$. The computation is elementary but tiresome. Try it if you like.

This behavior imitates the mutual repulsion of eigenvalues, illustrated in Exercise 12.6.1, by BM(1), conditioned not to hit the origin. The corresponding transition density is

$$(2\pi t)^{-1/2}\frac{1}{x}\left[e^{-(x-y)^2/2t} - e^{-(x+y)^2/2t}\right]y,$$

as I trust you found. This reduces to

$$(2\pi t)^{-3/2}2y^2 e^{-y^2/2t}$$

at $x = 0+$, exhibiting the same repulsive character as the spacing distribution. The same results are found in GOE – only the proofs are more complicated.

12.11 At the edge

The first question is: How to scale out there? The edge is located at $2\sqrt{n}$, more or less, though surely the largest eigenvalue is bigger still, and you need some idea of how to rescale $\lambda_n - 2\sqrt{n}$. The semi-circle can be used as a guide: for large n, the interval $ab \subset [-1, 1]$ contains about $n \times \frac{2}{\pi} \int_a^b \sqrt{1 - x^2}$ scaled eigenvalues, and if you want to catch just one of these near the edge, you should determine what small number h will make

$$n \times \frac{2}{\pi} \int_{1-h}^1 \sqrt{1 - x^2} \simeq n \times \frac{2}{\pi} \frac{\sqrt{2}}{n} h^{3/2} = 1.$$

Obviously, h must be comparable to $n^{-2/3}$, and if you now pretend that λ_n is something like $2\sqrt{n} \times (1 + Cn^{-2/3}) = 2\sqrt{n} + 2Cn^{-1/6}$, you will want to look at $(\lambda_n - 2\sqrt{n}) \times n^{1/6}$. This "demonstration" is, in fact, reliable: indeed, as Tracy–Widom [1994] found,

$$\mathbb{P}\left[\lambda_n \leq 2\sqrt{n} + 2Ln^{-1/6}\right]$$

$$= \mathbb{P}\left[\text{spec}\,\mathbf{x} \cap \left(2\sqrt{n} + 2Ln^{-1/6}, \infty\right) \text{ is void}\right]$$

$$= \det\left[I - \sqrt{n}\,\frac{H_n \otimes H_n' - H_n' \otimes H_n}{x - y}\,\sqrt{n} : x, y \geq 2\sqrt{n} + 2Ln^{-1/6}\right]$$

tends, as $n \uparrow \infty$, to

$$D(L) = \det(I - K : x, y \geq L)$$

where now $K(x, y) = (x - y)^{-1}(A \otimes A' - A' \otimes A)$, A being the (modified) Airy[17] function, so-called, solving $A'' = xA$ with

$$A(x) \simeq \frac{1}{2}\,\pi^{-1/2}x^{-1/4}\exp\left(-\frac{2}{3}\,x^{3/2}\right) \text{ at } \infty.$$

The computation is more delicate than in §12.8 since you must now look at $H_n(x)$ both for large degree n and large $x \simeq 2\sqrt{n}$, simultaneously, and this is just where its oscillatory behavior stops. The necessary information will be found in Bateman [1953(2):200(1)].

Tracy–Widom [1994] did not stop there but used the machinery of §12.9 to find a beautiful formula for $D = \det(I - K)$, *to wit*,

$$D(L) = \exp\left[\int_L^\infty (L - x)B^2(x)\,\mathrm{d}x\right],$$

---

[17] $A(x) = \frac{1}{2}\frac{1}{\sqrt{\pi}}\,x^{1/2}K_{-1/3}\left(\frac{2}{3}\,x^{3/2}\right)$ where $K_{-1/3}$ is the more familiar (modified) Bessel function solving $K'' + \frac{1}{x}K' + K = \frac{1}{9x^2}K$ with $K(x) \simeq x^{-1/3}$ at $0+$ and $K(x) \simeq x^{-1/2}\mathrm{e}^{-x}$ at $+\infty$. Too many Ks – sorry, but it's just this once.

in which B is the solution of a non-linear variant of $A'' = xA$, namely $B'' = xB + 2B^3$, with the same behavior as A at $+\infty$; see §12.12 for more about that. $D(\infty) = 1$, by inspection, and a simple computation produces

$$D'(L) = D(L) \times \int_L^\infty B^2 \simeq (8\pi L)^{-1} \exp\left(-\frac{2}{3}L^{3/2}\right),$$

i.e. a somewhat lighter tail than for spacing in the bulk. The same distribution is found at the edge in GOE; as for the bulk spacing, the proof is a bit more complicated.

12.12 Coda

What's been done in §§12.8–12.11 is a small part of a much wider picture. Here, I hope to give you some glimpses of that; see Baik–Deift–Suidan [2014] for a comprehensive view.

12.12.1 Some history old and new

There is an intriguing connection to a class of functions so to say *beyond* the classical "special" functions (Hermite, Bessel, Airy, and so on) and other "elementary" functions known in Riemann's day (rational and trigonometrical functions, elliptic functions, Abelian integrals, and the like). Informally speaking, these novel functions are the next simplest, not reducible to their special and/or elementary cousins, and there are six of them: P1, P2, P3, P4, P5, P6, so-named after P. Painlevé (1900) who noticed them first. I will not tell you what he was up to – that's a whole story in itself, for which see Its [2003] for the nicest introduction. What's relevant here is that nowadays they keep coming up in unexpected, I might even say unlikely places. P5 already entered into the bulk spacing, in the disguise of the 2×2 Hamiltonian of Fact 3, §12.9. P2 is precisely the function B of §12.11, descriptive of the edge. P2 also enters into the statistics of long increasing rows in a permutation of n letters[18] and (of all things) into the best protocol for the swift embarkation of flight passengers[19]. P3 appears in connection with the 2-dimensional Ising model of §10.5 (that's statistical mechanics). P5 has even an astonishing, *empirical* number-theoretical connection to the improvement Riemann (1859) hoped to make to Gauss's Prime Number Theorem. I cannot help but to describe it for you, it's so striking. That's the short list, but see Fokas–Its–Kapaev–Novokshenov [2006:17–29] for a much longer one where P1, P4, and P6 also appear.

[18] Baik–Deift–Johansson [1999].
[19] McCoy–Tracy–Wu [1977].

12.12.2 What's happening

This widening out of the picture is still so new and still so sparsely illustrated by particular examples that it's hard to say for sure, but it seems likely that P1–P6 are the building blocks of a class of limit theorems, both new and of common occurence, reminiscent of CLT with its simple rule of addition of independent copies of a variable $\mathbf{x}$ with $\mathbb{E}(\mathbf{x}) = 0$ and $\mathbb{E}(\mathbf{x}^2) = 1$ say, only now independence is lost, or at best off-stage, and addition is replaced by some more complicated (algebraic?) ways of combining variables, who knows how? Wouldn't it be nice to have some simple rule of thumb like CLT so you could say: Oh, P2 or P5 or whatever is coming out? Obviously, many more examples will be needed before such a rule could even be guessed at, but the evidence for it seems very persuasive.

12.12.3 Riemann and the prime numbers

Now comes the most remarkable aspect of the widening picture, discovered empirically by Odlyzko [2001]; compare Conrey [2003] for background and more information. The story begins with Riemann's (1859) not wholly successful improvement of Gauss's conjectured prime number theorem,

$$\#(p \le x) = \frac{x}{\log x} \times [1 + o(1)],$$

to the sharper form[20]

$$\#(p \le x) = \int_2^x \frac{dy}{\log y} + O\left(x^{\frac{1}{2}+}\right),$$

Riemann starts from his "zeta function" $\zeta(s) = 1 + 2^{-s} + 3^{-s} + \text{etc.}$ and its connection to the prime numbers via

$$\prod_{p=2}^{\infty} \frac{1}{1-p^{-s}} = \prod_{p=2}^{\infty}\sum_{k=0}^{\infty} p^{-ks} = \sum_{k_1,k_2,\text{etc.}\ge 0}\left(\frac{1}{2^{k_1}3^{k_2}5^{k_3}\text{etc.}}\right) = \sum_{1}^{\infty}\frac{1}{n^s},$$

s being taken in the half-plane $(1,\infty) \times \mathbb{R}$ so that product and sum make sense. Now ζ can be continued analytically into the whole complex plane, apart from a single simple pole at $s = 1$; it vanishes at $-2n$ ($n \ge 1$) – these are its "trivial" roots – and at no place else except for an infinite number of "non-trivial" roots in the open strip $(0,1) \times \mathbb{R}$.[21]

[20] Edwards [2001] explains in full what Riemann actually did.
[21] Ahlfors [1979:212–218] will take you nearly this far.

Riemann found a beautiful formula connecting the roots of ζ to the primes. A modern variant reads

$$\sum_{\substack{n=1 \\ p \geq 2 \\ p^n \leq x}}^{\infty} \sum \log p = x + \sum_1^{\infty} \frac{x^{-2n}}{2n} - \sum_{\omega} \frac{x^{1/2+\sqrt{-1}\,\omega}}{1/2 + \sqrt{-1}\,\omega} - \log 2\pi,$$

valid for $x \uparrow \infty$ up to tiny errors. This leads circuitously, with tears, to the improved error $O(x^{\frac{1}{2}+})$ provided that *without exception, the nontrivial roots lie on the vertical line* $L = \frac{1}{2} \times (0, \infty)$. That's Riemann's hypothesis, so-called. The reason for this hope (besides its undoubted truth) is not hard to understand. The roots off the line come in pairs $1/2 + \text{imaginary}\sqrt{-1}\,\omega \pm \text{real } h$, and if any such should appear, summands of amplitude $x^{1/2 \pm h}$ would come in, and the desired error $O(x^{\frac{1}{2}+})$ could not be expected. Now "undoubted truth" is one thing and proof is something else. But look: Hardy (1924) showed that *an infinite number* of the non-trivial roots lie on L; Levinson (1974) showed that *at least 1/3 of them do it, too*; and all the first 10^7 roots (listed by height) are known to lie there as well. In short, the evidence is very good indeed. Riemann's reasons are not known, though surely he had some deep idea.

Now comes the spooky part. Odlyzko computed some 7.8×10^7 non-trivial roots at heights near 10^{20} and made a scatter-plot of their (scaled) spacings. To the naked eye, these fall *plunk* on the bulk-spacing curve of Figure 12.12.1, reproduced here, except right at the top where a tiny

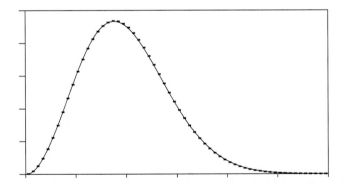

over-shoot may be perceived. Astonishing. And don't keep saying: But how can it be like that? Nobody knows how it can be like that. That's Feynman, out of context, but just the right thing to say.

Bibliography

Ahlfors, L. V. 1979. *Complex Analysis* (3rd ed.). McGraw-Hill, New York.

Andréief, C. 1883. Note sur une relation entre les intégrales définies des produits des fonctions. *Mem. Soc. Sci., Bordeaux*, **2**, 1–14.

Arnold, V.I. 1978. *Mathematical Methods of Classical Mechanics*. Springer-Verlag, Berlin.

Artin, E. 1924. Ein mechanisches System mit quasi-ergodischen Bahnen. *Abh. aus dem Math. Seminar Hamburg*, **3**, 170–175.

Ash, R. 1965. *Information Theory*. Interscience, J. Wiley & Sons, New York.

Baik, J., Deift P., and Johansson, K. 1999. On the distribution of the length of the longest increasing subsequence of random permutations, *J. Amer. Math. Soc.*, **12**, 1119–1178.

Baik J., Deift P., and Suidan T. 2014. *Some Combinatorial Problems and Random Matrix Theory*. AMS, Providence RI.

Berry, A.C. 1941. The accuracy of the Gaussian approximation to the sum of independent variates. *Trans. Amer. Math. Soc.*, **49**, 122–136.

Bateman, H., 1953. *Higher Transcendental Functions (2)*. McGraw-Hill, New York.

Billingsley, P. 1960. Hausdorff dimension in probability theory, *Illinois J. Math.*, **4**, 187–209.

Billingsley, P. 1965. *Ergodic Theory and Information*. J. Wiley & Sons, New York.

Billingsley, P. 1974. The probability theory of additive arithmetic functions. *Ann. Probability*, **2**, 749–791.

Billingsley, P. 1979. *Probability and Measure*. J. Wiley & Sons, New York.

Birkhoff, G.D. 1931. Proof of the ergodic theorem. *Proc. Nat. Acad. Sci. U.S.A.*, **17**, 656–660.

Boltzmann, L. 1872. Weitere Studien über das Wärmegleichgewicht unter Gasmolekülen. *Sitzungsberichter Akad. Wiss.*, **66**, 275–370.

Boltzmann, L. 1912. *Vorlesungen über Gastheorie*. Ambrosius Barth, Leipzig. See also *Lectures on Gas Theory*. U. Calif. Press, Berkeley, CA (1964) & Dover, New York (1995).

Breiman, L. 1957. The individual ergodic theorem of information theory. *Ann. Math. Stat.*, **28**, 809–811.

447

Breiman, L. 1967. *Introduction to Measure and Integral.* Addison-Wesley Co., Reading MA.

Breiman, L. 1968. *Probability.* Addison-Wesley Co., Reading MA.

Cameron, R.H. and Martin, W.T. 1944. A transformation of Wiener integrals. *Ann. Math.* **45**, 380–386.

Cameron, R.H. and Martin, W.T. 1947. The orthogonal development of non-linear functionals in series of Fourier–Hermite functionals. *Ann. Math.* **48**, 385–392.

Carleman, T. 1922. Sur les fonctions quasi-analytiques. *5ème Congrès des Math. Scand.*, Helsinfors, 181–196.

Carleman, T. 1957. Problèmes Mathématiques dans la Théorie Cinétique des Gaz. Almqvist-Wiksells, Uppsala, reprinted in *Édition Complète des Articles de Torsten Carleman.* Inst. Mittag-Leffler, Uppsala (1960).

Carlen, E., Carvalho, M.C., and Gabetta, E. 2000. Central limit theorem for Maxwellian molecules and truncation of the Wild expansion. *Comm. Pure Appl. Math.* **53**, 370–397.

Cercignani, C. 1969. *Mathematical Methods in Kinetic Theory.* Plenum Press, New York.

Cercignani, C., Illner, R., and Pulvirenti, M. 1994. *The Mathematical Theory of Dilute Gases.* Springer-Verlag, Berlin.

Chapman, S. 1916. On the law of distribution of molecular velocities. *Publ. Trans. Royal Soc. London*, **216**, 279–348.

Choquet, G. 1956. Existence des représentations intégrales au moyen des points extrémaux dans les cônes convexes. *C.R. Acad. Sci. Paris*, **243**, 699–702; 736–737.

Chung, K-L. 1973. Probabilistic approach in potential theory to the equilibrium problem. *Ann. Inst. Fourier (Grenoble)*, **23**, 313–322.

Ciesielski, Z. 1961. Hölder conditions for realizations of Gaussian processes. *Trans. Amer. Math. Soc.*, **99**, 403–413.

Conrey, J.B. 2003. The Riemann hypothesis. *Notices Amer. Math. Soc.*, **50**, 341–353.

Courant, R. and Robbins, A. 1951. *What is Mathematics?* Oxford University Press, Oxford/London/New York.

Cover, T.M. and Thomas, J.A. 1991. *Elements of Information Theory.* J. Wiley & Sons, New York.

Cramér, H. 1936. Über eine Eigenschaft der normalen Verteilungsfunktion. *Math. Zeit.*, **41**, 405–414.

Cramér, H. 1938. Sur un nouveau théorème-limite de la théorie des probabilités. *Actualités Scientifiques et Industrielles*, **736**, 5–33, Hermann & Cie, Paris.

Deift, P. 1999. *Orthogonal Polynomials and Random Matrices: a Riemann–Hilbert Approach.* Courant Lecture Notes in Mathematics, **3**, Courant Institute of Mathematical Sciences, New York, Amer. Math. Soc., Providence RI.

Deift, P., Kriecherbauer, T., McLaughlin, K.T.-R., Venakides, S., and Zhou, X. 1999. Uniform asymptotics for polynomials orthogonal with respect to varying exponential weights and applications to universality questions in random matrix theory, *Comm. Pure Appl. Math.*, **52**, 1335–1425.

Dobrushin, R.-L. 1965. Existence of a phase transition in the two-dimensional and three-dimensional Ising models, *Theory Probab. Appl.*, **10**, 193–213.

Donsker, M.D. 1951. An invariance principle for certain probability limit theorems. *Mem. Amer. Math. Soc.*, **6**, 1–12.

Donsker, M.D. and Varadhan, S.R.S. 1975. Asymptotic evaluation of certain Markov process expectations for large time, I & II. *Comm. Pure Appl. Math.*, **28**, 1–47; 279–301.

Donsker, M.D. and Varadhan, S.R.S. 1976. Asymptotic evaluation of certain Markov process expectations for large time, III. *Comm. Pure Appl. Math.*, **29**, 389–461.

Doob, J.L. 1949. Heuristic approach to the Kolmogorov–Smirnov theorems. *Ann. Math. Statistics*, **20**, 393–403.

Doob, J.L. 1953. *Stochastic Processes.* J. Wiley & Sons, New York.

Doyle, P.G. and Snell, J.L. 1984. *Random Walks and Electrical Networks.* Math. Assoc. of America, Washington, D.C.

Dresden, M. 1956. *Kinetic Theory Applied to Hydrodynamics.* Magnolia Petroleum Co., Dallas, TX.

Durrett, R. 1991. *Probability: Theory and Examples.* Wadsworth and Brooks/Cole Advanced Books & Software, Pacific Grove CA.

Dvoretsky, A. and Erdős, P. 1950. Some problems on random walk in space. *Proc. 2nd Berkeley Symp. on Math. Statist. and Prob.* (1950), California University Press, Berkeley CA, pp. 360–367.

Dym, H. and McKean, H.P. 1972. *Fourier Series and Integrals.* Academic Press, New York.

Dynkin, E.B. 1965. *Markov Processes.* Springer-Verlag, Berlin.

Dyson, F.J. 1962. A Brownian-motion model for the eigenvalues of a random matrix. *J. Math. Phys.*, **3**, 1191–1198.

Edwards, H.M. 2001. *Riemann's Zeta Function.* Dover, Mineola NY.

Eggleston, H.G. 1949. The fractional dimension of a set defined by decimal properties. *Quart. J. Math. Oxford Ser.*, **20**, 31–36.

Einstein, A. 1956. *Investigations on the Theory of the Brownian Movement.* Fürth, R. and Cowper, A.D., eds., Dover, New York.

Elias, P. 1961. Coding and decoding. In *Lectures on Communication System Theory*, E.J. Baghdady (ed), McGraw-Hill, Chapter 13, pp. 321–344.

Ellis, R.S. 1985. *Entropy, Large Deviations, and Statistical Mechanics.* Springer-Verlag, Berlin.

Enskog, D. 1917. *Kinetische Theorie der Vorgänge in mässig verdünnten Gasen.* Amlqvist-Wiksells, Uppsala.

Erdős, P. 1942. On the law of the iterated logarithm. *Ann. Math.* **43**, 419–436.

Erdős, P. 1949. On a new method in elementary number theory which leads to an elementary proof of the prime number theorem. *Proc. Nat. Acad. Sci. USA.*, **35**, 374–384.

Erdős, P. and Kac, M. 1940. The Gaussian law of error in the theory of additive number-theoretic functions. *Amer. J. Math.* **62**, 728–742.

Erdős, P. and Kac, M. 1947. On the number of positive sums of n independent random variables, *Bull. Amer. Math. Soc.* **53**, 1011–1020.

Esseen, C.-G. (1942). On the Liapunoff limit of error in the theory of probability. *Arkiv för Mat., Astr. & Fys*, **A28**, 1–19.

Fano, R.M. 1961. *Transmission of Information: A Statistical Theory of Communications.* MIT Press, Cambridge MA.

Feinstein, A. 1954. A new basic theorem of information theory. *Trans. Inst. Radio Eng.*, **4**, 2–22B.

Feller, W. 1948. On the Kolmogorov–Smirnov limit theorems for empirical distributions. *Ann. Math. Statistics*, **19**, 177–189.

Feller, W. 1966. *An Introduction to Probability Theory and its Applications, II* (2nd ed.). John Wiley & Sons, New York.

Feller, W. 1968. *An Introduction to Probability Theory and its Applications, I.* (2nd ed.). John Wiley & Sons, New York.

Fenchel, W. 1949. On conjugate convex functions. *Canadian J. Math.*, **1**, 73–77.

Fermi, E. 1956. *Thermodynamics.* Dover, New York.

Feynman, R. 1963. *Lectures on Physics*, vol. 1. Addison-Wesley Co., Reading MA.

Feynman, R. 1964. *Lectures on Physics*, vol. 2. Addison-Wesley Co., Reading MA.

Feynman, R. 1965. *Lectures on Physics*, vol. 3. Addison-Wesley Co., Reading MA.

Flanders, M. and Swan, D. 1963. The second law of thermodynamics, in *At the Drop of another Hat*, Angel Records, #PMC1216.

Fokas, A.S., Its, A.R., Kapaev, A.A., and Novokshenov, V.Y. 2006. *Painlevé Transcendents. The Riemann–Hilbert Approach.* Mathematical Surveys and Monographs, **128**, American Mathematical Society, Providence RI.

Garsia, A. 1965. A simple proof of E. Hopf's maximal ergodic theorem. *J. Math. Mech.*, **14**, 381–382.

Gauss, C.F. 1827. Arithmetisch Geometriches Mittel. *Werke*, Bd 3, 361–432, Teubner, Leipzig.

Gauss, C.F. 1884. *Allgemeine Lehrsätze in Beziehung auf die im verkehrten Verhältnisse des Quadrats der Entfernung wirkenden Anziehungs- und Abstossungskräfte.* Ostwalds Klassiker der exacten Wissenschaften, no 2, Teubner, Leipzig.

Gibbs, J.W. 1902. *Elementary Principles in Statistical Mechanics.* Yale University Press, New Haven CT, Dover, New York (1960).

Goldstein, S. 1951. Diffusion by discontinuous movement and on the telegraph equation. *Quart. J. Math. (Oxford)*, **4**, 129–156.

Glasser, M.L. and Zucker, I.J. 1977. Extended Watson integrals for the cubic lattices. *Proc. Nat. Acad. Sci. USA*, **74**, 1800–1801.

Gosnell, M. 2007. *Ice.* University of Chicago Press, Chicago.

Grad, H. 1958. *Principles of the Kinetic Theory of Gases.* Handbuch der Physik, **12**, Springer-Verlag, Berlin.

Griffiths, R.B. 1964. Peierls' proof of spontaneous magnetization in a two-dimensional Ising ferromagnet. *Phys. Rev.*, **136**, A437–A439.

Griffiths, R.B. 1967 (a). Correlations in Ising ferromagnets, I, II. *J. Math. Phys.*, **8**, 418–483; 484–489.

Griffiths, R.B. 1967 (b). Correlations in Ising ferromagnets, III. *Comm. Math. Phys.*, **6**, 121–127.

Grünbaum, F.A. 1971. Propagation of chaos for the Boltzmann equation. *Arch. Rat. Mech. & Anal.*, **42**, 323–345.

Grünbaum, F.A. 1972. Linearization for the Boltzmann equation. *Trans. Amer. Math. Soc.*, **165**, 425–449.

Halmos, P. and von Neumann, J. 1942. Operator methods in classical mechanics. II. *Ann. Math.*, **43**, 332–350.

Hardy, G.H. and Ramanujan, S. 1917. The normal number of prime factors of a number n. *Quart. J. Math.*, **48**, 76–92.

Hedlund, G.A. 1934. On the metrical transitivity of the geodesics on closed surfaces of constant negative curvature. *Ann. Math.*, **35**, 787–808.

Hedlund, G.A. 1937. A metrically transitive group defined by the modular groups. *Amer. J. Math.*, **57**, 668–678.

Hemmer, P., Kac, M. and Uhlenbeck, G.E. 1963. On the van der Waals theory of the vapor–liquid equilibrium. I, II. *J. Mathematical Phys.*, **4**, 216–228; 229–247.

Hemmer, P., Kac, M. and Uhlenbeck, G.E. 1964. On the van der Waals theory of the vapor–liquid equilibrium. III. *J. Mathematical Phys.*, **5**, 60–74.

Hewitt, E. & Savage, L.J. 1955. Symmetric measures on Cartesian products, *Trans. Amer. Math. Soc.*, **80**, 470–501.

Hilbert, D. 1912. *Grundzüge einer Allgemeinen Theorie der Linearen Integralgleichungen.* Teubner-Verlag, Leipzig.

Hopf, E. 1937. *Ergodentheorie.* Ergebnisse der Mathematik und ihrer Grenzgebiete, no. **5**, Springer, Berlin; Chelsea, New York (1948).

Householder, A.S. 1965. *The Theory of Matrices in Numerical Analysis.* Blaisdell, New York.

Ising, E. 1925. Beitrag zur Theorie des Ferromagnetismus. *Zeit. Phys.*, **31**, 253–258.

Itô, K. 1951. Multiple Wiener integral. *J. Math. Soc. Japan*, **3**, 157–169.

Itô, K. and McKean, H.P. 1960. Potentials and the random walk. *Illinois J. Math.*, **4**, 119–132.

Its, A. 2003. The Riemann–Hilbert problem and integrable systems. *Notices Amer. Math. Soc.*, **50**, 1389–1400.

Jain, N.C. and Pruitt, W.E. 1970. The range of recurrent random walk in the plane. *Z. Wahrschein. und Verw. Geb.*, **16**, 279–292.

Jellinek, R. 1968. *Probabilistic Information Theory.* McGraw-Hill, New York.

Jimbo, M., Miwa, T., Môri, Y. and Sato, M. 1980. Density matrix of an impenetrable Bose gas and the fifth Painlevé transcendent. *Physica D*, **1**, 80–158.

Kac, M. 1947. Random walk and the theory of Brownian motion. *Amer. Math. Monthly*, **54**, 369–391.

Kac, M. 1947. On the notion of recurrence in discrete stochastic processes. *Bull. AMS*, **53**, 1002–1010.

Kac, M. 1949. On the distribution of certain Wiener functionals. *Trans. Amer. Math. Soc.*, **65**, 1–13.

Kac, M. 1956. Foundations of kinetic theory. *Proc. Berkeley Symp. Math. Stat. Prob. (1954–1955)*, **3**, 171–197.

Kac, M. 1959a. *Probability and Related Topics in the Physical Sciences.* Interscience, London & New York.

Kac, M. 1959b. *Statistical Independence in Probability, Analysis, and Number Theory.* Carus Math. Monographs, no. 23, Math. Assoc. of America. J. Wiley & Sons, New York.

Kac, M. 1974. A stochastic model related to the telegrapher's equation. *Rocky Mountain Math. J.*, **4**, 491–509.

Kac, M. 1980. *Integration in Function Spaces and some of its Applications.* Lezioni Fermiane, Accademia Nazionale dei Lincei, Scuola Normale Superiore, Pisa.

Kac, M. and Ward, J.C. 1952. A combinatorial solution of the two-dimensional Ising model. *Phys. Rev.*, **88**, 1332–1337.

Kallianpur, G. and Robbins, H. 1953. Ergodic property of the Brownian motion. *Proc. Nat. Acad. Sci. USA*, **39**, 525–533.

van Kampen, N.G. 1964. Condensation of a classical gas with long-range attraction. *Phys. Rev.*, **135**, A362–A369.

Karlin, S. and McGregor, J. 1959. Coincidence probabilities. *Pacific. J. Math.*, **2**, 1141–1164.

Kauffman, B. 1949. Crystal statistics. II. Partition function evaluated by spinor analysis. *Phys. Rev.*, **76**, 1232–1243.

Kaufman, B. and Onsager, J. 1949. Crystal statistics. III. Short-range order in a binary Ising lattice, *Phys. Rev.*, **76**, 1244–252.

Kellogg, O.D. 1929. *Foundations of Potential Theory.* Grundlehren der Math. Wiss., no. **31**, Springer-Verlag, Berlin.

Kelly, D.E. and Sherman, S. 1968. General Griffiths's inequalities on correlations in Ising ferromagnets. *J. Math. Phys.*, **9**, 460–484.

Kelly, J. 1956. A new interpretation of information rate. *Bell Syst. Tech. J.*, **35**, 917–926.

Khinchine, A. 1923. Über dyadische Brüche. *Math. Zeit.*, **18**, 109–116.

Khinchine, A. 1924. Über einen Satz der Wahrscheinlichkeitsrechnung. *Fund. Math.*, **6**, 9–20.

Khinchine, A. 1933. *Asymptotische Gesetze der Wahrscheinlichkeitsrechnung.* Ergebnisse der Mathematik und ihrer Grenzgebiete, no. **2**, Springer, Berlin; Chelsea, New York (1948).

Knight, F. 1962. On the random walk and Brownian motion. *Trans. Amer. Math. Soc.*, **108**, 218–228.

Knight, F. 1963. Random walks and a sejourn density process of Brownian motion. *Trans. Amer. Math. Soc.*, **109**, 58–89.

Kolmogorov, A.N. 1933a. Sulla determinazione empirica di una legge di distribuzione. *Giorn. 1st. Ital. Attuari*, **4**, 83–91.

Kolmogorov, A.N. 1933b. *Grundbegriffe der Wahrscheinlichkeitsrechnung.* Ergebnisse der Mathematik und ihrer Grenzgebiete, no. **3**, Springer, Berlin; Chelsea, New York (1950).

Kramers, H.A. and Wannier, G.H. 1941. Statistics of the two-dimensional ferromagnet. *Phys. Rev.*, **60**, 252–276.

Landau, L. and Lifshitz, E. 1938. *Statistical Physics*. Oxford University Press, Oxford.

Lanford, O.E. 1975. Time evolution in large classical systems. *Dynamical Systems, Theory and Applications*, Lecture Notes in Phys., vol. **38**, Springer-Verlag, Berlin, pp. 1–111.

Lax, P. 2002. *Functional Analysis*. J. Wiley & Sons, New York.

Lebowitz, J. and Penrose, O. 1966. Rigorous treatment of the van der Waals–Maxwell theory of the liquid-vapor transition. *J. Math. Phys.*, **7**, 98–113.

Le Jan, Y. 1994. The central limit theorem for the geodesic flow on noncompact manifolds of constant negative curvature. *Duke Math. J.*, **74**, 159–175.

Levinson, N. 1974. At least one-third of zeroes of Riemann's zeta-function are on $\sigma = 1/2$. *Proc. Nat. Acad. Sci. USA*, **71**, 1013–1015.

Lévy, P. 1937. *Théorie de l'addition des variables aléatoires*. Gauthier-Villars, Paris.

Lévy, P. 1939. Sur certains processus stochastiques homogènes. *Compositio Math.*, **7**, 283–339.

Lévy, P. 1948. *Processus Stochastiques et Mouvement Brownien*. Gauthier-Villars, Paris.

Lévy, P. 1951. *Problèmes Concrets d'Analyse Fonctionnelle*. Gauthier-Villars, Paris.

Linnik, Yu.V. 1949. On the theory of nonuniform Markov chains. *Izv. Akad. Nauk SSSR Ser. Mat.*,**13**, 65–94.

Malliavin, P. 1997. *Stochastic Analysis*. Grundlehren der Math. Wiss., no. **137**, Springer-Verlag, Berlin.

Maxwell, J.C. 1861. On the dynamical theory of gases. *Phil. Trans. Royal Soc. London*, **157**, 49–88.

Maxwell, J.C. 1873. *The Scientific Letters and Papers of James Clerk Maxwell: 1862–1873, vol. 2*. Cambridge University Press, Cambridge (1995).

Maxwell, J.C. 1892. *A Treatise on Electricity and Magnetism*. Clarendon Press, London & New York.

McCoy, B., Tracy, C., and Wu, F. T. 1977. Painlevé functions of the third kind. *J. Math. Phys.*, **18**, 1058–1072.

McKean, H.P. 1960. The Bessel motion and a singular integral equation. *Mem. Coll. Sci. Kyoto*, **33**, 317–322.

McKean, H.P. 1963. Entropy is the only increasing functional of Kac's one-dimensional caricature of a Maxwellian gas. *Z. Wahrschein. und Verw. Geb.*, **2**, 167–172.

McKean, H.P. 1966a. Speed of approach to equilibrium for Kac's caricature of a Maxwellian gas. *Arch. Rat. Mech. Anal.*, **21**, 343–367.

McKean, H.P. 1966b. A class of Markov processes associated with nonlinear parabolic equations. *Proc. Nat. Acad. Sci. USA*, **56**, 1907–1911.

McKean, H.P. 1967. Chapman–Enskog–Hilbert expansion of a class of solutions of the telegraph equation. *J. Math. Phys.*, **8**, 547–552.

McKean, H.P. 1969. *Stochastic Integrals*. Academic Press, New York; AMS-Chelsea, Providence RI (2005).

McKean, H.P. 1973. Geometry of differential space. *Ann. Probability*, **1**, 197–206.

McKean, H.P. 1975. The central limit theorem for Carleman's equation. *Israel J. Math.*, **21**, 54–92.

McKean, H.P. 1975b. Fluctuations in the kinetic theory of gases. *Comm. Pure Appl. Math.*, **28**, 435–455.

McKean, H.P. 2011. Fredholm determinants. *Cent. Eur. J. Math.*, **9**, 205–243.

McKean, H.P. and Moll, V. 1997. *Elliptic Curves*. Cambridge University Press, Cambridge.

McMillan, B. 1952. Two inequalities implied by unique decipherability. *IEEE Trans. Information Theory*, **2**, 115–116.

Mehler, F.G. 1866. Über die Entwicklung einer Function von beliebig vielen Variabeln nach Laplaceschen Functionen höherer Ordnung. *J. Reine Angew. Math.*, **66**, 161–176.

Mehta, M.L. 1967. *Random Matrices*. Academic Press, Boston MA and Elsevier/Academic Press, Amsterdam (2004).

Milnor, J.W. (1963). *Topology from the Differentiable Viewpoint*. University of Virginia Press, Charlottesville, VA.

Mischler, S. and Mouhot, C. 2013. Kac's program in kinetic theory. *Invent. Math.* **193**, 1–147.

Munroe, M.E. 1953. *Introduction to Measure and Integration*. Addison-Wesley, Cambridge MA.

Münster, A. 1956. *Statistical Theromodynamics*. Springer-Verlag, Berlin.

Nagell, T. 1964. *Introduction to Number Theory*. Chelsea Publ. Co., New York.

Needham, T. 1997. *Visual Complex Analysis*. Oxford University Press, Oxford.

Nirenberg, L. 1959. On elliptic partial differential equations. *Ann. Scuola Norm. Sup. Pisa*, **13**, 115–162.

Odlyzko, A.M. 2001. The 10^{22}-nd zero of the Riemann zeta function. In *Dynamical, Spectral, and Arithmetic Zeta Functions*, van Frankenhuysen, M. and Lapidus, M.L., eds., Contemporary Math., **290**, Amer. Math. Soc., Providence RI, pp. 139–144.

Onsager, L. 1944. Crystal statistics. I. A two-dimensional model of an order-disorder transition. *Phys. Rev.*, **65**, 117–149.

Onsager, L. 1949. Discussion remark: spontaneous magnetization of the two-dimensional Ising model. *Nuovo Cimento Suppl.*, **6**, 261.

Ornstein, L.S. and Uhlenbeck, G.E. 1930. On the theory of Brownian motion. *Phys. Rev.*, **36**, 823–841.

Pólya, G. Über eine Aufgabe der Wahrscheinlichkeitsrechnung betreffend die Irrfahrt im Straßennetz. *Math. Ann.*, **84**, 149–160.

Peierls, R.F. 1936. On Ising's model of ferromagnetism. *Proc. Camb. Phil. Soc.*, **32**, 477–481.

Pogorelov, A.V. 1967. *Differential Geometry*. Nordhoff, Groningen.

Pollard, H. 1976. *Celestial Mechanics*. Carus Mathematical Monographs, no. **18**, Math. Assoc. America, J. Wiley & Sons, New York.

Ratner, M. 1973. The central limit theorem for geodesic flows on n-dimensional manifolds of negative curvature. *Israel J. Math.*, **16**, 181–197.

Ray, D. 1963. Sojourn times of a diffusion processes. *Illinois J. Math.*, **7**, 615–630.

Riesz, F. 1945. Sur la théorie ergodique. *Comm. Math. Helv.*, **17**, 221–239.

Rogers, L.C.G. and Williams, D. 1979. *Diffusions, Markov Processes, and Martingales.* Vol. 1. J. Wiley & Sons, Chichester.

Rogers, L.C.G. and Williams, D. 1987. *Diffusions, Markov Processes, and Martingales.* Vol. 2. J. Wiley & Sons, New York.

Rukeyser, M. 1942. *Willard Gibbs: American Genius*, Doubleday, New York. Reprinted 1988 Ox Bow Press, Woodbridge CT.

Ryll-Nardzewski, C. 1951. On the ergodic theorem II, *Studia Math.*, **12**, 74–79.

Seeley, A. 1966. *Introduction to Fourier Series and Integrals.* W.A. Benjamin, New York.

Selberg, A. 1949. An elementary proof of the prime-number theorem, *Ann. Math.* **50**, 305–313.

Shannon, C.E. 1948. A mathematical theory of communication. *Bell System Tech. J.*, **27**, 379–423; 623–656.

Shannon, C.E. and Weaver, W. 1949. *The Mathematical Theory of Communication.* The University of Illinois Press, Urbana IL.

Sinai, Y.G. 1960. The central limit theorem for geodesic flows on manifolds of constant negative curvature. *Dokl. Akad. Nauk.*, **133**, 1303–1306.

Smirnov, N.V. 1939. On the estimation of the discrepancy between empirical curves of distribution for two independent samples. *Mat. Sbornik*, **48**, 3–26; and *Bull. Math. Univ. Moscou* **2**, 3–14.

Sparre-Andersen, E. 1953. On the fluctuations of sums of random variables, I. *Math. Scand.* **1**, 263–285.

Sparre-Andersen, E. 1954. On the fluctuations of sums of random variables, II. *Math. Scand.* **2**, 195–223.

Spitzer, F. 1956. A combinatorial lemma and its applications to probability theory. *Trans. Amer. Math. Soc.*, **82**, 323–339.

Spitzer, F. 1957. The Wiener–Hopf equation whose kernel is a probability density. *Duke Math. J.*, **24**, 327–343.

Spitzer, F. 1964. Electrostatic capacity, heat flow, and Brownian motion. *Z. Wahrschein. und Verw. Geb.*, **3**, 110–121.

Strassen, V. 1964. An invariance principle for the law of the iterated logarithm. *Z. Wahrschein. und Verw. Gebiete*, **3**, 211–226.

Stroock, D. 1981. The Malliavin calculus and its applications. In *Stochastic integrals.* Lecture Notes in Math., **851**, pp. 394–432.

Stroock, D.W. 1993. *Probability Theory, an Analytic View.* Cambridge University Press, Cambridge.

Sullivan, D. 1982. Disjoint spheres, approximation by imaginary quadratic numbers, and the logarithm law for geodesics. *Acta Math.*, **149**, 215–237.

Sylvester, G.S. 1976. Continuous-spin inequalities for Ising ferromagnets. *J. Stat. Phys.*, **15**, 327–342.

Szegö, G. 1939. *Orthogonal Polynomials.* AMS Colloquium. Publ., vol. **23**, Amer. Math. Soc., Providence RI (1975).

Tanaka, H. 1973a. On Markov process corresponding to Boltzmann's equation of Maxwellian gas. *Lect. Notes Math.*, **330**, 478–489. Springer-Verlag, Berlin, Heidelberg, New York.

Tanaka, H. 1973b. An inequality for a functional of probability distributions and its applications to Kac's one-dimensional model of a Maxwellian gas. *Z. Wahrschein. und Verw. Geb.*, **27**, 47–52.

Tanaka, H. 1978. Probabilistic treatment of the Boltzmann equation of Maxwellian molecules. *Z. Wahrschein. und Verw. Gebiete*, **46**, 67–105.

Thompson, C.J. 1972. *Mathematical Statistical Mechanics.* Macmillan Co., New York.

Titchmarsh, E.C. 1959. *The Theory of the Riemann Zeta Function.* Oxford University Press, Oxford.

Tracy, C. and Widom, H. 1993. Introduction to random matrices. In *Geometric and Quantum Aspects of Integrable Systems (Scheveningen, 1992)*, Lecture Notes in Phys., **424**, Springer, Berlin, pp. 103–130.

Tracy, C. and Widom, H. 1994. Level spacing distributions and the Airy kernel. *Comm. Math. Phys.*, **159**, 151–174.

Trotter, H.F. 1958. Approximation of semi-groups of operators. *Pacific J. Math.*, **8**, 887–919.

Trotter, H.F. 1959. An elementary proof of the central limit theorem. *Arch. Math.*, **10**, 226–234.

Trotter, H.F. 1984. Eigenvalue distributions of large Hermitian matrices; Wigner's semi-circle law and a theorem of Kac, Murdock, and Szegö. *Adv. Math.*, **54**, 67–82.

Varadhan, S.R.S. 1984. *Large Deviations and Applications.* CBMS-NSF Regional Conference Series in Applied Mathematics, **46**, SIAM, Philadelphia PA.

Vedenyapin, V.V. 1988. Differential forms in spaces without a norm. A theorem on the uniqueness of Boltzmann's H-function. *Uspekhi Mat. Nauk*, **43**, 159–179, *Russian Math. Surveys*, **43**, 193–219.

Villani, C. 2002. A review of mathematical topics in collisional kinetic theory. *Handbook of Mathematical Fluid Dynamics*, vol. **I**, North-Holland, Amsterdam, pp. 71–305.

van der Waerden, B.L. 1941. Die lange Reichweite der regelmassigen Atomanordnung in Mischkristallen. *Z. Phys.*, **118**, 473.

Watson, G.N. 1939. Three triple integrals. *Quart. J. Math. (Oxford)*, **10**, 266–276.

Wendel, J.G. 1958. Spitzer's formula: a short proof. *Proc. Amer. Math. Soc.*, **9**, 905–908.

Wennberg, B. 1993. Stability and Exponential Convergence for the Boltzmann Equation. Thesis. Chalmers University of Technology, Sweden.

Weyl, H. 1916. Über die Gleichverteilung von Zahlen mod. Eins. *Math. Ann.*, **77**, 313–352.

Weyl, H. 1939. *The Classical Groups.* Princeton University Press, Princeton NJ.

Wielandt, H. 1950. Unzerlegbare, nicht-negative Matrizen. *Math. Zeit.*, **52**, 642–648.

Wiener, N. 1923. Differential space. *J. Math. Phys. Mass. Inst. Tech.*, **2**, 131–174.

Wiener, N. 1930. Generalized harmonic analysis. *Acta Math.*, **55**, 117–258.

Wiener, N. 1938. The homogeneous chaos. *Amer. J. Math.* **60**, 897–936.

Wiener, N. 1948. *Cybernetics.* J. Wiley & Sons, New York, and Hermann et Cie, Paris.

Wigner, E. 1958. On the distribution of roots of certain symmetric matrices, *Ann. Math.*, **67**, 325–327.

Wigner, E. 1967. Random matrices in physics. *SIAM Rev.*, **9**, 1–23.

Wild, E. 1951. On Boltzmann's equation in the kinetic theory of gases. *Proc. Camb. Phil. Soc.*, **41**, 602–609.

Wishart, J. 1928. Generalized product moment distribution in samples. *Biometrika*, **20A**, 32–52.

Wolfowitz, J. 1978. *Coding Theorems of Information Theory* (3rd ed.). Springer-Verlag, Berlin.

Yang, C.N. and Lee, T.D. 1952. Statistical theory of equations of state and phase transitions, I. Theory of condensation. *Phys. Rev.*, **87**, 404–409.

Yau, H.-T. 1998. Asymptotic solutions to dynamics of many body systems and classical continuum equations. *Current Developments in Mathematics.* Internat. Press, Somerville, MA, 155–236.

Index

1-sided stable density, 35

Airy function
modified, 443
André's reflection principle, 66, 69
Andréief's lemma, 424
annihilation operator, 223
arcsine law, 76, 152, 155
Brownian motion, 191
conditional, 81

backward equation, 212
Bernoulli trial, 8, 42, 60
Bessel function
modified, 443
Bessel operator
three-dimensional, 217
two-dimensional, 215
Bessel process, 215
binary symmetric channel, 306, 313
binomial distribution, 42
Birkhoff's theorem, 113, 263
Boltzmann's H-function, 295
Boltzmann's H-theorem, 359, 361
Boltzmann's collision integral, 360, 363

Boltzmann's constant, 161, 320
Boltzmann's equation, 356
2-speed gas, 367, 371
hydrodynamical approach, 362
Kac gas, 386
Borel–Cantelli lemma, 39
Breiman's estimate, 304
Brownian motion, 213
BM(1), 8, 9, 170, 206
Lévy's construction, 175
memoryless, 180
Wiener's construction, 178
absorbing, 189
circular, 216
higher dimension, 214
higher-dimensional, BM(d), 214
of Brownian motion, 226
reflecting, 188
relation to Hermite polynomials, 225
spherical, 218, 219, 227
tied, 203
Brownian scaling, 178
Brownian sheet, 227
bulk effect, 140, 168, 169

canonical density, 317
canonical ensemble, 263, 320
capacity, 110, 113
capacity of a channel, 306
Carathéodory's lemma, 1, 3, 174
Carleman's equation, 376
Carleman's test, 99
Cauchy density, 35
central limit theorem, 48, 61, 139, 147, 166, 200
  higher dimensions, 159
  Markov chains, 241
  RW(1), 99
chaos, 365, 400
  propagation, 367
charge, 109
Chebyshev inequality, 36, 43
  exponential-type, 47
Chung's formula, 106
circle bundle, 284
CLT, *see* central limit theorem
coding, 290
  block, 300
  noiseless, 299
conditional arcsine law, 81
conditional expectation, 11
  rules, 12
conditional probability, 11
  rules, 12
conditioned random walk, 121
conditioning, 128
conservation of energy, 318
continued fraction, 270, 285
convex function, 37
convolution, 33
correlation matrix, 157
Coulomb's law, 108
Cramér's estimate, 54
Cramér's rule, 128
creation operator, 223

crossing time, 419
curvature, 165, 280
  Gaussian, 280
Döblin's method of loops, 236
density
  1-sided stable, 35
  Cauchy, 35
  Gauss, 35
  stable, 103
Dirichlet's problem, 111
disorder, 147
distant passage, law of, 74
distribution function, 32
Doob's inequality, 37, 186
drift, 82, 228
  outward, 122
  simple, 120
Dyson's Coulomb gas, 415, 424
electrical circuits, 92
electricity, static, 106
electrostatic capacity, 109
electrostatic energy, 109
electrostatic potential, 108
elliptic equations, 112
empirical density, 364
empirical distribution, 45, 56
empirical frequency, 46
energy
  free, 321
  kinetic, 261, 318
  potential, 261, 318
entropy, 263, 292, 359
  change of, 322
  relative, 55, 56, 298
  statistical-mechanical, 294, 295
  thermodynamic, 147, 296
entropy production rate, 148
equation of state, 321

equidistribution, 72, 241
equilibrium Markov chain, 233
equipartition, 305
ergodicity, *see* metric intransitivity
Euler's equation, 362
event, 5
  tame, 38, 174
expectation, 6
  conditional, 11
    rules, 12
  probability
    rules, 12

fair game, 15, 68, 127, 183
Fano's inequality, 309
Fatou's lemma, 4
favorable game, 16, 183
Feynman–Kac formula, 192
field
  horizontal, 12–14
  independent, 7
field of events, 5
figure of merit, 146
Fisher's information, 148, 380, 391
fluctuations
  Gaussian, 140, 168, 169
focusing, 385
forward equation, 212
Fourier integral, 28
  higher-dimensional, 31
Fourier series
  for a smooth function, 27
  for the additive group of integers, 26
  higher-dimensional, 31
Fourier transform, 34
  inverse, 34
fractional dimension, 297

Fredholm determinants, 428
free energy
  convexity, 325
  Helmholtz, 322
Fubini's theorem, 5
fundamental cell, 282
Gambler's Ruin, 69, 73, 111, 118, 121, 122, 184
  Brownian motion, higher dimension, 216
gambling, 15
game
  fair, 15, 68, 127, 183
  favorable, 16, 183
Gauss density, 35
  circular, 86
Gauss's principle, 109
Gaussian approximation, *see* central limit theorem
Gaussian curvature, 280
Gaussian distribution, 146, 158, 160, 169, 171
  higher dimension, 157
Gaussian orthogonal ensemble, 403
Gaussian unitary ensemble, 422
Gaussian variables, 149
  higher dimension, 157
  uncorrelated, 158
generating function, 65
geodesic, 278
geodesic flow, 278
Gibbs's lemma, 55, 146, 263
gravitating bodies, 160
Green's function, 107, 111
  grounded, 111
grounding, 111

Hamilton's equations, 261, 318
hard ball gas, 327

Hardy–Littlewood estimate, 48
harmonic function, 111, 115
  non-negative, 115, 116, 118,
    120, 122, 123, 125
heat, 296
heat equation, 107, 170, 173,
    188, 211, 360, 392
heat flow, 130
heat flux, 173
Hermite polynomials, 223, 426
  relation to Brownian motion,
    225
Hewitt–Savage law, 38, 115, 135
Hilbert paradox, 362, 383
hitting time, higher dimension,
    105
holding time, 245
horizontal field, 12–14
hyperbolic   geometry,   *see*
  Poincaré's half-plane

ideal gas, 161, 319, 326
ideal gas law, 162
images, method of, 138, 189
independent events, 7
independent fields, 7
independent random variables,
    7
information, 147, 291, 297
information production rate,
    241, 301
  per letter, 278
invariance principle, 184
invariant distribution, 233, 234
irreversibility, 361
Ising model, 334
iterated logarithm, law of, 48,
    51, 61, 201
  general, 152
Itô calculus, 206

Itô's lemma, 208, 213
Itô's lemma, higher dimension,
    214
Itô's powers, 226
Itô's rule, 207

Jacobi identity, 30
Jacobi's complete elliptic inte-
  gral, $K(k)$, 91, 94
Jacobian, 229
Jensen's inequality, 37

Kac's gas, 385
Kelvin's method of images, 138,
    189
Kelvin's method of steepest de-
  scent, 24
Kelvin's principle, 109, 110
Kepler's laws, 260
Khinchine's constant, 273
Kolmogorov's 01 law, 38, 49,
    266
Kolmogorov's   inequality,   38,
    242
Kolmogorov–Smirnov statistic,
  D_N, 135, 202
Kramers–Wannier duality, 350
Kronecker's lemma, 134
Kubo's formula, 243

Lévy's rules, 18
Laplace transform, 35
Laplace's difference operator,
    106
Laplace's estimate, 25
Laplacian, 107
  infinite-dimensional, spheri-
    cal, 221
  spherical, 228
large deviations, 54, 323
  Markov chains, 250
large volume limit, 320

last-leaving probability, 106, 109, 110
law of distant passage, 103, 183
law of large numbers, 43, 49, 61, 132, 166, 266, 279
  Markov chain, 237
  non-equilibrium statistical mechanics, 365
  strong, 43
  weak, 43
Lebesgue's monotome convergence theorem, 4
Legendre dual, 58
Legendre duality, 257
Legendre–Fenchel duality, 58
likelihood ratio, 298
LLN, *see* law of large numbers
local limit theorem, 144
logarithmic Sobolev inequality, 148, 212
long runs, 74
loop, 71, 236
  duration, 71
  duration in high dimension, 99
  in higher dimension, 88
loop time, 62

magnetization, 341, 342
  spontaneous, 335, 338, 345, 350
    low temperature, 348
Markov chain, 232
  equilibrium, 233
  infinite, 237
  law of large numbers, 237
  Markov property, 245
  reversible, 233
Markov chains
  central limit theorem, 241

Markov process
  strict, 87
Markov property, 63
  simple, 61
    Brownian motion, 180
  strict, 62, 64
    Brownian motion, 180
Markov's method of moments, 99, 125, 169
Markov's methods of moments, 97
martingale, *see* fair game
Maxwell distribution, 161, 320, 359
Maxwell's equal area rule, 331
McMillan's theorem, 241
  continued fractions, 277
  information production rate, 302
mean field approximation, 345
mean value, *see* expectation
measurable function, 4
measurable set, 2
measure
  2-dimensional, 4
  Lebesgue, 1
  signed, 9
measure in infinite dimensions, 162
memory, 300
memoryless, 62
message source, 300
method of images, 138, 189
method of moments
  Markov's, 97, 99, 125, 169
metric intransitivity, 265
metric transitivity, 273
microcanonical ensemble, 263, 318, 319
Mills's ratio, 37, 40

mixing, 240, 266, 274
modular group, 282
modulus of continuity, 33
moments, problem of, 130
mountain pass lemma, 254

Newton's law of cooling, 173
Newton's second law of motion, 261
noise, 290, 305
  absence, 299
noisy channel, 305

Ornstein–Uhlenbeck process, 211
  stationary, 213, 227

Painlevé transcendents, 444
parabolic equations, 112
partition function, 320
passage time, 62, 64, 82
  Brownian motion, 181
passage, law of distant, 69
phase change, 325, 328, 336, 345
phase space, 262, 318
Poincaré recurrence, 267
Poincaré's half-plane, 278, 279
Poisson process, 248
Poisson summation, 29
Poisson walk
  tied, 204
Poisson's formula, 131
polynomial approximation
  Weierstraß theorem, 44
polynomial chaos of weight n, 223
pressure
  thermodynamic, 324
prime number theorem, 40, 166, 445
probability, 5
  conditional, 11

measure, 5
space, 11
problem of moments, 130
Pythagoras' rule, 27

Rademacher functions, 20
Radon–Nikodym derivative, 10
random variable, 6, 11
  independent, 7
random walk
  RW(1), 8, 60, 87
  RW(2), 87
  RW(3), 87, 159
  conditioned, 121, 125
  direction of, 92
  higher-dimensional, RW(d), 87
  number of repeat visits, 105
  speed of, 92
rate function, 251
Rayleigh's problem of random flights, 159
reflection principle of D. André, 66, 69, 137
return visits, 73
reversible Markov chain, 233
Riemann hypothesis, 446
Riemann zeta function, 445
Riesz's maximal ergodic lemma, 268
run, 46
runs
  long, 101

sample path, 8
sample space, 5, 42
scattering, 353, 354, 364
section of a cone, 115
shift-invariance, 21
signed measure, 9
Skorokhod embedding, 199

space-time walk, 126
Sparre-Andersen's identity, 152, 194
Spitzer's identity, 156
stable density, 103
stable law, 36, 69, 103
steepest descent, method of, 24
Stirling's approximation, 22
  poor man's, 22
stopping time, 62–64, 68
Stosszahlansatz
  Boltzmann's, 357, 361, 365
streaming, 353, 354
  2-speed gas, 374
submartingale, *see* favourable game
summable function, 4

tame, 38
  event, 174
  function, 221
telegrapher's equation, 373, 382
temperature, 296

absolute, 161, 320
  critical, 336, 344
thermodynamics
  second law, 295, 322
time average, 263
trinomial distribution, 89

uncertainty, 147, 291
upcrossing lemma, 16, 17

van der Waals gas, 328
  3-dimensions, 331
Vandermonde determinant, 405
Vieta's formulae, 21
volume
  of RW, 83
    higher-dimensional, 104, 112
  of RW(1), 188

Watson's integral, 90
Wiener's trick, 19
Wigner's semi-circle law, 403
Wigner's surmise, 433, 442
Wild's sum, 388

Printed in the United States
by Baker & Taylor Publisher Services